E. T. JAYNES:

PAPERS ON PROBABILITY, STATISTICS AND STATISTICAL PHYSICS

The Pallas Paperback Series is a natural outgrowth of Kluwer's scholarly publishing activities in the humanities and social sciences.
It is designed to accommodate original works in specialized fields which, by virtue of their broader applicability, deserve a larger audience and a lower price than the standard academic hardback. The series will also include books which have become modern classics in their fields, but which have not yet benefited from an appearance in a more accessible edition.

E. T. JAYNES:
PAPERS ON PROBABILITY, STATISTICS AND STATISTICAL PHYSICS

Edited by

R. D. ROSENKRANTZ
Department of Mathematics, Dartmouth College

KLUWER ACADEMIC PUBLISHERS
DORDRECHT / BOSTON / LONDON

Library of Congress Cataloging in Publication Data

Jaynes, E. T. (Edwin T.)
  E. T. Jaynes: papers on probability, statistics, and statistical physics.

  (Synthese library; v. 158)
  Bibliography: p.
  Includes index.
  1. Statistical mechanics–Collected works.   2. Entropy (Information theory)–Collected works.   3.   Probabilities–Collected works.
  4. Statistical physics–Collected works.   I.   Rosenkrantz, Roger D.,
  1938-   . II. Title.
  QC174.8.J38   1982   530.1'3   82-21623
  ISBN 90-277-1448-7
  ISBN 0-7923-0213-3 (Pallas Paperback)

Published by Kluwer Academic Publishers,
P.O. Box 17, 3300 AA Dordrecht, The Netherlands.

Sold and distributed in the U.S.A. and Canada
by Kluwer Academic Publishers Group,
101 Philip Drive, Norwell, MA 02061, U.S.A.

In all other countries, sold and distributed
by Kluwer Academic Publishers,
P.O. Box 322, 3300 AH Dordrecht, The Netherlands.

Reprinted 1989.
First published 1983 (Synthese
Library, Volume 158)

All Rights Reserved
© 1989 by Kluwer Academic Publishers
No part of the material protected by this copyright notice may be reproduced or
utilized in any form or by any means, electronic or mechanical,
including photocopying, recording or by any information storage and
retrieval system, without written permission from the copyright owner.

Printed in the Netherlands

TABLE OF CONTENTS

PREFACE	vii
ACKNOWLEDGEMENTS	xi
EDITOR'S INTRODUCTION	xiii
1. Introductory Remarks	1
2. Information Theory and Statistical Mechanics, I (1957)	4
3. Information Theory and Statistical Mechanics, II (1957)	17
4. Brandeis Lectures (1963)	39
5. Gibbs vs Boltzmann Entropies (1965)	77
6. Delaware Lecture (1967)	87
7. Prior Probabilities (1968)	114
8. The Well-Posed Problem (1973)	131
9. Confidence Intervals vs Bayesian Intervals (1976)	149
10. Where Do We Stand on Maximum Entropy? (1978)	210
11. Concentration of Distributions at Entropy Maxima (1979)	315
12. Marginalization and Prior Probabilities (1980)	337
13. What is the Question? (1981)	376
14. The Minimum Entropy Production Principle (1980)	401
SUPPLEMENTARY BIBLIOGRAPHY	425
INDEX	431

E. T. JAYNES

PREFACE

The first six chapters of this volume present the author's 'predictive' or information theoretic' approach to statistical mechanics, in which the basic probability distributions over microstates are obtained as distributions of maximum entropy (i.e., as distributions that are most non-committal with regard to missing information among all those satisfying the macroscopically given constraints). There is then no need to make additional assumptions of ergodicity or metric transitivity; the theory proceeds entirely by *inference* from macroscopic measurements and the underlying dynamical assumptions. Moreover, the method of maximizing the entropy is completely general and applies, in particular, to irreversible processes as well as to reversible ones.

The next three chapters provide a broader framework — at once Bayesian and objective — for maximum entropy inference. The basic principles of inference, including the usual axioms of probability, are seen to rest on nothing more than requirements of consistency, above all, the requirement that in two problems where we have the same information we must assign the same probabilities. Thus, statistical mechanics is viewed as a branch of a general theory of inference, and the latter as an extension of the ordinary logic of consistency. Those who are familiar with the literature of statistics and statistical mechanics will recognize in both of these steps a genuine 'scientific revolution' — a complete reversal of earlier conceptions — and one of no small significance. Indeed, the interplay between physics, probability and logic one finds here gives the work a wider import permeating down to the foundations of our knowledge in a way that is reminiscent of relativity or the quantum theory. But, unlike the quantum theory, which purports to set limits to what we can know, a major thrust of Jaynes' work is to liberate us from the imagined limits imposed by a frequency conception of probability.

Although Jaynes erects no artificial barriers to understanding — I know of no writer on technical subjects whose style is more to the point, incisive, or stimulating — some of the papers place heavy technical demands on the reader. And, in general, the rich ore in these hills is not to be tapped without some hard digging! The editor's introduction, which provides a brief overview of the leading ideas and some of the many interconnections between them, may help to guide those readers who have little previous acquaintance with these papers. The index — somewhere between a subject index and a 'synoptikon' — will also help you to locate ideas discussed in the introduction. Works referred to there are among those listed in the supplementary bibliography at the end of the book. The latter is divided into four sections as follows: background works, technical contributions and critical discussions of the entropy formalism and related methods of inference, a brief sampler (really, the tip of the iceberg) of papers applying the maximum entropy method in diverse fields of current research, and, finally, a short list of papers by Jaynes not included here, especially those dealing with the neoclassical theory of electrodynamics.

While the papers are reprinted in the order written, the better to portray the unfolding of an idea, there is no necessity that they be read in that order. Nearly everyone will find the first two chapters a good place to start, but readers primarily interested in statistics or the foundations of probability may want to pass directly to chapter seven. The M.I.T. paper (Chapter 10) is the pivotal piece of the collection, summing up all the papers that go before it and broaching new ideas taken up in the remaining chapters and still in process of development. This paper is an absolute leviathan! Reverberating with history and personal recollection and occasionally exploding with well-aimed critical bursts, it sweeps you up like a great tidal wave and carries you along for over one hundred pages at an accelerating tempo, leaving you at the end with a sense that its driving energy has still not spent itself (as, in fact, it had not). In the course of this voyage, Jaynes manages, among other things, to weave together the parallel histories of probability and statistical physics, to answer his critics, and to present, in the final section, the fullest account yet available of the Maxent treatment of irreversible processes.

I would like to express my gratitude to Mrs. J. C. Kuipers and others at D. Reidel for their help and patience in this enterprise and to the editor of this series, Jaakko Hintikka, for his encouragement to undertake it. Finally, I want to thank E. T. Jaynes for his splendid cooperation at every stage of the work. The introductory comments he has added to the volume and to the individual selections are especially valuable and welcome. But, even more than that, I want to thank Jaynes for writing these papers in the first place. The problems with which he has grappled — the Bertrand paradoxes, the marginalization paradoxes of statistical theory, and the seemingly intractable problems of irreversible thermodynamics — have withstood the efforts of many powerful minds (and given many others an attack of vertigo!). These are not puzzles that beckon one with a promise of easy gold at the hand of fay or elf. One must plunge into murky deeps and risk lying suspended indefinitely in an agony of confusion. Yet, in every case, Jaynes has managed to lay hold of a constructive principle that can steer us towards the light. The fruits of such hard-won gains always go beyond the mere harvesting of new scientific findings or the forging of powerful new instruments of inquiry; rather, advances at this fundamental level advance our understanding of thinking itself.

ACKNOWLEDGEMENTS

The editor wishes to thank the publishers mentioned below for permission to reprint the following papers:

'Information Theory and Statistical Mechanics, I, II', *Physical Review* **106** (1957), 620–630, **108** (1957), 171–190.
Reprinted by permission of the American Physical Society.

'Information Theory and Statistical Mechanics', 1962 Brandeis Summer Institute in Theoretical Physics, K. Ford, Editor, Benjamin, 1963.
Reprinted by permission of Benjamin-Cummings Publishing Company.

'Gibbs vs. Boltzmann Entropies', *American Journal of Physics* **33** (1965), 391–398.
Reprinted by permission of the *American Journal of Physics*.

'Foundations of Probability Theory and Statistical Mechanics', in *Delaware Seminar in the Foundations of Physics*, M. Bunge, Editor, Springer-Verlag, Berlin, 1967.
Reprinted by permission of Springer-Verlag, Inc., New York.

'Prior Probabilities', *IEEE Transactions on Systems Science and Cybernetics*, SSC-4, Sept. 1968, 227–241.
Reprinted by permission of the Institute of Electrical and Electronics Engineers, Inc., New York.

'The Well-Posed Problem', *Foundations of Physics* **3** (1973), 477–493.
Reprinted by permission of Plenum Press, New York.

'Confidence Intervals vs. Bayesian Intervals', in *Foundations of Probability Theory, Statistical Inference, and Statistical Theories of Science*, W. L. Harper and C. A. Hooker, Editors, D. Reidel Publishing Company, Dordrecht, Holland, 1976.
Reprinted by permission of D. Reidel Publishing Company.

'Where Do We Stand on Maximum Entropy?', in *The Maximum Entropy Formalism*, R. D. Levine and M. Tribus, Editors, M.I.T. Press, Cambridge, Massachusetts, 1978.
Reprinted by persmission of the M.I.T. Press.

ACKNOWLEDGEMENTS

'Marginalization and Prior Probabilities', in *Bayesian Analysis in Econometrics and Statistics*, A. Zellner, Editor, North-Holland Publishing Company, Amsterdam, 1980.

Reprinted by permission of North-Holland, Amsterdam.

'What is the Question?', in *Bayesian Statistics*, J. M. Bernardo *et al.*, Editors, Valencia University Press, Valencia, Spain, 1980.

Reprinted by permission of Valencia University Press.

'The Minimum Entropy Production Principle', in *Annual Review of Physical Chemistry*, S. Rabinovitch, Editor, Annual Reviews, Inc., Palo Alto, California, 1980.

Reprinted by permission of Annual Reviews, Inc., Palo Alto.

EDITOR'S INTRODUCTION

I

When methods successfully applied in science appear discredited, a creative tension is generated: one must either obtain the relevant results some other way or else place the methods in question on a more secure footing. Both sorts of effort have been visible throughout this century in the two related fields of statistics and statistical mechanics. For by 1900, the classical theory of probability, so fruitfully applied by Laplace to celestial mechanics and by the early workers in statistical mechanics, was faced with seemingly fatal paradoxes and contradictions – a situation not unlike that which then prevailed in set theory. There was a general turning away from the methods of Bayes and Laplace and attempts were made to ground both statistics and statistical mechanics on a frequency conception of probability. Although never wholly successful, these efforts dominated work in probability theory and its applications until quite recently. True, the important dissenting work of Harold Jeffreys (1939) showed how to derive the standard significance tests of the frequentist school almost effortlessly by Bayes' theorem, using an 'uninformative' prior probability distribution, and Jeffreys' methods were, moreover, of wider scope. Unfortunately, his uninformative priors, though intuitively appealing, lacked a compelling rationale, and were even widely believed to rest on nothing more solid than the discredited Laplacian principle of indifference.

The papers by E. T. Jaynes collected here stem, in part, from a desire to supply the missing rationale and, in part, from a growing realization that the methods of Jeffreys could be extended to provide a more satisfactory basis for statistical mechanics. Claude Shannon's development of information theory, which makes essential use of a measure of uncertainty he labelled 'entropy', pointed the way to the required extension of Jeffreys' methods. In attacking the problem of a most efficient encoding of English text, it was necessary to assign probabilities to all conceivable messages that might be sent. Of course, we can never know the 'true' probabilities in question; our statistical knowledge is both incomplete and unmanageably complex. Shannon is thus led to consider 'the source with the maximum entropy subject to the statistical conditions we wish to retain'. In writing down the solution to Shannon's problem, Jaynes found himself staring at Gibbs' canonical distribution. And it soon became evident that all of the distributions found by Gibbs could be derived in this way, without recourse to additional, ergodic assumptions. Jaynes' position at this point was like that of Jeffreys: he had a method that undoubtedly works, and one that is also simpler and of wider scope. In attempting to extend the method to continuous distributions, he was led back to the problem that Bayes and Laplace and Jeffreys had all run up against: how to represent ignorance of a parameter.

II

Jeffreys' uninformative priors are meant to provide "a formal way of expressing ignorance of the value of the parameter over the range permitted", and he argued that the same uninformative priors should be used for parameters with the same formal properties. Thus, he advocated a uniform prior for parameters ranging over the whole real line and a log-uniform prior (the logarithm of the parameter uniformly distributed) for positive parameters. But use of the uniform prior to represent 'complete ignorance' seemed open to the same charge of inconsistency once levelled at Bayes and Laplace. For if we are ignorant of θ, the argument runs, then, equally, we are ignorant of $T(\theta)$. But if T is a non-linear function, like $T(\theta) = \theta^k$, a uniform distribution of $T(\theta)$ induces a non-uniform distribution of θ, and we have an obvious contradiction.

Of course, one can deny the crucial premiss that *any* transformation of the parameter is admissible, but to make this convincing requires a systematic way of determining the group of admissible transformations. A short paper by R. T. Cox (1946) pointed the way. Cox obtained the basic 'addition' and 'multiplication' rules of probability calculus by imposing the requirement that different allowable ways of applying the calculus to a problem should yield the same answer. In particular, recasting the evidence in an equivalent form should not lead us to alter our probabilities. Jaynes then saw how to base Jeffreys' uninformative priors on a variant of the consistency principle, namely: *in two problems where we have the same information, we should assign the same probabilities.*

To illustrate, assume we are sampling a normal distribution of known (unit) variance. Then the density is a function, $h(X - \mu)$ of $X - \mu$, so that μ is a location parameter. A translation of coordinates, $X' = X + b, \mu' = \mu + b$, leaves the normal form and spread of the density unaltered, and so "we have the same information" in two versions of the problem that differ by a translation of coordinates. (Certain other reparametrizations, like a change of scale, would change either the form or spread of the distribution and so conflict with the information already at hand.) If f and g are the prior densities of μ and μ' in two equivalent versions of the problem, the consistency principle forces $f = g$, and this leads straight to the functional equation, $f(\mu) = f(\mu + b)$, expressing translation-invariance of f. The unique solution is, of course, the uniform prior advocated by Jeffreys.

One arrives similarly at the Jeffreys prior, $f(\sigma)d\sigma = d\sigma/\sigma$, for $\sigma > 0$, but the rationale Jaynes offers is, not that σ is positive, but that σ is a scale parameter, so that two versions of the problem which differ in the scale units employed are equivalent.

An amusing example is provided by "the anomalous law of first digits", often cited to illustrate the dangers of selecting an empirical distribution 'a priori' without recourse to observed frequencies. We are asked for the probability p_k that k ($k = 1, \ldots, 9$) is the first digit of a random entry from a table of numerical data. Naive application of the principle of indifference at the level of 'indifferent events' leads to a uniform distribution, $p_k = 1/9$, but the distribution obtained empirically is log-uniform: $p_k = \log(k+1) - \log(k)$, with logarithms to base 10. On the other hand, nothing has been said about the scale units employed, and if the problem has a definite solution, it must not depend on this unspecified information. And scale invariance does lead to a log-uniform distribution. Indeed, the present derivation shows *why* the logarithmic law holds for data that are ratio-scaled and that it need not hold for ordinal data, like street addresses.

The history of statistical physics is replete with empirically correct distributions obtained *perforce* by applying the principle of indifference, for one could not directly observe microstates, much less tabulate the frequencies with which a system enters different microstates. Jaynes' thesis is that successful applications of this sort are just the ones that can be rephrased as appealing to 'indifference between problems' rather than 'indifference between events'.

The consistency argument is by no means confined to location and scale parameters. In Chapter 8, Jaynes uses it to resolve the notorious Bertrand paradox, which asks for the probability that a random chord of a circle exceeds a side of the inscribed equilateral triangle. Applying the principle of indifference to different geometric constructions of the chord, different answers are obtained, and most writers who discuss the problem profess themselves unable to say which, if any, of these solutions would correctly describe a real experiment. Starting at the other end, Jaynes invites us to look at an actual experiment in which broom straws are tossed onto a circle from a height great enough to preclude even the skill needed to make the broom straws fall across the circle. This certainly gives a good sense to 'random chord'. But nothing has been said about the exact size and position of the circle, and the implied invariances are sufficient to single out a unique distribution for the center of a chord determined by a broom straw which does fall across the circle. We may not conclude, of course, that experimentally observed frequencies must agree with those predicted by the invariant distribution, but we can conclude that any exceptional experiment will produce different distributions on circles that differ only slightly in size or position from the given circle. Invariance under the admissible group of transformations is just a mathematical expression of the lack of skill or microscopic control needed to produce different frequencies when the initial conditions are slightly varied, and so we may even think of the invariant distribution as the 'objective chance distribution' of the 'random phenomenon' in question.

III

Uninformative priors are invariant under variation of unspecified details, but it is not always feasible to carry out the requisite group theoretical analysis. In such cases we do the next best thing: we find the distribution that is *maximally non-committal* with respect to missing information. (Any other distribution would pretend to knowledge we really lack.) So of all the distributions satisfying the given constraints (typically, mean value constraints), we choose the one that maximizes uncertainty, as measured by Shannon's entropy function. Entropy maximizing distributions thus enter the picture as a practical substitute for invariant distributions. They can also be used when the transformation group uncovered by our analysis is too sparse to single out a unique distribution, or to modify an uninformative prior in the light of experimentally given constraints.

This leaves open the question why maximize entropy, why not some other measure of uncertainty? Jaynes gives several answers. First, 'deductions made from any other information measure ... will eventually lead to contradictions' (see Chapter 2, appendix). Second, entropy maximizing distributions are obtained asymptotically by conditioning on given mean value constraints. And, third, the maximum entropy distribution is realized experimentally by an overwhelming majority of the trial sequences satisfying the given constraints. Its deep connections with consistency, conditionalization and observed frequency behavior all suggest that entropy maximization is a fundamental principle of probability theory.

Jaynes' first answer has been more fully articulated in the recent paper by Shore and Johnson (1980). They translate Shannon's desiderata for a measure of uncertainty into conditions of consistency for an abstract inferential operator that combines a prior p with mean value constraints D to yield a posterior distribution, $q = p \circ D$. Their axioms assert that the result should not depend on the coordinate system, or on whether we account for independent items of information separately or in terms of a joint density, etc.. In short, their axioms embody R. T. Cox's aforementioned requirement that different allowed ways of applying the inferential apparatus must lead to the same result. The only operator meeting their requirements assigns the pair (p, D) a posterior density, $q = p \circ D$, which among all those satisfying D, minimizes the *cross* (or *relative*) *entropy* with respect to p, defined in the discrete case by:

$$I(q, p) = \sum_i q_i \log(q_i/p_i)$$

with $p = (p_1, ..., p_n)$ and $q = (q_1, ..., q_n)$. The *entropy* of q is:

$$H(q) = -\sum_i q_i \log(q_i).$$

Obviously, an entropy maximizing q is cross entropy minimizing with respect to a uniform prior, and so entropy maximization is a special case of cross entropy minimization. (Actually, either may be viewed as a special case of the other.) Cross entropy is an interesting function in its own right. Kullback (1959) exhibited its power as a unifying principle in statistics, and it enters Jaynes' work in the problem of extending the entropy function invariantly to continuous distributions (Chapter 7, Section VI).

Suppose now that the new information D confines the outcome of an experiment to a proper subset E — the 'conditioning event'. Cross entropy minimization subject to $q(E) = 1$ yields the renormalization of the prior p to E, as required by Bayesian conditionalization. Similarly, if we impose the weaker (more realistic) constraint that the observation raises the probability of E to a value short of one, cross entropy minimization yields the generalization of Bayes' rule proposed by Jeffrey (1965). This rule depends on the order in which the observational inputs occur, but the modification Field (1978) has put forward to remove that blemish also pops right out as a special case of cross entropy minimization (see Domotor, 1980). Seen in this light, cross entropy minimization appears as a very general rule of minimal belief change, incorporating the various forms of Bayesian conditioning and entropy maximization as special cases.

Conversely, cross entropy minimization results asymptotically from Bayesian conditioning. As Jaynes notes (Chapter 10, p. 250), this is just the well-known Darin–Fowler method for obtaining the canonical distribution maximizing the entropy yields for a sample space S by conditioning on the product space S^n of repeated trials. The derivation is given in detail for both discrete and continuous distributions in van Campenhout and Cover (1981).

To see what is being asserted, consider n independent tosses of a die. We eliminate all possible outcome sequences whose associated 6-tuple of frequencies for the six faces fails to satisfy the given mean value constraint and then renormalize the original uniform distribution to the surviving 'admissible' sequences. The average of the frequency distributions $(f_1, \ldots, f_6)$ of these admissible sequences yields the conditioned distribution, $p^* = (p_1^*, \ldots, p_6^*)$, and as n increases, p^* tends to the maximum entropy distribution. This happens because an increasing proportion of the admissible sequences give rise to frequency 6-tuples whose entropy is close to the maximum (the 'concentration theorem' of Chapter 11 allows one to approximate this proportion), and all of these 'high entropy' frequency distributions differ inappreciably from the maximum entropy distribution. (The derivation of a cross entropy minimizing distribution is just the same.)

This is the 'correspondence with frequencies' alluded to earlier, and it is interesting to note that the first maximum entropy distributions were found in this way. Thus, Ludwig Boltzmann found his famous energy distribution for the molecules in a conservative force field by dividing the phase space into small cells R_k each small enough for the energy to be a constant E_k over the molecules within it and yet big enough to contain a large number N_k of molecules. The total number N of molecules and the total energy E are constants of the molecular motion, and Boltzmann argued that the 'most probable' distribution, $(N_1/N, \ldots, N_s/N)$, is the one, among all those satisfying the constraints $N_1 + \cdots + N_s = N$ and $N_1 E_1 + \cdots + N_s E_s = E$, which is realized by the greatest number of microstates. But this number is given by the multinomial coefficient, $W = N!/N_1! \ldots N_s!$, and this is a maximum when $N^{-1} \log W = -\Sigma (N_k/N) \log(N_k/N)$ is a maximum (where we have used Stirling's approximation to the log factorials). The latter expression is just the entropy of the frequency distribution, and so Boltzmann's 'most probable' distribution is the maximum entropy distribution satisfying the given constraints.

Boltzmann's law is a powerful one; it contains Clerk Maxwell's velocity distribution for the molecules of a gas at a given temperature as a special case. Yet, Boltzmann's derivation appears to ignore the dynamics altogether (to get something for nothing), and that is why Jaynes has taken such pains to explain clearly why the method delivers accurate predictions of thermodynamic quantities — *provided* the assumed mean value constraints and microscopic laws of motion embodied in the Hamiltonian of the system are empirically correct. (The point is discussed in Chapter 2, pp. 10–13, Chapter 3, pp. 19–20, and again in Chapter 10, pp. 227, 239–240, 281–282, 296–298.)

To begin with, Boltzmann's derivation does make use of the dynamics, both in the assumed energy conservation and in the fact that the volumes of the small cells R_k are invariants of the motion (Liouville's theorem). The empirical success of the law shows that none of the other myriad dynamical details are truly relevant. (This would not be so, of course, if the predictions were time-dependent.)

Since maximum entropy inference uses the broadest distribution compatible with the assumed constraints, the macroscopic quantities for which it yields sharp predictions must be characteristic of the great majority of the microstates to which it attaches appreciable probability, and the features of the system that are experimentally reproducible are precisely those which are characteristic of most of the states compatible with the conditions of the experiment. Sharp maximum entropy predictions can fail, then, only if there are new constraints not contained in the previously known laws of physics. A noteworthy historical case of this sort was the failure of Gibbs' canonical distribution to predict heat capacities and equations of state, a failure which pointed to the previously unsuspected discreteness of the energy levels. At a more mundane level, Jaynes uses Rudolf Wolf's data on dice to show how observed deviations from a maximum entropy distribution can suggest new physical constraints. Indeed, by comparing the entropy of the observed frequency distribution with the maximum entropy attainable with the assumed constraints, Jaynes is led to a new goodness-of-fit test that improves on the more usual chi squared test in several respects (see Chapter 11).

Jaynes' informational approach views the laws of statistical mechanics as *inferences* based entirely on the laws of mechanics; additional hypotheses of ergodicity or metric transitivity are not needed. Yet, it would be a mistake to think that the case for the information theoretic approach rests solely on its greater simplicity. Even complete success in proving the needed ergodic theorems would legitimize Gibbsian phase averaging only in the equilibrium case. Entropy maximization is subject to no such restriction; it applies equally well to irreversible processes, and it is here, most of all, that the theory generates new and testable predictions (see Chapter 10, Section D for the treatment of irreversible processes). In addition, Jaynes offers cogent criticism of the ergodic approach (Chapter 6, Section 2.3).

IV

Jaynes makes consistency the very cornerstone of an objectivist philosophy. The subjectivist or personalist school of Bayesian thought has also emphasized consistent (or 'coherent') degrees of belief, demonstrating that inconsistent betting odds open one to sure loss – a 'dutch book'. But within the wide limits set by coherence, subjectivists allow that two men in possession of the same information may reasonably differ in their probability assignments. While the accumulating data will *tend* to bring their subjective probabilities into alignment, logic does not *force* them to agree. Indeed, subjectivist doctrine does not even compel one to heed the evidence, for the thin conception of rationality subjectivists offer requires only that the beliefs held at any one time be coherent. Bayes' theorem enters solely as a means of ascertaining whether the beliefs one holds at different times are mutually consistent (de Finetti, 1972, pp. 144–145). But what are we to say of a man who alters his beliefs without any change in the evidence? We may think of him as conditioning on the null evidence (a tautology), and this will lead to credal probabilities no different from his initial probabilities. If now he finds himself with different beliefs, then, on the subjectivist's own showing, he is being inconsistent. And it is but a short step from this conclusion to the consistency principle by which Jaynes derives uninformative priors. In effect, Jaynes carries the consistency argument of the subjectivists a step farther and arrives at an 'impersonalism' that makes no use of anybody's personal opinions, but only of the data on which these are based (see Chapter 7, Section I).

Jaynes was impelled towards objectivism from another direction as well. If one obtains the canonical distribution of Gibbs by entropy maximization, one can no longer interpret that distribution as giving the frequencies with which a thermodynamic system enters different microstates. Instead, the distribution represents no more than the partial information about microstates contained in the macroscopic measurements. In this way, the distributions of statistical mechanics become credal or epistemic and entropy becomes a measure of ignorance rather than of disorder. As Jaynes recounts (Chapter 10, pp. 237–238) the reaction of Professor G. Uhlenbeck was to deny that entropy can measure amounts of ignorance, for different people have different amounts of ignorance, while entropy is a definite physical quantity that can be measured in the laboratory using thermometers and calorimeters. The answer to this little paradox is clear, but only if you view the credal probabilities in question as objective, for then entropy is a measure, not of this or that person's ignorance, but of the partial knowledge embodied in those laboratory measurements and objectively reflected in the assigned probability distribution. Nevertheless, entropy (like an uninformative prior) is anthropomorphic to this extent: it depends on what experiment we choose to perform on a given system (see Chapter 5). In the same way, the second law of thermodynamics becomes a statement about the loss of information regarding a system on which no new measurements are taken.

The objectivist finds unwonted support in the near coincidence of fiducial or confidence intervals based on a sufficient statistic with highest density intervals of a posterior distribution based on an uninformative prior, for the former attempt, in R. A. Fisher's words, 'to allow the data to speak for themselves'. At the same time, the Bayesian intervals are far easier to compute, for the (often *ad hoc*) choice of a test statistic and subsequent derivation of its sampling distribution are steps the Bayesian avoids. Mathematical equivalence of Bayesian and orthodox intervals cannot be achieved, however, where sufficient statistics are lacking, and it is Jaynes' contention (Chapter 9) that in such cases the differences can be magnified up to where common sense can clearly perceive the superiority of the Bayesian result (as when a confidence interval lies outside the allowed range of the parameter it estimates!). Jaynes shows how the information necessarily lost by a confidence interval in such cases (in his example, the half-range of two observations from a Cauchy distribution) can always be used to pick out a recognizable 'bad' subclass of samples in which the confidence interval fails to cover the true value as frequently as indicated by the stated confidence level, and a 'good' subclass on which the confidence interval is wider than it needs to be. Being of fixed width, the confidence interval is forced to rob Peter to pay Paul, making up for the 'bad' samples of very wide range at which it cannot possibly deliver the advertised reliability by giving us a needlessly wide interval in the great majority of 'good' samples.

Inevitably, some of Jaynes' points have been made by others, but his examples have a force that is wholly lacking in abstract discussions of these matters. I was present at the Western Ontario address on which Chapter 9 is based and can vividly recall the outcries it provoked. Jaynes' scrutiny of the actual *performance* of the rival methods should have appealed to orthodoxians of Neyman–Pearson persuasion, who base their entire approach on the demonstrable error characteristics of a test or a rejection rule, yet, to my knowledge, the only orthodox rebuttals that have since appeared are those by Margaret Maxfield and Oscar Kempthorne which follow Jaynes' paper in the Conference Proceedings. It did not seem appropriate to reprint the entire lengthy exchange here, but the reader is certainly urged to read it and judge for himself how many of the points raised by Maxfield and Kempthorne address the issue of actual performance.

I have reprinted, however, some portions of Jaynes' reply to Kempthorne, including his invaluable defense of 'impropriety'. The uninformative priors for location and scale parameters of unrestricted range are non-integrable, and such 'improper priors' have been thought to cause consistency problems. I can recall Dennis Lindley acknowledging this criticism at the same Western Ontario Conference and intoning solemn last rites for improper priors. That they come to grief is, perhaps, a conclusion not wholly unwelcome to those who remain sceptical of all attempts to objectify prior information. Yet, even subjectivists have a stake in impropriety, inasmuch as diffuse states of knowledge are often most naturally and most conveniently represented by an improper, uninformative prior (in particular, the improper beta prior for binomial sampling which Jaynes obtains group theoretically in Chapter 7 seems inescapably right). And we have already remarked on the near coincidence of Bayesian intervals based on improper priors with confidence intervals. In short, while Bayesian inference can doutbtless proceed without improper priors, their loss would rob the theory of much of its appeal, and so Jaynes has taken up the cudgels on their behalf.

He shows (Chapter 12) that the 'marginalization paradox' of Dawid, Stone, and Zidek, which occasioned all the fuss, has nothing to do with impropriety *per se*, but turns on a difference in the assumed prior knowledge, one that is hidden by a faulty notation. In characteristic fashion, Jaynes then uses the alleged paradox to initiate a new approach to uninformative priors, which yields, for example, Jeffreys' log-uniform prior for a scale parameter as the only prior that is 'uninformative' in the sense that 'it leads us to the same conclusion about other parameters θ as if the parameter σ had been removed from the model'. This argument appears to be of wider scope than the group theoretic method and appeals only to the usual axioms of probability.

The comments on impropriety in Jaynes' reply to Kempthorne (Chapter 9, pp. 205ff.) are also quite illuminating. Where the posterior distribution based on an improper prior integrates, we would obtain essentially the same posterior distribution using any truncated proper prior that is reasonably diffuse in the region of appreciable likelihood. (L. J. Savage called this the 'principle of stable estimation'.) Where, on the other hand, the posterior density based on an improper prior is itself improper, the theory is effectively warning us that the experiment is so uninformative that the exact limits of truncation do matter. Jaynes illustrates this with a reliability test in which the chosen test time is much longer than the desired lifetime of the machine. The failure of all tested items then conveys essentially no information, and the non-integrable posterior density expresses this by failing to impose a lower bound below which the mean lifetime is practically certain not to lie. Jaynes then notes that the improper prior in question gives inferences that agree with those of a standard significance test in the usual range of interest where the test time is relatively short and few failures are observed. But these significance tests break down completely when all tested items fail, rejecting the hypothesis that the mean lifetime exceeds any specified value at all levels of significance! And so he concludes, "it is the orthodoxian, and not the Bayesian, who is going to be in trouble in cases where 'improper priors' cannot be used".

Not all of the problems involving uninformative priors have been solved, as Jaynes himself is at pains to point out in several places. More needs to be said, in particular, about the sometimes disconcerting degree to which an uninformative prior depends on the experiment planned or the question posed. (On this, see Chapter 13 of this volume, pp. 379ff. and the paper by John Skilling listed in the supplementary bibliography.) It is, nevertheless, a safe prediction that further progress in the representation of prior information will follow the paths Jaynes has so clearly delineated.

The other great challenge his work poses is to separate substantive assumption from mere inference in the principles of any theory. Jaynes' own papers on the neo-classical theory of electrodynamics, a number of which are listed in the supplementary bibliography, pose this question for the quantum theory in an acute form (see p. 231). The technique of entropy maximization has definitely widened our inferential horizons and is proving invaluable whenever it is a question of extracting what is relevant from a complex mass of data. It is well known that many of the most useful models of applied probability — the normal and truncated normal, the Poisson and exponential, etc. — can be directly and easily obtained by maximizing entropy subject to distributional constraints. The small sample of papers listed in Section C of the supplementary bibliography will convey an idea of the vast and growing applications of the technique in diverse areas of current research. One might almost say that whenever a model seemingly based on substantive assumptions is delivering greater accuracy than the rather unrealistic character of the underlying assumptions would lead one to expect, there is probably an entropy argument lurking just under the surface.

1. INTRODUCTORY REMARKS

In a quantitative science, the mathematics exists as a kind of superstructure resting on a set of conceptual notions thought of as fundamental. Therefore, while the introduction of new mathematics takes place more or less continuously without much trauma, a major upheaval is required to replace one conceptual foundation with another. As a result, the acceptance of new concepts takes place only in widely separated jumps, preceded by controversy.

The writer does not know of any logical reason why science should be perceived in this way. Would it not be just as reasonable to take the opposite stance — dogmatic adherence to one mathematical system, while freely admitting new concepts? After all, the empirical success of a theory confirms only its mathematics, not the ideas you or I associate with it. Yet the history of science, from Ptolemy to Schwinger, shows that it is always a conceptual idea — even one having only the loosest of logical association with the successful mathematics — that gets elevated to the status of unassailable dogma.

This psychological phenomenon is just as much an obstacle to progress for us as for Copernicus. Progress in science requires the continual introduction and testing of new concepts, just as much as new mathematics. But in physics since Newton, a cyclic component in conceptual advances, with a period of about seventy years, is clearly discernible, as noted more fully in the *Delaware Lecture* (1967) reprinted here.

Thermodynamics and Statistical Mechanics occupy a rather special place in this scheme, because from the day Clausius discovered entropy there has been unremitting confusion, and resulting controversy, over conceptual questions that it raised. In 1951, a Century later, there was still unfinished business here; a viewpoint simple enough and broad enough to unify these fields and resolve their paradoxes had not been found, and various schools of thought — each subscribing to a different set of concepts — coexisted almost independently of each other. Yet all had adopted the same name, 'Statistical Mechanics', that Gibbs had coined for his program.

This book collects together a series of articles, written over a span of nearly thirty years, recording the gradual evolution of a 'new' and much more general viewpoint about these problems. The quotes seem appropriate, because the literature is so huge that almost every imaginable idea can be found, mentioned briefly and then forgotten, somewhere in the history of the subject. We are continually discovering old works that anticipate parts of what follows; doubtless, many more remain to be found.

Belatedly, it was realized that to put still another corpus of ideas under the umbrella of 'Statistical Mechanics' made an overcrowded terminology even more so. Accordingly, the name 'Predictive Statistical Mechanics' was coined for the particular line of thought expounded here, to distinguish it from all others. It will help to reduce confusion if others will adopt this name, which has a good basis in the current terminology of statistical inference. Indeed, our major break with the past is that our goal is not to 'explain irreversibility', but to predict observable facts.

It appeared that the following collection would be useful not only to scientists as a kind of textbook on Predictive Statistical Mechanics until something better is available; but also to philosophers as a case history for the development of new scientific concepts, in spite of the controversy they immediately stir up. And, critics need no longer attack article N ignorant of what is in article $N+1$.

Of course, Predictive Statistical Mechanics did not spring out fully formed in the first work. In most cases, the writing of article N was the stimulation of the further thinking that resulted in article $N+1$. For this reason, they are reprinted in chronological order; even though the subject matter may appear to wander back and forth between Predictive Statistical Mechanics and Statistical Inference, there is a logical thread connecting them in that order.

Indeed, I do not see Predictive Statistical Mechanics and Statistical Inference as different subjects at all; the former is only a particular realization of the latter, and it applies equally well in any 'generalized inverse' problem in which we have prior information about multiplicities that needs to be taken into account. There is no necessary connection with thermodynamics, other than the fact that thermodynamics provided, historically, the first example of such a problem. The recent major advances in Spectral Analysis (Childers, 1978; Currie, 1980) and Image Reconstruction (Frieden, 1980; Gull and Daniell, 1978, 1980) are straightforward applications of the 'Maximum Entropy Formalism' set forth in the first article below. Before long, we hope to have analyses of economic time series by methods developed in the later articles.

A brief commentary precedes each reprinted article, with background remarks, noting where critics or hindsight have functioned, relations to other works, etc. Not all of my articles in this field are reproduced here, only those which made some contribution to the development of the general thinking and whose writing had been completed by the end of 1980. Various articles by me and others, mentioned in commentaries but not referenced in the immediately following article, are referenced in the collection at the end of the book.

In summary, the following articles and commentaries trace the history of a conceptual innovation that started thirty years ago, only as a reinterpretation of Gibbs' canonical ensemble. But it survived the stern disapproval of the Establishment and the often bitter controversy that new ideas always provoke, and led finally to important new applications (Irreversible Statistical Mechanics, Image Reconstruction, Spectrum Analysis) far beyond the purview of 1951. Today, not only do Statistical Mechanics and Statistical Inference not appear as two different fields, even the term 'statistical' is not entirely appropriate. Both are special cases of a simple and general procedure that ought to be called, simply, 'inference'. Of course, the most important nontrivial applications do get involved in much technical detail, which will appear in further articles still being written.

I express my thanks to the Editor of this volume, Roger D. Rosenkrantz, who approached me with the suggestion that these works ought to be collected and reprinted; and who then performed the many necessary but thankless tasks dealing with the business side of this enterprise, which left me free to concentrate on the subject-matter. It is a major understatement to say that without this initiative and help this book would never have appeared.

St. Louis, MO E. T. JAYNES
December 1981

2. INFORMATION THEORY AND STATISTICAL MECHANICS I (1957)

The background of this work is explained in *Where Do We Stand?* (1978), also reprinted below. It started in 1951 as a private communication, the original purpose being only to convey the new ideas of Information Theory to Professor G. Uhlenbeck, in hope of enlisting his support and getting his constructive suggestions as to how these ideas might be implemented in Statistical Mechanics.

Just seventy years earlier – in nice correspondence with our theory of periodicity – a young man named Max Planck had tried to convey some new ideas about entropy to the Establishment Figures of that time, Kirchhoff and Clausius, with the same hope and the same success. Planck recalled the incident (which inspired his famous comment about the mechanism of scientific progress) in his Scientific Autobiography (1949).

In rereading ITSM (I) after many years, it does not appear to contain any actual misstatement of demonstrable fact, although the emphasis and language would be very different if I were writing it today. It is embarrassing to see the word 'bias' used in its colloquial sense, when it is also a technical term of statistics. This must have confused many readers.

I no longer subscribe to the views about 'subjectivist' and 'objectivist' probability theory on page 8. Further experience has taught me and other Bayesians that a single theory of probability suffices for all problems. Connections between probability and frequency appear automatically in Bayesian calculations whenever they are relevant. The remarks about 'objective statistical mechanics' on page 13 would, therefore, be deleted today.

It is now clear that the discussion of ergodicity and the 'principle of macroscopic uniformity' on pages 10–13 should have been amplified and emphasized much more strongly. This material is crucial to understanding why the method works, and why appeal to ergodicity would not help us in thermodynamic predictions. That is, a proof of ergodicity would not in any way affect the predictions that we have already by direct maximization of entropy. Only a proof of non-ergodicity, plus information about which subspace of our present one is actually used by Nature, could alter our predictions.

Conversely, while the success of maximum-entropy predictions does not constitute proof of ergodicity, failure of those predictions would give us strong evidence for non-ergodicity and a clue as to which subspace Nature is using. For this reason, entropy maximizers do not have the same 'fear of failure' that often inhibits other users of statistics — instead we look eagerly for it. It is only when our predictions fail that we obtain new evidence about Nature's workings. Most of the criticisms of this work are from persons who did not comprehend a word of these too brief attempts to explain the logical situation.

A further embarrassment is that in 1956 I knew about the Einstein-de Haas experiment, but not about Barnett's, and so predicted an effect that was already known. But amends were made in the review article with Steve Heims (1962), which extended this crude calculation to a realistic and unified treatment of gyromagnetic effects. It was a particular pleasure that we could vindicate completely Gibbs' rotationally canonical ensemble, which had been rejected fifty years earlier in the Ehrenfest review article (1912).

Information Theory and Statistical Mechanics

E. T. JAYNES
Department of Physics, Stanford University, Stanford, California
(Received September 4, 1956; revised manuscript received March 4, 1957)

Information theory provides a constructive criterion for setting up probability distributions on the basis of partial knowledge, and leads to a type of statistical inference which is called the maximum-entropy estimate. It is the least biased estimate possible on the given information; i.e., it is maximally noncommittal with regard to missing information. If one considers statistical mechanics as a form of statistical inference rather than as a physical theory, it is found that the usual computational rules, starting with the determination of the partition function, are an immediate consequence of the maximum-entropy principle. In the resulting "subjective statistical mechanics," the usual rules are thus justified independently of any physical argument, and in particular independently of experimental verification; whether or not the results agree with experiment, they still represent the best estimates that could have been made on the basis of the information available.

It is concluded that statistical mechanics need not be regarded as a physical theory dependent for its validity on the truth of additional assumptions not contained in the laws of mechanics (such as ergodicity, metric transitivity, equal *a priori* probabilities, etc). Furthermore, it is possible to maintain a sharp distinction between its physical and statistical aspects The former consists only of the correct enumeration of the states of a system and their properties; the latter is a straightforward example of statistical inference.

1. INTRODUCTION

THE recent appearance of a very comprehensive survey[1] of past attempts to justify the methods of statistical mechanics in terms of mechanics, classical or quantum, has helped greatly, and at a very opportune time, to emphasize the unsolved problems in this field.

[1] D. ter Haar, Revs. Modern Phys. 27, 289 (1955).

Although the subject has been under development for many years, we still do not have a complete and satisfactory theory, in the sense that there is no line of argument proceeding from the laws of microscopic mechanics to macroscopic phenomena, that is generally regarded by physicists as convincing in all respects. Such an argument should (a) be free from objection on mathematical grounds, (b) involve no additional arbitrary assumptions, and (c) automatically include an explanation of nonequilibrium conditions and irreversible processes as well as those of conventional thermodynamics, since equilibrium thermodynamics is merely an ideal limiting case of the behavior of matter.

It might appear that condition (b) is too severe, since we expect that a physical theory will involve certain unproved assumptions, whose consequences are deduced and compared with experiment. For example, in the statistical mechanics of Gibbs[2] there were several difficulties which could not be understood in terms of classical mechanics, and before the models which he constructed could be made to correspond to the observed facts, it was necessary to incorporate into them additional restrictions not contained in the laws of classical mechanics. First was the "freezing up" of certain degrees of freedom, which caused the specific heat of diatomic gases to be only $\frac{5}{7}$ of the expected value. Secondly, the paradox regarding the entropy of combined systems, which was resolved only by adoption of the generic instead of the specific definition of phase, an assumption which seems impossible to justify in terms of classical notions.[3] Thirdly, in order to account for the actual values of vapor pressures and equilibrium constants, an additional assumption about a natural unit of volume (h^{3N}) of phase space was needed. However, with the development of quantum mechanics the originally arbitrary assumptions are now seen as necessary consequences of the laws of physics. This suggests the possibility that we have now reached a state where statistical mechanics is no longer dependent on physical hypotheses, but may become merely an example of statistical inference

That the present may be an opportune time to re-examine these questions is due to two recent developments. Statistical methods are being applied to a variety of specific phenomena involving irreversible processes, and the mathematical methods which have proven successful have not yet been incorporated into the basic apparatus of statistical mechanics. In addition, the development of information theory[4] has been felt by many people to be of great significance for statistical mechanics, although the exact way in which it should be applied has remained obscure. In this connection it is essential to note the following. The mere fact that the same mathematical expression $-\sum p_i \log p_i$ occurs both in statistical mechanics and in information theory does not in itself establish any connection between these fields. This can be done only by finding new viewpoints from which thermodynamic entropy and information-theory entropy appear as the same *concept*. In this paper we suggest a reinterpretation of statistical mechanics which accomplishes this, so that information theory can be applied to the problem of justification of statistical mechanics. We shall be concerned with the prediction of equilibrium thermodynamic properties, by an elementary treatment which involves only the probabilities assigned to stationary states. Refinements obtainable by use of the density matrix and discussion of irreversible processes will be taken up in later papers.

Section 2 defines and establishes some of the elementary properties of maximum-entropy inference, and in Secs. 3 and 4 the application to statistical mechanics is discussed. The mathematical facts concerning maximization of entropy, as given in Sec. 2, were pointed out long ago by Gibbs. In the past, however, these properties were given the status of side remarks not essential to the theory and not providing in themselves any justification for the methods of statistical mechanics. The feature which was missing has been supplied only recently by Shannon[4] in the demonstration that the expression for entropy has a deeper meaning, quite independent of thermodynamics. This makes possible a reversal of the usual line of reasoning in statistical mechanics. Previously, one constructed a theory based on the equations of motion, supplemented by additional hypotheses of ergodicity, metric transitivity, or equal *a priori* probabilities, and the identification of entropy was made only at the end, by comparison of the resulting equations with the laws of phenomenological thermodynamics. Now, however, we can take entropy as our starting concept, and the fact that a probability distribution maximizes the entropy subject to certain constraints becomes the essential fact which justifies use of that distribution for inference.

The most important consequence of this reversal of viewpoint is not, however, the conceptual and mathematical simplification which results. In freeing the theory from its apparent dependence on physical hypotheses of the above type, we make it possible to see statistical mechanics in a much more general light. Its principles and mathematical methods become available for treatment of many new physical problems. Two examples are provided by the derivation of Siegert's "pressure ensemble" and treatment of a nuclear polarization effect, in Sec. 5.

2. MAXIMUM-ENTROPY ESTIMATES

The quantity x is capable of assuming the discrete values x_i, ($i=1,2 \cdots,n$). We are not given the corresponding probabilities p_i; all we know is the expectation

---

[2] J. W. Gibbs, *Elementary Principles in Statistical Mechanics* (Longmans Green and Company, New York, 1928), Vol. II of collected works.

[3] We may note here that although Gibbs (reference 2, Chap. XV) started his discussion of this question by saying that the generic definition "seems in accordance with the spirit of the statistical method," he concluded it with, "The perfect similarity of several particles of a system will not in the least interfere with the identification of a particular particle in one case with a particular particle in another. The question is one to be decided in accordance with the requirements of practical convenience in the discussion of the problems with which we are engaged."

[4] C. E. Shannon, Bell System Tech. J. 27, 379, 623 (1948); these papers are reprinted in C. E. Shannon and W. Weaver, *The Mathematical Theory of Communication* (University of Illinois Press, Urbana, 1949).

value of the function $f(x)$:

$$\langle f(x) \rangle = \sum_{i=1}^{n} p_i f(x_i). \qquad (2\text{-}1)$$

On the basis of this information, what is the expectation value of the function $g(x)$? At first glance, the problem seems insoluble because the given information is insufficient to determine the probabilities p_i.[5] Equation (2-1) and the normalization condition

$$\sum p_i = 1 \qquad (2\text{-}2)$$

would have to be supplemented by $(n-2)$ more conditions before $\langle g(x) \rangle$ could be found.

This problem of specification of probabilities in cases where little or no information is available, is as old as the theory of probability. Laplace's "Principle of Insufficient Reason" was an attempt to supply a criterion of choice, in which one said that two events are to be assigned equal probabilities if there is no reason to think otherwise. However, except in cases where there is an evident element of symmetry that clearly renders the events "equally possible," this assumption may appear just as arbitrary as any other that might be made. Furthermore, it has been very fertile in generating paradoxes in the case of continuously variable random quantities,[6] since intuitive notions of "equally possible" are altered by a change of variables.[7] Since the time of Laplace, this way of formulating problems has been largely abandoned, owing to the lack of any constructive principle which would give us a reason for preferring one probability distribution over another in cases where both agree equally well with the available information.

For further discussion of this problem, one must recognize the fact that probability theory has developed in two very different directions as regards fundamental notions. The "objective" school of thought[8,9] regards the probability of an event as an objective property of that event, always capable in principle of empirical measurement by observation of frequency ratios in a random experiment. In calculating a probability distribution the objectivist believes that he is making predictions which are in principle verifiable in every detail, just as are those of classical mechanics. The test of a good objective probability distribution $p(x)$ is: does it correctly represent the observable fluctuations of x?

On the other hand, the "subjective" school of thought[10,11] regards probabilities as expressions of human ignorance; the probability of an event is merely a formal expression of our expectation that the event will or did occur, based on whatever information is available. To the subjectivist, the purpose of probability theory is to help us in forming plausible conclusions in cases where there is not enough information available to lead to certain conclusions; thus detailed verification is not expected. The test of a good subjective probability distribution is does it correctly represent our state of knowledge as to the value of x?

Although the theories of subjective and objective probability are mathematically identical, the concepts themselves refuse to be united. In the various statistical problems presented to us by physics, both viewpoints are required. Needless controversy has resulted from attempts to uphold one or the other in all cases. The subjective view is evidently the broader one, since it is always possible to interpret frequency ratios in this way; furthermore, the subjectivist will admit as legitimate objects of inquiry many questions which the objectivist considers meaningless. The problem posed at the beginning of this section is of this type, and therefore in considering it we are necessarily adopting the subjective point of view.

Just as in applied statistics the crux of a problem is often the devising of some method of sampling that avoids bias, our problem is that of finding a probability assignment which avoids bias, while agreeing with whatever information is given. The great advance provided by information theory lies in the discovery that there is a unique, unambiguous criterion for the "amount of uncertainty" represented by a discrete probability distribution, which agrees with our intuitive notions that a broad distribution represents more uncertainty than does a sharply peaked one, and satisfies all other conditions which make it reasonable.[4] In Appendix A we sketch Shannon's proof that the quantity which is positive, which increases with increasing uncertainty, and is additive for independent sources of uncertainty, is

$$H(p_1 \cdots p_n) = -K \sum_i p_i \ln p_i, \qquad (2\text{-}3)$$

where K is a positive constant. Since this is just the expression for entropy as found in statistical mechanics, it will be called the entropy of the probability distribution p_i; henceforth we will consider the terms "entropy" and "uncertainty" as synonymous.

---

[5] Yet this is precisely the problem confronting us in statistical mechanics, on the basis of information which is grossly inadequate to determine any assignment of probabilities to individual quantum states, we are asked to estimate the pressure, specific heat, intensity of magnetization, chemical potentials, etc , of a macroscopic system. Furthermore, statistical mechanics is amazingly successful in providing accurate estimates of these quantities. Evidently there must be other reasons for this success, that go beyond a mere correct statistical treatment of the problem as stated above

[6] The problems associated with the continuous case are fundamentally more complicated than those encountered with discrete random variables, only the discrete case will be considered here

[7] For several examples, see E. P. Northrop, *Riddles in Mathematics* (D Van Nostrand Company, Inc , New York, 1944), Chap. 8

[8] H Cramer, *Mathematical Methods of Statistics* (Princeton University Press, Princeton, 1946)

[9] W Feller, *An Introduction to Probability Theory and its Applications* (John Wiley and Sons, Inc , New York, 1950)

[10] J M Keynes, *A Treatise on Probability* (MacMillan Company, London, 1921).

[11] H Jeffreys, *Theory of Probability* (Oxford University Press, London, 1939)

It is now evident how to solve our problem; in making inferences on the basis of partial information we must use that probability distribution which has maximum entropy subject to whatever is known. This is the only unbiased assignment we can make; to use any other would amount to arbitrary assumption of information which by hypothesis we do not have. To maximize (2-3) subject to the constraints (2-1) and (2-2), one introduces Lagrangian multipliers λ, μ, in the usual way, and obtains the result

$$p_i = e^{-\lambda - \mu f(x_i)}. \quad (2\text{-}4)$$

The constants λ, μ are determined by substituting into (2-1) and (2-2). The result may be written in the form

$$\langle f(x) \rangle = -\frac{\partial}{\partial \mu} \ln Z(\mu), \quad (2\text{-}5)$$

where
$$\lambda = \ln Z(\mu), \quad (2\text{-}6)$$

$$Z(\mu) = \sum_i e^{-\mu f(x_i)} \quad (2\text{-}7)$$

will be called the partition function

This may be generalized to any number of functions $f(x)$: given the averages

$$\langle f_r(x) \rangle = \sum_i p_i f_r(x_i), \quad (2\text{-}8)$$

form the partition function

$$Z(\lambda_1, \cdots, \lambda_m)$$
$$= \sum_i \exp\{-[\lambda_1 f_1(x_i) + \cdots + \lambda_m f_m(x_i)]\}. \quad (2\text{-}9)$$

Then the maximum-entropy probability distribution is given by

$$p_i = \exp\{-[\lambda_0 + \lambda_1 f_1(x_i) + \cdots + \lambda_m f_m(x_i)]\}, \quad (2\text{-}10)$$

in which the constants are determined from

$$\langle f_r(x) \rangle = -\frac{\partial}{\partial \lambda_r} \ln Z, \quad (2\text{-}11)$$

$$\lambda_0 = \ln Z. \quad (2\text{-}12)$$

The entropy of the distribution (2-10) then reduces to

$$S_{\max} = \lambda_0 + \lambda_1 \langle f_1(x) \rangle + \cdots + \lambda_m \langle f_m(x) \rangle, \quad (2\text{-}13)$$

where the constant K in (2-3) has been set equal to unity. The variance of the distribution of $f_r(x)$ is found to be

$$\Delta^2 f_r = \langle f_r^2 \rangle - \langle f_r \rangle^2 = \frac{\partial^2}{\partial \lambda_r^2} (\ln Z). \quad (2\text{-}14)$$

In addition to its dependence on x, the function f_r may contain other parameters $\alpha_1, \alpha_2, \cdots$, and it is easily shown that the maximum-entropy estimates of the derivatives are given by

$$\left\langle \frac{\partial f_r}{\partial \alpha_k} \right\rangle = -\frac{1}{\lambda_r} \frac{\partial}{\partial \alpha_k} \ln Z. \quad (2\text{-}15)$$

The principle of maximum entropy may be regarded as an extension of the principle of insufficient reason (to which it reduces in case no information is given except enumeration of the possibilities x_i), with the following essential difference. The maximum-entropy distribution may be asserted for the positive reason that it is uniquely determined as the one which is maximally noncommittal with regard to missing information, instead of the negative one that there was no reason to think otherwise. Thus the concept of entropy supplies the missing criterion of choice which Laplace needed to remove the apparent arbitrariness of the principle of insufficient reason, and in addition it shows precisely how this principle is to be modified in case there are reasons for "thinking otherwise."

Mathematically, the maximum-entropy distribution has the important property that no possibility is ignored; it assigns positive weight to every situation that is not absolutely excluded by the given information. This is quite similar in effect to an ergodic property. In this connection it is interesting to note that prior to the work of Shannon other information measures had been proposed[12,13] and used in statistical inference, although in a different way than in the present paper. In particular, the quantity $-\sum p_i^2$ has many of the qualitative properties of Shannon's information measure, and in many cases leads to substantially the same results. However, it is much more difficult to apply in practice. Conditional maxima of $-\sum p_i^2$ cannot be found by a stationary property involving Lagrangian multipliers, because the distribution which makes this quantity stationary subject to prescribed averages does not in general satisfy the condition $p_i \geq 0$. A much more important reason for preferring the Shannon measure is that it is the only one which satisfies the condition of consistency represented by the composition law (Appendix A). Therefore one expects that deductions made from any other information measure, if carried far enough, will eventually lead to contradictions.

3. APPLICATION TO STATISTICAL MECHANICS

It will be apparent from the equations in the preceding section that the theory of maximum-entropy inference is identical in mathematical form with the rules of calculation provided by statistical mechanics. Specifically, let the energy levels of a system be

$$E_i(\alpha_1, \alpha_2, \cdots),$$

where the external parameters α_i may include the volume, strain tensor applied electric or magnetic fields, gravitational potential, etc. Then if we know only the average energy $\langle E \rangle$, the maximum-entropy probabilities of the levels E_i are given by a special case of (2-10), which we recognize as the Boltzmann distribution. This observation really completes our derivation

---
[12] R. A Fisher, Proc Cambridge Phil Soc. **22**, 700 (1925)
[13] J L. Doob, Trans Am. Math. Soc **39**, 410 (1936).

of the conventional rules of statistical mechanics as an example of statistical inference; the identification of temperature, free energy, etc., proceeds in a familiar manner,[14] with results summarized as

$$\lambda_1 = (1/kT), \qquad (3\text{-}1)$$

$$U - TS = F(T, \alpha_1, \alpha_2, \cdots) = -kT \ln Z(T, \alpha_1, \alpha_2, \cdots), \qquad (3\text{-}2)$$

$$S = -\frac{\partial F}{\partial T} = -k \sum_i p_i \ln p_i, \qquad (3\text{-}3)$$

$$\beta_i = kT \frac{\partial}{\partial \alpha_i} \ln Z. \qquad (3\text{-}4)$$

The thermodynamic entropy is identical with the information-theory entropy of the probability distribution except for the presence of Boltzmann's constant.[15] The "forces" β_i include pressure, stress tensor, electric or magnetic moment, etc., and Eqs. (3-2), (3-3), (3-4) then give a complete description of the thermodynamic properties of the system, in which the forces are given by special cases of (2-15); i.e., as maximum-entropy estimates of the derivatives $(\partial E_i / \partial \alpha_k)$.

In the above relations we have assumed the number of molecules of each type to be fixed. Now let n_1 be the number of molecules of type 1, n_2 the number of type 2, etc. If the n_s are not known, then a possible "state" of the system requires a specification of all the n_s as well as a particular energy level $E_i(\alpha_1 \alpha_2 \cdots | n_1 n_2 \cdots)$. If we are given the expectation values

$$\langle E \rangle, \quad \langle n_1 \rangle, \quad \langle n_2 \rangle, \quad \cdots,$$

then in order to make maximum-entropy inferences, we need to form, according to (2-9), the partition function

$$Z(\alpha_1 \alpha_2 \cdots | \lambda_1 \lambda_2 \cdots, \beta) = \sum_{n_1 n_2 \cdots} \sum_i \exp\{-[\lambda_1 n_1 + \lambda_2 n_2 + \cdots + \beta E_i(\alpha_k | n_s)]\}, \qquad (3\text{-}5)$$

and the corresponding maximum-entropy distribution (2-10) is that of the "quantum-mechanical grand canonical ensemble;" the Eqs. (2-11) fixing the constants, are recognized as giving the relation between the chemical potentials

$$\mu_i = -kT\lambda_i, \qquad (3\text{-}6)$$

and the $\langle n_i \rangle$:

$$\langle n_i \rangle = \partial F/\partial \mu_i, \qquad (3\text{-}7)$$

where the free-energy function $F = -kT\lambda_0$, and $\lambda_0 = \ln Z$ is called the "grand potential."[16] Writing out (2-13) for this case and rearranging, we have the usual expression

$$F(T, \alpha_1 \alpha_2 \cdots, \mu_1 \mu_2 \cdots) = \langle E \rangle - TS + \mu_1 \langle n_1 \rangle + \mu_2 \langle n_2 \rangle + \cdots. \qquad (3\text{-}8)$$

It is interesting to note the ease with which these rules of calculation are set up when we make entropy the primitive concept. Conventional arguments, which exploit all that is known about the laws of physics, in particular the constants of the motion, lead to exactly the same predictions that one obtains directly from maximizing the entropy. In the light of information theory, this can be recognized as telling us a simple but important fact: *there is nothing in the general laws of motion that can provide us with any additional information about the state of a system beyond what we have obtained from measurement.* This refers to interpretation of the state of a system at time t on the basis of measurements carried out at time t. For predicting the course of time-dependent phenomena, knowledge of the equations of motion is of course needed. By restricting our attention to the prediction of equilibrium properties as in the present paper, we are in effect deciding at the outset that the only type of initial information allowed will be values of quantities which are observed to be constant in time. Any prior knowledge that these quantities would be constant (within macroscopic experimental error) in consequence of the laws of physics, is then redundant and cannot help us in assigning probabilities.

This principle has interesting consequences. Suppose that a super-mathematician were to discover a new class of uniform integrals of the motion, hitherto unsuspected. In view of the importance ascribed to uniform integrals of the motion in conventional statistical mechanics, and the assumed nonexistence of new ones, one might expect that our equations would be completely changed by this development. This would not be the case, however, unless we also supplemented our prediction problem with new experimental data which provided us with some information as to the likely values of these new constants. *Even if we had a clear proof that a system is not metrically transitive, we would still have no rational basis for excluding any region of phase space that is allowed by the information available to us.* In its effect on our ultimate predictions, this fact is equivalent to an ergodic hypothesis, quite independently of whether physical systems are in fact ergodic.

This shows the great practical convenience of the subjective point of view. If we were attempting to establish the probabilities of different states in the

---

[14] E Schrodinger, *Statistical Thermodynamics* (Cambridge University Press, Cambridge, 1948).

[15] Boltzmann's constant may be regarded as a correction factor necessitated by our custom of measuring temperature in arbitrary units derived from the freezing and boiling points of water Since the product TS must have the dimensions of energy, the units in which entropy is measured depend on those chosen for temperature It would be convenient in general arguments to define an "absolute cgs unit" of temperature such that Boltzmann's constant is made equal to unity Then entropy would become dimensionless (as the considerations of Sec 2 indicate it should be), and the temperature would be equal to twice the average energy per degree of freedom, it is, of course, just the "modulus" Θ of Gibbs.

[16] D ter Haar, *Elements of Statistical Mechanics* (Rinehart and Company, New York, 1954), Chap. 7.

objective sense, questions of metric transitivity would be crucial, and unless it could be shown that the system was metrically transitive, we would not be able to find any solution at all. If we are content with the more modest aim of finding subjective probabilities, metric transitivity is irrelevant. Nevertheless, the subjective theory leads to exactly the same predictions that one has attempted to justify in the objective sense. The only place where subjective statistical mechanics makes contact with the laws of physics is in the enumeration of the different possible, mutually exclusive states in which the system might be. Unless a new advance in knowledge affects this enumeration, it cannot alter the equations which we use for inference.

If the subject were dropped at this point, however, it would remain very difficult to understand why the above rules of calculation are so uniformly successful in predicting the behavior of individual systems. In stripping the statistical part of the argument to its bare essentials, we have revealed how little content it really has; the amount of information available in practical situations is so minute that it alone could never suffice for making reliable predictions. Without further conditions arising from the physical nature of macroscopic systems, one would expect such great uncertainty in prediction of quantities such as pressure that we would have no definite theory which could be compared with experiments. It might also be questioned whether it is not the most probable, rather than the average, value over the maximum-entropy distribution that should be compared with experiment, since the average might be the average of two peaks and itself correspond to an impossible value.

It is well known that the answer to both of these questions lies in the fact that for systems of very large number of degrees of freedom, the probability distributions of the usual macroscopic quantities determined from the equations above, possess a single extremely sharp peak which includes practically all the "mass" of the distribution. Thus for all practical purposes average, most probable, median, or any other type of estimate are one and the same. It is instructive to see how, in spite of the small amount of information given, maximum-entropy estimates of certain functions $g(x)$ can approach practical certainty because of the way the *possible* values of x are distributed. We illustrate this by a model in which the possible values x_i are defined as follows: let n be a non-negative integer, and ϵ a small positive number. Then we take

$$x_1^{n+1} = \epsilon, \quad x_{i+1} - x_i = \epsilon/x_i^n, \quad i = 1, 2, \cdots. \quad (3\text{-}9)$$

According to this law, the x_i increase without limit as $i \to \infty$, but become closer together at a rate determined by n. By choosing ϵ sufficiently small we can make the density of points x_i in the neighborhood of any particular value of x as high as we please, and therefore for a continuous function $f(x)$ we can approximate a sum as closely as we please by an integral taken over a corresponding range of values of x,

$$\sum_i f(x_i) \to \int f(x) \rho(x) dx,$$

where, from (3-9), we have

$$\rho(x) = x^n/\epsilon.$$

This approximation is not at all essential, but it simplifies the mathematics.

Now consider the problem: (A) Given $\langle x \rangle$, estimate x^2. Using our general rules, as developed in Sec. II, we first obtain the partition function

$$Z(\lambda) = \int_0^\infty \rho(x) e^{-\lambda x} dx = \frac{n!}{\epsilon \lambda^{n+1}},$$

with λ determined from (2-11),

$$\langle x \rangle = -\frac{\partial}{\partial \lambda} \ln Z = \frac{n+1}{\lambda}.$$

Then we find, for the maximum-entropy estimate of x^2,

$$\langle x^2 \rangle\{\langle x \rangle\} = Z^{-1} \int_0^\infty x^2 \rho(x) e^{-\lambda x} dx = \frac{n+2}{n+1} \langle x \rangle^2. \quad (3\text{-}10)$$

Next we invert the problem: (B) Given $\langle x^2 \rangle$, estimate x. The solution is

$$Z(\lambda) = \int_0^\infty \rho(x) \exp(-\lambda x^2) dx$$

$$= \frac{\epsilon^{\frac{1}{2}} n!}{2^{n+1}(n/2)!} \cdot \frac{1}{\epsilon \lambda^{\frac{1}{2}(n+1)}},$$

$$\langle x^2 \rangle = -\frac{\partial}{\partial \lambda} \ln Z = \frac{n+1}{2\lambda},$$

$$\langle x \rangle\{\langle x^2 \rangle\} = Z^{-1} \int_0^\infty \rho x(x) \exp(-\lambda x^2) dx$$

$$= \left(\frac{n+1}{2}\right)^{\frac{1}{2}} \frac{(\frac{1}{2}n)!}{[\frac{1}{2}(n+1)]!} \langle x^2 \rangle^{\frac{1}{2}}. \quad (3\text{-}11)$$

The solutions are plotted in Fig. 1 for the case $n=1$. The upper "regression line" represents Eq. (3-10), and the lower one Eq. (3-11). For other values of n, the slopes of the regression lines are plotted in Fig. 2. As $n \to \infty$, both regression lines approach the line at 45°, and thus for large n, there is for all practical purposes a definite functional relationship between $\langle x \rangle$ and $\langle x^2 \rangle$, independently of which one is considered "given," and which one "estimated." Furthermore, as n increases the distributions become sharper; in problem (A) we find for the variance of x,

$$\langle x^2 \rangle - \langle x \rangle^2 = \langle x \rangle^2/(n+1). \quad (3\text{-}12)$$

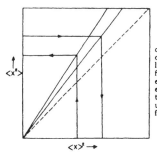

FIG 1 Regression of x and x^2 for state density increasing linearly with x. To find the maximum-entropy estimate of either quantity given the expectation value of the other, follow the arrows

Similar results hold in this model for the maximum-entropy estimate of any sufficiently well-behaved function $g(x)$. If $g(x)$ can be expanded in a power series in a sufficiently wide region about the point $x = \langle x \rangle$, we obtain, using the distribution of problem A above, the following expressions for the expectation value and variance of g:

$$\langle g(x) \rangle = g(\langle x \rangle) + g''(\langle x \rangle) \frac{\langle x \rangle^2}{2(n+1)} + O\left(\frac{1}{n^2}\right), \qquad (3\text{-}13)$$

$$\Delta^2(g) = \langle g^2(x) \rangle - \langle g(x) \rangle^2$$
$$= [g'(\langle x \rangle)]^2 \frac{\langle x \rangle^2}{n+1} + O\left(\frac{1}{n^2}\right). \qquad (3\text{-}14)$$

Conversely, a sufficient condition for x to be well determined by knowledge of $\langle g(x) \rangle$ is that x be a sufficiently smooth monotonic function of g. The apparent lack of symmetry, in that reasoning from $\langle x \rangle$ to g does not require monotonicity of $g(x)$, is due to the fact that the distribution of *possible* values has been specified in terms of x rather than g.

As n increases, the relative standard deviations of all sufficiently well-behaved functions go down like $n^{-\frac{1}{2}}$, it is in this way that definite laws of thermodynamics, essentially independent of the type of information given, emerge from a statistical treatment that at first appears incapable of giving reliable predictions The parameter n is to be compared with the number of degrees of freedom of a macroscopic system.

4. SUBJECTIVE AND OBJECTIVE STATISTICAL MECHANICS

Many of the propositions of statistical mechanics are capable of two different interpretations. The Maxwellian distribution of velocities in a gas is, on the one hand, the distribution that can be realized in the greatest number of ways for a given total energy; on the other hand, it is a well-verified experimental fact Fluctuations in quantities such as the density of a gas or the voltage across a resistor represent on the one hand the uncertainty of our predictions, on the other a measurable physical phenomenon. Entropy as a concept may be regarded as a measure of our degree of ignorance as to the state of a system; on the other hand, for equilibrium conditions it is an experimentally measurable quantity, whose most important properties were first found empirically. It is this last circumstance that is most often advanced as an argument against the subjective interpretation of entropy.

The relation between maximum-entropy inference and experimental facts may be clarified as follows. We frankly recognize that the probabilities involved in prediction based on partial information can have only a subjective significance, and that the situation cannot be altered by the device of inventing a fictitious ensemble, even though this enables us to give the probabilities a frequency interpretation. One might then ask how such probabilities could be in any way relevant to the behavior of actual physical systems A good answer to this is Laplace's famous remark that probability theory is nothing but "common sense reduced to calculation." If we have little or no infor-

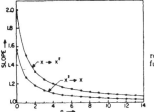

FIG 2 Slope of regression lines as a function of n

mation relevant to a certain question, common sense tells us that no strong conclusions either way are justified. The same thing must happen in statistical inference, the appearance of a broad probability distribution signifying the verdict, "no definite conclusion." On the other hand, whenever the available information is sufficient to justify fairly strong opinions, maximum-entropy inference gives sharp probability distributions indicating the favored alternative. Thus, *the theory makes definite predictions as to experimental behavior only when, and to the extent that, it leads to sharp distributions.*

When our distributions broaden, the predictions become indefinite and it becomes less and less meaningful to speak of experimental verification. As the available information decreases to zero, maximum-entropy inference (as well as common sense) shades continuously into nonsense and eventually becomes useless Nevertheless, at each stage it still represents the best that could have been done with the given information.

Phenomena in which the predictions of statistical mechanics are well verified experimentally are always those in which our probability distributions, for the macroscopic quantities actually measured, have enormously sharp peaks. But the process of maximum-entropy inference is one in which we choose the *broadest possible* probability distribution over the microscopic states, compatible with the initial data. Evidently, such sharp distributions for macroscopic quantities can emerge only if it is true that for *each* of the overwhelming majority of those states to which appreciable weight is assigned, we would have the *same* macroscopic behavior. We regard this, not merely as an interesting side remark, but as the essential fact without which statistical mechanics could have no experimental validity, and indeed without which matter would have no definite macroscopic properties, and experimental physics would be impossible. It is this principle of "macroscopic uniformity" which provides the objective content of the calculations, not the probabilities *per se*. Because of it, the predictions of the theory are to a large extent independent of the probability distributions over microstates. For example, if we choose at random one out of each $10^{10^{10}}$ of the possible states and arbitrarily assign zero probability to all the others, this would in most cases have no discernible effect on the macroscopic predictions.

Consider now the case where the theory makes definite predictions and they are not borne out by experiment. This situation cannot be explained away by concluding that the initial information was not sufficient to lead to the correct prediction; if that were the case the theory would not have given a sharp distribution at all. The most reasonable conclusion in this case is that the enumeration of the different *possible* states (i.e., the part of the theory which involves our knowledge of the laws of physics) was not correctly given. Thus, *experimental proof that a definite prediction is incorrect gives evidence of the existence of new laws of physics*. The failures of classical statistical mechanics, and their resolution by quantum theory, provide several examples of this phenomenon.

Although the principle of maximum-entropy inference appears capable of handling most of the prediction problems of statistical mechanics, it is to be noted that prediction is only one of the functions of statistical mechanics. Equally important is the problem of interpretation; given certain observed behavior of a system, what conclusions can we draw as to the microscopic causes of that behavior? To treat this problem and others like it, a different theory, which we may call objective statistical mechanics, is needed. Considerable semantic confusion has resulted from failure to distinguish between the prediction and interpretation problems, and attempting to make a single formalism do for both.

In the problem of interpretation, one will, of course, consider the probabilities of different states in the objective sense; i.e., the probability of state n is the fraction of the time that the system spends in state n. It is readily seen that one can never deduce the objective probabilities of individual states from macroscopic measurements. There will be a great number of different probability assignments that are indistinguishable experimentally; very severe unknown constraints on the possible states could exist. We see that, although it is now a relevant question, metric transitivity is far from necessary, either for justifying the rules of calculation used in prediction, or for interpreting observed behavior. Bohm and Schützer[17] have come to similar conclusions on the basis of entirely different arguments.

5. GENERALIZED STATISTICAL MECHANICS

In conventional statistical mechanics the energy plays a preferred role among all dynamical quantities because it is conserved both in the time development of isolated systems and in the interaction of different systems. Since, however, the principles of maximum-entropy inference are independent of any physical properties, it appears that in subjective statistical mechanics all measurable quantities may be treated on the same basis, subject to certain precautions. To exhibit this equivalence, we return to the general problem of maximum-entropy inference of Sec. 2, and consider the effect of a small change in the problem. Suppose we vary the functions $f_k(x)$ whose expectation values are given, in an arbitrary way; $\delta f_k(x_i)$ may be specified independently for each value of k and i. In addition we change the expectation values of the f_k in a manner independent of the δf_k; i.e., there is no relation between $\delta \langle f_k \rangle$ and $\langle \delta f_k \rangle$. We thus pass from one maximum-entropy probability distribution to a slightly different one, the variations in probabilities δp_i and the Lagrangian multipliers $\delta \lambda_k$ being determined from the $\delta \langle f_k \rangle$ and $\delta f_k(x_i)$ by the relations of Sec. 2. How does this affect the entropy? The change in the partition function (2-9) is given by

$$\delta \lambda_0 = \delta \ln Z = -\sum_k [\delta \lambda_k \langle f_k \rangle + \lambda_k \langle \delta f_k \rangle], \quad (5\text{-}1)$$

and therefore, using (2-13),

$$\delta S = \sum_k \lambda_k [\delta \langle f_k \rangle - \langle \delta f_k \rangle]$$
$$= \sum_k \lambda_k \delta Q_k. \quad (5\text{-}2)$$

The quantity

$$\delta Q_k = \delta \langle f_k \rangle - \langle \delta f_k \rangle \quad (5\text{-}3)$$

provides a generalization of the notion of infinitesimal heat supplied to the system, and might be called the "heat of the kth type." If f_k is the energy, δQ_k is the heat in the ordinary sense. We see that the Lagrangian multiplier λ_k is the integrating factor for the kth type of heat, and therefore it is possible to speak of the kth type of temperature. However, we shall refer to λ_k as the quantity "statistically conjugate" to f_k, and use the terms "heat" and "temperature" only in their conventional sense. Up to this point, the theory is completely symmetrical with respect to all quantities f_k.

---

[17] D. Bohm and W. Schutzer, Nuovo cimento, Suppl. II, 1004 (1955).

In all the foregoing discussions, the idea has been implicit that the $\langle f_k \rangle$ on which we base our probability distributions represent the results of measurements of various quantities. If the energy is included among the f_k, the resulting equations are identical with those of conventional statistical mechanics. However, in practice a measurement of energy is rarely part of the initial information available; it is the temperature that is easily measurable. In order to treat the experimental measurement of temperature from the present point of view, it is necessary to consider not only the system σ_1 under investigation, but also another system σ_2. We introduce several definitions:

A *heat bath* is a system σ_2 such that

(a) The separation of energy levels of σ_2 is much smaller than any macroscopically measurable energy difference, so that the possible energies E_{2i} form, from the macroscopic point of view, a continuum.

(b) The entropy S_2 of the maximum-entropy probability distribution for given $\langle E_2 \rangle$ is a definite monotonic function of $\langle E_2 \rangle$; i.e., σ_2 contains no "mechanical parameters" which can be varied independently of its energy.

(c) σ_2 can be placed in interaction with another system σ_1 in such a way that only energy can be transferred between them (i.e., no mass, momentum, etc.), and in the total energy $E = E_1 + E_2 + E_{12}$, the interaction term E_{12} is small compared to either E_1 or E_2. This state of interaction will be called *thermal contact*.

A *thermometer* is a heat-bath σ_2 equipped with a pointer which reads its average energy. The scale is, however, calibrated so as to give a number T, called the *temperature*, defined by

$$1/T \equiv dS_2/d\langle E_2 \rangle. \quad (5\text{-}4)$$

In a measurement of temperature, we place the thermometer in thermal contact with the system σ_1 of interest. We are now uncertain not only of the state of the system σ_1 but also of the state of the thermometer σ_2, and so in making inferences, we must find the maximum-entropy probability distribution of the total system $\Sigma = \sigma_1 + \sigma_2$, subject to the available information. A state of Σ is defined by specifying simultaneously a state i of σ_1 and a state j of σ_2 to which we assign a probability p_{ij}. Now however we have an additional piece of information, of a type not previously considered; we know that the interaction of σ_1 and σ_2 may allow transitions to take place between states (ij) and (mn) if the total energy is conserved:

$$E_{1i} + E_{2j} = E_{1m} + E_{2n}.$$

In the absence of detailed knowledge of the matrix elements of E_{12} responsible for these transitions (which in practice is never available), we have no rational basis for excluding the possibility of any transition of this type. Therefore all states of Σ having a given total energy must be considered equivalent; the probability p_{ij} in its dependence on energy may contain only $(E_{1i} + E_{2j})$, not E_{1i} and E_{2j} separately.[18] Therefore, the maximum-entropy probability distribution, based on knowledge of $\langle E_2 \rangle$ and the conservation of energy, is associated with the partition function

$$Z(\lambda) = \sum_{ij} \exp[-\lambda(E_{1i} + E_{2j})] = Z_1(\lambda) Z_2(\lambda), \quad (5\text{-}5)$$

which factors into separate partition functions for the two systems

$$Z_1(\lambda) = \sum_i \exp(-\lambda E_{1i}), \quad Z_2(\lambda) = \sum_j \exp(-\lambda E_{2j}), \quad (5\text{-}6)$$

with λ determined as before by

$$\langle E_2 \rangle = -\frac{\partial}{\partial \lambda} \ln Z_2(\lambda); \quad (5\text{-}7)$$

or, solving for λ by use of (2-13), we find that the quantity statistically conjugate to the energy is the reciprocal temperature:

$$\lambda = dS_2/d\langle E_2 \rangle = 1/T. \quad (5\text{-}8)$$

More generally, this factorization is always possible if the information available consists of certain properties of σ_1 by itself and certain properties of σ_2 by itself. The probability distribution then factors into two independent distributions

$$p_{ij} = p_i(1) p_j(2), \quad (5\text{-}9)$$

and the total entropy is additive:

$$S(\Sigma) = S_1 + S_2. \quad (5\text{-}10)$$

We conclude that the function of the thermometer is merely to tell us what value of the parameter λ should be used in specifying the probability distribution of system σ_1. Given this value and the above factorization property, it is no longer necessary to consider the properties of the thermometer in detail when incorporating temperature measurements into our probability distributions; the mathematical processes used in setting up probability distributions based on energy or temperature measurements are exactly the same but only interpreted differently.

It is clear that any quantity which can be interchanged between two systems in such a way that the total amount is conserved, may be used in place of energy in arguments of the above type, and the fundamental symmetry of the theory with respect to such quantities is preserved. Thus, we may define a "volume bath," "particle bath," "momentum bath," etc., and the probability distribution which gives the most unbiased representation of our knowledge of the state of a system is obtained by the same mathematical procedure whether the available information consists of a measurement of $\langle f_k \rangle$ or its statistically conjugate quantity λ_k.

---

[18] This argument admittedly lacks rigor, which can be supplied only by consideration of phase coherence properties between the various states by means of the density matrix formalism. This, however, leads to the result given.

We now give two elementary examples of the treatment of problems using this generalized form of statistical mechanics.

The pressure ensemble.—Consider a gas with energy levels $E_i(V)$ dependent on the volume. If we are given macroscopic measurements of the energy $\langle E \rangle$ and the volume $\langle V \rangle$, the appropriate partition function is

$$Z(\lambda,\mu) = \int_0^\infty dV \sum_i \exp[-\lambda E_i(V) - \mu V],$$

where λ, μ are Lagrangian multipliers. A short calculation shows that the pressure is given by

$$P = -\langle \partial E_i(V)/\partial V \rangle = \mu/\lambda,$$

so that the quantity statistically conjugate to the volume is

$$\mu = \lambda P = P/kT.$$

Thus, when the available information consists of either of the quantities $(T,\langle E \rangle)$, plus either of the quantities $(P/T,\langle V \rangle)$, the probability distribution which describes this information, without assuming anything else, is proportional to

$$\exp\left\{-\left[\frac{E_i(V) + PV}{kT}\right]\right\}. \quad (5\text{-}11)$$

This is the distribution of the "pressure ensemble" of Lewis and Siegert.[19]

A nuclear polarization effect.—Consider a macroscopic system which consists of σ_1 (a nucleus with spin I), and σ_2 (the rest of the system). The nuclear spin is very loosely coupled to its environment, and they can exchange angular momentum in such a way that the total amount is conserved; thus σ_2 is an angular momentum bath. On the other hand they cannot exchange energy, since all states of σ_1 have the same energy. Suppose we are given the temperature, and in addition are told that the system σ_2 is rotating about a certain axis, which we choose as the z axis, with a macroscopically measured angular velocity ω. Does that provide any evidence for expecting that the nuclear spin I is polarized along the same axis? Let m_2 be the angular momentum quantum number of σ_2, and denote by n all other quantum numbers necessary to specify a state of σ_2. Then we form the partition function

$$Z_2(\beta,\lambda) = \sum_{n,m_2} \exp[-\beta E_2(n,m_2) - \lambda m_2], \quad (5\text{-}12)$$

where $\beta = 1/kT$, and λ is determined by

$$\langle m_2 \rangle = -\frac{\partial}{\partial \lambda} \ln Z_2 = \frac{B\omega}{\hbar}, \quad (5\text{-}13)$$

where B is the moment of inertia of σ_2. Then, our most unbiased guess is that the rotation of the molecular

[19] M. B. Lewis and A. J. F. Siegert, Phys. Rev. **101**, 1227 (1956).

surroundings should produce on the average a nuclear polarization $\langle m_1 \rangle = \langle I_z \rangle$, equal to the Brillouin function

$$\langle m_1 \rangle = -\frac{\partial}{\partial \lambda} \ln Z_1(\lambda), \quad (5\text{-}14)$$

where

$$Z_1(\lambda) = \sum_{m=-I}^{I} e^{-\lambda m}. \quad (5\text{-}15)$$

In the case $I = \frac{1}{2}$, the polarization reduces to

$$\langle m_1 \rangle = -\tfrac{1}{2} \tanh(\tfrac{1}{2}\lambda). \quad (5\text{-}16)$$

If the angular velocity ω is small, (5-12) may be approximated by a power series in λ:

$$Z_2(\beta,\lambda) = Z_2(\beta,0)[1 - \lambda \langle m_2 \rangle_0 + \tfrac{1}{2}\lambda^2 \langle m_2^2 \rangle_0 + \cdots],$$

where $\langle \ \rangle_0$ stands for an expectation value in the nonrotating state. In the absence of a magnetic field $\langle m_2 \rangle_0 = 0$, $\hbar^2 \langle m_2^2 \rangle_0 = kTB$, so that (5-13) reduces to

$$\lambda = -\hbar\omega/kT. \quad (5\text{-}17)$$

Thus, the predicted polarization is just what would be produced by a magnetic field of such strength that the Larmor frequency $\omega_L = \omega$. If $|\lambda| \ll 1$, the result may be described by a "dragging coefficient"

$$\langle m_1 \rangle = \frac{\hbar^2 I(I+1)}{3kTB} \langle m_2 \rangle. \quad (5\text{-}18)$$

There is every reason to believe that this effect actually exists, it is closely related to the Einstein-de Haas effect. It is especially interesting that it can be predicted in some detail by a form of statistical mechanics which does not involve the energy of the spin system, and makes no reference to the mechanism causing the polarization. As a numerical example, if a sample of water is rotated at 36 000 rpm, this should polarize the protons to the same extent as would a magnetic field of about 1/7 gauss. This should be accessible to experiment. A straightforward extension of these calculations would reveal how the effect is modified by nuclear quadrupole coupling, in the case of higher spin values.

6. CONCLUSION

The essential point in the arguments presented above is that we accept the von-Neumann—Shannon expression for entropy, very literally, as a measure of the amount of uncertainty represented by a probability distribution; thus entropy becomes the primitive concept with which we work, more fundamental even than energy. If in addition we reinterpret the prediction problem of statistical mechanics in the subjective sense, we can derive the usual relations in a very elementary way without any consideration of ensembles or appeal to the usual arguments concerning ergodicity or equal *a priori* probabilities. The principles and mathematical methods of statistical mechanics are seen to be of much more general applicability than conventional arguments would lead one to suppose. In the problem of prediction, the maximization of entropy is not an application of a law of physics, but merely a method of reasoning which ensures that no unconscious arbitrary assumptions have been introduced.

APPENDIX A. ENTROPY OF A PROBABILITY DISTRIBUTION

The variable x can assume the discrete values $(x_1, \cdots x_n)$. Our partial understanding of the processes which determine the value of x can be represented by assigning corresponding probabilities $(p_1, \cdots, p_n)$. We ask, with Shannon,[4] whether it is possible to find any quantity $H(p_1 \cdots p_n)$ which measures in a unique way the amount of uncertainty represented by this probability distribution. It might at first seem very difficult to specify conditions for such a measure which would ensure both uniqueness and consistency, to say nothing of usefulness Accordingly it is a very remarkable fact that the most elementary conditions of consistency, amounting really to only one composition law, already determines the function $H(p_1 \cdots p_n)$ to within a constant factor. The three conditions are:

(1) H is a continuous function of the p_i.
(2) If all p_i are equal, the quantity $A(n) = H(1/n, \cdots, 1/n)$ is a monotonic increasing function of n.
(3) The composition law. Instead of giving the probabilities of the events $(x_1 \cdots x_n)$ directly, we might group the first k of them together as a single event, and give its probability $w_1 = (p_1 + \cdots + p_k)$; then the next m possibilities are assigned the total probability $w_2 = (p_{k+1} + \cdots + p_{k+m})$, etc. When this much has been specified, the amount of uncertainty as to the composite events is $H(w_1 \cdots w_r)$. Then we give the conditional probabilities $(p_1/w_1, \cdots, p_k/w_1)$ of the ultimate events $(x_1 \cdots x_k)$, given that the first composite event had occurred, the conditional probabilities for the second composite event, and so on. We arrive ultimately at the same state of knowledge as if the $(p_1 \cdots p_n)$ had been given directly, therefore if our information measure is to be consistent, we must obtain the same ultimate uncertainty no matter how the choices were broken down in this way. Thus, we must have

$$H(p_1 \cdots p_n) = H(w_1 \cdots w_r) + w_1 H(p_1/w_1, \cdots, p_k/w_1) + w_2 H(p_{k+1}/w_2, \cdots, p_{k+m}/w_2) + \cdots. \quad \text{(A-1)}$$

The weighting factor w_1 appears in the second term because the additional uncertainty $H(p_1/w_1, \cdots, p_k/w_1)$ is encountered only with probability w_1. For example, $H(1/2, 1/3, 1/6) = H(1/2, 1/2) + \frac{1}{2} H(2/3, 1/3)$.

From condition (1), it is sufficient to determine H for all rational values

$$p_i = n_i / \sum n_i,$$

with n_i integers. But then condition (3) implies that H is determined already from the symmetrical quantities $A(n)$. For we can regard a choice of one of the alternatives $(x_1 \cdots x_n)$ as a first step in the choice of one of

$$\sum_{i=1}^{n} n_i$$

equally likely alternatives, the second step of which is also a choice between n_i equally likely alternatives. As an example, with $n = 3$, we might choose $(n_1, n_2, n_3) = (3, 4, 2)$. For this case the composition law becomes

$$H\left(\frac{3}{9}, \frac{4}{9}, \frac{2}{9}\right) + \frac{3}{9} A(3) + \frac{4}{9} A(4) + \frac{2}{9} A(2) = A(9).$$

In general, it could be written

$$H(p_1 \cdots p_n) + \sum_i p_i A(n_i) = A(\sum n_i). \quad \text{(A-2)}$$

In particular, we could choose all n_i equal to m, whereupon (A-2) reduces to

$$A(m) + A(n) = A(mn). \quad \text{(A-3)}$$

Evidently this is solved by setting

$$A(n) = K \ln n, \quad \text{(A-4)}$$

where, by condition (2), $K > 0$. For a proof that (A-4) is the only solution of (A-3), we refer the reader to Shannon's paper.[4] Substituting (A-4) into (A-2), we have the desired result,

$$H(p_1 \cdots p_n) = K \ln(\sum n_i) - K \sum p_i \ln n_i,$$
$$= -K \sum_i p_i \ln p_i. \quad \text{(A-5)}$$

3. INFORMATION THEORY AND STATISTICAL MECHANICS II (1957)

This sequel had the primary purpose of extending the *Maxent* formalism to the density matrix, but it also went much more deeply into conceptual matters. Its weakness was in trying to do too much. A single article communicates best if it confines itself to a single topic, but this one touched lightly on a dozen matters, each of which needs a full article to do it justice. In particular, all the ramifications of the 'Information Game' and its relation to stochastic theory, could easily fill a book. The article remains, to this day, a source of unfinished fragments of ideas in need of development.

The factorization property of the density matrix, Eq. (10.18), was at the time new and startling; for me this consistency in the handling of several independent pieces of information, in an area undreamt of in the original Information Theory of Shannon, was important evidence that we were on the right track.

The material on array probabilities and the hierarchy of unitary transformations under which the density matrix is invariant, has implications for quantum theory beyond statistical mechanics, not yet brought out. In fact, the remarks of Sections 7–9 do not hint at the great ferment of thought that went on in the years 1951–1956 when this article was being written. Some of it is recorded in the Stanford Thesis of Ray Nelson (1956).

The discussion of irreversible processes at the end is incomplete. The intention was to rectify this with a third article, built in the image of quantum electrodynamics; but my views on that topic started to change, and there followed a long search for a better basis for nonequilibrium theory, in which the partition functional generalization emerged as the only satisfactory approach. Finally, with the theory of macroscopic sources of Wm. C. Mitchell (1967), the original plan seemed altogether crude and unnecessary, and the third article was never written.

The reaction of some readers to my use of the word 'subjective' in these articles, was astonishing. Thereafter I derived a certain amount of malicious pleasure from sitting back to enjoy the spectacle. There is something so patently ridiculous in the sight of a grown man recoiling in horror from something so harmless as a three-syllable word. 'Subjective' must surely be the most effective scare word yet invented. Yet it was used in what still seems a valid sense: 'depending on the observer'.

In Euclidean geometry the coordinates of a point are 'subjective' in the sense that they depend on the orientation of the observer's coordinate system; while the distance between two points is 'objective' in the sense that it is independent of the observer's orientation. That is all I ever meant by the term; yet twenty-five years later the shock waves are still arriving.

While this was being written there appeared the article of Denbigh (1981), attacking my statement that thermodynamic entropy is 'subjective' in the sense that it depends on which macroscopic coordinates we use to define the thermodynamic state. Of course, he does not deny that entropy does so depend. As far as I can see, there is no disagreement between us on any question of fact, and I am reduced to conjecturing what the word 'subjective' must mean to him; evidently it is something too terrible to divulge.

Information Theory and Statistical Mechanics. II

E. T. JAYNES
Department of Physics, Stanford University, California
(Received March 15, 1957)

Treatment of the predictive aspect of statistical mechanics as a form of statistical inference is extended to the density-matrix formalism and applied to a discussion of the relation between irreversibility and information loss A principle of "statistical complementarity" is pointed out, according to which the empirically verifiable probabilities of statistical mechanics necessarily correspond to incomplete predictions A preliminary discussion is given of the second law of thermodynamics and of a certain class of irreversible processes, in an approximation equivalent to that of the semiclassical theory of radiation.

It is shown that a density matrix does not in general contain all the information about a system that is relevant for predicting its behavior In the case of a system perturbed by random fluctuating fields, the density matrix cannot satisfy any differential equation because $\dot{\rho}(t)$ does not depend only on $\rho(t)$, but also on past conditions The rigorous theory involves stochastic equations in the type $\rho(t) = \mathcal{G}(t,0)\rho(0)$, where the operator $\mathcal{G}$ is a functional of conditions during the entire interval $(0 \to t)$ Therefore a general theory of irreversible processes cannot be based on differential rate equations corresponding to time-proportional transition probabilities However, such equations often represent useful approximations.

INTRODUCTION

IN a previous paper[1] the prediction of equilibrium thermodynamic properties was developed as a form of statistical inference, based on the Shannon[2] concept of entropy as an information measure, and the subjective interpretation of probabilities. The guiding principle is that the probability distribution over microscopic states which has maximum entropy subject to whatever is known, provides the most unbiased representation of our knowledge of the state of the system. The maximum-entropy distribution is the broadest one compatible with the given information; it assigns positive weight to every possibility that is not ruled out by the initial data.

This method of inference is extended in the following sections (numbered consecutively from those of I), to the density-matrix formalism, which makes possible the treatment of time-dependent phenomena. It is then applied to a discussion of the relation of information loss and irreversibility, and to a treatment of relaxation processes in an approximation equivalent to that of the semiclassical theory of radiation. The more rigorous treatment, corresponding to quantum electrodynamics, will be taken up in a later paper.

Our picture of a prediction process is as follows. At the initial time $t=0$ certain measurements are made. In practice, these will always represent far less than the maximum observation which would enable us to determine a definite pure state. Therefore, we must have recourse to maximum-entropy inference in order to represent our degree of knowledge about the system in a way free of arbitrary assumptions with regard to missing information.[3] As time goes on, each state of the maximum-entropy distribution changes due to perturbations that are in general unknown; thus it "spreads out" into several possibilities, and our initial knowledge as to the state of the system is gradually lost. In the "semiclassical" approximation considered here, the final state of affairs is usually one in which the initial information is completely lost, the density matrix relaxing into a multiple of the unit matrix. The prediction of thermal equilibrium, in which the limiting form of the density matrix is that of the Boltzmann distribution with finite temperature, is found only by using a better approximation which takes into account the quantum nature of the surroundings.

It is of the greatest importance to recognize that in all of this semiclassical theory it is possible to maintain the view that the system is at all times in some definite but unknown pure state, which changes because of definite but unknown external forces, the probabilities represent only our ignorance as to the true state. With such an interpretation the expression "irreversible process" represents a semantic confusion; it is not the physical process that is irreversible, but rather our ability to follow it. The second law of thermodynamics then becomes merely the statement that although our information as to the state of a system may be lost in a variety of ways, the only way in which it can be gained is by carrying out further measurements. Essential for this is the fact, analogous to Liouville's theorem, that in semiclassical approximation the laws of physics do not provide any tendency for systems initially in different states to "accumulate" in certain final states in preference to others; i.e., the time-development matrix is unitary.

In opposition to the foregoing views, one may assert

---
[1] E T Jaynes, Phys Rev **106**, 620 (1957) Hereinafter referred to as I
[2] C E. Shannon, Bell System Tech J **27**, 379, 623 (1948) These papers are reprinted in C E Shannon and W Weaver, *The Mathematical Theory of Communication* (University of Illinois Press, Urbana, 1949)
[3] A very interesting quotation from J W Gibbs [*Collected Works* (Longmans, Green and Company, New York, 1928), Vol II, p 180] suggests the same basic idea. In discussing the interaction of a body and a heat-bath, he says "The series of phases through which the whole system runs in the course of time may not be entirely determined by the energy, but may depend on the initial phase in other respects In such cases the ensemble obtained by the microcanonical distribution of the whole system, which includes all possible time-ensembles combined in the proportion which seems least arbitrary, will better represent than any one time-ensemble the effect of the bath "

that irreversibility is not merely a loss of human information; it is an experimental fact, well recognized long before the development of statistical mechanics. Furthermore, the relaxation times calculated below are not merely measures of the rate at which we lose information; they are experimentally measurable quantities expressing the rate at which physical systems approach equilibrium. Therefore, the probabilities involved in our calculations must be ascribed some objective meaning independent of human knowledge.

Objections of this type have already been answered in large part in I, particularly Sec. 4. However, we wish to indicate briefly how those arguments apply to the case of time-dependent phenomena. The essential fact is again the "principle of macroscopic uniformity." In the first place, it has been shown that the only quantities for which maximum-entropy inference makes definite predictions are those for which we obtain sharp probability distributions. Since maximum-entropy inference uses the broadest distribution compatible with the initial data, the predictable properties must be characteristic of the great majority of those states to which appreciable weight is assigned. Maximum-entropy inference can never lead us astray, for any quantity which it is incapable of predicting will betray that fact by yielding a broad probability distribution.

We can, however, say much more than this. We take it as self-evident that the features of irreversible processes which are experimentally reproducible are precisely those characteristic of most of the states compatible with the conditions of the experiment. Suppose that maximum-entropy inference based on knowledge of the experimentally imposed conditions makes a definite prediction of some phenomenon, and it is found experimentally that no such phenomenon exists. Then the predicted property is characteristic of most of the states appearing in the subjective maximum-entropy distribution, but it is not characteristic of most of the states physically allowed by the experimental conditions. Consider, on the other hand, the possibility that a phenomenon might be found which is experimentally reproducible but *not* predictable by maximum-entropy inference. This phenomenon must be characteristic of most of the states allowed by the experimental conditions, but it is not characteristic of most of the states in the maximum-entropy distribution. In either case, there must exist new physical states, or new *constraints* on the physically accessible states, not contained in the presently known laws of physics.

In summary, we assert that *if it can be shown that the class of phenomena predictable by maximum-entropy inference differs in any way from the class of experimentally reproducible phenomena, that fact would demonstrate the existence of new laws of physics, not presently known.* Assuming that this occurs, and the new laws of physics are eventually worked out, then maximum-entropy inference based on the new laws will again have this property.

From this we see that adoption of subjective probabilities in no way weakens the theory in its ability to give reliable and useful results. On the contrary, the full power of statistical mechanics cannot be seen until one makes this distinction between its subjective and objective aspects. Once this is done, its mathematical rules become a methodology for a very general type of scientific reasoning.

7. REPRESENTATION OF A QUANTUM-MECHANICAL SYSTEM

We now develop a method of representing any state of knowledge of a quantum-mechanical system, leaving aside for the moment any consideration of how this knowledge might have been obtained. Suppose that on the basis of the information available we conclude that the system may be in the "pure state" ψ_1 with probability w_1, or it may be in the state ψ_2 with probability w_2, etc. The various alternative possibilities ψ_i are not necessarily mutually orthogonal, but each may be expanded in terms of a complete orthonormal set of functions u_k:

$$\psi_i = \sum_k u_k a_{ki}. \qquad (7.1)$$

This state of knowledge may be visualized in a geometrical fashion by considering a complex function space, whose dimensionality may be finite or infinite, in which the state ψ_i is represented by a point P_i with coordinates a_{ki}, $k=1, 2, \cdots$. At P_i, place a weight w_i; thus the state of knowledge is described by a set (which may be discrete or continuous) of weighted points; such a set will be called an *array*. Since each of the possible wave functions is normalized to unity,

$$(\psi_i, \psi_i) = \int |\psi_i|^2 d\tau = 1,$$

we have

$$\sum_k |a_{ki}|^2 = 1, \qquad (7.2)$$

and all points P_i are at unit "distance" from the origin, on the surface of the unit hypersphere.

If each of the possible states ψ_i satisfies the same Schrödinger equation,

$$i\hbar \dot\psi = H\psi,$$

then as time goes on the function space as a whole is subjected to a unitary transformation, so that all "distances" and scalar products

$$(\psi_i, \psi_j) = \int \psi_i{}^* \psi_j d\tau$$

remain invariant, and the entire motion of the array may be visualized as a "rigid rotation" of the hypersphere. An array with this behavior will be called *simple*. A simple array is conceptually somewhat like a microcanonical ensemble; it consists of points lying on a closed surface which are subjected, in consequence of the equations of motion, to a measure-preserving transformation which continually unfolds as t increases.

The transformation with time may be of a different type; much more interesting is the case where the initial information is of the form: "The system may be in state ψ_i with probability w_i, and in this case the Hamiltonian will be H_i." Then different parts of the array are subjected to different rotations, and separations or interpenetrations occur. Such an array will be called *compound*. It arises, for example, when we have a system consisting of two coupled spins in a strong magnetic field, and we wish to describe our knowledge of the state of one of them.

Consider a measurable quantity represented by a Hermitian operator F; in state ψ_i its expectation value is

$$\langle F \rangle_i = (\psi_i, F\psi_i) = \sum_{kn} a_{ki} a_{ni}^* F_{nk}, \quad (7.3)$$

where $F_{nk} = (u_n, Fu_k)$ are the matrix elements of F in the u_k representation. The average of (7.3) over the array is

$$\langle F \rangle = \sum_i w_i \langle F \rangle_i = \text{Tr}(\rho F), \quad (7.4)$$

where

$$\rho_{kn} = \sum_i w_i a_{ki} a_{ni}^* = \langle a_k a_n^* \rangle \quad (7.5)$$

is the density matrix. The probability $p(f)$ that a measurement of F will yield the particular eigenvalue f, is also expressible as an expectation value; define the projection operator O by $O\psi = (\varphi, \psi)\varphi$, where φ is the corresponding normalized eigenfunction of $F: F\varphi = f\varphi$. Then

$$p(f) = \langle O \rangle = \text{Tr}(\rho O). \quad (7.6)$$

From (7.5) it is seen that in general an infinite number of different arrays, representing different mixtures of pure states, all lead to the same density matrix. The most general discrete array which leads to a given density matrix ρ corresponds to the most general matrix A (not necessarily square) for which

$$\rho = AA^\dagger, \quad (7.7)$$

the dagger denoting the Hermitian conjugate. An array is uniquely determined by A, for from (7.2) and (7.5) we have

$$A_{ki} = a_{ki} w_i^{\frac{1}{2}}, \quad \sum_k |A_{ki}|^2 = w_i.$$

To find another array with the same density matrix, insert a matrix U:

$$\rho = (AU)(U^{-1}A^\dagger).$$

This has the form $BB^\dagger$ with $B = AU$ if and only if U is unitary; thus the group of transformations from one array of n states to another of n states is isomorphic with the group of unitary transformations in n dimensions. These are not, however, transformations of the wave functions ψ_i, but of the *probability-normalized* wave functions

$$\Psi_i = \psi_i w_i^{\frac{1}{2}}. \quad (7.8)$$

If we carry out the unitary transformation

$$\Phi_j = \sum_i \Psi_i U_{ij}, \quad (7.9)$$

and write

$$\Phi_j = \varphi_j p_j^{\frac{1}{2}},$$

where φ_j is normalized to unity, then the array in which state φ_j has probability p_j, leads to the same density matrix as the original array $\{\psi_i, w_i\}$. Evidently an array is determined uniquely by specifying a set $\{\Psi_i\}$ of probability-normalized states.

From an array $\{\Psi_i\}$ of n states we can construct new arrays of $(n+1)$ states. Define $\Psi_{n+1} \equiv 0$, then new transformations of the form (7.9) are possible, in which U is a unitary matrix of dimensionality $(n+1)$. These generate an infinite number of new arrays for which, in general, all $(n+1)$ states Φ_j are different from each other and from zero. The inverse process of *contracting* an array to one of fewer states is possible if any linear combination of the ψ_i vanishes.

An array of n states will be called *minimal* with respect to its density matrix ρ if no array exists which leads to ρ with fewer than n states. The states of an array are linearly independent if and only if the array is minimal.

In general, a given density matrix can be represented in only one way as a mixture of orthogonal states. Since ρ is Hermitian, there always exists a unitary matrix U which diagonalizes it,

$$d = U\rho U^{-1}, \quad (7.10)$$

with $d_{mn} = d_m \delta_{mn}$. If the eigenvalues d_m of ρ are non-degenerate, only one such matrix U exists. The basis functions of the new representation in which ρ is diagonal,

$$v_m = \sum_k u_k U_{km}^{-1}, \quad (7.11)$$

are the orthogonal states which, when mixed with probabilities d_m, lead to the given density matrix.

Suppose we have a density matrix ρ and a state φ which is considered a "candidate" for inclusion in a minimal array which will lead to ρ. What is the probability $p_A(\varphi)$ which should be assigned to φ in such an array? To answer this, we first construct the orthogonal array $\{v_m, d_m\}$, and expand

$$\varphi = \sum_m v_m C_m$$

If this is to be equivalent to one of the columns of (7.9), it is necessary that

$$\frac{1}{p_A} = \sum_m \frac{|C_m|^2}{d_m}. \quad (7.12)$$

This is uniquely determined by the density matrix and the state φ, regardless of which other states φ_j might also appear in the array. The array probability p_A is in general different from the measurement probability (7.6), which is equal to

$$p_M(\varphi) = \sum_m d_m |C_m|^2. \quad (7.13)$$

It is readily shown that $p_M \geq p_A$, with equality if and only if φ is an eigenstate of ρ.

The representation in terms of orthogonal states is important in connection with the entropy which measures our knowledge of the system. It might be thought that for an array $\{\psi_i, w_i\}$ we could define an entropy by

$$S_A = -\sum_i w_i \ln w_i. \qquad (7.14)$$

This, however, would not be satisfactory because the w_i are not in general the probabilities of mutually exclusive events. According to quantum mechanics, if the state is known to be ψ_i, then the probability of finding it upon measurement to be ψ_j, is $|(\psi_j, \psi_i)|^2$. Thus, the probabilities w_i refer to independent, mutually exclusive events only when the states ψ_i of the array are orthogonal to each other, and only in this case is the expression (7.14) for entropy satisfactory. This array of orthogonal states has another important property; consider the totality of all possible arrays which lead to a given density matrix, and the corresponding expressions (7.14). The array for which (7.14) attains its minimum value is the orthogonal one, which therefore provides, in the sense of information content, the most economical description of the freedom of choice implied by a density matrix (Appendix A).

For the orthogonal array, the w_i in (7.14) are identical with the eigenvalues d_i of the density matrix, so for numerical calculation of entropy given ρ, one would find the eigenvalues and use the formula

$$S = -\sum_i d_i \ln d_i. \qquad (7.15)$$

In general discussions it is convenient to express this

$$S = -\text{Tr}(\rho \ln \rho). \qquad (7.16)$$

Since this could also be written as $S = -\langle \ln \rho \rangle$, it is the natural extension to quantum mechanics of the Gibbs definition of entropy.

Equation (7.16) assigns zero entropy to any pure state, whether stationary or not. It has been criticized on the grounds that according to the Schrödinger equation of motion it would be constant in time, and thus one could not account for the second law of thermodynamics; this has led some authors[4,5] to propose instead the expression

$$S = -\sum_k \rho_{kk} \ln \rho_{kk}, \qquad (7.17)$$

which involves only diagonal elements of ρ in the energy representation, for which a "quantum-mechanical spreading" phenomenon can be demonstrated. It will be shown in detail below how the objections to (7.16) may be answered. With regard to (7.17), we note that it does not assign the same entropy to all pure states, but von Neumann[6] has shown that any

[4] R. C. Tolman, *The Principles of Statistical Mechanics* (Clarendon Press, Oxford, 1938).
[5] D. ter Haar, *Elements of Statistical Mechanics* (Rinehart and Company, Inc., New York, 1954).
[6] J. von Neumann, *Mathematische Grundlagen der Quantenmechanik* (Dover Publications, New York, 1943), Chap V.

pure state may be converted reversibly and adiabatically into any other pure state.

Since, according to (7.4), knowledge of ρ enables one to calculate the expectation value of any Hermitian operator, it is tempting to conclude that the density matrix contains all of our information as to the objective state of the system. Thus, although many different arrays would all lead to the same density matrix, the differences between them would be considered physically meaningless, only their second moments (7.5) corresponding to any physical predictions. The concept of any array as something separate and distinct from a density matrix might then appear superfluous. That this is not the case, however, will be seen in Sec. 13 below, where it is shown that the resolution of a compound array into independent simple arrays may represent useful information which cannot be expressed in terms of the resultant density matrix.

8. SUFFICIENCY AND COMPLETENESS OF THE DENSITY MATRIX

If a density matrix provides a definite probability assignment for each possible outcome of a certain experiment, in a way that makes full use of all of the available relevant information, we shall say that ρ is *sufficient* for that experiment. A density matrix that is sufficient for all conceivable experiments on a system will be called *complete* for that system. Strictly speaking, we should always describe a density matrix as sufficient or complete *relative* to certain initial information.

The assertion that complete density matrices exist involves several assumptions, in particular that all measurable quantities may be represented by Hermitian operators, and that all experimental measurements may be expressed in terms of expectation values. We do not wish to go into these questions, but only to note the following. Even if it be granted that it is always possible in principle to operate with a complete density matrix, it would often be extremely awkward and inconvenient to do so in practice, because it would require us to consider the density matrix and dynamical quantities as operators in a much larger function space than we wish to use.

To see this by a simple example, consider a "molecular beam" experiment in which particles of spin $\tfrac{1}{2}$ are prepared by apparatus A, then sent into a detection system B which determines whether the spin is up or down with respect to some chosen z axis. Assume, for simplicity, that only one particle at a time is processed in this way. A particle thus has, for our purposes, two possible states u_+ and u_-; our knowledge of the nature of the apparatus A could be incorporated into an array and its corresponding (2×2) density matrix, from which we can calculate the probability of finding the spin aligned in any particular direction. Thus, the (2×2) density matrix adequately represents our state of knowledge as to the outcome of any spin measurement made on a single particle; i.e., it is a sufficient statistic for any such measurement. The question is, does it also adequately represent our knowledge of the *ensemble* of particles (assuming that the apparatus A is "stationary," so that each particle, considered by itself, would be represented by the same density matrix). More specifically, is it possible for apparatus A to produce a physical situation which can be measured in our detection apparatus, but for which the (2×2) density matrix gives no probability assignment? One such property is easily found; the detecting apparatus tells us not only the fraction of spins aligned along the $+z$ axis, but also the *order* in which spin up and spin down occurred, so that correlations between spin states of successive particles can be observed. Now all possible such correlations can be described only by considering the entire ensemble of N particles as a single quantum-mechanical system with 2^N possible states, and therefore a density matrix which is a sufficient statistic for all conceivable measurements on the spin system must have 2^N rows and columns.[7] This, however, would completely destroy the simplicity of the theory, and in practice we would probably prefer to retain the original (2×2) density matrix for predicting the results of measurements on single particles, while recognizing its insufficiency for other measurements which the same apparatus could perform.

9. SUBJECTIVE AND OBJECTIVE INTERPRETATIONS

The topic of Sec. 8 is closely related to some of the most fundamental questions in physics. According to quantum mechanics, if a system is known to be in state ψ_i, then the probability that measurement of the quantity F will result in the particular eigenvalue f, is $\langle O \rangle_i$, where O is the projection operator of Eq. (7.6). Are we to interpret this probability in the objective or subjective sense, i.e., are the probability statements of quantum mechanics expressions of empirically verifiable laws of physics or merely expressions of our incomplete ability to predict, whether due to a defect in the theory or to incomplete initial information? The current interpretation of quantum mechanics favors the first view, but it is important to note that the whole content of the theory depends critically on just what we mean by "probability." In calling a probability objective, we do not mean that it is necessarily "correct," but only that a conceivable experiment exists by which its correctness or incorrectness could be empirically determined. In calling a probability assignment subjective, we mean that it is not a physical property of any system, but only a means of describing our information about the system; therefore it is meaningless to speak of verifying it empirically.

Is there any operational meaning to the statement

[7] This is a very conservative statement. It would be more realistic to assume that all the coordinates of apparatus A must also be included in the space upon which this complete density matrix operates.

that the probabilities of quantum mechanics are objective? If so, we should be able to devise an experiment which will measure these probabilities, for example the probability that a measurement of the quantity F will give the result f. In order to do this, we will need to repeat a measurement of F an indefinitely large number N of times, with systems that have all been prepared in exactly the same way, and record the fraction of cases in which the particular result f was obtained. Which density matrix should we use to predict the result of this experiment? In principle, we should always use the one which contains the greatest amount of information about the ensemble of N systems; i.e., which is complete. The apparatus which prepares them may be producing correlations, thus the ensemble of N systems should be considered as a single large quantum-mechanical system. The probability statements which come from the theory are then of the form, "the probability that system 1 will yield the result f_1, *and* system 2 will yield the value f_2, $\cdots$, is $p(f_1 \cdots f_N)$." But then measurement of F in each of the N small systems is not N repetitions of an experiment; it is only a single experiment from the standpoint of the total system Clearly, no probability assignment can be verified by a single measurement. Note that the question whether correlations were in fact present between different systems is irrelevant to the question of principle involved; even if the distribution factors

$$p(f_1 \cdots f_N) = p_1(f_1) p_2(f_2) \cdots p_N(f_N) \qquad (9.1)$$

it remains a joint distribution, not one for a single system. We can, of course, always obtain the single-system probabilities by summation:

$$p_1(f_1) = \sum_{f_2} \cdots \sum_{f_N} p(f_1, f_2 \cdots f_N), \qquad (9.2)$$

but $p_1(f_1)$ now refers specifically to system 1, and the results of measurements on the other systems are irrelevant to the question whether $p_1(f_1)$ was verified. We cannot avoid the difficulty by repeating all this M times, because for that experiment the complete density matrix would refer to all NM systems, and we would be in exactly the same situation. Thus, the probability statements obtained from a complete density matrix cannot be verified.

In practice, of course, one will never bother with such considerations, but will find a density matrix which operates only on the space of a single system and incorporates as much information as possible subject to that limitation. The probability $p(f)$ computed from this density matrix is presumably equal to $p_1(f)$ in (9.2). If the result f is obtained approximately $Np(f)$ times, one says that the predictions have been verified, and $p(f)$ is correct in an objective sense. This result is obtained, however, only by renouncing the possibility of predicting any mutual properties of different systems, and the record of the experiment contains some information about those mutual properties.

Thus, we enunciate as a general principle: *Empirical verifiability of a probability assignment, and completeness of the density matrix from which the probabilities were obtained, are incompatible conditions.* Whenever we use a density matrix whose probabilities are verifiable by certain measurements, we necessarily renounce the possibility of predicting the results of other measurements which can be made on the same apparatus.

This principle of "statistical complementarity" is not restricted to quantum mechanics, but holds in any application of probability theory; in a very fundamental sense no experiment can ever be repeated, and the most comprehensive probability assignments are necessarily incapable of verification

If an operational viewpoint[8–10] is to be upheld consistently, it appears that the probabilities computed from a complete density matrix must be interpreted in the subjective sense Since this complete density matrix might be a projection operator corresponding to a pure state, one is led very close to the views of Einstein[11] and Bohm[12] as to the interpretation of quantum mechanics

Entirely different considerations suggest the same conclusion. A density matrix represents a fusion of two different statistical aspects; those inherent in a pure state and those representing our uncertainty as to which pure state is present. If the former probabilities are interpreted in the objective sense, while the latter are clearly subjective, we have a very puzzling situation Many different arrays, representing different combinations of subjective and objective aspects, all lead to the same density matrix, and thus to the same predictions However, if the statement, "only certain specific aspects of the probabilities are objective," is to have any operational meaning, we must demand that some experiment be produced which will distinguish between these arrays

10. MAXIMUM-ENTROPY INFERENCE

The methods of maximum-entropy inference described in I may be generalized immediately to the density-matrix formalism. Suppose we are given the expectation values of the operators $F_1 \cdots F_m$; then the density matrix which represents the most unbiased picture of the state of the system on the basis of this much information is the one which maximizes the entropy subject to these constraints. As before, this is accomplished by finding the density matrix which unconditionally maximizes

$$S - \lambda_1 \langle F_1 \rangle - \cdots - \lambda_m \langle F_m \rangle, \qquad (10.1)$$

in which the λ_i are Lagrangian multipliers The result may be described in terms of the partition function

$$Z(\lambda_1 \cdots \lambda_m) = \mathrm{Tr}[\exp(-\lambda_1 F_1 - \cdots - \lambda_m F_m)], \qquad (10.2)$$

with the λ_k determined by

$$\langle F_k \rangle = -\frac{\partial}{\partial \lambda_k} \ln Z. \qquad (10.3)$$

The maximum-entropy density matrix is then

$$\rho = \exp[-\lambda_0 1 - \lambda_1 F_1 - \cdots - \lambda_m F_m] \qquad (10.4)$$

which is correctly normalized ($\mathrm{Tr}\rho = 1$) by setting

$$\lambda_0 = \ln Z, \qquad (10.5)$$

and the corresponding entropy becomes

$$S = \lambda_0 + \lambda_1 \langle F_1 \rangle + \cdots + \lambda_m \langle F_m \rangle. \qquad (10.6)$$

Use of (10.5) and (10.6) enables us to solve (10.3) for the λ_k.

$$\lambda_k = \partial S / \partial \langle F_k \rangle. \qquad (10.7)$$

If the operator F_k contains parameters α_i, we find as before that the maximum-entropy estimates of the derivatives are given by

$$\left\langle \frac{\partial F_k}{\partial \alpha_i} \right\rangle = -\frac{1}{\lambda_k} \frac{\partial}{\partial \alpha_i} \ln Z. \qquad (10.8)$$

For an infinitesimal change in the problem, λ_k is the integrating factor for the kth analog of infinitesimal heat;

$$\delta S = \sum_k \lambda_k \delta Q_k, \qquad (10.9)$$

with

$$\delta Q_k = \delta \langle F_k \rangle - \langle \delta F_k \rangle. \qquad (10.10)$$

All of these relations except (10.2) and (10.4) are formally identical with those found in I, the F_k now being interpreted as matrices instead of functions of a discrete variable x.

The definitions of heat bath and thermometer given in I remain applicable, and the discussion of experimental measurement of temperature proceeds as before with the difference that maximization of entropy of the combined system now automatically takes care of the question of phase relations. We have two systems σ_1 and σ_2, with complete orthonormal basis functions $u_n(1)$, $v_k(2)$, respectively. A state ψ_i of the combined system $\sigma = \sigma_1 \times \sigma_2$ is then some linear combination

$$\psi_i(1,2) = \sum_{nk} u_n(1) v_k(2) a_{nki}.$$

If ψ_i occurs with probability w_i, the density matrix is

$$(nk|\rho|n'k') = \sum_i w_i a_{nki} a_{n'k'i}{}^* = \langle a_{nk} a_{n'k'}{}^* \rangle.$$

---

[8] P W Bridgman, *The Logic of Modern Physics* (The Macmillan Company, New York, 1927)
[9] P A M Dirac, *The Principles of Quantum Mechanics* (Clarendon Press, Oxford, 1935), second edition, Chap I
[10] Hans Reichenbach, *Philosophic Foundations of Quantum Mechanics* (University of California Press, Berkeley, 1946)
[11] *Albert Einstein Philosopher-Scientist*, edited by P A Schilpp (Library of Living Philosophers, Inc , Evanston, 1949), pp 665–684
[12] D Bohm, Phys Rev **85**, 166, 180 (1952); **89**, 458 (1953)

An operator $\mathcal{G}(1,2)$ has matrix elements

$$(nk|\mathcal{G}|n'k') = \int\int u_n^*(1)v_k^*(2)\mathcal{G}(1,2)u_{n'}(1)v_{k'}(2)d\tau_1 d\tau_2$$

and its expectation value is

$$\langle\mathcal{G}\rangle = \mathrm{Tr}(\rho\mathcal{G}) = \sum_{nn'kk'}(nk|\rho|n'k')(n'k'|\mathcal{G}|nk).$$

An operator F_1 which operates only on the coordinates of system 1 is represented in the space of the combined system by a direct product matrix,[13] $\tilde{\mathfrak{F}}_1 = F_1 \times 1$, with matrix elements

$$(nk|\tilde{\mathfrak{F}}_1|n'k') = (n|F_1|n')\delta_{kk'}.$$

Similarly, for an operator F_2 of system 2, we obtain $\tilde{\mathfrak{F}}_2 = 1 \times F_2$, and

$$(nk|\tilde{\mathfrak{F}}_2|n'k') = \delta_{nn'}(k|F_2|k').$$

Consider, as before, the system of interest σ_1, and a thermometer σ_2. Let their Hamiltonians be H_1, H_2, respectively. In the function space of the combined system σ, these Hamiltonians are represented by

$$\mathfrak{H}_1 = H_1 \times 1, \quad \mathfrak{H}_2 = 1 \times H_2. \quad (10.11)$$

The available information now consists of a given (measured) value of $\langle H_2 \rangle$, and the knowledge that energy may be transferred between σ_1 and σ_2 in such a way that the total amount is conserved. In practice we never have detailed knowledge of the weak-interaction Hamiltonian $\mathfrak{H}_{12}$ of a type that would tell us which transitions may in fact take place and which will not. Therefore we have no rational basis for excluding the possibility of any transition between states of σ with a given total energy, and the most unbiased representation of our state of knowledge must treat all such states as equivalent, in their dependence on energy. Any other procedure would amount to arbitrarily favoring some states at the expense of others, in a way not warranted by any of the available information. Therefore only the total energy may appear in our density matrix, and we have to find that matrix which maximizes

$$S - \lambda\langle\mathfrak{H}_1 + \mathfrak{H}_2\rangle, \quad (10.12)$$

subject to the observed value of $\langle H_2 \rangle$. The matrix involved in (10.2) and (10.4) now factors into a direct product:

$$\exp[-\lambda(\mathfrak{H}_1 + \mathfrak{H}_2)] = (e^{-\lambda H_1}) \times (e^{-\lambda H_2}), \quad (10.13)$$

so that the partition function reduces to

$$Z(\lambda) = Z_1(\lambda)Z_2(\lambda), \quad (10.14)$$

with

$$Z_1(\lambda) = \mathrm{Tr}\exp(-\lambda H_1),$$
$$Z_2(\lambda) = \mathrm{Tr}\exp(-\lambda H_2). \quad (10.15)$$

[13] P. R. Halmos, *Finite Dimensional Vector Spaces* (Princeton University Press, Princeton, 1948), Appendix II.

Similarly, the density matrix (10.4) is the direct product

$$\rho = \left[\frac{\exp(-\lambda H_1)}{Z_1(\lambda)}\right] \times \left[\frac{\exp(-\lambda H_2)}{Z_2(\lambda)}\right] = \rho_1 \times \rho_2. \quad (10.16)$$

Because of the absence of correlations between the two systems, it is true once again that the function of the thermometer is merely to tell us the value of the parameter λ in ρ_1, and the properties of the thermometer need not be considered in detail when incorporating temperature measurements into our theory.

An important feature of this theory is that measurement of averages of several noncommuting quantities may be treated simultaneously without interference. Consider, for example, three interacting systems $\sigma = \sigma_1 \times \sigma_2 \times \sigma_3$, where σ_1 is the system of interest, and σ_2 is a thermometer. Some physical quantity F, represented in the space of σ_1 by the operator F_1, and in σ_3 by F_3, can be transferred between σ_1 and σ_3 in such a way that the total amount is conserved. F_1 could stand for angular momentum, volume, etc., and need not commute with H_1. In addition suppose that a quantity $\langle G_1 \rangle$ is measured directly in σ_1, where G_1 does not necessarily commute with either H_1 or F_1. Now the available information consists of the measured values of $\langle G_1 \rangle$, $\langle H_2 \rangle$, and $\langle F_3 \rangle$, plus the conservation laws of F and H. The various operators are now represented in the space σ by direct product matrices as follows:

$$\mathfrak{H}_1 = H_1 \times 1 \times 1, \quad \tilde{\mathfrak{F}}_1 = F_1 \times 1 \times 1,$$
$$\mathfrak{H}_2 = 1 \times H_2 \times 1, \quad \tilde{\mathfrak{F}}_3 = 1 \times 1 \times F_3,$$
$$\mathfrak{G}_1 = G_1 \times 1 \times 1,$$

and the density matrix that provides the most unbiased picture of the state of the total system is the one that maximizes

$$S - \lambda\langle\mathfrak{H}_1 + \mathfrak{H}_2\rangle - \mu\langle\tilde{\mathfrak{F}}_1 + \tilde{\mathfrak{F}}_3\rangle - \nu\langle\mathfrak{G}_1\rangle. \quad (10.17)$$

We now find the factorization property

$$\exp[-\lambda(\mathfrak{H}_1 + \mathfrak{H}_2) - \mu(\tilde{\mathfrak{F}}_1 + \tilde{\mathfrak{F}}_3) - \nu\mathfrak{G}_1]$$
$$= [e^{-\lambda H_1 - \mu F_1 - \nu G_1}] \times [e^{-\lambda H_2}] \times [e^{-\mu F_3}], \quad (10.18)$$

so that once again the partition function and density matrix factor into independent parts for the three systems

$$Z(\lambda,\mu,\nu) = Z_1(\lambda,\mu,\nu)Z_2(\lambda)Z_3(\mu), \quad \rho = \rho_1 \times \rho_2 \times \rho_3,$$

and the pieces of information obtained from σ_2, σ_3 are transferred into ρ_1 without interference.

11. INFORMATION LOSS AND IRREVERSIBILITY

In classical statistical mechanics the appearance of irreversibility can always be traced either to the replacement of a fine-grained probability distribution by a coarse-grained one, or to a projection of a joint probability distribution of two systems onto the subspace of one of them. Both processes amount to a loss, whether voluntary or not, of some of the information which is in principle available. The former is often justified by the very persuasive argument that the mathematics would otherwise be too complicated. But mathematical difficulties, however great, have no bearing on matters of principle, and this way of looking at it causes one to lose sight of a much more important positive reason for discarding information. After sufficient "stirring" has occurred, two different fine-grained distributions will lead to predictions that are macroscopically the same, differing only in microscopic details. Thus, even if we were good enough mathematicians to deal with a fine-grained distribution, its replacement by a coarse-grained one would still be the elegant method of treating the prediction of macroscopic properties, because in this way one eliminates irrelevant details at an early stage of the calculation.

In quantum mechanics, as in classical theory, the increase in entropy characteristic of irreversibility always signifies, and is identical with, a loss of information. It is important to realize that the tendency of entropy to increase is not a consequence of the laws of physics as such, for the motion of points of an array is a unitary transformation prescribed by the Schrödinger equation in a manner just as "deterministic" as is the motion of phase points in classical theory. An entropy increase may occur unavoidably, due to our incomplete knowledge of the forces acting on a system, or it may be an entirely voluntary act on our part. In the latter case, an entropy increase is the means by which we simplify a prediction problem by discarding parts of the available information which are irrelevant, or nearly so, for the particular predictions desired. It is very similar to the statistician's practice of "finding a sufficient statistic." The price we must pay for this simplification is that the possibility of predicting other properties with the resulting equations is thereby lost.

The natural way of classifying theories of irreversible processes is according to the mechanism by which information is lost or discarded. In most of the existing theories we find that this consists of the repetition, at regular intervals, of one of the following procedures. Suppose we wish to find the expectation value of the quantity F; in the representation in which F is diagonal it reduces to

$$\langle F \rangle = \text{Tr}(\rho F) = \sum_n \rho_{nn} F_{nn}. \quad (11.1)$$

Since only the diagonal elements of ρ contribute, $\langle F \rangle$ can be calculated as well by using the density matrix ρ', where

$$\rho_{nk}' = \rho_{nn}\delta_{nk} \quad (11.2)$$

The process of replacing ρ by ρ' will be called *removing coherences*, and is clearly permissible whenever all the quantities which we wish to calculate are diagonal simultaneously. It is readily verified that removal of coherences represents loss of information: $S(\rho') \geq S(\rho)$, with equality if and only if $\rho = \rho'$.

The second procedure by which information may be discarded is an invariant operation, exactly analogous to its classical counterpart. Consider two interacting systems σ_1 and σ_2. As already noted, an operator F_1 which operates only on the variables of σ_1 is represented in the space of the combined system $\sigma = \sigma_1 \times \sigma_2$ by the direct product matrix $\mathfrak{F}_1 = F_1 \times 1$. The expectation value of any such operator reduces to a trace involving only the space of σ_1:

$$\langle F_1 \rangle = \text{Tr}(\rho \mathfrak{F}_1) = \text{Tr}(\rho_1 F_1), \quad (11.3)$$

where ρ_1 is the "projection" of the complete density matrix ρ onto the subspace σ_1, with matrix elements

$$(n|\rho_1|n') = \sum_k (nk|\rho|n'k). \quad (11.4)$$

Similarly, we can project ρ onto σ_2, with the result

$$(k|\rho_2|k') = \sum_n (nk|\rho|nk')$$

and for any operator F_2 of system 2 we can define $\mathfrak{F}_2 = 1 \times F_2$, whereupon $\langle F_2 \rangle = \text{Tr}(\rho \mathfrak{F}_2) = \text{Tr}(\rho_2 F_2)$.

In the projection onto σ_1, the parts of ρ that are summed out contain information about the state of system σ_2 and about correlations between possible states of σ_1 and σ_2, both of which are irrelevant for predicting the average of F_1.

The operation of *removing correlations* consists of replacing ρ by the direct product $\rho_1 \times \rho_2$, with matrix elements

$$(nk|\rho_1 \times \rho_2|n'k') = (n|\rho_1|n')(k|\rho_2|k'), \quad (11.5)$$

and the expectation value of any operator composed additively of terms which operate on σ_1 alone or on σ_2 alone, is found as well from $(\rho_1 \times \rho_2)$ as from ρ. The removal of correlations also involves a loss of information, the entropy after removal of correlations is additive and never less than the original entropy.

$$S(\rho_1 \times \rho_2) = S(\rho_1) + S(\rho_2) \geq S(\rho), \quad (11.6)$$

with equality if and only if $\rho = \rho_1 \times \rho_2$.

These remarks generalize in an obvious way to the case of any number of subsystems, to remove correlations from a density matrix ρ operating on the space of three systems $\sigma = \sigma_1 \times \sigma_2 \times \sigma_3$, project it onto each of the σ_i, and replace ρ by the direct product of the projections.

$$\rho \to \rho_1 \times \rho_2 \times \rho_3.$$

If an operator F_2 operates only on the space of σ_2, its matrix representation in the σ space and expectation value are given by

$$\mathfrak{F} = 1 \times F_2 \times 1, \quad \langle F_2 \rangle = \text{Tr}(\rho \mathfrak{F}) = \text{Tr}(\rho_2 F_2)$$

Most treatments of irreversible processes in the past have been based on the removal of coherences in the energy representation, and the resulting concept of "occupation numbers" N_k, proportional to the diagonal elements ρ_{kk} in this representation. One then introduces a transition probability per unit time λ_{kn}, which usually, but not always,[14,15] conforms to the assumption of "microscopic reversibility" $\lambda_{kn}=\lambda_{nk}$, and equations of the form

$$dN_k/dt = \sum_m (\lambda_{km} N_m - \lambda_{mk} N_k) \quad (11.7)$$

are the starting point of the theory. The existence of time-proportional transition probabilities is not, however, a general consequence of quantum mechanics, but involves assumptions about the type of perturbing forces responsible for the transitions, and mathematical approximations which represent a loss of information. That information is lost somewhere is seen from the fact the entropy, as calculated from (11.7), is in general an increasing function of the time, while that obtained from rigorous integration of a Schrodinger equation is necessarily constant The nature of the information-discarding process in (11.7), as well as a clear statement of the type of physical problems to which equations of this form are applicable, can be appreciated only by starting from a more fundamental viewpoint.

12. SUBJECTIVE H THEOREM

In the remainder of this paper, we consider a certain approximation, which might be called the "semiclassical theory of irreversible processes," since it is related to a complete theory in the same way that the semiclassical theory of radiation[16] is related to quantum electrodynamics. The system of interest σ is treated as a quantum-mechanical system, but outside influences are treated classically, their effect on σ being represented by perturbing terms in the Hamiltonian which are considered definite if unknown functions of the time. It is of interest to see which aspects of irreversible processes are found in this approximation, and which ones depend essentially on the quantum nature of the surroundings.

Let the Hamiltonian of the system be

$$H = H_0 + V(t), \quad (12.1)$$

where H_0 is stationary and defines the "energy levels" of the system, and $V(t)$ represents the perturbing effect of the environment. Suppose that at time t' we are given information which leads (by maximum-entropy inference, if needed) to the density matrix $\rho(t')$. At other times, the effect of the Hamiltonian (12.1) is to carry out a unitary transformation

$$\rho(t) = U(t,t')\rho(t')U(t,t')^{-1}$$
$$= U(t,t')\rho(t')U(t',t), \quad (12.2)$$

where the time-development matrix $U(t,t')$ is determined from the Schrödinger equation (with $\hbar=1$)

$$i\frac{\partial}{\partial t}U(t,t') = H(t)U(t,t'), \quad (12.3)$$

[14] J S Thomsen, Phys Rev **91**, 1263 (1953)
[15] R T Cox, *Statistical Mechanics of Irreversible Change* (Johns Hopkins Press, Baltimore, 1955)
[16] L I Schiff, *Quantum Mechanics* (McGraw Hill Book Company, Inc, New York, 1949)

with $U(t,t)=1$. The entropy

$$S(t) = -\text{Tr}[\rho(t)\ln\rho(t)] \quad (12.4)$$

is unchanged by a unitary transformation, and therefore remains constant regardless of the magnitude or time variations of $V(t)$. Consider, however, the circumstance that $V(t)$ may not be known with certainty; during the time interval $(t' \to t)$ it may have been the operator $V^{(1)}(t)$ with probability P_1, or it may have been $V^{(2)}(t)$ with probability P_2, $\cdots$, etc. Then our state of knowledge of the system may be represented by a compound array, which is a fusion of several simple arrays corresponding to the different $V^{(\alpha)}(t)$, and which are subject to different rotations. At time t, the density matrix will be the average of the matrices that would result from each of the possible interactions:

$$\rho(t) = \sum_\alpha P_\alpha U^{(\alpha)}(t,t')\rho(t')U^{(\alpha)}(t',t), \quad (12.5)$$

and the transformation $\rho(t') \to \rho(t)$ is no longer unitary. We might also have a continuous distribution of unknown interactions, and therefore an integration over α, or more generally there might be several parameters $(\alpha_1 \cdots \alpha_n)$ in $V(t)$, with probability distribution $P(\alpha_1 \cdots \alpha_n) d\alpha_1 \cdots d\alpha_n$. We will understand the notation in (12.5) to include such possibilities. Our uncertainty as to $V(t)$ will be reflected in increased uncertainty, as measured by the entropy, in our knowledge of the state of system σ. It is shown in Appendix A that, in case α is discrete, there is an upper limit to this increase, given by the following inequality:

$$S(t') \leq S(t) \leq S(t') + S(P_\alpha), \quad (12.6)$$

where

$$S(P_\alpha) \equiv -\sum_\alpha P_\alpha \ln P_\alpha \quad (12.7)$$

Equation (12.6) has an evident intuitive content, the entropy of a system is a measure of our uncertainty as to its true state, and by applying an unknown signal to it, this uncertainty will increase, but not by more than our uncertainty as to the signal The maximum increase in entropy can occur only in the following rather exceptional circumstances. The totality of all possible states of the system forms a function space S. Suppose that our initial state of knowledge is that the system is in a certain subspace S_0 of S If the perturbation $V^{(\alpha)}(t)$ is applied, this is transformed into some other subspace

$$S_\alpha = U^{(\alpha)} S_0,$$

and the maximum increase of entropy can occur only if the different subspaces S_α are disjoint, i.e., every state in S_α must be orthogonal to every state in S_β if $\alpha \neq \beta$ From this we see two reasons why the increase is usually less than the maximum possible amount, (a) it may be that even though $V^{(\alpha)}(t)$ and $V^{(\beta)}(t)$ are different functions, they nevertheless produce the same, or nearly the same, net transformation U in time $(t-t')$, so that our knowledge of the final state does not suffer from the uncertainty in the perturbation,

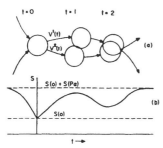

Fig 1 Illustration of the subjective H theorem (a) The array (b) The resulting entropy curve

and (b) our initial uncertainty may be so great that no such disjoint subspaces exist regardless of the nature of the $V^{(\alpha)}(t)$. The extreme case is that of complete initial ignorance; $\rho(t')$ is a multiple of the unit matrix. Then, no matter what is done to the system we cannot acquire any additional uncertainty, and the entropy does not change at all.

Equation (12 6) corresponds closely to relations that have been used to demonstrate the second law of thermodynamics in the past, and it will be called the "subjective H theorem." The inequalities hold for all times, positive or negative, given the density matrix at time $t'=0$, our uncertainty as to the perturbing signal $V(t)$ affects our knowledge of the past state of the system just as much as it does the future state. We cannot conclude from (12.6) that "entropy always increases." It may fluctuate up and down in any way as long as it remains within the prescribed bounds On the other hand, it is true *without exception* that the entropy can at no time be less than its value at the instant t' for which the density matrix was given

Figure 1 represents an attempt to illustrate several of the foregoing remarks by picturing the array. The diagram represents a portion of the surface of the unit hypersphere upon which all points of the array lie [17] The interior of a circle represents a certain subspace $S_i(t)$ which moves in accordance with the Schrodinger equation. Separated circles represent disjoint subspaces, while if two circles overlap, the subspaces have a certain linear manifold of states in common. The information given to us at time $t'=0$ locates the system somewhere in the subspace S_0 The two possible interactions $V^{(1)}(t)$, $V^{(2)}(t)$ would induce rigid rotations of the hypersphere which would carry S_0 along two different trajectories as shown. The lower part of the diagram represents the resulting entropy curve $S(t)$. If the subspaces S_1, S_2 coincide at some time t_1, then $S(t_1)$

[17] The representation is necessarily very crude, since a continuous 1:1 mapping of a region of high dimensionality onto a region of lower dimensionality is topologically impossible Nevertheless such diagrams represent enough of the truth to be very helpful, and there seems to be little danger of drawing fundamentally incorrect conclusions from them

$=S(0)$. At times when they are completely separated, we have $S(t)=S(0)+S(P_\alpha)$, and in case of partial overlapping the entropy assumes intermediate values.

13. INFORMATION GAME

A typical process by which the subjective H theorem can lead to a continual increase of entropy, and which illustrates the essential nature of irreversibility, may be described in terms of a game. We have a sequence of observers $\mathcal{O}_1$, $\mathcal{O}_2$, $\mathcal{O}_3$, $\cdots$, who play as follows. At the beginning of the game they are given the possible Hamiltonians $H_\alpha = H_0 + V^{(\alpha)}(t)$ and the corresponding probabilities P_α. At time t_1, observer $\mathcal{O}_1$ is given a density matrix $\rho_1(t_1)$. He computes from (12.5) the density matrix $\rho_1(t)$ which represents his state of knowledge at all other times on this basis, and the corresponding entropy curve $S_1(t)$. He then tells observer $\mathcal{O}_2$ the value which the density matrix $\rho_1(t)$ assumes at time t_2, and gives no other information.

$\mathcal{O}_2$ now computes a density matrix $\rho_2(t)$ which represents *his* state of knowledge at all times, on the basis of the information given him, and a corresponding entropy curve $S_2(t)$. He will, of course, have $\rho_2(t_2) = \rho_1(t_2)$, but in general there will be no other time at which these density matrices are equal. The reason for this is seen in Fig. 2, in which we assume that there are only two possible perturbations $V^{(1)}$, $V^{(2)}$. The information given to $\mathcal{O}_1$ locates the system somewhere in the subspace S_0 at time t_1. At a different time t_2, this will be separated into two subspaces $S_1(t_2)$ and $S_2(t_2)$, corresponding to the two possible perturbations. For simplicity of the diagram, we assume that they are disjoint. At any other time t_3, the array of $\mathcal{O}_1$ is still represented by two possible subspaces $S_1(t_3)$, $S_2(t_3)$. Observer $\mathcal{O}_2$, however, is not in as advantageous a position as $\mathcal{O}_1$; although he is given the same density matrix at time t_2, and therefore can locate the subspaces $S_1(t_2)$ and $S_2(t_2)$, he does *not* know that $S_1(t_2)$ is associated only with the perturbation $V^{(1)}$, $S_2(t_2)$ only with $V^{(2)}$. Therefore, he can only assume that either perturbation may be associated with either subspace, and the array representing the state of knowledge of $\mathcal{O}_2$ for general times consists of four subspaces.

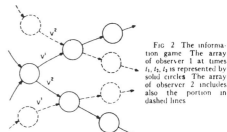

Fig 2 The information game The array of observer 1 at times t_1, t_2, t_3 is represented by solid circles The array of observer 2 includes also the portion in dashed lines

The game continues; $\mathfrak{O}_2$ tells $\mathfrak{O}_3$ what the density matrix $\rho_2(t_2)$ is, and $\mathfrak{O}_3$ calculates *his* density matrix $\rho_3(t)$ (which, at general times other than t_2, must be represented by eight possible subspaces), and the entropy curve $S_3(t)$, $\cdots$, and so on.

The subjective H theorem applied to the nth observer gives

$$S_n(t_n) \leq S_n(t) \leq S_n(t_m) + S(P_a), \quad (13.1)$$

while from the rules of the game,

$$S_{n-1}(t_n) = S_n(t_n). \quad (13.2)$$

Therefore, we have

$$S_1(t_1) \leq S_2(t_2) \leq S_3(t_3) \leq \cdots. \quad (13.3)$$

Note that no such inequality as $t_1 \leq t_2 \leq t_3 \leq \cdots$ need be assumed, since the subjective H theorem works as well backwards as forwards; *the order of increasing entropy is the order in which information was transferred, and has nothing to do with any temporal order.*

An important conclusion from this game is that a density matrix does not in general contain all of the information about a system which is relevant for predicting its behavior; even though $\mathfrak{O}_1$ and $\mathfrak{O}_2$ had the same knowledge about possible perturbations, and represented the system by the same density matrix at time t_2, they were nevertheless in very different positions as regards the ability to predict its behavior at other times. The information which was lost when $\mathfrak{O}_1$ communicated with $\mathfrak{O}_2$ consisted of correlations between possible perturbing forces and the different simple arrays which are contained in the total compound array. The effect of this information loss on an observer's knowledge of the system was not immediate, but required time to "develop." Thus, it is not only the entire density matrix, but also the particular resolution (12.5) into parts arising from different simple arrays, that is relevant for the prediction problem.

For these and other reasons, an array must be considered as a more fundamental and meaningful concept than the density matrix; even though many different arrays lead to the same density matrix, they are not equivalent in all respects. In problems where the entropy varies with time, the array which at each instant represents the density matrix as a mixture of orthogonal states is difficult to obtain, and without any particular significance. The one which is resolved into simple arrays, each representing the unfolding of a possible unitary transformation, provides a clearer picture of what is happening, and may contain more information relevant to predictions.

The density matrices $\rho_n(t)$ determined by the successive observers in the information game may be represented in a compact way as follows. Consider first the case where there is only a single possible perturbation, and therefore ρ undergoes a unitary transformation

$$\rho(t) = U(t,t')\rho(t')U^{-1}(t,t'). \quad (13.4)$$

This could also be written in another kind of matrix notation as

$$\rho_{nn'}(t) = \sum_{kk'}(nn'|G(t,t')|kk')\rho_{kk'}(t'), \quad (13.5)$$

or,

$$\rho(t) = G(t,t')\rho(t'), \quad (13.6)$$

where

$$(nn'|G(t,t')|kk') = U_{nk}(t,t')U_{n'k'}{}^*(t,t') \quad (13.7)$$

is the direct product matrix

$$G = U \times U^*. \quad (13.8)$$

In (13.4) ρ is considered as an $(N \times N)$ matrix, while in (13.6) it is a vector with N^2 components, and G is an $(N^2 \times N^2)$ matrix. It is readily verified that G has the group property

$$G(t,t')G(t',t'') = G(t,t'') \quad (13.9)$$

in consequence of the same property possessed by U.

The advantage of writing the transformation law in the form (13.6) is that, in the case where there are several possible perturbations $V^{(a)}(t)$, the transformation with time (12.5) cannot be written as a similarity transformation with any "averaged U matrix," but it is expressible by a G matrix averaged over the distribution P_a:

$$\rho(t) = \mathcal{G}(t,t')\rho(t'), \quad (13.10)$$

where

$$\mathcal{G}(t,t') = \sum_a P_a G^{(a)}(t,t'). \quad (13.11)$$

The essential feature of the irreversibility found in the information game is that $\mathcal{G}(t,t')$ does *not* possess the group property (13.9).

$$\mathcal{G}(t,t')\mathcal{G}(t',t'') \neq \mathcal{G}(t,t''), \quad (13.12)$$

for on one side we have the product of two averages, on the other the average of a product. If (13.12) were an equality valid for all times, it would imply that $\mathcal{G}$ has an inverse $\mathcal{G}^{-1}(t,t') = \mathcal{G}(t',t)$, whereupon (13.10) could be solved for $\rho(t')$,

$$\rho(t') = \mathcal{G}(t',t)\rho(t). \quad (13.13)$$

But then, the subjective H theorem would give

$$S(t) \geq S(t'), \text{ from } (13.10);$$
$$S(t') \geq S(t), \text{ from } (13.13).$$

In the general case $\mathcal{G}(t,t')$ may be singular.

The density matrices of the successive observers are now given by

$$\rho_1(t) = \mathcal{G}(t,t_1)\rho_1(t_1),$$
$$\rho_2(t) = \mathcal{G}(t,t_2)\mathcal{G}(t_2,t_1)\rho_1(t_1), \quad (13.14)$$
$$\rho_3(t) = \mathcal{G}(t,t_3)\mathcal{G}(t_3,t_2)\mathcal{G}(t_2,t_1)\rho_1(t_1),$$

in which the information game is exhibited as a Markov

chain,[18,19] the ordering index giving the sequence of information transfer rather than a time sequence.

14. STEP-RELAXATION PROCESS

In the preceding section, the information game was interpreted in the "passive" sense; i.e., we assumed that a certain one of the perturbations $V^{(\alpha)}(t)$ was the one in fact present, and this same one persisted for all time. The different observers then represent different ways of looking at what is in reality only one physical situation, their increasing uncertainty as to the true state being due only to the incomplete transmission of information from one observer to the next.

The game may equally well be interpreted in the "active" sense, in which there is only one observer, but at each of the times $t_1, t_2, t_3, \cdots$, the perturbation is interrupted and a new choice of one of the $V^{(\alpha)}(t)$ made in accordance with the probability distribution P_α. Although it is not required by the equations, it is perhaps best at this point, merely to avoid certain teleological distractions, to assume that

$$t_1 \leq t_2 \leq t_3 \leq \cdots. \qquad (14.1)$$

At each of these times the observer loses exactly the same information that was lost in the communication process of the passive interpretation, and his knowledge of the state of the system progressively deteriorates according to the same Eqs. (13.14) as before. The density matrix which represents the best physical predictions he is able to make is then

$$\rho(t) = \begin{cases} \rho_1(t), & t_1 \leq t \leq t_2 \\ \rho_2(t), & t_2 \leq t \leq t_3 \\ \cdots \\ \rho_n(t), & t_n \leq t \leq t_{n+1}. \end{cases} \qquad (14.2)$$

This is a continuous function of time, since

$$\rho_n(t_n) = \rho_{n-1}(t_n).$$

In the following we consider only the case where ρ operates on a function space σ of finite dimensionality N. The maximum possible entropy of such a system is

$$S_{\max} = \ln N, \qquad (14.3)$$

which is attained if and only if ρ is a multiple of the unit matrix:

$$\rho_{nk} = N^{-1}\delta_{nk}. \qquad (14.4)$$

From this fact and (13.3), it follows that the sequence of values $S(t_n)$ must converge to some definite final entropy:

$$\lim_{n\to\infty} S(t_n) = S_\infty \leq S_{\max}. \qquad (14.5)$$

To investigate the limiting form of the density matrix as $t \to \infty$, some spectral properties of the transformation matrices are needed. Let $\mathcal{G}$ stand for any one of the

[18] J L Doob, Anr Math 43, 351 (1942)
[19] W Feller, *An Introduction to Probability Theory and its Applications* (John Wiley and Sons, Inc, New York, 1950).

$(N^2 \times N^2)$ step transformations $\mathcal{G}(t_n, t_{n-1})$ operating in the direct product space $\sigma \times \sigma = \sigma^2$, and x, y be any vectors of N^2 components upon which $\mathcal{G}$ can operate. Instead of denoting the components of x, y by a single index running from 1 to N^2, we use two indices each running from 1 to N, so that x, y may also be interpreted as $(N \times N)$ matrices operating in the space σ. We define inner products in the usual way by

$$(x,y) = \sum_{n,k=1}^{N} x_{nk}^* y_{nk} = \text{Tr}(x^\dagger y). \qquad (14.6)$$

Since $\mathcal{G}$ is not a normal matrix (i.e., it does not commute with its Hermitian conjugate), we may not assume the orthogonality, or even the existence of a complete set, of its eigenvectors. However, every square matrix has at least one eigenvector belonging to each eigenvalue, so that as x varies over all possible directions, the set of numbers

$$g(x) \equiv (x, \mathcal{G}x)/(x,x)$$

includes all the eigenvalues of $\mathcal{G}$. Writing

$$x_\alpha = U^{(\alpha)} x U^{(\alpha)-1}$$

it is readily shown that $(x_\alpha, x_\alpha) = (x, x)$. From (12.5) we have

$$\mathcal{G}x = \sum_\alpha P_\alpha x_\alpha,$$

and therefore

$$|(x, \mathcal{G}x)| = |\sum P_\alpha (x, x_\alpha)| \leq \sum P_\alpha |(x, x_\alpha)|$$
$$\leq \sum P_\alpha [(x,x)(x_\alpha, x_\alpha)]^{\frac{1}{2}} = (x,x),$$

where the Schwarz inequality has been used. We conclude that for all x,

$$|g(x)| \leq 1, \qquad (14.7)$$

with equality if and only if $x_\alpha = x$ for all α. This is evidently the case if x is a multiple of the unit matrix; thus (14.4) is always an eigenvector of $\mathcal{G}$ with the eigenvalue unity. Only in exceptional circumstances could $\mathcal{G}$ have any other eigenvalue of magnitude unity; this would require that some x other than (14.4) must exist which is invariant under all the unitary transformations $U^{(\alpha)}$.

By a similar argument, one can derive a slightly weaker inequality than (14.7):

$$(\mathcal{G}x, \mathcal{G}x) \leq (x,x), \qquad (14.8)$$

which shows that $\text{Tr}[\rho^2(t_n)]$ is a non-increasing function of n, which must converge to some definite final value.

From these relations several features of the long-time behavior may be inferred. First consider $\mathcal{G}$ to be brought, by similarity transformations, to the canonical form

$$T\mathcal{G}T^{-1} = \begin{pmatrix} A_1 & & & \\ & A_2 & & \\ & & \cdot & \\ & & & \cdot \\ & & & & A_r \end{pmatrix}, \qquad (14.9)$$

where each A_i contains all those, and only those, terms which arise from the eigenvalue λ_i. If λ_i is nondegenerate, A_i is simply the number λ_i. If λ_i is an m-fold multiple root of $|\mathcal{G}-\lambda 1|=0$, then A_i may be the $(m\times m)$ diagonal matrix $\lambda_i 1$, or it may have one or more "superdiagonal" terms[20]

$$A_i = \begin{bmatrix} \lambda_i & 1 & 0 & 0 & \cdot \\ 0 & \lambda_i & 1 & 0 & \cdot \\ 0 & 0 & \lambda_i & 0 & \cdot \\ \cdot & \cdot & \cdot & \cdot & \cdot \end{bmatrix}. \quad (14.10)$$

The simplest type of step-relaxation process to describe is the one in which all of the matrices $\mathcal{G}(t_n, t_{n-1})$ are equal; i.e., $t_n = n\tau$, and each of the possible perturbations $V^{(\alpha)}(t)$ is periodic with period τ. The general conclusions will be the same regardless of whether this is the case. We now have

$$\rho(t_n) = \mathcal{G}^n \rho(0), \quad (14.11)$$

and those parts of the canonical form $T\mathcal{G}^n T^{-1}$ arising from the eigenvalue $\lambda = 0$ are annihilated in a finite number of steps, while the sections A_i^n for which $0 < |\lambda_i| < 1$ are exponentially attenuated. Thus, the situation as $n \to \infty$ depends only on those A_i^n for which $|\lambda_i| = 1$. There are two possibilities:

(a) The ergodic case. If $\mathcal{G}$ has only one eigenvalue with $|\lambda_i| = 1$ [which must therefore correspond to the eigenvector (14.4)], the sequence $\{\mathcal{G}^n\}$ converges to the projection onto (14.4); i.e.,

$$\lim_{n \to \infty} \rho(t_n) = N^{-1} 1, \quad (14.12)$$

independently of $\rho(0)$. The information contained in the initial distribution becomes completely lost, and the limiting entropy is the maximum possible value (14.3) In practice, this would be the usual situation.

(b) If $\mathcal{G}$ has more than one eigenvalue with $|\lambda_i| = 1$, the density matrix does not necessarily approach any fixed limit. Nevertheless, the entropy $S(t_n)$ must do so Therefore, by an argument like that of Appendix A, the ultimate behavior must be one in which a certain similarity transformation is repeated indefinitely. For example, this ultimate transformation could consist of a permutation of the rows and columns of ρ. In this case, traces of the initial information are never lost, and the limiting entropy is less than $\ln N$.

These results correspond closely to those of the theory of long-range order in crystals,[21,22] in which one introduces a stochastic matrix which relates the probability distribution of one crystal layer to that of an adjacent one The existence or nonexistence of probability influences over arbitrarily long distances depends on the degeneracy (in magnitude) of the greatest eigenvalue of this matrix

---

[20] S. Lefschetz, *Lectures on Differential Equations* (Princeton University Press, Princeton, 1946), Chap I
[21] J Ashkin and W E Lamb, Jr, Phys Rev **64**, 159 (1943)
[22] G F Newell and E W Montroll, Revs Modern Phys **25**, 353 (1953)

15. PERTURBATION BY A STATIONARY STOCHASTIC PROCESS

We now investigate the change in our knowledge of the state of a system for which the perturbing Hamiltonian $V(t)$ is a stationary random function of time. Certain aspects of irreversible processes may be described in terms of such a model, although we will find that other essential features, such as the mechanism by which thermal equilibrium is established, require better approximations in which the quantum nature of the perturbing forces is taken into account

In classical statistical mechanics an ergodic hypothesis facilitated the mathematics by allowing one to replace time averages by ensemble averages. We now find the reverse situation; that calculation of $\mathcal{G}(t,t')$ is facilitated by an ergodic principle that enables us to replace the "ensemble average" (13.11) by a time average, and then to make use of correlation functions and the Wiener-Khintchine theorem. In Eq. (13 10), $G^{(\alpha)}(t,t')$ may be regarded as a certain functional $F[V^{(\alpha)}(t)]$ of $V^{(\alpha)}(t)$, which depends on the values assumed by this operator in the time interval $(t' \to t)$. The statement that $V(t)$ is a stationary stochastic process implies that the average of this functional

$$\bar{F}_0{}^\alpha = \sum_\alpha P_\alpha F[V^{(\alpha)}(t)] \quad (15.1)$$

is not affected by which particular sample of the function $V^{(\alpha)}(t)$ is involved in (15 1), i e, if we were to insert instead the values assumed by $V^{(\alpha)}(t)$ in some other equal time interval $(t' + \tau \to t + \tau)$, the average

$$\bar{F}_\tau{}^\alpha = \sum_\alpha P_\alpha F[V^{(\alpha)}(t+\tau)] \quad (15.2)$$

would be independent of τ Conversely, if

$$\bar{F}_\tau{}^\alpha = \bar{F}_0{}^\alpha$$

for all functionals and all values of τ, this implies that $V(t)$ has exactly the same statistical properties after any time translation, so that $V(t)$ must be a stationary stochastic process Under these conditions the expression (15.1) will not be affected by averaging it over all time translations,

$$\bar{F}^\alpha = \bar{F}^{\alpha\tau} = \lim_{T \to \infty} \frac{1}{2T} \int_{-T}^{T} \sum P_\alpha F[V^{(\alpha)}(t+\tau)] d\tau \quad (15.3)$$

Our ergodic assumption is that in this formula the averaging over P_α is redundant, i e,

$$\bar{F}^\alpha = \bar{F}^\tau = \lim_{T \to \infty} \frac{1}{2T} \int_{-T}^{T} F[V(t+\tau)], \quad (15.4)$$

in which the parameter α may be dropped.

The preceding paragraph was written in a conventional kind of language which has made it appear that a substantial assumption has been introduced, one whose correctness should be demonstrated if the resulting theory is to be valid. Such conventional modes of expression, however, do not do full justice to the situation as it is presented to us in practice. To see this, we need only ask, "What do we really mean by the functions $V^{(a)}(t)$ and the probabilities P_a?" In most cases there is only one function $V(t)$. Knowledge of the statistical properties of V cannot then be obtained by observing the frequency with which the particular function $V^{(a)}(t)$ appears in an ensemble of similar situations, because no such ensemble exists. By the probability P_a we could mean only the average frequency, over long periods of time, with which a configuration locally like $V^{(a)}$ occurs in the *single* function $V(t)$. *The means by which the probabilities P_a are defined already involve a time-averaging procedure.* Therefore (15.4) is not an assumption at all; it is merely the natural way of stating a fact which is expressed only awkwardly by (15.1). Equation (15.4) carries out in a single step both the averaging procedure in (15.1) and the process by which the $V^{(a)}$ and P_a are determined.

The problem is thus reduced to a calculation of $\mathcal{G}(t,t') = \mathcal{G}(t-t')$, where

$$\mathcal{G}(t) = \lim_{T \to \infty} \frac{1}{2T} \int_{-T}^{T} [U(t+\tau,\tau) \times U^*(t+\tau,\tau)] d\tau. \quad (15.5)$$

The exact evaluation of $\mathcal{G}(t)$ would require a rigorous solution of the Schrödinger equation (12.3) for arbitrary $V(t)$. In practice one must resort to approximate solutions at this point, and it is fortunate that in many practical situations $\mathcal{G}(t)$ is determined to a good approximation by the use of second-order perturbation theory. The characteristic feature of such problems is found by noting that although $\mathcal{G}(t,t')$ does not in general possess the group property (13.12), an equality of this form may be approximately correct for certain choices of times, provided the perturbation is weak and has a short correlation time. Thus, suppose that $t'' < t' < t$, and we try to represent $\mathcal{G}(t,t'')$ by a product

$$\mathcal{G}(t,t'') \simeq \mathcal{G}(t,t') \mathcal{G}(t',t'') \quad (15.6)$$

The approximation involved in (15.6) consists of the discarding, at time t', of mutual correlations which were built up in the time interval $(t'' \to t')$ between possible functions $V(t)$ and the corresponding simple arrays. If $V(t)$ is a weak perturbation, it can change the state of the system only slowly, and a long time is required for any strong correlations to develop. However, if the time τ_c over which appreciable autocorrelations persist in $V(t)$ is very short compared to $(t'-t'')$, the mutual correlations discarded were actually accumulated only during an interval τ_c just prior to t', and will be relatively unimportant; thus (15.6) may be a very good approximation. On the other hand, it will never be an exact equality, because the values of $V(t)$ just prior to t' will necessarily have some influence on its behavior just after t', whose effect is lost in the approximation.

These considerations lead to a means for approximate calculation of $\mathcal{G}(t-t')$. Divide the time interval $(t' \to t)$ into n equal intervals: $(t-t') = n\tau$, and set

$$\mathcal{G}(t-t') \simeq [\mathcal{G}(\tau)]^n. \quad (15.7)$$

If $\tau \gg \tau_c$, this is a good approximation, and if in addition it is possible to choose τ short enough so that the change of state during time τ is given adequately by second-order perturbation theory, this leads to a feasible method of calculation. With this approximation, the theory is reduced in its essentials to that of the step-relaxation process of the preceding section.

The most important feature of the final solution can be seen directly from (15.7). The change of state with time has a simple "stroboscopic" property: if we observe the density matrix only at the instants $t_m = m\tau$, we see the approach to equilibrium take place in a stepwise exponential fashion, describable by relaxation times. This result is already guaranteed by the nature of the approximation in (15.7) quite independently of any further details, and in particular independently of any assumptions concerning the level spacings of the system. However, the level spacings are important in determining the appropriate form of the solution. For example, if the correlation time τ_c is extremely short compared to all characteristic times of the system, we may, while satisfying the condition $\tau \gg \tau_c$, still have $|\omega_{kl}|\tau \ll 1$ for all transitions frequencies ω_{kl}. In this case, the change in ρ during time τ is very small, and (15.7) may be replaced by a linear differential equation with constant coefficients. Thus, defining K_1 by

$$K_1 = [\mathcal{G}(\tau) - 1]/\tau, \quad (15.8)$$

we have approximately

$$d\rho/dt \simeq K_1 \rho. \quad (15.9)$$

K_1 has N^2 eigenvalues λ_i, one of which must be zero since K_1 annihilates (14.4). By an argument like that leading to (14.7) one shows that $\text{Re}(\lambda_i) \leq 0$. Thus each element of ρ will relax to a final state according to a superposition of exponentials $\exp(\lambda_i t)$, with several different relaxation times in general.

The right-hand side of (15.9) is generally a poor approximation to the instantaneous time derivative of ρ, but gives only the average rate of change over the period τ. Similarly, the matrix K_1 resembles a time derivative of $\mathcal{G}$; in the following section we present reasons for expecting that a slightly different definition of K_1 will render (15.9) more accurate as far as giving the long-term drift is concerned.

16. EXACTLY SOLUBLE CASE

In the case where the perturbation $V(t)$ commutes with H_0, it is possible to evaluate (15.5) exactly without use of perturbation theory. This case is a very special one, since the perturbation causes no transitions but only a loss of coherences; nevertheless it has found some applications in the theory of pressure-broadening of spectral lines[23,24] and exchange narrowing[25] in paramagnetic resonance.

The perturbing forces represented by $V(t)$ often arise as a superposition of many small independent effects, and in this case the central limit theorem of probability theory shows that the distribution of $V(t)$ will be Gaussian. Furthermore, in most applications one will not have enough information about $V(t)$ to determine any unique objective probability distribution; we may know, for example, only the average energy density, therefore the mean-square value, of the perturbing fields, plus a few features of their spectral density. Maximum-entropy inference would then be needed in order to represent our knowledge of $V(t)$ in a way free of arbitrary assumptions. Since a Gaussian distribution has maximum entropy for a given variance, one should always use a Gaussian distribution if the available information consists only of the first and second moments. In the following we consider only the Gaussian case

The Hamiltonian has matrix elements

$$H_{kl}(t) = [\omega_k + V_k(t)]\delta_{kl}. \quad (16.1)$$

The solution of (12.3) for the time-development matrix is substituted into (15.4) to give

$$(kk'|\mathcal{G}(t,t')|ll') = \delta_{kl}\delta_{k'l'}e^{i\omega_{k'k}(t-t')}\left\langle \exp\left[i\int_{t'}^{t}f_{k'k}(t'')dt''\right]\right\rangle, \quad (16.2)$$

where $\omega_{k'k} = \omega_{k'} - \omega_k$, and

$$f_{k'k}(t) = V_{k'}(t) - V_k(t) \quad (16.3)$$

is a real Gaussian random function with mean value zero (by definition, since any constant part of V may be included in H_0). So also, therefore, is the function

$$g(t) = \int_0^t f(t')dt', \quad (16.4)$$

where we have dropped the subscripts for brevity. The probability distribution of $g(t)$ is determined by its second moment

$$\sigma(t) = \langle g^2(t) \rangle = \int_0^t dt' \int_0^t dt'' \langle f(t')f(t'') \rangle$$

$$= \int_0^t dt' \int_0^t dt'' \varphi(t'-t''), \quad (16.5)$$

where

$$\varphi(\tau) = \lim_{T\to\infty}\frac{1}{2T}\int_{-T}^{T} f(t+\tau)f(t)dt \quad (16.6)$$

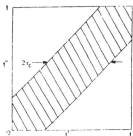

FIG. 3. Region of integration in Eq. (16.5). Appreciable contributions to the integral come only from shaded part.

is the autocorrelation function of $f(t)$. A short calculation shows that for a Gaussian function with variance $\sigma(t)$, the average required in (16.2) is

$$\langle e^{ig} \rangle = e^{-\frac{1}{2}\sigma(t)}, \quad (16.7)$$

and thus the exact solution (13.10) of the relaxation problem is

$$\rho_{kk'}(t) = e^{i\omega_{k'k}t}\rho_{kk'}(0)e^{-\frac{1}{2}\sigma_{k'k}(t)}. \quad (16.8)$$

Since $\sigma_{kk}=0$, the diagonal elements of ρ are unchanged, but the off-diagonal elements relax to zero in a manner described by (16.5).[26]

We assume that there exists a correlation time τ_c such that the correlation function (16.6) is essentially zero whenever $|\tau| > \tau_c$. The region of integration in (16.5) may be represented by a square as in Fig. 3, and it is seen that although $\sigma(t)$ necessarily starts out proportional to t^2 for small t, it approximates a linear function of time when $t > \tau_c$. The function $\sigma(t)$ therefore has the form of Fig. 4, and for $t > \tau_c$ it reduces to

$$\sigma(t) \cong 2\pi I(0)[t-t_1]. \quad (16.9)$$

The quantity

$$I(\omega) = \frac{1}{2\pi}\int_{-\infty}^{\infty} \varphi(t)e^{-i\omega t}dt \quad (16.10)$$

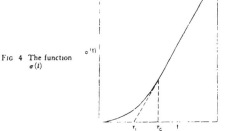

FIG. 4. The function $\sigma(t)$

---

[23] P. W. Anderson, Phys Rev **76**, 647 (1949). Earlier references are given in this paper
[24] S. Bloom and H Margenau, Phys Rev **90**, 791 (1953)
[25] P W Anderson and P R Weiss, Revs Modern Phys **25**, 269 (1953)

[26] In some cases it may be possible to evaluate (16.7) directly even though $\langle g^2 \rangle$ does not exist. For example, we may have $f(t) = $ constant, with probability distribution $p(f)df$. Then (16.7) is a Fourier transform, and with Lorentzian $p(f)$ we obtain a decay law exactly exponential for all times

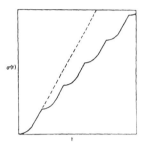

Fig 5 Slip effect caused by discarding correlations. The approximate solution is represented by the solid line, while the dashed line is the exact solution

is the spectral density of $f(t)$ for frequency ω, and τ_1 is a short time somewhat less than τ_c, indicated on Fig. 4. Thus when $t > \tau_c$, the relaxation process goes into an exponential damping, the element $\rho_{kk'}$ having a relaxation time $T_{kk'}$, where

$$1/T_{kk'} = \pi I_{k'k}(0). \quad (16.11)$$

Note that although the final formulas involve only the spectral density at zero frequency, the condition that $\varphi(t)$ should be very small for $|t| > \tau_c$ implies certain conditions on $I(\omega)$ at other frequencies It is required not only that $I(\omega)$ be large over a band width $\sim \tau_c^{-1}$ of frequencies, but also that it be a sufficiently smooth function of frequency. Discontinuities in $I(\omega)$ produce oscillations in $\varphi(t)$ and $\sigma(t)$ which may persist for long periods, rendering (16.9) inaccurate.

It is of interest to compare the exact solution (16.8) with the one which would be obtained using the approximation of (15 7). Here we stop the integration process of (16 5) after each interval τ, throw away mutual correlations between ρ and $V(t)$, and use the density matrix thus obtained as the initial condition for the next period. The resulting $\sigma(t)$ is illustrated in Fig 5 It is seen that the approximation "slips behind" the exact solution by a time delay τ_1 each time the mutual correlations are discarded

There is an apparent paradox in this result It seems natural to suppose that any mathematical approximation must "lose information," and therefore increase the entropy However, we find the relaxation process taking place more rapidly in the exact treatment than in the approximate one. $S_{\text{exact}}(t) \geq S_{\text{approx}}(t)$ Thus, the approximation has not "lost information," but has "injected false information." The reason for this can be visualized as follows Suppose that at time $t = 0$ the array consisted of a single point, i e , a pure state. At later times it will consist of a continuous distribution of points filling a certain volume, which continually expands as t increases It is very much like an expanding sphere of gas, where strong correlations will develop between position and velocity, a molecule near the edge of the sphere is very likely to be moving away from the center. This corresponds roughly to the correlations between different states of the array and

different possible perturbing signals $V(t)$. Now suppose that in an expanding gas sphere these correlations are suddenly lost; the set of velocities existing at time τ is suddenly redistributed among the molecules at random. Then a molecule near the edge is equally likely to be moving toward or away from the center. The general expansion is momentarily interrupted, but soon resumes its former rate.

This paradox shows that "information" is an unfortunate choice of word to describe entropy expressions. Furthermore, one can easily invent situations where acquisition of a new piece of information (that an event previously considered improbable had in fact occurred) can cause an increase in the entropy. The terms "uncertainty" or "apparent uncertainty" come closer to carrying the right connotations.

Note that, if we were to use the slope of the approximate curve in Fig. 5 just before time τ, instead of the average drift over period τ, to calculate the relaxation time, we would obtain a more accurate value whenever $\tau > \tau_c$.

17. PERTURBATION THEORY APPROXIMATION

Returning to the general case, we conjecture that a similar situation to that just found will occur: i.e., that the differential equation

$$d\rho/dt = K_2\rho, \quad (17.1)$$

where

$$K_2 = \left(\frac{d\mathcal{G}}{dt}\right)_{t \to \tau_-}, \quad (17.2)$$

will give a slightly more accurate long-term solution than will (15.9). The evaluation of $\mathcal{G}(\tau)$ using perturbation theory is in essence identical with the treatments of nuclear spin relaxation given by Wangsness and Bloch,[27] Fano,[28] Ayant,[29] and Bloch.[30,31] Only a brief sketch of the calculations is given here, although we wish to point out certain limitations on the applicability of previous treatments.

One solves the equation of motion (12 3) by use of time-dependent perturbation theory, retaining terms through the second order. The result of substituting this solution into (15.5) is expressed compactly as follows. Define a matrix $\varphi(t)$ whose elements consist of all correlation functions of V_{kl}, $V_{k'l'}$:

$$(kk'|\varphi(t-t')|ll') = \langle V_{kl}(t) V_{k'l'}{}^*(t') \rangle, \quad (17.3)$$

in which the average is taken over all time translations. $\varphi(t)$ has the symmetry properties

$$(kk'|\varphi(t)|ll') = (ll'|\varphi(t)|kk')^* = (l'l|\varphi(-t)|k'k) \quad (17.4)$$

We assume again that there exists a correlation time

---

[27] R K Wangsness and F Bloch, Phys Rev **89**, 728 (1953)
[28] U Fano, Phys Rev **96**, 869 (1954)
[29] Y Ayant, J phys radium **16**, 411 (1955)
[30] F Bloch, Phys Rev **102**, 104 (1956)
[31] F Bloch, Phys Rev **105**, 1206 (1957)

τ_c such that all components of $\varphi(t)$ are essentially zero whenever $t > \tau_c$. In this case the "partial Fourier transforms" of φ, defined by

$$\Phi(\omega) = \int_0^\tau e^{-i\omega t}\varphi(t)dt \quad (17.5)$$

are independent of τ. Finally, we introduce the symbols

$$(kk'|nn') = (kk'|\Phi(\omega_{n'k'})|nn') = (nn'|kk')^*. \quad (17.6)$$

In terms of these quantities, we obtain

$$(kk'|\mathcal{G}(\tau)|nn') = e^{-i\omega_{kk'}\tau}\{\delta_{kn}\delta_{k'n'} \\ -\delta_{kn}q(\omega_{n'k'})\sum_p(pp|k'n') - \delta_{k'n'}q(\omega_{kn})\sum_p(kn|pp) \\ + q(\omega_{kn} - \omega_{k'n'})[(kk'|nn') + (n'n|k'k)]\}, \quad (17.7)$$

where
$$q(\omega) = (e^{i\omega\tau} - 1)/i\omega.$$

In the case of extremely short correlation time, so that $|\omega_{kn}\tau| \ll 1$, as assumed in (15.9) and (17.1), $q(\omega_{kn}) = \tau$ for all transition frequencies ω_{kn}, and (17.7) leads to the differential equation

$$\dot{\rho}_{kk'} + i\omega_{kk'}\rho_{kk'} = \sum_{n,n'}\{[(kk'|nn') + (n'n|k'k)]\rho_{nn'} \\ - (nn|k'n')\rho_{kn'} - (kn|n'n')\rho_{nk'}\}. \quad (17.8)$$

This case of perturbation by extremely wide-band "white noise" applies to many cases of nuclear spin relaxation in liquids,[32] its condition of validity being that the correlation time (roughly, period of molecular rotation) is short compared to the Larmor precession periods.

In the approximation of (17.8) the quantities $(kk'|nn')$ are real if $\varphi(t)$ is real, as will usually be the case:

$$(kk'|nn') \simeq \int_0^\infty \cos(\omega_{n'k'}t)(kk'|\varphi(t)|nn')dt. \quad (17.9)$$

The neglected term is small, since by hypothesis $\varphi(t)$ is very small before $\sin(\omega_{n'k'}t)$ attains an appreciable magnitude. Equation (17.9) is π times the "mixed spectral density," at frequency $\omega_{n'k'}$, of $V_{kn}(t)$ and $V_{k'n'}(t)$. To interpret (17.8) we transfer all terms containing $\rho_{kk'}$ to the left-hand side

$$\dot{\rho}_{kk'} + \left(\frac{1}{T_{kk'}} + i\omega_{kk'}\right)\rho_{kk'} = \text{"driving forces."} \quad (17.10)$$

The relaxation times $T_{kk'}$ are given by

$$1/T_{kk'} = \gamma_k + \gamma_{k'} - \gamma_{kk'}, \quad (17.11)$$

where
$$\gamma_k = \sum_p(kk|pp),$$
$$\gamma_{kk'} = (kk'|kk') + (k'k|k'k). \quad (17.12)$$

If the correlation time τ_c is not short compared to the periods $(\omega_{kn})^{-1}$, then the time of integration τ must

[32] Bloembergen, Purcell, and Pound, Phys. Rev. **73**, 679 (1948).

be chosen so long that the formulation (17.8) in terms of a differential equation breaks down. In this case a different approach, used by Wangsness and Bloch,[27] may be attempted. Here one removes the rapid time variations of ρ due to H_0 by transforming to the interaction representation, in which the density matrix is

$$\bar{\rho}(t) = e^{iH_0 t}\rho(t)e^{-iH_0 t}, \quad (17.13)$$

and attempts to describe the relaxation process by a linear differential equation with constant coefficients, satisfied by the slowly varying $\bar{\rho}(t)$. This is not always possible, however, for Eqs (15.5) and (15.7) hold only in the original Schrödinger representation. If H_0 is diagonal, the matrix $\mathcal{G}_I$ which gives the change of state in the interaction representation,

$$\bar{\rho}(t) = \mathcal{G}_I(t, t')\bar{\rho}(t'), \quad (17.14)$$

is related to the previous $\mathcal{G}$ by

$$(kk'|\mathcal{G}_I(t+\tau, t)|nn') \\ = e^{i(\omega_{kn} - \omega_{k'n'})t}e^{i\omega_{kk'}\tau}(kk'|\mathcal{G}(\tau)|nn'), \quad (17.15)$$

so that although $\mathcal{G}$ is a function only of $(t-t')$, this is not in general true of $\mathcal{G}_I$. Consequently an approximation of the form (15.7) cannot be valid in general for $\mathcal{G}_I$. However, it is seen that those elements of $\mathcal{G}_I$ for which

$$\omega_{kn} = \omega_{k'n'} \quad (17.16)$$

depend only on $(t-t')$. Therefore, if by any means one can justify discarding elements of $\mathcal{G}_I$ not satisfying (17.16), this method will work. Referring to (17.7), it is seen that the elements which satisfy (17.16) are just the "secular terms" which increase proportional to τ, while the unwanted terms are the oscillating ones. Therefore if the time τ is sufficiently long, and the level spacings are such, that the quantities

$$|\omega_{kn} - \omega_{k'n'}|\tau$$

are either large compared to unity, or zero, for all combinations of levels, the secular terms will be much larger than the oscillating ones and we obtain the approximate differential equation

$$\frac{\partial \bar{\rho}_{kk'}}{\partial t} = \sum_{n,n'} \{\delta(\omega_{kn} - \omega_{k'n'})[(kk'|nn') \\ + (n'n|k'k)]\bar{\rho}_{nn'} - \delta(\omega_{k'n'})(nn|k'n')\bar{\rho}_{kn'} \\ - \delta(\omega_{kn})(kn|n'n')\bar{\rho}_{nk'}\}. \quad (17.17)$$

If there is no degeneracy and the density matrix is initially diagonal, (17.17) reduces to

$$\partial \bar{\rho}_{kk}/\partial t = 2\pi \delta_{kk'} \sum_n I_{kn}(\omega_{nk})(\bar{\rho}_{nn} - \bar{\rho}_{kk}), \quad (17.18)$$

where

$$I_{kn}(\omega) = \frac{1}{2\pi}\int_{-\infty}^{\infty} e^{-i\omega t}(kk|\varphi(t)|nn)dt \quad (17.19)$$

is the spectral density, at frequency ω, of $V_{kn}(t)$. Equation (17.18) is to be compared to (11.7); we have a time-proportional transition probability satisfying the condition of microscopic reversibility. Note, however, that this result depends entirely on the assumptions as to spectral properties of $V(t)$ and the various approximations made, which ensured that off-diagonal elements of ρ would not appear. From the definition (15.5) of $\mathcal{G}$ it follows that, in the case that $\rho(0)$ is diagonal, the rigorous expression for diagonal elements at time t is

$$\rho_{kk}(t) = \sum_n \langle |U_{kn}(t,0)|^2 \rangle \rho_{nn}(0)$$
$$= \sum_n \lambda_{kn}(t) \rho_{nn}(0), \quad (17.20)$$

so that in general the transition probabilities $\lambda_{kn}(t)$ are neither time proportional nor symmetric.[33] On the other hand, the so-called λ-hypothesis,[14] if stated in the form

$$\sum_k \lambda_{kn}(t) = \sum_n \lambda_{kn}(t) = 1,$$

is always satisfied in this semiclassical theory, in consequence of the unitary character of U.[34]

In (17.17) we may again transfer all terms containing $\bar{\rho}_{kk'}$ to the left-hand side[35]:

$$\frac{\partial \bar{\rho}_{kk'}}{\partial t} + \left[\frac{1}{T_{kk'}} + i(\delta\omega_k - \delta\omega_{k'})\right] \bar{\rho}_{kk'}$$
$$= \text{"driving forces,"} \quad (17.21)$$

where (17.11) holds, but in place of (17.12) we now have

$$\gamma_k + i\delta\omega_k = \sum_p (kk|pp). \quad (17.22)$$

The quantities γ_k and $\delta\omega_k$ are defined to be real. We interpret these relations as follows. In consequence of the random perturbations, the energy of state k is uncertain by an amount γ_k (in frequency units), and in addition its average position is shifted by an amount $\delta\omega_k$. Because of this uncertainty in energy, different possible states of the array drift out of phase with each other, and the off-diagonal element $\bar{\rho}_{kk'}$ tends to relax to zero with a relaxation time $T_{kk'}$. The term

$$\gamma_{kk'} = (kk'|kk') + (k'k|k'k)$$
$$= \int_{-\infty}^\infty \langle V_{kk}(t) V_{k'k'}(0) \rangle dt \quad (17.23)$$

corrects for the fact that there may be correlations between the "instantaneous level shifts" $V_{kk}(t)$, $V_{k'k'}(t)$

---

[33] A trivial exception occurs if the system has only two linearly independent states, for a (2×2) unitary matrix necessarily satisfies $|U_{12}|^2 = |U_{21}|^2$. This is not true in any higher dimensionality.

[34] The possibility that λ_{kn} is not proportional to t may lead in some cases to a differential equation for $\bar{\rho}$ with time-dependent coefficients, analogous to Eq (2 24) of reference 31

[35] If there is no degeneracy and the level spacing is the most general type for which there is no relation of the form $\omega_{kn} = \omega_{k'n'}$ for $k \neq k'$, the right-hand side of (17 21) is zero for all off-diagonal elements $\rho_{kk'}$.

so that the contributions of the level widths γ_k, $\gamma_{k'}$ to the rate of relaxation are not independent. Due to the terms $\gamma_{kk'}$ the uncertainty in energy γ_k is different from the reciprocal of the mean lifetime of state k against transitions. The predicted line widths are, of course, the reciprocals of the relaxation times $T_{kk'}$.

The symbols $(kk|pp)$ may be expressed in terms of the spectral density of $V_{kp}(t)$. Inverting the Fourier transform (17.19) and substituting the result into (17.5), (17.6), we obtain

$$(kk|pp) = \pi I_{kp}(\omega_{pk}) + iP\int_{-\infty}^{\infty} \frac{I_{kp}(\omega)d\omega}{\omega - \omega_{pk}}, \quad (17.24)$$

where P stands for the Cauchy principal value. Thus the level widths depend on the spectral density at the transition frequencies, while the level shifts depend mainly on the manner in which the spectral density varies near the transition frequencies. This can be stated in simpler form in the usual case where $V_{kp}(t) = Q_{kp}f(t)$, where Q_{kp} is constant, and $f(t)$ is a real random function. Let $\varphi(t)$ be the autocorrelation function of $f(t)$; then the level widths and level shifts are proportional to the cosine and sine transforms of $\varphi(t)$:

$$\gamma_k = \sum_p |Q_{kp}|^2 \int_0^\infty \cos(\omega_{kp}t)\varphi(t)dt,$$
$$\delta\omega_k = \sum_p |Q_{kp}|^2 \int_0^\infty \sin(\omega_{kp}t)\varphi(t)dt. \quad (17.25)$$

From this we see that the level shifts will be small compared to the level widths if $\varphi(t)$ becomes vanishingly small before $\sin(\omega_{kp}t)$ reaches its first maximum. This, however, is just the condition for validity of (17.8). Thus, whenever the correlation time τ_c is so long that (17.17) is required instead of (17.8) one may expect appreciable level shifts.

If the quantities $\omega_{kn}\tau$ are of order unity, neither of the differential Eqs. (17.17), (17.8) is applicable. In fact, it is clear already from the rigorous expression $\rho(t) = \mathcal{G}(t,t')\rho(t')$ that in general a relaxation process cannot be described by any differential equation, for the rate of change of ρ does not depend only on its momentary value, but is a functional of past conditions during the entire interval $(t' \to t)$. Thus, the formulation in terms of differential equations is fundamentally inappropriate. It is convenient in those special cases where it can be justified, because of the easy interpretation in terms of relaxation times and level shifts. However, the quantities necessary for comparison with experiment can always be inferred directly from (17.7), the validity of which does not depend on the magnitudes of the quantities $\omega_{kn}\tau$.[34]

The symmetry of the transition probabilities given by (17.18) arises only because the $V_{kn}(t)$ are here considered numbers. If in better approximation one takes into account the quantum nature of the surroundings, they must be considered as operators which operate on the state vector of the perturbing system σ_2 (the "heat bath"). Then, as shown by Ayant,[29] the definition of correlation functions (17.5) remains valid, provided the brackets are now interpreted as standing for the expectation value taken over the system σ_2, and the differential Eq. (17.8) or (17.17) then represents an approximation in which mutual correlations between the two systems are discarded at intervals τ, in the manner of (11.5). One now finds that the probabilities of upward and downward transitions are no longer equal. In the treatment of Ayant, the question of equality of these transition probabilities is reduced to the question whether the spectral density of the perturbing forces is the same at frequencies $(+\omega)$ and $(-\omega)$. This is correct provided one always defines the perturbing terms to be real, as in (17.25); note, however, that the symmetry of transition probabilities in (17.18) does not require that the spectral density of $V_{kn}(t)$ be an even function of frequency. It is sufficient if the spectral density of V_{kn} at frequency $(+\omega)$ is equal to that of V_{nk} at $(-\omega)$, and this is always the case if V is Hermitian.

If one assumes a Boltzmann distribution for the heat bath and neglects the effect of the system of interest σ_1 in modifying this distribution, the solution of (17.17) tends to another Boltzmann distribution corresponding to the same temperature.[27,30] Treatment of this case and that of "secular equilibrium" from the subjective point of view will be considered in a later paper.

18. CONCLUSION

The foregoing represents the first stage of an attempt to provide a new foundation for the predictive aspect of statistical mechanics, in which a single basic principle and method applies to all cases, equilibrium or otherwise.

The phenomenon of nuclear spin relaxation is a particularly good one to serve as a guide to a general theory of irreversible processes. It is complicated enough to require most of the techniques of a general theory, but at the same time it is simple enough so that in many cases the calculations can be carried out explicitly. Nuclear induction experiments, in which the predictions of the Bloch-Wangsness theory[27,30,31] are verified down to fine details,[26] provide a good illustration of many of the above remarks. Here the experiments are performed on samples containing of the order of 10^{20} nuclei, and one measures the time dependence of their total magnetic moment when subject to various applied fields. In the theory, however, one usually calculates a density matrix $\rho_1(t)$ which operates only in the function space of a single spin, or of some small aggregate of spins such as those attached to a single molecule. The possibility of predicting mutual properties of different spin units is therefore lost.

[38] J. T. Arnold, Phys. Rev. 102, 136 (1956); W. A. Anderson, Phys Rev. 102, 151 (1956).

It would, however, always be better in principle to adopt the "global" view in which the entire assemblage of spins in the sample is the system treated To the extent that different molecular units behave independently, the complete density matrix ρ thus obtained would be a direct product of a very large number of matrices. However, this would hardly ever be true because some correlations between different spin units would be expected. Thus, the question is raised whether, and to what extent, predictions made only from ρ_1 can be trusted. At first glance it seems that they could not be, for in most cases the density matrix $\rho_1(t)$ differs only very slightly from a multiple of the unit matrix, and thus represents a very "broad" probability distribution. According to the discussions of maximum-entropy inference in I and the introduction to the present paper, it would appear that this is a case where the theory fails to make any definite predictions, so that unless the probabilities in ρ_1 could be established in the objective sense, the calculations of Sec. 17 would be devoid of physical content.

The thing which rescues us from this situation is, of course, the fact that the experiments refer not to a single spin unit, but to a very large number of them. We must not, however, jump to the obvious conclusion that the "law of large numbers," or the central limit theorem,[19] automatically restores reliability to our predictions To do so would be to make the logical error of the experimenter who thought that he could add three significant figures to his measurements merely by repeating them a million times The correctness of the usual calculations can be demonstrated without explicit reference to the laws of large numbers, by application of the principles of Sec. 11. This is, in fact, the example *par excellence* of how much a prediction problem can be simplified by discarding irrelevant information.

Suppose that we had solved the problem from the global viewpoint, obtained the complete density matrix $\rho(t)$, and demonstrated that it gave a sharp distribution, and therefore reliable predictions, for the total magnetic moment $\mathbf{M} = \mathbf{M}_1 + \mathbf{M}_2 + \cdots + \mathbf{M}_N$. Then the only thing of further interest would be the value of $\langle \mathbf{M} \rangle$. According to Sec. 11, this can be calculated as well from the direct product matrix

$$\rho_1 \times \rho_2 \times \cdots \times \rho_N,$$

where ρ_k is the projection of ρ onto the space of the kth system. If the small systems are equivalent, the $\langle \mathbf{M}_k \rangle$ are all equal, and thus we obtain

$$\langle \mathbf{M} \rangle = \mathrm{Tr}(\rho \mathbf{M}) = N \, \mathrm{Tr}(\rho_1 \mathbf{M}_1).$$

This equation is exact regardless of whether correlations exist. Thus, *if ρ_1 embodies all of the available information about a single spin system, the predictions of total moment of N systems obtained from it are just as reliable as are those obtained from the global density matrix ρ*. We cannot estimate this reliability from ρ_1 alone; loss of that information is part of the price we had to pay for simplification of the problem. If correlations between different spin units are strong, it will of course be very difficult to obtain ρ_1 without first solving a larger problem. Thus, in practice one will obtain only an approximate value of ρ_1; however, a one percent error in the calculated value of $\langle M_1 \rangle$ leads only to a one percent error in $\langle M \rangle$.

APPENDIX A. SUBJECTIVE H THEOREM

Consider the density matrix (12.5) with $t'=0$; at any particular time there exists a unitary matrix $V(t)$ which diagonalizes $\rho(t)$, so that (12.5) may be written in terms of the diagonal matrices,

$$d(t) = \sum_a P_a W_a d(0) W_a^{-1}, \qquad (A.1)$$

where

$$W_a = V(t) U^{(a)}(t,0) V^{-1}(0) \qquad (A.2)$$

is a unitary matrix. The eigenvalues $d_m(t)$ of $\rho(t)$ are thus related to the eigenvalues of $\rho(0)$ by

$$d_m(t) = \sum_n B_{mn} d_n(0), \qquad (A.3)$$

where the quantities B_{mn} form a doubly stochastic matrix:

$$\sum_m B_{mn} = \sum_n B_{mn} = 1. \qquad (A.4)$$

The first of the inequalities (12.6) is then proved as follows:

$$\begin{aligned} S(t) - S(0) &= \sum_n d_n(0) \ln d_n(0) - \sum_m d_m(t) \ln d_m(t) \\ &= \sum_{mn} B_{mn} d_n(0) \ln [d_n(0)/d_m(t)] \\ &\geq \sum B_{mn} [d_n(0) - d_m(t)] = 0. \end{aligned} \qquad (A.5)$$

Here use has been made of the fact that $\ln x \geq (1-x^{-1})$, with equality if and only if $x=1$. Thus, the equality sign in (A.5) holds if and only if $B_{mn}=0$ for each combination of m, n for which $d_n(0) \neq d_m(t)$. If $\rho(0)$ is nondegenerate, this means that the eigenvalues $d_m(t)$ must be a permutation of the $d_n(0)$.

The second of the inequalities (12.6) follows from the fact that for any given density matrix ρ, the "array entropy" S_A of Eq. (7.14) attains its minimum value, equal to $S = -\text{Tr}(\rho \ln \rho)$ for the orthogonal array. To prove this, let the orthogonal array be the one with N states, where the state v_n has probability d_n, and let $\{\psi_m, w_m\}$ be any other array with M states, where $M \geq N$, which leads to the same density matrix. The two arrays are related by a transformation of the form (7.9)

$$\psi_m w_m^{\frac{1}{2}} = \sum_n v_n d_n^{\frac{1}{2}} U_{nm},$$

where U_{nm} is an $(M \times M)$ unitary matrix, and we define $d_n \equiv 0$, $N < n \leq M$. From this and the orthogonality of the v_n it follows that

$$w_m = \sum_n C_{mn} d_n, \qquad (A.6)$$

where $C_{mn} = |U_{mn}|^2$ is a doubly stochastic matrix, and thus by the previous argument (A.5),

$$S \leq S_A. \qquad (A.7)$$

Now in the case considered here, let $\rho(0)$ be represented by its orthogonal array $\{v_n(0), d_n(0)\}$. At time t, the density matrix (12.5) is represented by the array in which the state

$$\psi_{an}(t) = U^{(a)}(t,0) v_n(0)$$

has probability $w_{an} = P_a d_n(0)$. The array entropy is thus

$$S_A(t) = -\sum_{an} w_{an} \ln w_{an} = S(0) + S(P_a) = \text{const}, \qquad (A.8)$$

which, together with (A.7), proves the theorem.

4. BRANDEIS LECTURES (1963)

These lectures, delivered at Brandeis University in July 1962, mark the end of an evolutionary phase in which, still misguided by the thinking of the past, I believed that the treatment of irreversible processes must be fundamentally different from equilibrium theory. But there is an advance in that the distinction between information entropy S_I and experimental entropy S_E, not yet clearly seen in the 1957 papers, is now recognized and stressed. Of course, this step was crucial for any rational discussion of irreversibility and the second law. The beginning of the next phase is recalled in 'Where do we Stand?' reprinted elsewhere in this volume.

The Brandeis lectures have also the recognition that the continuous information measure as given by Shannon was not derived by him from any desideratum, but only written down by analogy with the discrete measure; and if we derive it by a limiting process from the discrete case there is an extra term $m(x)$. Of course, expressions of this type had been given three years earlier by Kullback, and sixty years earlier by Gibbs; but they were not given this motivation. Recognition of $m(x)$ restored the invariance of the theory under parameter changes, which had been a minor problem for Shannon but a major one for us.

The most important result of this work is the demonstration that, contrary to what had been asserted endlessly in the literature of Statistical Mechanics, the constancy of von Neumann's entropy expression $S_I = -\text{Tr}(\rho \log \rho)$ does not conflict with the Clausius adiabatic form of the second law, $S_{\text{final}} \geqslant S_{\text{initial}}$. Quite the opposite; the second law is an elementary consequence of that constancy. It is only after it has been maximized that the von Neumann information entropy S_I corresponds to the experimental entropy S_E of Clausius. To the best of my knowledge, however, this demonstration has been totally ignored, and the traditional false statement continues to be repeated throughout the literature of Statistical Mechanics.

While the important result was being ignored, the trivial illustrative example of dice in the opening remarks acquired a world-wide notoriety. About a dozen articles have now been written attacking or defending this example, and they are still appearing. We return to the topic in 'Where do we Stand?' and in 'Concentration of Distributions at Entropy Maxima'.

CONTENTS

1. Introduction

2. The General Maximum-Entropy Formalism

3. Application to Equilibrium Thermodynamics

4. Generalization
   a. Density Matrix
   b. Continuous Distributions

5. Distribution Functions

6. Entropy and Probability

7. Conclusion

   References

1. INTRODUCTION

At the beginning of every problem in probability theory, there arises a need to assign some initial probability distribution; or what is the same thing, to "set up an ensemble." This is a problem which cannot be evaded, and for which the laws of physics give us no help. For example, the laws of physics tell us that a density matrix $\rho(t)$ must vary with time according to $i\hbar\dot{\rho} = [H, \rho]$, but they do not tell us what function $\rho(0)$ should be put in at the start. Assignment of $\rho(0)$ is, of course, a matter of free choice on our part — it is for us to say which problem we want to solve.

The assignment of initial probabilities must, in order to be useful, agree with the initial information we have (i.e., the results of measurements of certain parameters). For example, we might know that at time $t = 0$, a nuclear spin system having total (measured) magnetic moment $M(0)$, is placed in a magnetic field H, and the problem is to predict the subsequent variation $M(t)$, which presumably tends to an equilibrium value $M(\infty) = \chi_0 H$ after a long time. What initial density matrix for the spin system $\rho(0)$, should we use? Evidently, we shall want it to satisfy, at the very least,

$$\text{Tr}\,(\rho(0)M_{op}) = M(0) \tag{1}$$

where M_{op} is the operator corresponding to total magnetic moment. But Eq. (1) is very far from uniquely specifying $\rho(0)$. Out of the infinite number of density matrices satisfying (1), which should we choose as the starting point of our calculation to predict $M(t)$?

Conventional quantum theory has provided an answer to the problem of setting up initial state descriptions only in the limiting case where measurements of a "complete set of commuting observables" have been made, the density matrix $\rho(0)$ then reducing to the projection operator onto a pure state $\psi(0)$ which is the appropriate simultaneous eigenstate of all the measured quantities. But there is almost no experimental situation in which we really have all this information, and before we have a theory able to treat actual experimental situations, existing quantum theory must be supplemented with some principle that tells us how to translate, or encode, the results of measurements into a definite state description $\rho(0)$. Note that the problem is not to find the $\rho(0)$ which correctly describes the "true physical situation." That is unknown, and always remains so, because of incomplete information. In order to have a usable theory we must ask the much more modest question: "What $\rho(0)$ best describes our state of knowledge about the physical situation?"

In order to emphasize that this problem really has nothing to do with the laws of physics (and, as a corollary, that its solution will have applications outside the field of physics), consider the following problem. A die has been tossed a very large number N of times, and we are told that the average number of spots up per toss was not 3.5, as we might expect from an honest die, but 4.5. Translate this information into a probability assignment P_n, $n = 1, 2, \ldots, 6$, for the n-th face to come up on the next toss.

To explain more fully what is meant by this, note that we are not asking for an estimate of the fraction (i.e., the relative frequency) of tosses which give n spots. There is, indeed, a connection between the probability and the frequency, which we will derive later. But the problem stated is to reason as best we can about the individual case. The probability P_n must therefore be interpreted in the so-called "subjective" sense; it is only a means of describing how strongly we believe that the n-th face will come up in the next toss.

To state the problem more drastically, imagine that we are offered several bets, at various odds, on various values of n, and we are compelled to accept one of these bets. The probabilities

P_n are the basic raw material from which we decide which one to accept. This is typical of many practical problems faced by the scientist, the engineer, the statistician, the politician; and indeed all of us. We are continually faced with situations where some definite decision must be made now, even though we do not have all the information we might like.

Conventional probability theory does not provide any principle for assigning the probabilities P_n; so let us think about it a little. We must, evidently, choose the P_n such that

$$\sum_{n=1}^{6} P_n = 1 \qquad (2)$$

$$\sum_{n=1}^{6} nP_n = 4.5 \qquad (3)$$

where (3) is analogous to (1). A possible solution of (2) and (3) is indicated in Fig. 1; we could take $P_4 = P_5 = 1/2$, all other $P_n = 0$. This agrees with all the given data. But our common sense tells us it is not a reasonable assignment. The assignment of Fig. 2 is

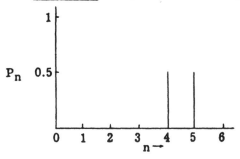

Fig. 1

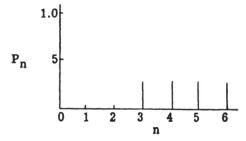

Fig. 2

evidently a more honest description of what we know. But even this is not reasonable—nothing in the data tells us that n = 1, 2 are impossible events. In Fig. 2, we are still jumping to conclusions not warranted by the available evidence. Evidently, it is unreasonable to assign probability zero to any situation unless our data really rules out that case. If we assign $P_1 > 0$, $P_2 > 0$, then in order to keep the average at 4.5, we shall have to give some increased weight to the cases n = 5, 6. Figure 3 shows an assignment that agrees with the data and does not ignore any possibility. But it still seems unreasonable to give the case n = 6 such exceptional treatment. Figure 4 represents what we should probably

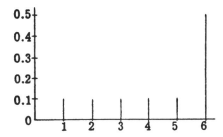

Fig. 3

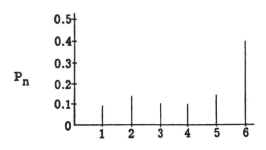

Fig. 4

call a backward step—nothing in the data of the problem indicates any reason for such an uneven treatment. A reasonable assignment P_n must not only agree with the data and must not ignore any possibility—but it must also not give undue emphasis to any possibility. The P_n should vary as smoothly as possible, in some sense. One criterion of "smoothness" might be that adjacent differences $P_{n+1} - P_n$ should be constant; and, indeed, there is a solution with that property. It is given by $P_n = (12n - 7)/210$ and is shown in Fig. 5. This is evidently the most reasonable probabili-

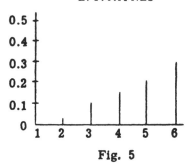

Fig. 5

ty assignment so far. But there is a limit to how high an average you can get with this linear variation of P_n. If we took the extreme case, $P_n =$ (const.)$(n - 1)$, we should again violate one of our principles because $P_1 = 0$, and the average would be only $\sum nP_n = 70/15 = 4.67$. Suppose the data of the problem had been changed so that the average is to be 4.7 instead of 4.5. Then there is no straight-line solution satisfying $P_n \geq 0$. The P_n must lie on some concave curve, as in Fig. 6. But the principles by which we reason surely

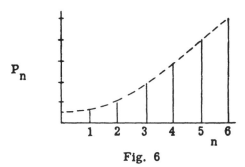

Fig. 6

are the same whether the data specify 4.5 or 4.7; so it appears that a result qualitatively such as Fig. 6 should be used also when $n = 4.5$.

This is about as far as qualitative reasoning can take us, and I have carried the argument through on that basis in order to show how ordinary common sense leads us to a result that has all the important features of the quantitative solution given below. The probability assignment P_n which most honestly describes what we know is the one that is as smooth and "spread out" as possible subject to the data. It is the most conservative assignment in the sense that it does not permit one to draw any conclusions not warranted by the data.

This suggests that the problem is a variational one; we need a measure of the "spread" of a probability distribution which we can maximize, subject to constraints which represent the available information. It is by now amply demonstrated by many workers that the "information measure" introduced by Shannon[1] has special properties of consistency and uniqueness which make it the correct measure of "amount of uncertainty" in a probability distribution. This is, of course, the expression

$$S_I = -\sum_i p_i \log p_i \qquad (4)$$

which, for some distributions and in some physical situations, has long been recognized as representing entropy. However, we have to emphasize that "information-theory entropy" S_I and the experimental thermodynamic entropy S_e are entirely different concepts. Our job cannot be to postulate any relation between them; it is rather to deduce whatever relations we can from known mathematical and physical facts. Confusion about the relation between entropy and probability has been one of the main stumbling blocks in developing a general theory of irreversibility.

2. THE GENERAL MAXIMUM-ENTROPY FORMALISM

To generalize the above problem somewhat, suppose that the quantity x can take on the values $(x_1, x_2, \ldots, x_n)$ where n can be finite or infinite, and that the average values of several functions $f_1(x), f_2(x), \ldots, f_m(x)$ are given, where m < n. The problem is to find the probability assignment $p_i = p(x_i)$ which satisfies the given data: $p_i \geq 0$,

$$\sum_{i=1}^{n} p_i = 1 \qquad (5)$$

$$\sum_{i=1}^{n} p_i f_k(x_i) = \langle f_k(x) \rangle = F_k \qquad k = 1, 2, \ldots, m \qquad (6)$$

and, subject to (5) and (6), maximizes the information theory entropy

$$S_I = -\sum_{i=1}^{n} p_i \log p_i \qquad (7)$$

The solution to this mathematical problem can be found immediately by the method of Lagrangian multipliers, and special cases are given in every statistical mechanics textbook. This method has the merit that it leads immediately to the answer, but the weakness that it does not make it obvious whether one obtains a true absolute maximum of S_I. The following argument establishes this important result more rigorously.

Let $(p_1 \ldots p_n)$ and $(u_1 \ldots u_n)$ be any two possible probability distributions over the x_i; i.e., $p_i \geq 0$, $u_i \geq 0$, $i = 1, 2, \ldots n$ and

$$\sum_{i=1}^{n} p_i = \sum_{i=1}^{n} u_i = 1 \tag{8}$$

Then, by using the fact that $\log x \geq (1 - x^{-1})$, with equality if and only if $x = 1$, we find the following:

Lemma

$$\sum_{i=1}^{n} p_i \log \frac{p_i}{u_i} \geq \sum_{i=1}^{n} p_i (1 - \frac{u_i}{p_i}) = 0 \tag{9}$$

with equality if and only if $p_i = u_i$, $i = 1, 2, \ldots n$. Now make the choice

$$u_i \equiv \frac{1}{Z(\lambda_1 \ldots \lambda_n)} \exp\{-\lambda_1 f_1(x_i) - \ldots - \lambda_m f_m(x_i)\} \tag{10}$$

where $\lambda_1 \ldots \lambda_m$ are fixed constants, and

$$Z(\lambda_1 \ldots \lambda_m) \equiv \sum_{i=1}^{n} \exp\{-\lambda_1 f_1(x_i) - \ldots - \lambda_m f_m(x_i)\} \tag{11}$$

will be called the "partition function." Substituting (10) into (9) results in the inequality

$$\sum_{i=1}^{n} p_i \log p_i \geq \sum_{i=1}^{n} p_i \log u_i = -\sum_{i=1}^{n} p_i [\lambda_1 f_1(x_i) + \ldots + \lambda_m f_m(x_i)] - \log Z(\lambda_1 \ldots \lambda_m)$$

or

$$S_I \le \log Z(\lambda_1\ldots\lambda_m) + \sum_{k=1}^{m} \lambda_k \langle f_k \rangle \tag{12}$$

Now let the distribution p_i vary over the class of all possible distributions that satisfy (6). The right-hand side of (12) remains fixed, and (12) shows that S_I attains its maximum possible value

$$(S_I)_{max} = \log Z + \sum_{k=1}^{m} \lambda_k \langle f_k \rangle \tag{13}$$

if and only if p_i is taken as the generalized canonical distribution (10). It only remains to choose the unspecified constants λ_k so that (6) is satisfied. This is the case, as one readily verifies, if the λ_k are determined in terms of the given data $F_k = \langle f_k \rangle$ by

$$\langle f_k \rangle = -\frac{\partial}{\partial \lambda_k} \log Z(\lambda_1\ldots\lambda_m) \quad k = 1, 2, \ldots, m \tag{14}$$

We now survey rapidly the main formal properties of the distribution found. The maximum attainable entropy (13) is some function of the given data:

$$(S_I)_{max} = S(\langle f_1 \rangle, \ldots \langle f_m \rangle) \tag{15}$$

and, by using (13) and (14), we find

$$\frac{\partial S}{\partial \langle f_k \rangle} = \lambda_k \quad k = 1, 2, \ldots, m \tag{16}$$

Regarding, in (14), the $\langle f_k \rangle$ expressed as functions of $(\lambda_1 \ldots \lambda_m)$, we find, on differentiating, the reciprocity law

$$\frac{\partial \langle f_k \rangle}{\partial \lambda_j} = \frac{\partial \langle f_j \rangle}{\partial \lambda_k} = -\frac{\partial^2}{\partial \lambda_k \partial \lambda_j} \log Z = A_{jk} \tag{17}$$

while by the same argument, if we regard λ_k in (16) expressed as a function of $\langle f_1 \rangle \ldots \langle f_m \rangle$, we find a corresponding law

$$\frac{\partial \lambda_k}{\partial \langle f_j \rangle} = \frac{\partial \lambda_j}{\partial \langle f_k \rangle} = \frac{\partial^2 S}{\partial \langle f_j \rangle \partial \langle f_k \rangle} = B_{jk} \qquad (18)$$

Comparing (17) and (18) and remembering the chain rule for differentiating,

$$\frac{\partial \langle f_j \rangle}{\partial \langle f_k \rangle} = \sum_\ell \frac{\partial \langle f_j \rangle}{\partial \lambda_\ell} \frac{\partial \lambda_\ell}{\partial \langle f_k \rangle} = \delta_{jk}$$

we see that the second derivatives of S and of log Z yield inverse matrices:

$$A = B^{-1} \qquad (19)$$

The functions $\log Z(\lambda_1 \ldots \lambda_n)$ and $S(\langle f_1 \rangle \ldots \langle f_n \rangle)$ are equivalent in the sense that each gives full information about the probability distribution; indeed (13) is just the Legendre transformation that takes us from one representative function to the other.

The reciprocity law (17) acquires a deeper meaning when we consider the "fluctuations" in our probability distribution. Using the distribution (10), a short calculation shows that the second central moments of the distribution of the $f_k(x)$ are given by

$$\langle (f_k - \langle f_k \rangle)(f_\ell - \langle f_\ell \rangle) \rangle = \langle f_k f_\ell \rangle - \langle f_k \rangle \langle f_\ell \rangle$$
$$= \frac{\partial^2}{\partial \lambda_k \partial \lambda_\ell} \log Z \qquad (20)$$

and so, comparing with (17), there is a universal relation between the "fluctuations" of the f_k and the "compliance coefficients" $\partial \langle f_k \rangle / \partial \lambda_\ell$:

$$\langle f_k f_\ell \rangle - \langle f_k \rangle \langle f_\ell \rangle = -\frac{\partial \langle f_k \rangle}{\partial \lambda_\ell} = -\frac{\partial \langle f_\ell \rangle}{\partial \lambda_k} \qquad (21)$$

Likewise, higher derivatives of $\log Z(\lambda_1 \ldots \lambda_n)$ yield higher central moments of the f_k, in a manner analogous to (20), and a hierarchy of fluctuation laws similar to (21).

In addition to their dependence on x, the functions f_k may depend on another parameter, α. The partition function will then also have an explicit dependence on α:

$$Z(\lambda_1 \ldots \lambda_m; \alpha) \equiv \sum_{i=1}^{n} \exp\{-\lambda_1 f_1(x_i; \alpha) - \ldots - \lambda_m f_m(x_i; \alpha)\} \quad (22)$$

and a short calculation shows that the expected derivatives

$$\left\langle \frac{\partial f_k}{\partial \alpha} \right\rangle$$

satisfy the relations

$$\sum_{k=1}^{m} \lambda_k \left\langle \frac{\partial f_k}{\partial \alpha} \right\rangle = -\frac{\partial}{\partial \alpha} \log Z = -\frac{\partial S}{\partial \alpha} \quad (23)$$

If several parameters $\alpha_1 \ldots \alpha_r$ are present, a relation of this form will hold for each of them.

Finally, we note an important variational property which generalizes (16) to the case where we have also variations in the parameters $\alpha_1 \ldots \alpha_r$. Let $Z = Z(\lambda_1 \ldots \lambda_m; \alpha_1 \ldots \alpha_r)$, and consider an arbitrary small change in the problem, where the given data $\langle f_k \rangle$ and the parameters α_j are changed by small amounts $\delta \langle f_k \rangle$, $\delta \alpha_j$. This will lead to a change $\delta \lambda_k$ in λ_k. From (13), the maximum attainable entropy is changed by

$$\delta S = \sum_{k=1}^{m} \frac{\partial \log Z}{\partial \lambda_k} \delta \lambda_k + \sum_{j=1}^{r} \frac{\partial \log Z}{\partial \alpha_j} \delta \alpha_j$$

$$+ \sum_{k=1}^{m} \langle f_k \rangle \delta \lambda_k + \sum_{k=1}^{m} \lambda_k \delta \langle f_k \rangle \quad (24)$$

The first and third terms cancel by virtue of (14). Then, using (23), we have

$$\delta S = -\sum_{j=1}^{r} \sum_{k=1}^{m} \lambda_k \left\langle \frac{\partial f_k}{\partial \alpha_j} \right\rangle \delta \alpha_j + \sum_{k=1}^{m} \lambda_k \delta \langle f_k \rangle \quad (25)$$

Now we can write

$$\sum_{j=1}^{r}\left\langle\frac{\partial f_k}{\partial \alpha_j}\right\rangle \delta\alpha_j = \left\langle\sum_{j=1}^{r}\frac{\partial f_k}{\partial \alpha_j}\delta\alpha_j\right\rangle = \langle\delta f_k\rangle \qquad (26)$$

and so finally

$$\delta S = \sum_{k=1}^{m}\lambda_k[\delta\langle f_k\rangle - \langle\delta f_k\rangle] \qquad (27)$$

or

$$\delta S = \sum_{k=1}^{m}\lambda_k \delta Q_k \qquad (28)$$

where

$$\delta Q_k \equiv \delta\langle f_k\rangle - \langle\delta f_k\rangle \qquad (29)$$

In general δQ_k is not an exact differential; i.e., there is no function $Q_k(\lambda_1\ldots\lambda_m;\alpha_1\ldots\alpha_r)$ which yields δQ_k by differentiation. But (28) shows that λ_k is an integrating factor such that $\sum_k \lambda_k \delta Q_k$ is the exact differential of some "state function" $S(\lambda_1\ldots\lambda_m;\alpha_1\ldots\alpha_r)$.

All the above relations, (10) to (29), are elementary consequences of maximizing the information theory entropy subject to constraints on average values of certain quantities. Although they bear a strong formal resemblance to the rules of calculation provided by statistical mechanics, they make no reference to physics, and, therefore, they must apply equally well to any problem, in or out of physics, where the situation can be described by (1) enumerating a discrete set of possibilities and by (2) specifying average values of various quantities. The above formalism has been applied also to problems in engineering[2] and economics.[3]

In most problems, interest centers on making the best possible predictions for a specific situation, and we are not really interested in properties of any ensemble, real or imaginary. (For example, we want to predict the magnetization M(t) of the particular spin system that exists in the laboratory.) In this case, as already emphasized, the maximum-entropy probability assignment p_i cannot be regarded as describing any objectively existing state of affairs; it is only a means of describing a state of knowledge in a way that is "maximally noncommittal" by a certain criterion.

The above equations then represent simply the best predictions we are able to make on the given information. We are not entitled to assert that the predictions must be "right," only that to make any better ones, we should need more information than was given. However, in cases where it makes sense to imagine x_i as being the result of some random experiment which can be repeated many times, a somewhat more "objective" interepretation of this formalism is possible, which in its essentials was given already by Boltzmann. We are given the same average values $\langle f_k(x) \rangle$ as before, but we are now asked a different question. If the random experiment is repeated N times, the result x_i will be obtained m_i times, $i = 1, 2, \ldots, n$. We are to make the best estimates of the numbers m_i on the basis of this much information. The knowledge of average values tells us that

$$\sum_{i=1}^{n} \frac{m_i}{N} f_k(x_i) = \langle f_k \rangle \qquad k = 1, 2, \ldots, m \tag{30}$$

and, of course,

$$\sum_{i=1}^{n} \frac{m_i}{N} = 1 \tag{31}$$

Equations (30) and (31) do not uniquely determine the m_i if $m < n - 1$, and so again it is necessary to introduce some additional principle, which now amounts to stating what we mean by the "best" estimate. The following criterion seems reasonable. In N repetitions of the random experiment, there are a priori n^N conceivable results, since each trial could give independently any of the results $(x_1, x_2, \ldots, x_n)$. But for given m_i, there are only W of these possible, where

$$W \equiv \frac{N!}{m_1! \ldots m_n!} = \frac{N!}{(Ng_1)!(Ng_2)! \ldots (Ng_n)!} \tag{32}$$

and

$$g_i = \frac{m_i}{N} \qquad i = 1, 2, \ldots, n \tag{33}$$

is the relative frequency with which the result x_i is obtained.

Which choice of the g_i can happen in the greatest number of ways? If we have to guess the frequencies on the basis of no more information than (30), it seems that a reasonable criterion is to ask what choice will maximize (32) while agreeing with (30). Now in the limit of large N, we have by the Stirling formula,

$$\lim_{N\to\infty} \frac{1}{N} \log W = \lim_{N\to\infty} \frac{1}{N} \log \left[\frac{N!}{(Ng_1)!\ldots(Ng_n)!} \right]$$
$$= - \sum_{i=1}^{n} g_i \log g_i \qquad (34)$$

and so, if we are to estimate limiting frequencies in an indefinitely large number of trials, we have in (30) and (34) formulated exactly the same mathematical problem as in (6) and (7). The same solution (10) and formal properties, Eqs. (11) to (29), follow immediately, and we have an alternative interpretation of the maximum-entropy formalism: the probability p_i which information theory assigns to the event x_i at a single trial is numerically equal to an estimate of the relative frequency g_i of this result in an indefinitely large number of trials, obtained by enumerating all cases consistent with our knowledge, and placing our bets on the situation that can happen in the greatest number of ways. Thus, for example, the fluctuation laws (21) describe, on the one hand, our uncertainty as to the unknown true values of $f_k(x)$ in a specific instance; on the other hand, they give the best estimates we can make of the average departures from $\langle f_k \rangle$ in many repetitions of the experiment, by the criterion of placing our bets on the situation that can happen in the greatest number of ways.

Two points about these interpretations should be noted:

1. In most practical problems, repeated repetition of the experiment is either impossible or not relevant to the real problem, which is to do the best we can with the individual case. Thus if one were to insist, as has sometimes been done, that only the second interpretation is valid, the result would be to deny ourselves the use of this formalism in most of the problems where it is helpful.

2. The argument leading from the averages (30) to the estimate of frequencies g_i was not deductive reasoning, but only plausible reasoning. Consequently, we are not entitled to assert that the estimates g_i must be right; only that, in order to make any better estimates, we should need more information. Thus the apparently greater "objectivity" of the second interpretation is to a large extent illusory.

BRANDEIS LECTURES 53

3. APPLICATION TO EQUILIBRIUM THERMODYNAMICS

We apply the formalism of the preceding section to the following situation: $m = 1$, $f_1(x_i, \alpha) = E_i(V)$. The parameter V (volume) and the expectation value of the energy of the system $\langle E \rangle$ are given. The partition function is

$$Z(\lambda, V) \equiv \sum_{i=1}^{\infty} e^{-\lambda E_i(V)} \qquad (35)$$

Then, by (14), λ is determined from

$$\langle E \rangle = -\frac{\partial}{\partial \lambda} \log Z \qquad (36)$$

and, as a special case of (23), we have

$$\lambda \left\langle \frac{\partial E}{\partial V} \right\rangle = -\frac{\partial}{\partial V} \log Z \qquad (37)$$

But $-\langle \partial E/\partial V \rangle = \langle P \rangle$ is the maximum-entropy estimate of pressure, and so the predicted equation of state is

$$\langle P \rangle = \frac{1}{\lambda} \frac{\partial}{\partial V} \log Z \qquad (38)$$

To identify the temperature and entropy, we use the general variational property (28). A small change δV in volume will change the energy levels by $\delta E_i = (\partial E_i/\partial V) \delta V$, and if this is carried out infinitely slowly (i.e., reversibly), the "adiabatic theorem" of quantum mechanics tells us that the probabilities p_i will not be changed. So, the maximum-entropy estimate of the work done is

$$\delta W = -\langle \delta E \rangle \qquad (39)$$

Of course, the given $\langle E \rangle$ is interpreted as the thermodynamic energy function U. In addition to the change δV, we allow a small reversible heat flow δQ, and by the first law, the net change in energy is $\delta U = \delta Q - \delta W$, or

$$\delta Q = \delta \langle E \rangle - \langle \delta E \rangle \qquad (40)$$

Thus, if f_k is the energy, then the δQ_k defined by (29) is the predicted heat flow in the ordinary sense. Equation (28) shows that for <u>any</u> quantity f_k, there is a quantity δQ_k formally analogous to heat.

In the present case (28) reduces to

$$\delta S(\langle E \rangle, V) = \lambda \, \delta Q \tag{41}$$

Now the Kelvin temperature is defined by the condition that $(1/T)$ is the integrating factor for infinitesimal reversible heat in closed systems and the experimental entropy S_e is defined as the resulting state function. So from (41) the predicted temperature T' and experimental entropy S_e' are given by

$$\lambda = \frac{1}{kT'} \tag{42}$$

$$S_e' = kS(\langle E \rangle, V) = k(S_I)_{max} \tag{43}$$

The presence of Boltzmann's constant k merely indicates the particular practical units in which we choose to measure temperature and entropy. For theoretical discussions, we may as well adopt units such that $k \equiv 1$.

All that we have shown so far is that the general maximum-entropy formalism leads automatically to definitions of quantities <u>analogous</u> to those of thermodynamics. This is, of course, as far as any mathematical theory can go; no amount of mathematics can prove anything about experimental facts. To put it differently, before we can establish any connection between our theoretical entropy S_e' and the experimentally measured quantity S_e, we have to introduce some physical assumption about what the result of an experiment would in fact be:

> <u>Physical assumption</u>: The equilibrium thermodynamic properties of a system, as measured experimentally, agree with the results calculated by the usual methods of statistical mechanics; i.e., from the canonical or grand canonical ensemble appropriate to the problem. (44)

This assumption has proved correct in every case where one has succeeded in carrying out the calculations, and its universal validity is taken so much for granted nowadays that authors of textbooks no longer list it as an assumption. But strictly speaking, all we can prove here is that systems conforming to this assumption will also conform to various other statements made below.

If we accept (44), then the identification of entropy is complete, and connection between information theory entropy and experimental entropy for the present problem can be stated as a theorem.

Theorem: Let $p_i \equiv \text{prob}(E_i)$ be any probability assignment which conforms to the data in the sense that $\langle E \rangle = \sum_i p_i E_i$ is the measured energy. Let $S_I \equiv -\sum p_i \log p_i$ be the corresponding information theory entropy, and S_e be the experimentally measured entropy for the system. The additive constant is chosen so that at zero temperature $S_e = \log n$, where n is the degeneracy of the ground state, and let S_e be expressed in units such that Boltzmann's constant $k = 1$. Then

$$S_I \leq S_e \tag{45}$$

with equality if and only if p_i is chosen as the canonical distribution

$$p_i = \frac{1}{Z} \exp\{-\lambda E_i(V)\} \tag{46}$$

This is the physical meaning, for the present problem, of the general inequality (12). Obviously, the above statement can be greatly generalized; we can introduce more degrees of freedom in addition to V, we can consider open systems, where the number of molecules can change, and we can use the grand canonical ensemble, etc. The corresponding statement will still hold; over all probability assignments that agree with the data in the aforementioned sense, the information theory entropy attains an absolute maximum, equal to the experimental entropy, if and only if p_i is taken as the appropriate canonical or grand canonical distribution.

Remarks: 1. We have taken $\langle E \rangle$ as the given quantity. In practice, it is usually the temperature that is measured. To treat the temperature as the observable, one must regard the system of interest to be in contact with a heat reservoir, with which it may exchange energy and which acts as a thermometer. Detailed analysis of the resulting system (given in reference[4]) leads to the same probability assignments as we have found with $\langle E \rangle$ as the given datum.

2. If not only $\langle E \rangle$ is known, but also the accuracy of the measurement, as given for example by $\langle E^2 \rangle$, then this information may be incorporated into the problem by taking $f_1(x_i, \alpha) = E_i(V)$, $f_2(x_i, \alpha) = E_i^2(V)$. The partition function (11) becomes

$$Z(\lambda_1, \lambda_2, V) = \sum_i \exp[-\lambda_1 E_i(V) - \lambda_2 E_i^2(V)] \qquad (47)$$

and from (14),

$$\langle E \rangle = -\frac{\partial}{\partial \lambda_1} \log Z \qquad \langle E^2 \rangle = -\frac{\partial}{\partial \lambda_2} \log Z \qquad (48)$$

The fluctuation theorem (21) then gives the relation

$$\langle E^3 \rangle - \langle E \rangle \langle E^2 \rangle = -\frac{\partial \langle E \rangle}{\partial \lambda_2} = -\frac{\partial \langle E^2 \rangle}{\partial \lambda_1} \qquad (49)$$

In principle, whenever information of this sort is available, it should be incorporated into the problem. In practice, however, we find that for the macroscopic systems that exhibit reproducible thermodynamic properties, the variance $\langle E^2 \rangle - \langle E \rangle^2$ as calculated from (46) is already very small compared to any reasonable mean-square experimental error, and so the additional information about accuracy of the measurement did not lead to any difference in the predictions. This is, of course, the basic reason for the success of the Gibbs canonical ensemble formalism.

3. The theory as developed here has, in principle, an additional freedom of choice not present in conventional statistical mechanics. The statement that a system has a definite, reproducible equation of state means, for example, that if we fix experimentally any two of the parameters P, V, T, then the third is determined. Correspondingly, in the theory it should be true that information about any two of these quantities should suffice to enable us to predict the third; there is no basic reason for constructing our ensembles always in terms of energy rather than any other measurable quantities. Use of energy has the mathematical convenience that energy is a constant of the motion, and so the statement that the system is in equilibrium (i.e., measurable parameters are not time-dependent) requires no additional constraint. With an ensemble based on some quantity, such as pressure or magnetization, which is not an intrinsic constant of the motion, if we wish to predict equilibrium properties we need to incorporate into the theory an additional statement, involving the equations of motion, which specifies that these quantities are constant. To do this requires no new principles of reasoning beyond those given above; we merely include the values of such a quantity $f(t_i)$ at many different times (or in the limit, at all times) into the set of quantities f_k whose expectation values are given. In the limit, the

partition function thus becomes a partition functional:

$$Z[\lambda(t)] = \sum_i \exp[-\int \lambda(t) f(x_i, t)\, dt] \qquad (50)$$

and the relations (14) determining the λ's go into the corresponding functional derivative relations

$$\langle f(t) \rangle = -\frac{\delta}{\delta \lambda(t)} \log Z[\lambda(t)] \qquad (51)$$

which determine the function $\lambda(t)$.

We have not found any general proof that the predicted equation of state is independent of the type of information used, but a special case is proved in the 1961 Stanford thesis of Dr. Douglas Scalapino. There it is shown that the same equation of state of a paramagnetic substance with spin-spin interaction is obtained whatever the input information. We conjecture that this is true for any system that exhibits an experimentally reproducible equation of state.

It is doubtful whether this new degree of freedom in applying the theory will prove useful in calculations pertaining to the equilibrium state, since it is more complicated than the usual procedure. However, it is just this extra freedom that makes it possible to develop a general formalism for irreversible processes; indeed, prediction of time-dependent phenomena is obviously impossible as long as our probability distributions depend only on constants of the motion. Equations (50) and (51) form the starting point for a general theory of the nonequilibrium steady state, the Scalapino thesis providing an example of the calculation of transport coefficients from them.

4. GENERALIZATION

For most applications of interest, the foregoing formalism needs to be generalized to the case of (a) systems described by a density matrix or (b) continuous probability distributions as occur in classical theory. We indicate briefly how this is done.

a. **Density Matrix**

The expectation value of an operator F_k of a system described by the density matrix ρ is

$$\langle F_k \rangle = \text{Tr}\,(\rho F_k) \tag{52}$$

where Tr stands for the trace. The information theory entropy corresponding to ρ is

$$S_I = -\text{Tr}\,(\rho \log \rho) \tag{53}$$

(See reference[5] for the arguments that lead to this definition of S_I and discussion of other expressions which have been proposed.) Maximizing S_I subject to the constraints imposed by knowledge of the $\langle F_k \rangle$ yields

$$\rho = \frac{1}{Z(\lambda_1 \ldots \lambda_m)} \exp\{-\lambda_1 F_1 - \ldots - \lambda_m F_m\} \tag{54}$$

where

$$Z(\lambda_1 \ldots \lambda_m) \equiv \text{Tr}\,\exp\{-\lambda_1 F_1 - \ldots - \lambda_m F_m\} \tag{55}$$

To prove (54), use the lemma

$$\text{Tr}\,(\rho \log \rho) \geq \text{Tr}\,(\rho \log \sigma) \tag{56}$$

analogous to (9). Here ρ is any density matrix satisfying (52), and σ is the canonical density matrix (54). All the formal relations (12) to (29) still hold, except that when the F_k do not all commute, the fluctuation law (21) must be generalized to

$$-\frac{\partial \langle F_k \rangle}{\partial \lambda_j} = -\frac{\partial \langle F_j \rangle}{\partial \lambda_k} = \int_0^1 \langle e^{xA} F_k e^{-xA} F_j \rangle \, dx - \langle F_k \rangle \langle F_j \rangle \tag{57}$$

where

$$A \equiv \sum_{k=1}^{m} \lambda_k F_k \tag{58}$$

For all ρ that agree with the data in the sense of (52), we have $S_I(\rho) \leq S_e$, with equality if and only if ρ is the canonical matrix (54).

b. Continuous Distributions

Shannon's fundamental uniqueness theorem (reference,[1] theorem 3) which establishes $-\sum p_i \log p_i$ as the correct information measure, goes through only for discrete probability distributions. At the present time, the only criterion we have for finding the analogous expression for the continuous case is to pass to the limit from a discrete one; presumably, future study will give a more elegant approach. The following argument can be made as rigorous as we please, but at considerable sacrifice of clarity. In the discrete entropy expression

$$S_I^{(d)} = -\sum_{i=1}^{n} p_i \log p_i \qquad (59)$$

we suppose that the discrete points x_i, $i = 1, 2, \ldots, n$, become more and more numerous, in such a way that, in the limit $n \to \infty$, the density of points approaches a definite function $m(x)$:

$$\lim_{n \to \infty} \frac{1}{n}(\text{number of points in } a < x < b) = \int_a^b m(x)\, dx \qquad (60)$$

If this passage to the limit is sufficiently well behaved, it will also be true that adjacent differences $(x_{i+1} - x_i)$ in the neighborhood of any particular value of x will tend to zero so that

$$\lim_{n \to \infty} [n(x_{i+1} - x_i)] = [m(x_i)]^{-1} \qquad (61)$$

The discrete probability distribution p_i will go over into a continuous probability density $w(x)$, according to the limiting form of

$$p_i = w(x_i)(x_{i+1} - x_i)$$

or, from (61),

$$p_i \to w(x_i)[nm(x_i)]^{-1} \qquad (62)$$

Consequently, the discrete entropy (59) goes over into the integral

$$S_I^{(d)} \to -\int w(x)\, dx\, \log\left[\frac{w(x)}{nm(x)}\right]$$

In the limit, this contains an infinite term log n; but if we subtract this off, the difference will, in the cases of interest, approach a definite limit which we take as the continuous information measure:

$$S_I^{(c)} \equiv \lim [S_I^{(d)} - \log n] = -\int w(x) \log\left[\frac{w(x)}{m(x)}\right] dx \qquad (63)$$

The expression (63) is invariant under parameter changes; i.e., instead of x another quantity y(x) could be used as the independent variable. The probability density and measure function m(x) transform as

$$w_1(y)\, dy = w(x)\, dx$$
$$m_1(y)\, dy = m(x)\, dx$$

so that (63) goes into

$$S_I^{(c)} = -\int w_1(y)\, dy\, \log\left[\frac{w_1(y)}{m_1(y)}\right] \qquad (64)$$

To achieve this invariance it is necessary that the "measure" m(x) be introduced. I stress this point because one still finds, in the literature, statements to the effect that the entropy of a continuous probability distribution is <u>not</u> an invariant. This is due to the historical accident that in his original papers, Shannon[1] assumed, without calculating, that the analog of $\sum p_i \log p_i$ was $\int w \log w\, dx$, and got into trouble for lack of invariance. Only recently have we realized that mathematical deduction from the uniqueness theorem, instead of guesswork, yields the invariant information measure (63).

In many cases it is more natural to pass from the discrete distribution to a continuous distribution of several variables, $x_1 \ldots x_r$; in this case the results readily generalize to

$$S_I^{(c)} = -\int \ldots \int w(x_1 \ldots x_r) \log\left[\frac{w(x_1 \ldots x_r)}{m(x_1 \ldots x_r)}\right] dx_1 \ldots dx_r \qquad (65)$$

We apply this to the Liouville function of classical mechanics. For a system of N particles, $W_N(x_1 p_1 \ldots x_2 p_x; t)\, d^3x_1 \ldots d^3p_N$ is the probability that at time t the system is in the element $d^3x_1 \ldots d^3p_N$ of 6N-dimensional phase space. Before we can set up the information measure for this case, we must decide on a basic measure $m(x_1 \ldots p_N)$ for phase space. In classical statistical mechanics, one has always taken uniform measure: m = const., largely because one couldn't think of anything else to do. However, the more careful writers have all stressed the fact that within the context of classical theory, no real justification of this has ever been produced. For the present, I propose to dodge this issue by regarding classical statistical mechanics merely as a limiting form of the (presumably more fundamental) discrete quantum statistical mechanics. In other words, the well-known proposition that each discrete quantum state corresponds to a volume h^{3N} of classical phase space, will determine our uniform measure as resulting from equal weighting of all orthogonal quantum states, and passing to the limit h→0. Thus, apart from an irrelevant additive constant which we drop, our information measure will be just the negative of the Gibbs H-function, H_G:

$$-S_I = H_G = \int W_N \log W_N \, d\tau \tag{66}$$

where $d\tau = d^3x_1 \ldots d^3p_N$.

With this continuous probability distribution, we are able to incorporate into the theory a more detailed kind of macroscopic information than we have considered up till now. Suppose we are given the macroscopic density $\rho(x)$ as a function of position. We interpret this as specifying at each point of space, the expectation value of a certain quantity:

$$\langle f_1(x_1 p_1 \ldots x_N p_N; x) \rangle = \int W_N f_1 \, d\tau = \rho(x) \tag{67}$$

where the phase function f_1 is given by

$$f_1(x_1 p_1 \ldots x_N p_N; x) = \sum_{i=1}^{N} m\, \delta(x_i - x) \tag{68}$$

The position x now plays the same role as the index k in the elementary version of the formalism, Eqs. (10) to (29), and so in place of the sum $\sum \lambda_k f_k(x_i)$ in the exponent of the probability distribution, this information will place the integral

$$\int \lambda(x) f_1 \, d^3x$$

into the exponent of W_N. The partition function then becomes a partition functional of the function $\lambda(x)$.

In general, we might have several phase functions of this kind, whose expectation values are given at each point of space:

$$\langle f_1(x_1 \ldots p_N; x) \rangle = \int W_N f_1 \, d\tau$$
$$\cdots \qquad \cdots \qquad (69)$$
$$\langle f_m(x_1 \ldots p_N; x) \rangle = \int W_N f_m \, d\tau$$

Maximization of S_I subject to these constraints gives the partition functional

$$Z[\lambda_1(x), \ldots, \lambda_m(x)] = \int d\tau \, \exp\left\{-\sum_{k=1}^{m} \int \lambda_k(x) \right. \qquad (70)$$
$$\left. \times f_k(x_1 \ldots p_N; x) \, d^3x \right\}$$

The Lagrange multiplier functions $\lambda_k(x)$ are determined by relations analogous to (14), but now involving the functional derivatives:

$$\langle f_k(x_1 \ldots p_N; x) \rangle = -\frac{\delta}{\delta \lambda_k(x)} \log Z[\lambda_1(x), \ldots, \lambda_m(x)] \qquad (71)$$

and the other properties, Eqs. (16) to (29), are likewise easily generalized.

Example: Suppose the macroscopic density of mass, momentum, and kinetic energy are given at the initial time. This corresponds to expectation values of (68), and

$$\langle f_2(x_1 \ldots p_N; x) \rangle = \left\langle \sum_{i=1}^{N} p_i \delta(x_i - x) \right\rangle = P(x) \qquad (72)$$

$$\langle f_3(x_1 \ldots p_N; x) \rangle = \left\langle \sum_{i=1}^{N} \frac{p_i^2}{2m} \delta(x_i - x) \right\rangle = K(x) \qquad (73)$$

Since all the given data are formed additively from contributions of each particle, the maximum-entropy Liouville function factors:

$$W_N = w_1(x_i, p_i) \tag{74}$$

(this would not be the case if the given information concerned mutual properties of different particles, such as the potential energy), and the exponential in the partition functional (70) reduces to

$$-\int d^3x \, [\lambda_1(x) \sum_i m \, \delta(x_i - x) + \lambda_2(x) \cdot \sum_i p_i \, \delta(x_i - x)$$
$$+ \lambda_3(x) \sum_i \frac{p_i^2}{2m} \delta(x_i - x)]$$
$$= -\sum_{i=1}^N [m\lambda_1(x_i) + p_i \cdot \lambda_2(x_i) + \frac{p_i^2}{2m}\lambda_3(x_i)]$$

so that

$$\log Z = N \log \int \{\exp[-m\lambda_1(x) - p \cdot \lambda_2(x) \\ - \frac{p^2}{2m}\lambda_3(x)]\} \, d^3x \, d^3p \tag{75}$$

Application of (71) now yields the physical meaning of the Lagrange multipliers: defining the "mass velocity" $u(x)$ by $P(x) = \rho(x)u(x)$, and the "local temperature" $T(x)$ by the mean-square velocity as seen by an observer moving at velocity $u(x)$, we find

$$\lambda_3(x) = \frac{1}{kT(x)} = \beta(x)$$
$$\lambda_2(x) = \beta(x)u(x) \tag{76}$$
$$m\lambda_1(x) = 1/2 \, mu^2(x)\beta(x) - 3/2 \log \beta(x) - \log \rho(x) + (\text{const.})$$

and the single-particle distribution function w_1 of (74) reduces to

$$w_1(x, p) = \frac{\rho(x)}{mN[2\pi mkT(x)]^{3/2}} \exp\left\{-\frac{[p - mu(x)]^2}{2mkT(x)}\right\} \tag{77}$$

In this rather trivial example we merely recover a well-known result; but from a different viewpoint than the usual one, which leads us to interpret (77) differently, and regard it as a very special case. The method used enables us to translate other kinds of macroscopic information into definite probability distributions. In other words, we suggest that the maximum-entropy formalism provides the general solution to the problem of "setting up an ensemble" to describe an arbitrary macroscopic situation, equilibrium or nonequilibrium.

The distributions found in the above way, of course, describe the situation only at the initial time for which the macroscopic information is given. For predictions referring to other times, one should, in principle, solve the equations of motion, or Liouville equation,

$$W_N + [W_N, H] = 0 \tag{78}$$

where H is the Hamiltonian and $[W_N, H]$, the Poisson bracket. In practice, the history of irreversible statistical mechanics has been one of unceasing efforts to replace this impossibly difficult calculation by a simpler one, in which we try to reduce (78) to an "irreversible" equation variously termed Boltzmann equation, rate equation, or master equation. Although considerable progress has been made in this direction in recent years, we are still far from really bridging the gap between these two methods of description.

As a preliminary step in this direction, it is necessary that we understand clearly the physical meaning of the Liouville function W_N and the various reduced distribution functions derived from it. The following section surveys these questions.

5. DISTRIBUTION FUNCTIONS

A recent review of transport theory by Dresden[6] (hereafter referred to as MD) illustrates that attempts to bridge the gap between phenomenological rate equations and fundamentals (equations of Liouville and Gibbs) have been largely frustrated because basic conceptual difficulties, dating from the time of Boltzmann, are still unresolved. This section is intended as a supplement to the discussion of these problems given to MD, Sec. I.B.

Early attempts to base transport theory on the BBGKY hierarchy of distribution functions made no distinction between the Boltzmann distribution function $f(x, p, t)$ and the single-particle function $w_1(xpt)$ of the hierarchy. In MD this distinction is pointed out without, however, stating any precise relation between them. To do this requires, first of all, precise definitions of f and the Liouville function W_N. Boltzmann originally defined f as giving the actual number of particles in various cells of six-dimensional phase space; thus if R is the set of phase points comprising a cell, the number of particles in R is *

$$n_R = \int_R f(x, p, t) \, d^3x \, d^3p \qquad (79)$$

The well-known paradoxes involving the H-theorem led to a feeling that this definition should be modified; but the exact way seems never to have been stated. Here we retain the definition (79), which has at least the merit of being a precise statement, and accept the consequence that the Boltzmann collision equation cannot be strictly correct, for reasons given by Zermelo and Loschmidt.

From (79) it is immediately clear that Boltzmann's f is not a probability distribution at all, but a "random variable." In other words, instead of saying that f gives the probability of various conditions, we should ask, "What is the probability that f takes on various values?"

Establishment of a precise connection between Boltzmann's f and the single-particle function of the hierarchy,

$$w_1(x_1, p_1, t) = \int W_N \, d^3x_2 \ldots d^3p_N \qquad (80)$$

requires no coarse-graining, time-smoothing, or any other mutilation of the hierarchy. If we agree that a particle will be considered "in R" if its center of gravity is in R, and that the Liouville function W_N is symmetric under permutations of particle labels, then from (79) and (80) the exact connection between them is simply,

$$\langle f \rangle = N w_1 \qquad (81)$$

where the angular brackets denote an average over the Liouville function W_N. The only "statistical notion" which needs to be adjoined to it is the usual one that $W_N \, d\tau$ shall be interpreted as the probability that the individual system is in the phase region $d\tau$. To say that W_N refers to number density in a fictitious ensemble is only to say the same thing in different words; this cannot be emphasized too strongly. Indeed, the notion of an ensemble is merely a device that enables us to speak of probabilities on the Gibbs, or global level, as if they were frequencies, in some larger system

---

* Same as Uhlenbeck, in M. Kac, "Probability and Related Topics in Physical Sciences", Appendix I, p. 192 (Interscience, 1959).

which is defined for just that purpose.

The reason why it was felt necessary to introduce the notion of an ensemble is that the development of equilibrium statistical mechanics took place entirely in a period when the frequency theory of probability was the only one considered respectable. It has been taken for granted that any probability distributions used must be, in principle, empirically measurable frequencies, and that the fundamental problem of statistical mechanics is to justify these distributions in the frequency sense.

The statistical practice of physicists has tended to lag about 20 years behind current developments in the field of basic probability and statistics. I hope to shorten that gap to about 10 years by pointing out that a revolution in statistical thought has recently taken place, brought about largely by the development of statistical decision theory. Two brief summaries of these developments have been published[5,7] and a detailed analysis of the present situation[8] will soon be available. The net result is a vindication of the viewpoint of Laplace, and of Jeffreys,[9] that probability theory is properly regarded as an extension of logic to the case of inductive, or plausible, reasoning, the probabilities denoting basically a "degree of reasonable belief," rather than limiting frequencies. This does not mean that there are no longer any connections between probability and frequency; the situation is rather that every connection between probability and frequency which is actually used in applications is deducible as a mathematical consequence of the "inductive logic" theory of probability.[8] Equation (81), and others given below, provide examples of the kind of connections that exist.

Use of probability in this "modern" (actually the original) sense is, of course, essential to the maximum-entropy formalism; for the frequencies with which different microscopic states are occupied are manifestly not given, in general, by a distribution canonical in the observed quantities; indeed, for a time-dependent problem the notion of occupation frequency is meaningless. Nevertheless, in a problem where frequencies are meaningful, if our job is to estimate those frequencies, our best estimate on the basis of the information available will be numerically equal to the probabilities. One example of this was given in the "objective" interpretation of the maximum-entropy formalism in Sec. 2, and we now give another example which clarifies the meaning of the distribution functions.

From Eqs. (79) and (81) one sees that the single-particle function w_1 does not contain full information about the distribution of particles in six-dimensional phase space. Integrating (81) over the cell R, we see that it determines only the expectation value of particle occupation numbers:

$$\langle n_R \rangle = N \int_R w_1(x, p, t) \, d^3x \, d^3p \tag{82}$$

In words: the integral in (82) represents the probability that any specified particle is in the phase cell R. This is not the same as the fraction of particles in that cell but represents only the expectation value of that fraction, over the Liouville distribution W_N. Before we are justified in the usual interpretation which identifies (82) with the actual number of particles in R, it must be shown that the variance of the n_R distribution is small:

$$\frac{\langle n_R^2 \rangle - \langle n_R \rangle^2}{\langle n_R \rangle^2} \ll 1 \tag{83}$$

Unless (83) is satisfied, the Liouville function is making no definate prediction about the number of particles in R. But we are not allowed to postulate (83) on the grounds of any "law of large numbers" even for a cell R of macroscopic size, because the two-particle distribution function of the hierarchy,

$$w_2(x_1 p_1, x_2 p_2, t) = \int W_N \, d^3x_3 \ldots d^3p_N \tag{84}$$

completely determines whether (83) is or is not satisfied. To see this, introduce the characteristic function of the set R:

$$M(x, p) \equiv \begin{cases} 1, & x, p \text{ in R} \\ 0, & \text{otherwise} \end{cases} \tag{85}$$

Then

$$\langle n_R^2 \rangle = \sum_{i,j=1}^{N} \langle M(x_i, p_i) M(x_j, p_j) \rangle = N I_1 + N(N-1) I_2 \tag{86}$$

where

$$I_1 \equiv \int_R w_1(x, p) \, d^3x \, d^3p \tag{87}$$

$$I_2 \equiv \int_R d^3x_1 \, d^3p_1 \int_R d^3x_2 \, d^3p_2 \, w_2(x_1 p_1, x_2 p_2) \tag{88}$$

The measure of dispersion (83) then reduces to

$$\frac{I_2 - I_1^2}{I_1^2} + \frac{I_1 - I_2}{N I_1^2} \tag{89}$$

Thus, when $N \gg 1$ and $\langle n_R \rangle \gg 1$, the necessary and sufficient condition for validity of (83) becomes

$$\left| \frac{I_2}{I_1^2} - 1 \right| \ll 1 \tag{90}$$

Usually one omits gravitational forces from the Hamiltonian and chooses a Liouville function which makes w_1 independent of position. If we then describe thermal equilibribrium by $W_N \sim \exp(-\beta H)$ and choose a cell R consisting of all of momentum space, and a region V_R of ordinary space of macroscopic size, Eq. (90) becomes the necessary and sufficient condition that the Liouville function makes a sharp prediction of the density of the fluid; i.e., it predicts that only one phase is present in V_R. Thus the condition for condensation, or more precisely for the coexistence of more than one phase, is that (90) fails to hold. Equation (82) then gives only a weighted average of the density of the various possible phases.

Similarly, in the problem of deriving the laws of hydrodynamics from the Liouville equation, one needs to find the predicted momentum density. In terms of the Boltzmann distribution function, the total momentum in any phase cell R is

$$\mathbf{P} = \int_R \mathbf{p} f(x, p, t) \, d^3x \, d^3p \tag{91}$$

and we choose R to consist of all momentum space plus a cell S' of ordinary space that is "microscopically large but macroscopically small." Again, the single-particle function gives only the expectation value,

$$\langle \mathbf{P} \rangle = N \int_R \mathbf{p} w_1(x, p, t) \, d^3x \, d^3p \tag{92}$$

but w_1 gives no information at all as to whether this is a <u>reliable</u> prediction. To answer this, we must appeal to the two-particle function:

$$\langle P^2 \rangle = N \int_R p^2 w_1 \, dx \, dp + N(N-1) \int_R dx \, dp \int_R dx' \, dp' \\ \times \mathbf{p} \cdot \mathbf{p}' \, w_2(x, p, x', p') \tag{93}$$

If the variance of P is everywhere small, then the Liouville function is making a definite prediction of a flow pattern; i.e., it predicts laminar flow. But if the last term of (93) is large, the single-particle function gives only a weighted average of several possible flows. In this case, the information put into the Liouville function was not sufficient to determine any definite mass motion of the fluid. But if we incorporated into W_N all the information about the experimentally imposed conditions, the theory is now telling us that under these conditions the flow will not be experimentally reproducible. In other words, the theory is predicting turbulent flow.

These examples show that the proper physical interpretation of the distributions (i.e., their exact relation to physical quantities) is not an obscure philosophical point. Failure to distinguish between w_1 and f as given in (79) means failure to distinguish between expectation values and actual values, and amounts to the same thing as simply postulating that ensemble averages are equal to observed values of physical quantities. This is not only unjustified because of the probability nature of W_N; it would mean loss of the correct criterion for phase changes and of the criterion which distinguishes between laminar and turbulent flow.

On the other hand, we can see no basis for any distinction between equilibrium and nonequilibrium situations here. One of the most elementary theorems of probability theory assures us that, for any phase function Q and any probability assignment W_N whatsoever, the expectation value $\langle Q \rangle$, denoted by Q_{obs} in MD, is the best estimate of Q in the sense that it minimizes the expected square of the error. Whether the information put into W_N permits an <u>accurate</u> estimate (i.e., whether the expected square of the error is small), can be neither postulated nor denied arbitrarily; it is determined by W_N. In all cases, equilibrium or otherwise, the test is to calculate $\langle Q^2 \rangle = \int Q^2 W_N \, dv$, and see whether it is sufficiently close to $\langle Q \rangle^2$ in the sense of (83). If calculation of $\langle Q \rangle$ requires knowledge of the function w_s of the hierarchy, but not w_{s+1}, and 2s < N, then information about the reliability of the ensemble average $\langle Q \rangle$ as an estimate of Q appears for the first time in the function w_{2s}, and is, of course, retained in all higher-order functions.

Any system of "kinetic equations," such as the Boltzmann or Bogoliubov scheme, which attempts to write the higher-order functions in terms of w_1, throws away information about the reliability of the predictions. This, however, may represent a net advantage if it simplifies the mathematics without greatly affecting the actual predictions; consequently the search for such kinetic equations is a major objective of current theoretical effort. If the particles move under the influence of a potential energy function $V(x_1 \ldots x_N)$, the exact differential equation satisfied by $w_1(x_1, p_1, t)$ may be written compactly

$$\frac{\partial w_1}{\partial t} + \frac{p_{1\alpha}}{m} \frac{\partial w_1}{\partial x_\alpha} + \frac{\partial}{\partial p_{1\alpha}} [\langle F_\alpha \rangle w_1] = 0 \qquad (94)$$

where

$$\langle F_\alpha \rangle = -\int \frac{\partial V}{\partial x_{1\alpha}} (x_2 \ldots p_N | x_1 p_1) \, d^3x_2 \ldots d^3p_N \qquad (95)$$

is the conditional expectation value of the force seen by particle 1, given that it has position and momentum (x_1, p_1). Here $(x_2 \ldots p_N | x_1 p_1)$ is the conditional probability density for the other particles, defined by $W_N(x_1 \ldots p_N) = (x_2 \ldots p_N | x_1 p_1) w_1(x_1 p_1)$.

Although direct calculation of $\langle F_\alpha \rangle$ would be very difficult, the form of (94) should prove useful in two respects. In the first place, it shows that, although the basic ideas may be stated in entirely different terms, any proposed equation for w_1, such as the Boltzmann, the Fokker-Planck, or the Bogoliubov equation, is equivalent to some assumption about the expected force $\langle F_\alpha \rangle$. The physical reasonableness of any proposed equation may, therefore, be judged by comparing it to (94), and seeing what explicit assumption it makes about $\langle F_\alpha \rangle$. Second, (94) shows that all the complications of this subject reduce ultimately to the determination of one quantity, $\langle F_\alpha \rangle$. Therefore, a phenomenological theory should be feasible in which $\langle F_\alpha \rangle$ is determined from appropriate experiments. In situations close to equilibrium, one finds in this way that in first approximation $\langle F_\alpha \rangle$ is proportional to the density gradient, and independent of p_1. The condition for condensation, which is a particular kind of hydrodynamic instability, is then that this proportionality coefficient exceeds a certain critical value.

6. ENTROPY AND PROBABILITY

Now we turn to what is perhaps the most serious confusion of all in current irreversible statistical mechanics—the interpretation of entropy in terms of probability distributions. As recent literature gives ample testimony, even the issue of Boltzmann's versus Gibbs' H functions to represent entropy has not been resolved in any commonly agreed way. For example, in MD it is stated that the Boltzmann H,

$$H_B = \int f \log f \, d^3x \, d^3p \qquad (96)$$

---

* G. E. Uhlenbeck, *Physics Today*, July 1960, pp. 17-21, together with remark in Kac book (see p. 65 here). Think of the problem of "setting up an ensemble" solved by Bogoliubov.

is "directly related" to the entropy, whereas the Gibbs expression

$$H_G = \frac{1}{N} \int W_N \log W_N \, dv \qquad (97)$$

is rejected with the statement: "There is, however, no possibility of identifying or relating H_G to the macroscopic entropy, for one proves directly from (23) and (18) that H_G is constant in time, whereas the macroscopic entropy always increases in a nonequilibrium situation." Similar statements appeared in the Ehrenfest[10] review article of 1912, when the work of Gibbs had not yet been understood. From the frequency with which this objection to Gibbs' H has been repeated in the literature since then, it is clear that the nature of Gibbs' contribution has not been fully appreciated to this day.

We wish to point out that the mathematical relations proved by Gibbs, plus one physical assumption which is universally accepted today (although it had hardly been formulated at the time of the Ehrenfest article) are sufficient to prove, on the contrary, the following four statements:

(I) The Gibbs H has a simple and universally valid connection with the entropy; for all probability assignments that agree with the measured thermodynamic parameters we have $S \geq -kH_G$, with equality if and only if H_G is computed from the appropriate canonical or grand canonical probability assignment.

(II) The Boltzmann H is related to the entropy in only one case, the nonexistent ideal Boltzmann (i.e., not Bose or Fermi) gas. In general, $H_B \leq H_G$, and the entropy can be either greater or less than $-kH_B$.

(III) The constancy of Gibbs' H, far from conflicting with the increase of entropy, is the sole dynamical property needed to demonstrate that increase.

(IV) The Gibbs H provides a generalized definition of entropy for nonequilibrium cases, in such a way that the usual statement of the second law remains valid. It gives, therefore, a new rule telling which nonequilibrium states are accessible from others in adiabatic processes.

The fourth statement is a nontrivial extension of the second law which is capable of being tested experimentally, and whose finding required only a careful reading of Gibbs. Since the second law is a statement of experimental fact, it cannot be "proved" mathematically without some assumption about what the result of an experiment would be. The assumption we need is just the statement (44) which we appealed to before.

Before turning to the proofs, some preliminary remarks are needed. We are still faced with the ambiguity in the definition of f.

The function defined by (79) is singular in such a way that the integral (96) diverges; thus before we can introduce a Boltzmann H at all, we have to abandon Boltzmann's definition of f in favor of some other, unspecified one. In MD it is stated that f gives an "average" occupation number, and that this can be made more precise by reference to an equation which is indeed an average over an undefined probability distribution P. If we suppose that, in going to fundamentals, this would eventually become an average over the Liouville function W_N, we have a definition of H_B for which exact relations can be proved. In other words, we mean to use the single-particle function w_1 of the hierarchy to define a Boltzmann H:

$$H_B = \int w_1 \log w_1 \, d^3x \, d^3p \qquad (98)$$

There is really no other way of doing it if we are ever to prove precise statements about Boltzmann's H, because eventually this will have to depend on precise properties of the dynamics, and the Liouville hierarchy is just the precise expression of the dynamics.

Another point is that, strictly speaking, all this should be restated in terms of quantum theory using the density matrix formalism. This will introduce the N! permutation factor, a natural zero for entropy, alteration of numerical values if discreteness of energy levels becomes comparable to kT, etc. But there seems to be complete agreement as to how this transcription is to be made, and it will affect the Boltzmann and Gibbs expressions in the same way. We shall first attempt to define the Boltzmann H as $H' = \text{Tr}(\sigma \log \sigma)$, where σ is the "molecular" density matrix operating in the Hilbert space of a single molecule and gives occupation numbers. The Gibbs H will become $H_G' = N^{-1}\text{Tr}(\rho \log \rho)$, where ρ is the "global" density matrix with an enormously greater number of rows and columns, operating in the entire Hilbert space of the system. On closer examination, we shall wonder whether the diagonal elements of σ are to represent the actual values, probable values, average values, etc. of the occupation numbers, and H' will peter out in ambiguities until we note that, if it is to have any precisely provable properties, it must be precisely related to the dynamics; i.e., out of all possible definitions of σ, we decide to use ρ_1, the projection of ρ onto the subspace of a single molecule, as defined in reference,[5] Sec. 11. Its diagonal elements are expectation values, over the global density matrix ρ, of occupation fractions. Then with H_G' and $H_B' = \text{Tr}(\rho_1 \log \rho_1)$ we can prove exactly the same inequalities as for the classical case. Thus, the issue of Boltzmann versus Gibbs entropy expressions does not involve quantum theory, and we continue to use classical terminology for brevity.

Statement (I) is now just the theorem (45) already proved, if one grants the physical assumption (44), for the quantum theory case.

Statement (II) quotes a well-known mathematical theorem,

$$H_G \geq H_B \tag{99}$$

with equality if and only if the Liouville function factors "almost everywhere"

$$W_N(x_1 \ldots p_N) = \prod_{i=1}^{N} w_1(x_i, p_i) \tag{100}$$

which corresponds, in quantum theory, to the condition that the global density matrix is a direct product[5]

$$\rho = \rho_1 \times \rho_2 \times \ldots \times \rho_N \tag{101}$$

where ρ_i is the projection of ρ onto the Hilbert space of the i-th molecule. The final part of statement II then follows from the fact that the canonical distribution $W_N \sim \exp(-\beta H)$ has the factorized form (100) only in the case of an ideal Boltzmann gas. In this case the "Boltzmann entropy," $S_B = -k\bar{H}_B$, is equal to the experimental entropy; in all other cases, if w_1 is constructed from the appropriate canonical distribution W_N, we shall have $S_B > S_e$.

Statement III is likewise an immediate consequence of statement I and the well-known fact that H_G is, in consequence of the equations of motion, constant in time in either classical or quantum theory. To make this clearer, consider the following experiment. At time $t = 0$, we measure the values of various parameters $X_1 \ldots X_n$ adequate to determine the state of a thermodynamic system of n degrees of freedom. The experimental entropy is, of course, some function $S_e(X_1 \ldots X_n)$ of the measured quantities; and not primarily related to any probability distribution. But we have shown that the maximum attainable information theory entropy S_I, corresponding to the appropriate canonical distribution based on the values of $X_1 \ldots X_n$, is equal to S_e. At some later time t, a new measurement of the thermodynamic state yields different values, $X'_1, \ldots, X'_n$, and a different experimental entropy $S_e(X'_1 \ldots X'_n)$. But the inequality $S_I \leq S_e$ still holds; and so the statement that S_I (or what is the same thing, H_G) is constant, then gives us $S'_e \geq S_e$.

There is still an apparent paradox hiding here; for suppose we choose t negative. It looks as if this argument then says that the experimental entropy in the past was greater than at the time of the measurements $X_1 \ldots X_n$. Actually, the explanation of this paradox has been given before.[5] We have, of course, assumed in the above that forward integration of the equations of motion does, in fact, yield the correct predictions at time t; i.e., the measured X_i' are equal to ensemble averages calculated from the time-developed Liouville function obeying (78), or the time-developed global density matrix obeying $i\hbar\dot\rho = [H, \rho]$. In reference,[5] it is shown that this is the case <u>if the observed change $X_i \to X_i'$ is an experimentally reproducible one</u>. But we know that many past macroscopic states X_i'' would all relax into the same state X_i at time t = 0. Thus, we suggest that the correct statement of the second law is that spontaneous decreases in the experimental entropy, although not absolutely prohibited by the laws of physics, <u>cannot occur in an experimentally reproducible process</u>.

Statement IV now follows from the fact that nothing in the above reasoning restricts us to equilibrium states. In conventional thermodynamics, the experimental entropy is defined only for equilibrium states; however, our definition $S_e \equiv [\max S_I$ over all probability distributions that agree with the data in the sense of (52)] defines a function $S_e(X_1 \ldots X_n)$ of the experimentally measured parameters for the equilibrium or nonequilibrium case, which by the above arguments cannot spontaneously decrease in an experimentally reproducible process. It can no longer be found by numerical integration of dQ/T over a reversible path; but the content of statement IV is that a function S_e still <u>exists</u>, such that the usual statement of the second law remains valid. It requires a great deal more analysis, to be given elsewhere, before we can reduce this to a suggestion of a definite experiment that could test statement IV; I am trying here only to point out in the briefest terms why it is that an extension of the second law is predicted by theory as soon as we have understood everything revealed by Gibbs about the connection between entropy and probability.

Finally, we note that the Boltzmann H-theorem, whether correct or not, cannot have any real relevance to the second law. For, summarizing the above inequalities,

$$-kH_B \geq -kH_G \leq S_e \tag{102}$$

where the first inequality becomes an equality if and only if there are no interparticle correlations (i.e., ideal Boltzmann gas), the second if and only if H_G is computed from the appropriate canonical distribution. Obviously, whether H_B increases or decreases allows us to infer nothing about S_e. The situation is even worse than that; for the Boltzmann H-theorem was based on incorrect equations of motion, and whether H_B increases or decreases depends on the form of the distribution and the force law. To see this, note that from (98) and the exact equation of motion (94), the exact rate of change of H_B is just the negative of the expected divergence in momentum space of the molecular force $\langle F_\alpha \rangle$:

$$\dot{H}_B = -\left\langle \frac{\partial \langle F_\alpha \rangle}{\partial p_\alpha} \right\rangle \tag{103}$$

and this can have either sign. For example, if $\langle F_\alpha \rangle$ is dominated by a "dragging" term as in the Langevin equation: $\langle F_\alpha \rangle = -Kp_\alpha + \ldots$, then we find that the exact equations give us an "anti-H-theorem," $\dot{H}_B > 0$.

7. CONCLUSION

We have seen that the principle of maximum entropy leads immediately to the same final rules of calculation that conventional statistical mechanics had provided only after long and inconclusive discussion of phase space, ergodicity, metric transitivity, etc.; and then only for the equilibrium case. The viewpoint advocated here thus represents, from the pedagogical standpoint, a considerable simplification of the subject. But this agreement also means that, from a pragmatic standpoint, if there is any new content in this principle, we must look for it in the extension to the statistical mechanics of irreversible processes, where there does not exist at present any general formal theory, and ask whether the principle of maximum entropy provides such a basis. Over the past several years, my students and I have verified that all the commonly accepted principles of irreversible statistical mechanics can be derived from this formalism; that is, of course, a minimum requirement that any proposed new theory must pass. The real test of these ideas can come only through their application to problems that have resisted solution by older methods. Although a few results along this line are now in,[11] and a few others have been hinted at in these talks, a final settlement of the questions raised still lies rather far in the future.

References

1. C. E. Shannon, Bell System Tech. J., <u>27</u>, 379, 623 (1948). Reprinted in C. E. Shannon and W. Weaver, "The Mathematical Theory of Communication," Univ. of Illinois Press, Urbana, Illinois, 1949.

2. E. T. Jaynes, Note on unique decipherability, IRE Trans. Inform. Theory, Sept. 1959.
3. E. T. Jaynes, New engineering applications of information theory, in "Symposium on Engineering Applications of Random Function Theory and Probability" (J. L. Bogdanoff and F. Kozin, eds.), Wiley, New York, 1963.
4. E. T. Jaynes, Phys. Rev., 106, 620 (1957).
5. E. T. Jaynes, Phys. Rev., 108, 171 (1957).
6. M. Dresden, Rev. Mod. Phys., 33, 265 (1961).
7. E. T. Jaynes, Am. J. Phys., 31, 66 (1963).
8. E. T. Jaynes, "Probability Theory in Science and Engineering," McGraw-Hill, New York, in press.
9. H. Jeffreys, "Theory of Probability," Oxford Univ. Press, London, 1938.
10. P. and T. Ehrenfest, Encykl. Math. Wiss., IV 2, II, Heft 6 (1912). Reprinted in "Paul Ehrenfest, Collected Scientific Papers" (M. J. Klein, ed.), North-Holland, Amsterdam, 1959. (English translation by M. J. Moravcsik, Cornell Univ. Press, Ithaca, New York, 1959.)
11. S. Heims and E. T. Jaynes, Rev. Mod. Phys., 34, 143 (1962).

5. GIBBS vs BOLTZMANN ENTROPIES (1965)

In the 1950's and early 1960's there was a great deal of activity trying to develop a theory of irreversible processes in terms of the notion of local entropy production. For a time it appeared that this conception might be justified in statistical theory. However, these arguments (de Groot and Mazur, 1962) used the Boltzmann H-function definition of entropy, $S_B = -kH_B$ based on the single-particle distribution function. Recognition of the difficulties caused by this was slow in coming.

Dresden (1961) asserted that S_B was 'directly related to the entropy', while the Gibbs S_G based on the full N-particle distribution function was rejected with the statement that there was 'no possibility of relating it to the entropy'. Prigogine (1963) questioned the validity of the Gibbs relation $T\,dS = dU + P\,dV - \Sigma\,\mu_i\,dn_i$ and the Kubo expressions for transport coefficients because derivations based on S_B did not support them.

However, to others it appeared that the Gibbs relation, properly used, was not a physical hypothesis, but the definition of μ_i; and the Kubo formulas had been derived directly from first principles. Failure to confirm either could signify only faulty reasoning or faulty premises, and it seemed that the difficulty lay in use of an incorrect entropy expression.

Spurred by this discrepancy and following a lengthy discussion with E. P. Wigner on these problems, I wrote the following article analyzing the entropy question in detail. It demonstrated explicitly what had long been known implicitly by some; for arbitrary interparticle forces the Gibbs entropy $S_G = -kH_G$ using the canonical ensemble gives the correct thermodynamic relations, while S_B is correct only for an ideal gas at thermal equilibrium.

The point was amplified further in a note (Jaynes, 1971) pointing out that when attractive forces are present, use of S_B would make some well-known experimental facts appear to be violations of the second law (if the approach to equilibrium takes place via conversion of kinetic energy into potential energy, S_B decreases rather than increases).

The demonstration of the Clausius adiabatic form of the second law in the Brandeis lectures was repeated in classical form here, but again to no avail. It remains unnoticed.

As a former student of Eugene Wigner, I had been invited to contribute an article to the October 1962 issue of Reviews of Modern Physics, planned as a kind of Festschrift for his 60'th birthday. I wrote a paper entitled 'Conceptual Problems in Statistical Mechanics' but at the last minute felt that it was not of high enough quality for that occasion and withdrew it, a decision deeply regretted on seeing some of the articles that did appear. However, some portions of it that appeared sound enough to stand the test of time (the 'anthropomorphic' remarks) were retrieved and built into the following article. More portions appear in the Delaware lecture reprinted below, but most of it is still unpublished.

But it is just those anthropomorphic remarks that have aroused the most ire, most recently in the article of Denbigh (1981). Accordingly, let me state here that I stand by those remarks, and believe they are correct as presented below. Indeed, they state only what had been well recognized by Boltzmann, Gibbs, G. N. Lewis, J. von Neumann, and E. P. Wigner long before.

Denbigh objects that my arguments on the nature of entropy would apply as well to energy. Not so! The difference is that energy is a property of the microstate, and so all observers, whatever macroscopic variables they may choose to define their thermodynamic states, must ascribe the same energy to a system in a given microstate. But they will ascribe different entropies to that microstate, because entropy is not a property of the microstate, but rather of the reference class in which it is embedded. As we learned from Boltzmann, Planck, and Einstein, the entropy of a thermodynamic state is a measure of the number of microstates compatible with the macroscopic quantities that you or I use to define the thermodynamic state.

It is a sad commentary on the state of conceptual progress here that these facts are still not grasped by prominent workers in the field, although they were demonstrated very cogently over 100 years ago by Gibbs (1876) in his discussion of gas diffusion. We can hardly blame students for being as confused as their teachers, in a field that ought to have become, long ago, just as clear and rational as mechanics or optics. In my opinion, this would have happened if Gibbs and Einstein had been read more assiduously.

Needless to say, it was just this property of entropy – that it measures our degree of information about the microstate, that is conveyed by data on the macroscopic thermodynamic variables – that made information theory such a powerful tool in showing us how to generalize Gibbs' equilibrium ensembles to nonequilibrium ones. The generalization could never have been found by those who thought that entropy was, like energy, a physical property of the microstate.

GIBBS VS BOLTZMANN ENTROPIES

Gibbs vs Boltzmann Entropies*

E. T. JAYNES
Department of Physics, Washington University, St Louis, Missouri
(Received 27 March 1964, in final form, 5 November 1964)

The status of the Gibbs and Boltzmann expressions for entropy has been a matter of some confusion in the literature. We show that: (1) the Gibbs H function yields the correct entropy as defined in phenomenological thermodynamics; (2) the Boltzmann H yields an "entropy" that is in error by a nonnegligible amount whenever interparticle forces affect thermodynamic properties; (3) Boltzmann's other interpretation of entropy, $S = k \log W$, is consistent with the Gibbs H, and derivable from it, (4) the Boltzmann H theorem does not constitute a demonstration of the second law for dilute gases, (5) the dynamical invariance of the Gibbs H gives a simple proof of the second law for arbitrary interparticle forces; (6) the second law is a special case of a general requirement for any macroscopic process to be experimentally reproducible. Finally, the "anthropomorphic" nature of entropy, on both the statistical and phenomenological levels, is stressed.

I. INTRODUCTION

IN the writer's 1962 Brandeis lectures[1] on statistical mechanics, the Gibbs and Boltzmann expressions for entropy were compared briefly, and it was stated that the Gibbs formula gives the correct entropy, as defined in phenomenological thermodynamics, while the Boltzmann H expression is correct only in the case of an ideal gas. However, there is a school of thought which holds that the Boltzmann expression is directly related to the entropy, and the Gibbs' one simply erroneous. This belief can be traced back to the famous Ehrenfest review article,[2] which severely criticized Gibbs' methods.

* Supported by the National Science Foundation Grant NSF G23778.
[1] *Statistical Physics* (1962 Brandeis Theoretical Physics Lectures, Vol 3), edited by K. W. Ford (W. A. Benjamin, Inc , New York, 1963), Chap. 4. Note that typographical errors occur in Eqs. 20, 49, 74, 78, 94, and the inequality preceding Eq. 90.
[2] P. Ehrenfest and T. Ehrenfest, Encykl Math. Wiss., IV 2, II, Issue 6 (1912). Reprinted in *Paul Ehrenfest, Collected Scientific Papers*, edited by M. J Klein (North-Holland Press, Amsterdam, 1959). English translation by M. J. Moravcsik, *The Conceptual Foundations of the Statistical Approach in Mechanics* (Cornell University Press, Ithaca, New York, 1959).

While it takes very little thought to see that objections to the Gibbs H are immediately refuted by the fact that the Gibbs canonical ensemble does yield correct thermodynamic predictions, discussion with a number of physicists has disclosed a more subtle, but more widespread, misconception. The basic inequality of the Gibbs and Boltzmann H functions, to be derived in Sec. II, was accepted as mathematically correct; but it was thought that, in consequence of the "laws of large numbers" the difference between them would be practically negligible in the limit of large systems.

Now it is true that there are many different entropy expressions that go into substantially the same thing in this limit; several examples were given by Gibbs. However, the Boltzmann expression is not one of them; as we prove in Sec. III, the difference is a direct measure of the effect of interparticle forces on the potential energy and pressure, and increases proportionally to the size of the system.

Failure to recognize the fundamental role of the Gibbs H function is closely related to a much deeper confusion about entropy, probability, and irreversibility in general. For example, the Boltzmann H theorem is almost universally equated to a demonstration of the second law of thermodynamics for dilute gases, while ever since the Ehrenfest criticisms, it has been claimed repeatedly that the Gibbs H cannot be related to the entropy because it is constant in time.

Closer inspection reveals that the situation is very different. Merely to exhibit a mathematical quantity which tends to increase is not relevant to the second law unless one demonstrates that this quantity is related to the entropy as measured experimentally. But neither the Gibbs nor the Boltzmann H is so related for any distribution other than the equilibrium (i.e., canonical) one. Consequently, although Boltzmann's H theorem does show the tendency of a gas to go into a Maxwellian velocity distribution, this is not the same thing as the second law, which is a statement of experimental fact about the direction in which the observed *macroscopic* quantities (P,V,T) change.

Past attempts to demonstrate the second law for systems other than dilute gases have generally tried to retain the basic idea of the Boltzmann H theorem. Since the Gibbs H is dynamically constant, one has resorted to some kind of coarse-graining operation, resulting in a new quantity $\bar{H}$, which tends to decrease. Such attempts cannot achieve their purpose, because (a) mathematically, the decrease in $\bar{H}$ is due only to the artificial coarse-graining operation and it cannot, therefore have any physical significance; (b) as in the Boltzmann H theorem, the quantity whose increase is demonstrated is not the same thing as the entropy. For the fine-grained and coarse-grained probability distributions lead to just the same predictions for the observed macroscopic quantities, which alone determine the experimental entropy; the difference between H and $\bar{H}$ is characteristic, not of the macroscopic state, but of the particular way in which we choose to coarse-grain. Any really satisfactory demonstration of the second law must therefore be based on a different approach than coarse-graining.

Actually, a demonstration of the second law, in the rather specialized situation visualized in the aforementioned attempts, is much simpler than any H theorem. Once we accept the well-established proposition that the Gibbs canonical ensemble does yield the correct equilibrium thermodynamics, then there is logically no room for any assumption about which quantity represents entropy; it is a question of mathematically demonstrable fact. But as soon as we have understood the relation between Gibbs' H and the experimental entropy, Eq. (17) below, it is immediately obvious that the constancy of Gibbs' H, far from creating difficulties, is precisely the dynamical property we need for the proof.

It is interesting that, although this field has long been regarded as one of the most puzzling and controversial parts of physics, the difficulties have not been mathematical. Each of the above assertions is proved below or in the Brandeis lectures, using only a few lines of elementary mathematics, all of which was given by Gibbs. It is the enormous *conceptual* difficulty of this field which has retarded progress for so long. Readers not familiar with recent developments may, I hope, be pleasantly surprised to see how clear and basically simple these problems have now become, in several respects. However, as we will see, there are still many complications and unsolved problems.

Inspection of several statistical mechanics textbooks showed that, while most state the formal relations correctly, their full implications are never noted. Indeed, while all textbooks give extensive discussions of Boltzmann's H, some recent ones fail to mention even the existence of the Gibbs H.[3] I was unable to find any explicit mathematical demonstration of their difference. It appeared, therefore, that the following note might be pedagogically useful.

II. THE BASIC INEQUALITY

We consider, as usual, a monoatomic fluid of N particles. The ensemble is defined by the N-particle distribution function, or Liouville function, $W_N(x_1,p_1; x_2,p_2; \cdots ; x_N,p_N; t)$ which gives the probability density in the full phase space of

---

[3] A notable exception is the monumental work of R. C. Tolman, *The Principles of Statistical Mechanics* (Oxford University Press, London, 1938) Tolman repeatedly stresses the superiority of Gibbs' approach, although he still attempts to base the second law on coarse-graining.

the system. The Gibbs H is then

$$H_G = \int W_N \log W_N d\tau \quad (1)$$

and the corresponding Boltzmann H is

$$H_B = N \int w_1 \log w_1 d\tau_1, \quad (2)$$

where $w_1(x_1, p_1; t)$ is the single-particle probability density

$$w_1(x_1, p_1; t) = \int W_N d\tau_{-1}. \quad (3)$$

Here and in the following, we use the notation: $d\tau \equiv d^3x_1 \cdots d^3p_N$, $d\tau_1 \equiv d^3x_1 d^3p_1$, $d\tau_{-1} \equiv d^3x_2 \cdots d^3p_N$ to stand for phase-volume elements in the full phase space, the space of one particle, and the space of all particles except one, respectively.

Both the Gibbs and Boltzmann H functions are often defined in slightly different ways, in which one uses distribution functions with different normalizations. This changes the numerical values by additive constants which, for fixed N, are independent of the thermodynamic state and therefore not relevant to the present discussion. These additive constants are important, however, in connection with the "Gibbs paradox" about entropy of mixing, and the resolution of this paradox by quantum statistics is well known. The distribution functions used above are understood to be *probability densities*; i.e., normalized according to $\int W_N d\tau = \int w_1 d\tau_1 = 1$.

Using (3) and the fact that W_N is symmetric under permutations of particle labels, we can write H_B in a more symmetrical form

$$H_B = N \int W_N \log w_1(x_1, p_1) d\tau$$

$$= \int W_N \log[w_1(1) \cdots w_1(N)] d\tau,$$

where we use the abbreviation: $(i) \equiv (x_i, p_i)$. We have, then,

$$H_B - H_G = \int W_N \log\left[\frac{w_1(1) \cdots w_1(N)}{W_N(1 \cdots N)}\right] d\tau. \quad (4)$$

Now on the positive real axis, $\log x \leq (x-1)$, with equality if and only if $x = 1$. Therefore

$$H_B - H_G \leq \int W_N \left[\frac{w_1(1) \cdots w_1(N)}{W_N(1 \cdots N)} - 1\right] d\tau = 0,$$

and we have proved

Theorem 1: The Gibbs and Boltzmann H functions satisfy the inequality

$$H_B \leq H_G, \quad (5)$$

with equality if and only if W_N factors "almost everywhere" into a product of single-particle functions

$$W_N(1 \cdots N) = w_1(1) \cdots w_1(N).$$

III. CANONICAL ENSEMBLE

Theorem 1 holds for any symmetrical W_N. The magnitude of the difference $(H_G - H_B)$ depends on the distribution function, and we are particularly interested in the case of thermal equilibrium, represented by the canonical distribution $W_N \sim \exp(-\beta H)$, where $\beta = (kT)^{-1}$ and H is the Hamiltonian, taken of the form

$$H = \sum_{i=1}^{N} \frac{p_i^2}{2m} + V(x_1 \cdots x_N), \quad (6)$$

where the potential-energy function $V(x_1 \cdots x_N)$ is a symmetrical function of the particle coordinates, which we suppose for simplicity depends only the relative coordinates (relaxing this restriction by adding gravitational potential energy leads to a number of interesting results, but does not change the conclusions of this section). More explicitly, we have

$$W_N = \left(\frac{\beta}{2\pi m}\right)^{3N/2} Q^{-1}$$

$$\times \exp\left\{-\beta V(x_1 \cdots x_N) - \beta \sum_i \frac{p_i^2}{2m}\right\}, \quad (7)$$

where

$$Q(\beta, \Omega) \equiv \int_\Omega \exp(-\beta V) d^3x_1 \cdots d^3x_N$$

$$= \Omega \int_\Omega \exp(-\beta V) d^3x_2 \cdots d^3x_N \quad (8)$$

is the "configuration integral," and in the last

expression we have made use of the fact that V depends only on relative coordinates, and supposed the range of interparticle forces negligibly small compared to the size of the container, so that the final integration supplies only a factor Ω. From (3), the corresponding single-particle function is then

$$w_1(x,p) = (\beta/2\pi m)^{3/2}\Omega^{-1}\exp(-\beta p^2/2m). \quad (9)$$

We therefore have

$$[w_1(1)\cdots w_1(N)]/W_N(1\cdots N) = Q\Omega^{-N}e^{\beta V},$$

and (4) reduces to

$$H_B - H_G = \log Q - N\log\Omega + \beta\langle V\rangle, \quad (10)$$

where the angular brackets $\langle\ \rangle$ denote the canonical ensemble average. It is also true that

$$\langle V\rangle = -\partial\log Q/\partial\beta, \quad (11a)$$

$$\beta\langle P\rangle = \partial\log Q/\partial\Omega, \quad (11b)$$

where P is the pressure; Eq. (11) are well-known identities of the canonical ensemble. From (10), (11), we thus find that on an infinitesimal change of state,

$$d(H_B - H_G) = \beta d\langle V\rangle + \beta[\langle P\rangle - P_0]d\Omega, \quad (12)$$

where $P_0 \equiv NkT/\Omega$ is the pressure of an ideal gas with the same temperature and density. Introducing the "entropies" $S_i = -kH_i$ and integrating (12) over a reversible path (i.e., a locus of equilibrium states), we see that the difference varies according to

$$(S_G - S_B)_2 - (S_G - S_B)_1$$

$$= \int_1^2 \frac{d\langle V\rangle + [\langle P\rangle - P_0]d\Omega}{T}. \quad (13)$$

Now from (9), using $\langle p^2\rangle = 3mkT$, we find that

$$S_B = \tfrac{3}{2}Nk\log(2\pi mkT) + Nk\log\Omega + \tfrac{3}{2}Nk,$$

from which

$$\left(\frac{\partial S_B}{\partial T}\right)_\Omega dT = \frac{3}{2}\frac{NkdT}{T} = \frac{d\langle K\rangle}{T},$$

$$\left(\frac{\partial S_B}{\partial\Omega}\right)_T d\Omega = \frac{Nk}{\Omega}d\Omega = \frac{P_0 d\Omega}{T},$$

where $\langle K\rangle = \tfrac{3}{2}NkT$ is the total kinetic energy.

Over the reversible path (13) the Boltzmann entropy therefore varies according to

$$(S_B)_2 - (S_B)_1 = \int_1^2 \frac{d\langle K\rangle + P_0 d\Omega}{T}, \quad (14)$$

and from (13), (14) we finally have for the Gibbs entropy

$$(S_G)_2 - (S_G)_1 = \int_1^2 \frac{d\langle K+V\rangle + \langle P\rangle d\Omega}{T}$$

$$= \int_1^2 \frac{dQ}{T}. \quad (15)$$

Equations (14), (15) are the main results sought. From them it is clear that (a) the "Boltzmann entropy" is the entropy of a fluid with the same density and temperature, but without interparticle forces; it completely neglects both the potential energy and the effect of interparticle forces on the pressure; (b) the Gibbs entropy is the correct entropy as defined in phenomenological thermodynamics, which takes into account all the energy and the total pressure, and is therefore equally valid for the gas or condensed phases; (c) the difference between them is *not* negligible for any system in which interparticle forces have any observable effect on the thermodynamic properties. If the system exhibits an equation of state or heat capacity different from those of an ideal gas, the Boltzmann entropy will be in error by a corresponding amount.

IV. THE SECOND LAW

We can now demonstrate the second law very easily, for the specialized case usually considered. The following argument can be greatly generalized, although we do not do so here.

It is well known[1] that the canonical distribution (7) is uniquely determined by a variational property; over all distributions W_N that agree with the experimental energy U, in the sense that the mean value of the Hamiltonian is

$$\langle H\rangle \equiv \int W_N H d\tau = U, \quad (16)$$

the Gibbs H attains an absolute minimum for the canonical distribution. For this case, we have just shown that, if the arbitrary additive constant is properly adjusted at a single point, then the Gibbs entropy $S_G = -kH_G$ will be the same as the experimental entropy at all points. Therefore, the general relation between S_G and the experimental entropy S_e is: over all distributions W_N that agree with the experimental energy in the sense of (16), we have

$$S_G \leq S_e \qquad (17)$$

with equality if, and only if, S_G is computed from the canonical distribution (7).

At time $t=0$, let our system be in complete thermal equilibrium so that all its reproducible macroscopic properties are represented by the canonical distribution; then the equality holds in (17). Now force the system to carry out an adiabatic change of state (i.e., one involving no heat flow to or from its environment), by appylying some time-dependent term in the Hamiltonian (such as moving a piston or varying a magnetic field). It is well known that the N-particle distribution function varies according to the Liouville equation $\dot{W}_N = \{H(t), W_N\}$ where the right-hand side is the Poisson bracket; and in consequence H_G remains constant.

At a later time t', the system is allowed to come once more, but still adiabatically, to equilibrium (which means experimentally that macroscopic quantities such as pressure or magnetization are no longer varying), so that a new experimental entropy S_e' can be defined. If the time-developed distribution function $W_N(t')$ leads to a correct prediction of the new energy U' in the sense of (16), then the inequality (17) still holds. The fact that H_G is a constant of the motion then gives $S_e \leq S_e'$, which is the second law.

V. INTUITIVE MEANING OF THE SECOND LAW

The above proof has the merit of being almost unbelievably short, but partly for that reason, the physical basis of the second law is not made clear. In the following we are not trying to give a rigorous mathematical demonstration; that has just been done. We are trying rather to exhibit the basic intuitive reason for the second law. We recall Boltzmann's original conception of entropy as measuring the logarithm of phase volume associated with a macroscopic state. If Boltzmann's interpretation $S = k \log W$ is to be compatible with Gibbs' $S = -kH_G$, it must be true that the quantity $W \equiv \exp(-H_G)$ measures, in some sense, the phase volume of "reasonably probable" microstates.

Such a connection can be established as follows. Define a "high-probability" region R of phase space, consisting of all points where $W_N \geq C$, and choose the constant C so that the total probability of finding the system somewhere in this region is $(1-\epsilon)$, where $0 < \epsilon < 1$. Call the phase volume of this region $W(\epsilon)$; in equations,

$$\int_R W_N d\tau = 1 - \epsilon,$$

$$\int_R d\tau = W(\epsilon).$$

Evidently, with a continuously varying probability density W_N, it is not strictly meaningful to speak of the "phase volume of an ensemble," without qualifications; but the "minimum phase volume of 50% probability" or the "minimum phase volume of 99% probability" do have precise meanings.

A remarkable limit theorem first noted by Shannon[5] shows that for most purposes the particular probability level ϵ is unimportant. We quote the result without proof; it is an adaptation of the fundamental "asymptotic equipartition property" (AEP) of Information Theory.[6] We suppose that the distribution function W_N from which H_G and $W(\epsilon)$ are computed is either a canonical distribution or a time-developed version of one resulting from some dynamical perturbation; and that the system is such that the canonical ensemble predicts relative fluctuations in energy which tend to zero as $N^{-1/2}$ in the "thermodynamic limit" as $N \to \infty$ at constant density. The Gibbs H per particle, H_G/N, then approaches a definite limit, and

$$\lim_{N \to \infty} \{[H_G + \log W(\epsilon)]/N\} = 0 \qquad (18)$$

---

[4] E. T. Jaynes, Phys. Rev. **108**, 171 (1957).
[5] C. E. Shannon, Bell Syst. Tech. J. **27**, 379, 623 (1948); reprinted in C. E. Shannon and W. Weaver, *The Mathematical Theory of Communication* (University of Illinois Press, Urbana, Illinois, 1949). See, particularly, Sec. 21.
[6] A. Feinstein, *Foundations of Information Theory* (McGraw-Hill Book Company, Inc., New York, 1958), Chap. 6.

provided ϵ is not zero or unity. The principal feature of this theorem, at first sight astonishing, is that the result is independent of ϵ. Changing ϵ does, of course, change $W(\epsilon)$; and generally by an enormous factor. But the change in $\log W(\epsilon)$ grows less rapidly than N, and in the limit it makes no difference.

The intuitive meaning of this theorem is that the Gibbs H *does* measure the logarithm of phase volume of reasonably probable microstates and, remarkably, for a large system the amount per particle, $\log W(\epsilon)/N$, becomes independent of just what we mean by "reasonably probable." We are thus able to retain Boltzmann's original formula, $S = k \log W$, which is seen to be precisely related to the Gibbs H, not the Boltzmann one.

With this interpretation of entropy, let us reconsider the above experiment. At time $t = 0$, we measure a number of macroscopic parameters $\{X_1(0), \cdots, X_n(0)\}$ adequate to define the thermodynamic state. The corresponding canonical distribution determines a high-probability region R_0, of phase volume W_0. The aforementioned variational property of the canonical ensemble now implies that, of all ensembles agreeing with this initial data in the sense of (16), the canonical one defines the *largest* high-probability region. The phase volume W_0 therefore describes the full range of possible initial microstates; and not some arbitrary subset of them; this is the basic justification for using the canonical distribution to describe partial information.

On the "subjective" side, we can therefore say that W_0 measures our *degree of ignorance* as to the true unknown microstate, when the only information we have consists of the macroscopic thermodynamic parameters; a remark first made by Boltzmann.

But, and perhaps more pertinent, we can also say on the "objective" side, that W_0 measures the *degree of control of the experimenter over the microstate*, when the only parameters he can manipulate are the usual macroscopic ones. On successive repetitions of the experiment, the initial microstate will surely not be repeated; it will vary at random over the high-probability region R_0.

When we carry out an adiabatic change of state, the region R_0 is transformed, by the equations of motion, into a new region R_t. From either the constancy of H_G, or directly from Liouville's theorem, the phase volume remains unchanged; $W_t = W_0$. Each possible initial microstate in R_0 uniquely determines a possible final one in R_t, and on successive repetitions of the experiment, the final state varies over R_t at random.

At the end of this experiment, under the new equilibrium conditions, we note the new values $\{X_1(t), \cdots, X_n(t)\}$ of the thermodynamic quantities. Now consider the region R', consisting of all microstates that are compatible with these new $X_i(t)$, whether or not they could have resulted from the experiment just described; i.e., whether or not they also lie in R_t. By (17) and (18), the final experimental entropy is $S_e' = k \log W'$, where W' is the phase volume of R'; the experimental entropy is a measure of all conceivable ways in which the final macrostate can be realized, and not merely of all ways in which it could be produced in one particular experiment.

Now it is obvious that, if the observed change of state $X_i(0) \to X_i(t)$ is to be experimentally reproducible, the region R_t resulting from the experiment must be totally contained in R'. But this is possible only if the phase volumes satisfy $W_t \leq W'$, which is again the second law!

At this point, we finally see the real reason for the second law; since phase volume is conserved in the dynamical evolution, *it is a fundamental requirement on any reproducible process that the phase volume W' compatible with the final state cannot be less than the phase volume W_0 which describes our ability to reproduce the initial state.*

But this argument has given us more than the second law; in the past the notion "experimental entropy" has been defined, in conventional thermodynamics, only for *equilibrium* states. It is suddenly clear that the second law is only a very special case of a general restriction on the direction of any reproducible process, whether or not the initial and final states are describable in the language of thermodynamics; the expression $S = k \log W$ gives a generalized definition of entropy applicable to arbitrary nonequilibrium states, which still has the property that it can only increase in a reproducible experiment. This can be shown directly from Liouville's theorem, without any consideration of canonical distributions or the asymptotic equipartition theorem.

Finally, it is clear that this extension of the second law can be subjected to experimental tests.

Returning to the case of equilibrium thermodynamics, these considerations (which are easily extended[1] to quantum statistics) lead us to state the conventional second law in the form: *The experimental entropy cannot decrease in a reproducible adiabatic process that starts from a state of complete thermal equilibrium.*

The necessity of the last proviso is clear from a logical standpoint in our derivation of the second law in Sec. IV; for if the preparation of the system just before $t=0$ imposes any constraints other than those implied by the canonical distribution, the manifold of possible initial states will be reduced below W_0, and we shall not have an equality in Eq. (17) initially. This necessity is also shown strikingly from an experimental standpoint in the phenomenon of spin echoes,[7,8] which is a gross violation of any statement of the second law that fails to specify anything about the past history of the system. This proviso has not been particularly emphasized before, but it has always been obvious that some such condition would be needed before we had a really air-tight statement of the second law, which could not be violated by a clever experimenter. The future behavior of the system is uniquely determined, according to the laws of mechanics, only when one has specified perhaps 10^{24} microscopic coordinates and momenta; it could not possibly be determined merely by the values of the three or four quantities measured in typical thermodynamic experiments.

Specifying "complete thermal equilibrium" is still not as precise a statement as we might wish. Experimentally, the only criterion as to whether it is satisfied seems to be that the system is "aged," i.e., that it is quiescent, the macroscopic quantities X_i unchanging, for a sufficiently long time; and only experience can tell the experimenter how long is "sufficiently long."

Theoretically, we can understand this requirement as meaning that, for purposes of prediction, lack of knowledge of the present microstate can be, in part, compensated by knowledge of the past history of the macroscopic state. As we

[7] E. L. Hahn, Phys. Rev **80**, 580 (1950)
[8] A. L. Bloom, Phys. Rev. **98**, 1104 (1955).

observe the system to be quiescent for a longer and longer time, we become more and more confident that it is not in an atypical microstate that will lead to "abnormal" behavior in the future. In Hahn's experiment[7] the spin system, having no observable net magnetization at time $t=0$, is nevertheless able to develop, spontaneously and almost magically, a large and reproducible magnetization at a later time only because it "remembers" some very atypical things that were done to it before $t=0$.

In this observation lies the clue that shows how to extend the mathematical methods of Gibbs to a general formalism for predicting irreversible phenomena, we must learn how to construct ensembles which describe not only the present values of macroscopic quantities, but also whatever information we have about their past behavior. The details of this generalization will be given elsewhere.

VI. THE "ANTHROPOMORPHIC" NATURE OF ENTROPY

After the above insistence that any demonstration of the second law must involve the entropy as measured experimentally, it may come as a shock to realize that, nevertheless, thermodynamics knows of no such notion as the "entropy of a physical system." Thermodynamics does have the concept of the entropy of a *thermodynamic* system; but a given physical system corresponds to many different thermodynamic systems.

Consider, for example, a crystal of Rochelle salt. For one set of experiments on it, we work with temperature, pressure, and volume. The entropy can be expressed as some function $S_e(T,P)$. For another set of experiments on the same crystal, we work with temperature, the component e_{xy} of the strain tensor, and the component P_z of electric polarization; the entropy as found in these experiments is a function $S_b(T,e_{xy},P_z)$. It is clearly meaningless to ask, "What is the entropy of the crystal?" unless we first specify the set of parameters which define its thermodynamic state.

One might reply that in each of the experiments cited, we have used only part of the degrees of freedom of the system, and there is a "true" entropy which is a function of all these parameters simultaneously. However, we can always introduce as many new degrees of freedom as we please. For example, we might expand each element of the strain tensor in a complete orthogonal set of functions $\varphi_k(x,y,z)$

$$e_{ij}(x,y,z) = \sum_k a_{ij,k} \varphi_k(x,y,z)$$

and by a sufficiently complicated system of levels, we could vary each of the first 1000 expansion coefficients $a_{ij,k}$ independently. Our crystal is now a thermodynamic system of over 1000 degrees of freedom; but we still believe that the laws of thermodynamics would hold. So, the entropy must be a function of over 1000 independent variables. There is no end to this search for the ultimate "true" entropy until we have reached the point where we control the location of each atom independently. But just at that point the notion of entropy collapses, and we are no longer talking thermodynamics!

From this we see that entropy is an anthropomorphic concept, not only in the well-known statistical sense that it measures the extent of human ignorance as to the microstate. *Even at the purely phenomenological level, entropy is an anthropomorphic concept.* For it is a property, not of the physical system, but of the particular experiments you or I choose to perform on it.

This points up still another qualification on the statement of the second law without which it is, strictly speaking, no law at all. If we work with a *thermodynamic* system of n degrees of freedom, the experimental entropy is a function $S_e(X_1 \cdots X_n)$ of n independent variables. But the *physical* system has any number of additional degrees of freedom X_{n+1}, X_{n+2}, etc. We have to understand that these additional degrees of freedom are not to be tampered with during the experiments on the n degrees of interest; otherwise one could easily produce apparent violations of the second law.

For example, the engineers have their "steam tables," which give measured values of the entropy of superheated steam at various temperatures and pressures. But the H_2O molecule has a large electric dipole moment; and so the entropy of steam depends appreciably on the electric field strength present. It must always be understood implicitly (because it is never stated explicitly) that this extra thermodynamic degree of freedom was not tampered with during the experiments on which the steam tables are based; which means, in this case, that the electric field was not inadvertently varied from one measurement to the next.

Recognition that the "entropy of a physical system" is not meaningful without further qualifications is important in clarifying many questions concerning irreversibility and the second law. For example, I have been asked several times whether, in my opinion, a biological system, say a cat, which converts inanimate food into a highly organized structure and behavior, represents a violation of the second law. The answer I always give is that, until we specify the set of parameters which define the *thermodynamic state* of the cat, no definite question has been asked!

It seems apparent, in view of complications which we have encountered in the attempt to give a complete statement of the second law, that much more work needs to be done in this field. Glib, unqualified statements to the effect that "entropy measures randomness" are in my opinion totally meaningless, and present a serious barrier to any real understanding of these problems. A full resolution of all the questions that can be raised requires a much more careful analysis than any that has been attempted thus far. Perhaps the most difficult problem of all is to learn how to state clearly *what is the specific question we are trying to answer?* However, I believe that in the above arguments we have been able to report some progress in this direction.

VII. ACKNOWLEDGMENTS

I have profited from discussions of these problems, over many years, with Professor E. P. Wigner, from whom I first heard the remark, "Entropy is an anthropomorphic concept." It is a pleasure to thank Professor Wm. C. Band for reading a preliminary draft of this article, and suggesting an important clarification of the argument.

6. DELAWARE LECTURE (1967)

The invitation to participate in the Delaware Seminar provided an opportunity to put on record quite a collection of thoughts that had accumulated in my notes for several Colloquium talks given at various Universities since the Brandeis lectures. So they were assembled into one manuscript for the following lecture delivered at the University of Delaware on March 9, 1965.

On first reading, it appears to be three lectures on three different topics: the theory of Periodicity in Scientific Creation, the interpretation of Quantum Theory, and the foundations of Statistical Mechanics. If any common thread can be said to connect them, it is that we are observing three different consequences of the standard psychological reaction to conceptual problems of science: to try to sweep them under the rug instead of bringing them out into the open and discussing them.

As we now realize, Statistical Mechanics was held up for decades by conceptual misunderstandings, leading to a misplaced emphasis in research. That is, almost everybody took it for granted that the Gibbs rules of calculation must be justified as an application of the laws of mechanics; whereas the Gibbs rules were expressing only the laws of inference. Today, this is so clear to anyone who uses those rules in maximum entropy image reconstruction or spectral analysis, that it is hard to understand how the confusion could have persisted so long in Statistical Mechanics.

The unceasing confusion that swirls about the Copenhagen interpretation of Quantum Theory is, in my opinion, the direct result of a very similar, but more subtle, misplaced emphasis. The mathematical rules of present Quantum Theory, like the Gibbs rules, are highly succesful and clearly contain a great deal of very fundamental truth. But nobody knows what they mean; they are in part expressions of laws of Nature, in part expressions of principles of human inference, and we have not yet learned how to disentangle them. The positivist Copenhagen philosophy has prevented solution of the problem by denying that there is any distinction between reality and our knowledge of reality. That this leads to such absurdities as prediction of psychokinesis was recognized by Schrödinger and Einstein; a specific example of this arising in current Quantum Optics is pointed out in Jaynes (1980).

The historical analysis of the work of Gibbs, and the remarks about the relation of ensembles to real systems, were taken from the earlier work planned for the Wigner Festschrift. The main theme of that discussion — that imprecisely defined concepts place a limit on the development of a theory, that no amount of mathematical prowess can overcome, found another confirmation in the attempts to extend Onsager's irreversible thermodynamics, as noted in 'The Minimum Entropy Production Principle' reprinted here.

Foundations of Probability Theory
and Statistical Mechanics

EDWIN T. JAYNES

Department of Physics, Washington University
St. Louis, Missouri

1. What Makes Theories Grow?

Scientific theories are invented and cared for by people; and so have the properties of any other human institution — vigorous growth when all the factors are right; stagnation, decadence, and even retrograde progress when they are not. And the factors that determine which it will be are seldom the ones (such as the state of experimental or mathematical techniques) that one might at first expect. Among factors that have seemed, historically, to be more important are practical considerations, accidents of birth or personality of individual people; and above all, the general philosophical climate in which the scientist lives, which determines whether efforts in a certain direction will be approved or deprecated by the scientific community as a whole.

However much the "pure" scientist may deplore it, the fact remains that military or engineering applications of science have, over and over again, provided the impetus without which a field would have remained stagnant. We know, for example, that ARCHIMEDES' work in mechanics was at the forefront of efforts to defend Syracuse against the Romans; and that RUMFORD'S experiments which led eventually to the first law of thermodynamics were performed in the course of boring cannon. The development of microwave theory and techniques during World War II, and the present high level of activity in plasma physics are more recent examples of this kind of interaction; and it is clear that the past decade of unprecedented advances in solid-state physics is not entirely unrelated to commercial applications, particularly in electronics.

Another factor, more important historically but probably not today, is simply a matter of chance. Often, the development of a field of knowledge has been dependent on neither matters of logic nor practical applications. The peculiar vision, or blindness, of individual persons can be decisive for the direction a field takes; and the views of one man can persist for centuries whether right or wrong. It seems incredible to us today that the views of Aristotle and Ptolemy could have dominated thought in mechanics and astronomy for a millenium, until GALILEO and others pointed out that we are all surrounded daily by factual evidence to the contrary; and equally incredible that, although thermometers (or rather, thermoscopes) were made by GALILEO before 1600, it required another 160 years before the distinction between temperature and heat was clearly recognized, by JOSEPH BLACK. (Even here, however, the practical applications were never out of sight; for GALILEO's thermoscopes were immediately used by his colleagues in the medical school at Padua for diagnosing fever; and JOSEPH BLACK's prize pupil was named JAMES WATT). In an age averse to any speculation, FRESNEL was nevertheless able, through pure speculation about elastic vibrations, to find the correct mathematical relations governing the propagation, reflection, and refraction of polarized light a half-century before MAXWELL's electromagnetic theory; while at the same time the blindness of a few others delayed recognition of the first law of thermodynamics for forty years.

Of far greater importance than these, however, is the general philosophical climate that determines the "official" views and standards of value of the scientific community, and the degree of pressure toward conformity with those views that the community exerts on those with a tendency to originality. The reality and effectiveness of this factor are no less great because, by its very nature, individual cases are more difficult to document; its effects "in the large" are easily seen as follows.

If you make a list of what you regard as the major advances in physical theory throughout the history of science, look up the date of each, and plot a histogram showing their distribution by decades, you will be struck immediately by the fact that advances in theory do not take place independently and randomly; they have a strong tendency to appear in small close clusters, spaced about sixty to seventy years apart. What we are observing here is the result of an interesting social phenomenon; this pressure toward conformity with certain officially proclaimed views, and away from free speculation, is subject to large periodic fluctuation. The last three cycles can be followed very easily, and the pressure maxima and minima can be dated rather precisely.

At the point of the cycle where the pressure is least, conditions are ideal for the creation of new theories. At these times, no one feels very sure just where the truth lies, and so free speculation is encouraged. New ideas of any kind are welcomed, and judged as all theories ought to be judged; on grounds of their logical consistency and agreement with experiment. Of course, we are only human; and so we also have a strong preference for theories which have a beautiful simplicity of concept. However, as stressed by many thinkers from OCCAM to EINSTEIN, this instinct seldom leads us away from the truth, and usually leads us toward it.

Eventually, one of these theories proves to be so much more successful than its competitors that, in a remarkably short time the pressure starts rising, all effective opposition ceases, and only one voice is heard. A well-known human frailty — overeagerness of the fresh convert — rides rough-shod over all lingering doubts, and the successful theory hardens into an unassailable official dogma, whose absolute, universal, and final validity is proclaimed independently of the factual evidence that led to it. We have then reached the peak of the pressure cycle; a High Priesthood arises whose members believe very sincerely that they are, at last, in possession of Absolute Truth, and this gives them the right and duty to combat errors of opinion with all the forces at their command. Exactly the same attitude was responsible, in still earlier times, for the Spanish Inquisition and the burning of witches.

At times of a pressure maximum, all free exercise of the imagination is frowned upon, and if one persists, severely punished. New ideas are judged, not on grounds of logic or fact, but on grounds of ideological conformity with the official dogma. To openly advocate ideas which do not conform is to be branded a crackpot and to place one's professional career in jeopardy; and very few have the courage to do this. Those who are students at such a time are taught only one view; and they miss out on the give and take, the argument and rational counter-argument, which is an essential ingredient in scientific progress. A tragic result is that many fine talents are wasted, through the misfortune of being born at the wrong time.

This high-pressure phase starts to break up when new facts are discovered, which clearly contradict the official dogma. As soon as one such fact is known, then we are no longer sure just what the range of validity of the official theory is; and we usually have enough clues by then so that additional disconcerting facts can be found without difficulty. The voice of the High Priests fades, and soon we have again reached a pressure minimum, in which nobody feels very sure where the truth lies and new suggestions are again given a fair hearing, so that creation of new theories is again socially possible.

Let us trace a few cycles of this pressure fluctuation (see Fig. 1). The pressure minimum that occurred at the end of the eighteenth century is now known as the "Age of Reason".

During a fairly short period many important advances in physical theory were made by such persons as LAPLACE, LAGRANGE, LAVOISIER, and FOURIER. Then a pressure maximum occurred in the first half of the nineteenth century, which is well described in some thermodynamics textbooks, particularly that of EPSTEIN [1]. This period of hostility toward free speculation seems to have been brought about, in part, by the collapse of SCHELLING's *Naturphilosophie*, and its chief effect was to delay recognition of the first law of thermodynamics for several decades. As already noted, FRESNEL was one of the very few physicists who escaped this influence sufficiently to make important advances in theory.

Another pressure minimum was reached during the third quarter of the nineteenth century, when a new spurt of advances took place in a period of only fifteen years (1855—1870), in the hands of MAXWELL, KELVIN, HERTZ, HELMHOLTZ, CLAUSIUS, BOLTZMANN, and several

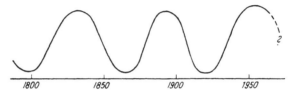

Fig. 1. Some recent fluctuations in social pressure in science

others. During this short period thermodynamics, electromagnetic theory, and kinetic theory were developed nearly to their present form; but the very success of these efforts led to another of the inevitable pressure maxima, which we recognize as being in full flower in the period 1885—1900. One of the tragedies (at least from the standpoint of physics) caused by this was the virtual loss of the talents of POINCARÉ. While his contributions to physical theory are considerable, still they are hardly commensurate with what we know of his enormous abilities. This was recognized and explained by E. T. BELL [2] in these words: "He had the misfortune to be in his prime just when physics had reached one of its recurrent periods of senility." The official dogma at that time was that all the facts of physics are to be explained in terms of Newtonian mechanics; particularly that of particles interacting through central forces. Herculean efforts were made to explain away MAXWELL's electromagnetic theory by more and more complicated mechanical models of the ether — efforts which remind us very much of the earlier single-minded insistence that all the facts of astronomy must be explained by adding more and more Ptolemaic epicycles.

An interesting manifestation toward the end of this period was the rise of the school of "Energetics", championed by MACH and OSTWALD, which represents an early attempt of the positivist philosophy to limit the scope of science. This school held that, to use modern terminology, the atom was not an "observable", and that physical theories should not, therefore, make use of the concept. The demise of this school was brought about rapidly by PERRIN's quantitative measurements on the Brownian motion, which verified EINSTEIN's predictions and provided an experimental value for AVOGADRO's number.

The last "Golden Age of Theory" brought about by the ensuing pressure minimum, lasted from about 1910 to 1930, and produced our present general realitivity and quantum theories. Again, the spectacular success of the latter — literally thousands of quantitatively correct predictions which could not be matched by any competing theory — brought about the inevitable pressure rise, and for twenty-five years (1935—1960) theoretical physics was paralyzed by one of the most intense and prolonged high-pressure periods yet recorded. During this period the official dogma has been that all of physics is now to be explained by prescribing initial and final state vectors in a Hilbert space, and computing transition matrix elements between them. Any attempt to find a more detailed description than this stood in conflict with the official ideology, and was quickly suppressed without any attempt to exhibit a logical inconsistency or a conflict with experiment; this time, a few individual cases can be documented [3].

There are now many signs that the pressure has started down again; several of the supposedly universal principles of quantum theory have been confronted with new facts, or new investigations, which make us unsure of their exact range of validity. In particular, one of the fundamental tasks of any theory is to prescribe the class of physical states allowed by Nature. In MAXWELL's electromagnetic theory, for example, any mathematical solution of MAXWELL's equations is held to represent a possible physical state, which could in principle be produced in the laboratory. In quantum theory, we were taught for many years that the class of possible physical states is in 1:1 correspondence with solutions of the Schrödinger equation that are either symmetric or antisymmetric under permutations of identical particles. Our confidence in the universal validity of this rule has, recently, been shaken in two respects. In the first place, study of "parastatistics" has shown that much more general types of symmetry in configuration space can also be described by the machinery of quantized wavefunctions, and these new possibilities are not ruled out by experimental evidence. Secondly, the superposition principle (which may be regarded as a consequence of the above-mentioned rule, although it is usually considered in a still more general sense) holds that, if ψ_1 and ψ_2 are any two possible physical states, then any linear combination $\psi \equiv a_1 \psi_1 + a_2 \psi_2$ is also a possible physical state. But with the appearance of superselection rules, we are no longer sure what the range of validity of the superposition principle is.

The discovery of parity nonconservation was a great psychological shock; a principle which had been taught to a generation of physicists as a universally valid physical law, so firmly established that it could be used to rule out *a priori* certain theoretical possibilities, such as WEYL'S twocomponent relativistic wave equation, was found not to be universally valid after all; and again we are unsure as to its exact range of validity, and WEYL's equation has been resurrected.

Several quantum mechanics textbooks assure us that the phenomenon of spontaneous emission places a fundamental irreducible minimum value on the width of spectral lines. Such statements are now confronted with the laser, which — in instruments now commercially available, and as simple to operate as a sixty-watt light bulb — produce spectral lines over a million times narrower than the supposedly fundamental limit! Thus, all around the edges of quantum theory we see the familiar kind of crumbling which, historically, has always signalled the incipient breakdown of the theory itself.

I hasten to add that, of course, none of these developments affects the basic "hard core" of quantum theory in any way; they show only that certain gratuitous additions to quantum theory (which had, however, become very closely associated with the basic theory) were unsound in the sense that they were not of *universal* validity. But it is inevitable that, faced these developments, more and more physicists will ask themselves how many other principles are destined to crumble a little at the edges, so that they can again be considered valid objects for inquiry; and not articles of faith to be asserted dogmatically for the purpose of discouraging inquiry.

In particular, the uncertainty principle has stood for a generation, barring the way to more detailed descriptions of nature; and yet, with the lesson of parity still fresh in our minds, how can anyone be quite so sure of its universal validity when we note that, to this day, it has never been subjected to even one direct experimental test?

Today, elementary particle theorists are busily questioning and reexamining all the foundations of quantum field theory, in a way that would have been regarded as utter heresy ten years ago; and some have suggested that perhaps the whole apparatus of fields and Hamiltonians ought to be simply abandoned in favor of more abstract approaches. It would be quite inconsistent with the present mood of theoretical physics if we failed to question and re-examine *all* of the supposedly sacred principles of quantum theory.

For all these reasons, I think we are going to see a rapid decrease in pressure in the immediate future, and another period of great theoretical advances will again be socially possible in perhaps ten years. And I think we can predict with confidence that some of the clues which will lead to the next round of advances are to be found in the many suggestions already made by dissenters from the Copenhagen theory — suggestions which have, thus far, been met only by sneers and attacks, which no attempt to study their real potentialities.

2. Statistical Mechanics

At this point, I see that you are looking about anxiously and wondering if you are in the right room; for the announced title of this talk was, "Foundations of Probability Theory and Statistical Mechanics". What has all this to do with statistical mechanics? Well, I wanted to say a few things first about general properties of physical theories because statistical mechanics is, in several respects, an exceptional case. Statistical methods exist independently of physical theories, and so statistical mechanics is subject to additional outside interactions from other fields. The field of probability and statistics is also subject to periodic fluctuations, but they are not in phase with the fluctuations taking place in physics (they are right now at a deep pressure minimum); and so the history of statistical mechanics is more complicated.

In particular, statistical mechanics missed out on the latest pressure minimum in physics, because it coincided with a pressure maximum in statistics; the transition to quantum statistics took place quietly and uneventfully without any real change in the basic formalism of GIBBS, and without any extension of the range of applicability of the theory. There was no advance in understanding, as witnessed by the fact that debates about irreversibility continue to this day, repeating exactly the same arguments and counter-arguments that were used in the time of BOLTZMANN; and the newest and oldest textbooks you can find hardly differ at all in their presentation of fundamentals. In short, statistical mechanics has suffered a period of stagnation and decadence that makes it unique in the recent history of science.

A new era of active work in statistical mechanics started, however, about 1955, in phase with a revolution in statistical thought but not at first directly influenced by it. This was caused, in part, by practical needs; an understanding of irreversible processes became increasingly necessary in chemical and mechanical engineering as one demanded more efficient industrial processing plants, stronger and more reliable materials, and bigger and better bombs. There is always a movement of scientific talent into areas where generous financial support is there for the taking. Another cause was the appearance of a few people who were genuinely interested in the field for its own sake; and perhaps it helped to reflect that, since it had been virtually abandoned for decades, one might be able to work in this field free of the kind of pressure noted above, which was paralyzing creative thought in other areas of physics.

Regardless of the reasons for this renewed activity, we have now made considerable progress in theoretical treatment of irreversible processes; at least in the sense of successful calculation of a number of particular cases. It is an opportune time to ask whether this has been accompanied by any better understanding, and whether the foundations of the subject can now be put into some kind of order, in contrast to the chaos that has persisted for almost a century. I hope to show now that the answer to both of these questions is yes; and that recent developments teach us an important lesson about scientific methodology in general.

Let me state the lesson first, and then illustrate it by examples from statistical mechanics. It is simply this: *You cannot base a general mathematical theory on imprecisely defined concepts. You can make some progress that way; but sooner or later the theory is bound to dissolve in ambiguities which prevent you from extending it further.* Failure to recognize this fact has another unfortunate consequence which is, in a practical sense, even more disastrous: *Unless the conceptual problems of a field have been clearly resolved, you cannot say which mathematical problems are the relevant ones worth working on; and your efforts are more than likely to be wasted.* I believe that, in this century, thousands of man-years of our finest mathematical talent have been lost through failure to understand this simple principle of methodology; and this remark applies with equal force to physics and to statistics.

2.1. Boltzmann's Collision Equation

Let us consider some case histories. Boltzmann sought to describe the approach to equilibrium in a gas in terms of the distribution $f(x, p, t)$. In his first work, this function was defined as giving the actual number of particles in various cells of phase space; thus if R denotes the set of points comprising a region of six-dimensional phase space, the number of particles in R is to be computed from

$$n_R = \int_R f(x, p, t) d^3x \, d^3p. \tag{1}$$

After some physical arguments which need not concern us here, Boltzmann concluded that the time evolution of the gas should be described by his famous "collision equation",

$$\frac{\partial f}{\partial t} + \sum_\alpha \left[\frac{P_\alpha}{m} \frac{\partial f}{\partial x_\alpha} + F_\alpha \frac{\partial f}{\partial P_\alpha} \right] = \int \partial^3 P' \int \partial \Omega (\bar{f}\bar{f}' - ff') \sigma \tag{2}$$

where F_α is the α-component of external force acting on a particle; and the right-hand side represents the effects of collisions in redistributing particles in phase space, in a way familiar to physicists. As a consequence of this equation, it is easily shown that the quantity

$$H_B \equiv \int f \log f \, d^3x \, d^3p \tag{3}$$

can only decrease (in this equation we integrate over all the accessible phase space); and so BOLTZMANN sought to identify the quantity

$$S_B \equiv -kH_B \tag{4}$$

with the entropy, making the second law of thermodynamics a consequence of the dynamical laws, as expressed by (2). As we know, this was challenged by ZERMELO and LOSCHMIDT who produced two counter-examples, based on time-reversal and on the POINCARÉ recurrence theorem, showing that Eq. (2) could not possibly be an exact expression of the dynamical equations of motion, *and thereby placing the range of validity of Boltzmann's theory in doubt.*

At this point, confusion entered the subject; and it has never left it. For BOLTZMANN then retreated from his original position, and said that he did not intend that $f(x, p, t)$ should represent necessarily the *exact* number of particles in various regions [indeed, it is clear that the only function f which has exactly the property of Eq. (1) is a sum of delta-functions: $f(x, p, t) = \Sigma_i \, \delta(x - x_i) \, \delta(p - p_i)$, where $x_i(t)$, $p_i(t)$ are the position and momentum of the i-th particle]. It represents only the *probable* number of particles; or perhaps the *average* number of particles; or perhaps it gives the *probability* that a given particle is to be found in various regions. The decrease in H_B is then not something which must happen *every* time; but only what will *most probably* happen; or perhaps what will happen *on the average*, etc.

Unfortunately, neither BOLTZMANN nor anybody else has ever become more explicit than this about just what BOLTZMANN's f; and therefore BOLTZMANN's H-theorem, means. When our concepts are not precisely defined, they are bound to end up meaning different things to different people, thus creating rooom for endless and fruitless debate, of exactly the type that has been going on ever since. Furthermore, when we debate about imprecise concepts, we can never be sure whether we are arguing about a question of fact; or only a question about the meaning of words. From BOLTZMANN's day to this, the debate has never been able to rise above this level.

If you think my characterization of the situation has been too laconic, and unfair to many honest seekers after the truth, I invite you to examine a recent review article on transport theory [4]. On page 271, the author states that "The Boltzmann distribution function — is the (probable) number of particles in the positional range d^3x and the velocity range d^3v — ". On page 274 this is altered to: "The quantity f, the Boltzmann distribution function — is, roughly speaking, the average number of particles in a cell in the $x-v$ space (the μ-space). f refers to a single system. A more precise definition of f can be obtained through the use of the master function P." Consulting this master function, we find that neither the definition of P, nor its connection with f, is ever given. This, furthermore, is not a particularly bad example; it is typical of what one finds in discussions of BOLTZMANN's theory.

Let us note some of the difficulties that face the practical physicist because of this state of utter confusion with regard to basic concepts. Suppose we try to assess the validity of BOLTZMANN's equation (2) for some particular problem; or we try to extend it to higher powers in the density, where higher order collisions will become important in addition to the binary ones that are taken into account, in some sense, in (2). If we agree that f represents an *average* number of particles, we must still specify what this average is to be taken over. Is it an average over the particles, an average over time for a single system, an average over many copies of the single system, or an average over some probability distribution? Different answers to this question are going to carry different implications about the range of validity of (2), and about the correct way of extending it to more general situations. Even without answering it at all, however, we can still see the kind of difficulties that are going to face us. For if $f(x, p, t)$ is an average over something, then the left-hand side of (2) is also an average over this same something. So also, therefore, is the right-hand side if the equation is correct. But on the right-hand side we see the product of two f's; the product of two averages.

If you meditate about this for a moment, I think you will find it hard to avoid concluding that, if f is an average, then the right-hand side ought to contain the average of a product, not the product of the averages. These quantities are surely different; but we cannot say how different until we say what we are averaging over. *Until this ambiguity in the definition of Boltzmann's f is cleared up, we cannot assess the range of validity of Eq. (2), and we cannot say how it should be extended to more general problems.* Because of imprecise concepts, the theory reaches an impasse at the stage where it has barely scratched the surface of any real treatment of irreversible processes!

2.2. Method of GIBBS

For our second case history, we turn to the work of GIBBS. This was done some thirty years after the aforementioned work of BOLTZMANN, and the difficulties noted above, plus many others for which we do not have time here, were surely clear to GIBBS, who was extremely careful in matters of logic, detail, and definitions.

All important advances have their precursors, the full significance of which is realized only later; and the innovations of GIBBS were not entirely new. For example, considerations of the full phase space (Γ-space) appear already in the works of MAXWELL and BOLTZMANN; and GIBBS' canonical ensemble is clearly only a small step removed from the distribution laws of MAXWELL and BOLTZMANN. However, GIBBS applied these ideas in a way which was unprecedented; so much so that his work was almost totally rejected ten years later in the famous Ehrenfest review article [5], which has had a dominating influence on thought in statistical mechanics for fifty years. In this article, the methods of GIBBS are attacked repeatedly, and the physical superiority of BOLTZMANN's approach is proclaimed over and over again. For example, GIBBS' canonical and grand canonical ensembles are dismissed as mere "analytical tricks", which do not solve the problem; but only enable GIBBS to *evade* what the authors consider to be real problems of the subject!

Since then, of course, the mathematical superiority of GIBBS' methods for calculating equilibrium thermodynamic properties has become firmly established; and so statistical mechanics has become a queer hybrid, in which the practical calculations are always based on the methods of GIBBS; while in the pedagogy virtually all one's attention is given to repeating the arguments of BOLTZMANN.

This hybrid nature — the attempt to graft together two quite incompatible philosophies — is nowhere more clearly shown than in the fact that the "official" commentary on GIBBS' work [6] devotes a major amount of space to discussion of ergodic theories. Now, it is a curious fact that if you study GIBBS' work, you will not find the word "ergodic" or the concept of ergodicity, at any point. Recalling that ergodic theorems, or hypotheses, had been actively discussed by other writers for over thirty years, and recalling GIBBS' extremely meticulous attention to detail, I think the only possible conclusion we can draw is that GIBBS simply *did not consider ergodicity as relevant to the foundations of the subject*. Of course, he was far too polite a man to say so openly; and so he made the point simply by developing his theory without making any use of it. Unfortunately, this tactic was too subtle to be appreciated by most readers; and the few who did notice it took it to be a defect in GIBBS' presentation, in need of correction by others.

This situation has had very unfortunate consequences, in that the work of GIBBS has been persistently misunderstood; and in particular, the full power and generality of the methods he introduced have not yet been recognized in any existing textbook. However, it is not a question of placing blame on anyone; for we can understand and sympathize with the position of everyone involved. I think that a historical study will convince you, as it has convinced me, that all of this is the more or less inevitable result of the fact that GIBBS did not live long enough to complete his work. The principle he had discovered was so completely new, and the method of thinking so completely different from what had gone before, that it was not possible to explain it fully, or to explore its consequences for irreversible phenomena, in the time that was granted to him.

GIBBS was in rapidly failing health at the time he wrote his work on statistical mechanics, and he lapsed into his final illness very soon after the manuscript was sent to the publisher. In studying his book, it is clear that it was never really finished; and we can locate very accurately the place where time and energy ran out on him. The first eleven chapters are written in his familiar style — extremely meticulous attention to detail, while unfolding a carefully thought out logical development. At Chapter 12, entitled, "On the Motion of Systems and Ensembles of Systems Through Long Periods of Time", we see an abrupt change of style; the treatment becomes sketchy, and amounts to little more than a random collection of observations, trying to state in words what he had not yet been able to reduce to equations. On pages 143—144 he tries to explain the methodology which led him to his canonical and grand canonical ensembles, as well as the ensemble canonical in the angular momenta which was presented in Chapter 4 but not applied to any problem [7]. However, he devotes only two sentences to this; and the principle he states is what we would recognize today as the principle of maximum entropy! To the best of my knowledge, this passage has never been noted or quoted by any other author (it is rather well hidden among discussions of other topics); and I discovered it myself only by accident, three years after I had written some papers [8] advocating this principle as a general foundation for statistical mechanics. This discovery convinced me that there was much more to the history of this subject than one finds in any textbook, and induced me to study it from the original sources; some of the resulting conclusions are being presented in this talk.

GIBBS' discussion of irreversibility in this chapter does not advance beyond pointing to a qualitative analogy with the stirring of colored ink in water; and this forms the basis for another of the EHRENFEST's criticisms of his work. I think that, had GIBBS been granted a few more years of vigorous health, this would have been replaced by a simple and rigorous demonstration of the second law based on other ideas. For it turns out that all the clues necessary to point the way to this, and all the mathematical material needed for the proof, were already present in the first eleven chapters of his book; it requires only a little more physical reasoning to see that introduction of coarse-grained distributions does not advance our understanding of irreversibility and the second law, for the simple reason that the latter are experimentally observed *macroscopic* properties; and the fine-grained and coarse-grained distributions lead to just the same predictions for all macroscopic quantities. Thus, the difference between the fine-grained and coarse-grained H-functions has nothing to do with the experimentally observed entropy; it depends only on the particular way in which we choose to coarse-grain.

On the other hand, the variational (maximum entropy) property noted by GIBBS does lead us immediately to a proof, not only of the second law, but of an extension of the second law to nonequilibrium states. I have recently pointed this out [9] and supplied the very simple proof, which I think is just the argument GIBBS would have given if he had been able to complete his work. However, this is not the main point I wish to discuss tonight, so let us turn back to other topics.

In defense of the EHRENFEST's position, it has to be admitted that, through no fault of his own, GIBBS did fail to present any clear description of the motivation behind his work. I believe that it was virtually impossible to understand what GIBB's methods amounted to, *and therefore how great was their generality and range of validity*, until the appearance of SHANNON's work on Information Theory, in our own time [10]. Finally, until recently the situation in probability theory itself, which was in a high-pressure phase completely dominated by the frequency theory, which only sneers and attacks on the theories of LAPLACE and JEFFREYS, has made it impossible even to discuss, much less publish, the viewpoint and approach which I believe has now solved these problems.

Now, in order to lend a little more substance to these remarks, let's examine some equations, the net result of GIBBS' work. Considering a closed system (i.e., no particles enter or leave), the thermodynamic properties are to be calculated from the Hamiltonian $H(q_i, p_i)$ as follows. First, we define the *partition function*

$$Z(\beta, V) \equiv \int \exp(-\beta H) dq_1 dp_1 \ldots dq_n dp_n, \qquad (5)$$

where we integrate over all the accessible phase space, and the dependence on the volume V arises because the range of integration over the coordinates q_i depends on V. If we succeed in evaluating this function, then all thermodynamic properties are known; for the energy function (which determines the thermal properties) is given by

$$U = -\frac{\partial}{\partial \beta} \log Z \qquad (6)$$

in which we interpret β as $(kT)^{-1}$, where k is BOLTZMANN's constant and T the KELVIN temperature; and the equation of state is

$$P = \frac{1}{\beta} \frac{\partial}{\partial V} \log Z. \tag{7}$$

Now, isn't this a beautifully simple and neat prescription? For the first time in what has always been a rather messy subject, one had a glimpse of the kind of formal elegance that we have in mechanics, where a single equation (HAMILTON's principle) summarizes everything that needs to be said. Of all the founders of statistical mechanics, only GIBBS gives us this formal simplicity, generality, and as it turned out, a technique for practical calculation which the labors of another sixty years have not been able to improve on. The transition to quantum statistics took place so quietly and uneventfully because it consisted simply in the replacement of the integral in (5) by the corresponding discrete sum; and nothing else in the formalism was altered.

In the history of science, whenever a field has reached such a stage, in which thousands of separate details can be summarized by, and deduced from, a single formal rule — then an extremely important synthesis has been accomplished. Furthermore, by understanding the basis of this rule it has always been possible to extend its application far beyond the original set of facts for which it was designed. And yet, this did not happen in the case of GIBBS' formal rule. With only a few exceptions, writers on statistical mechanics since GIBBS have tried to snatch away this formal elegance by grafting GIBBS' method onto the substrate of BOLTZMANN's ideas, for which GIBBS himself had no need. However, a few, including TOLMAN and SCHRÖDINGER, have seen GIBBS' work in a different light — as something that can stand by itself without having to lean on unproved ergodic hypotheses, intricate but arbitrarily defined cells in phase space, Z-stars, and the like. Thus, while a detailed study will show that there are as many different opinions as to the reason for GIBBS' rules as there are writers on the subject, a more coarse-grained view shows that these writers are split into two basic camps; those who hold that the ultimate justification of GIBBS' rules must be found in ergodic theorems; and those who hold that a principle for assigning *a priori* probabilities will provide a sufficient justification. Basically, the confusion that still exists in this field arises from the fact that, while the *mathematical content* of GIBBS' formalism can be set forth in a few lines, as we have just seen, the *conceptual basis* underlying it has never been agreed upon.

Now, while GIBBS' formalism has a great generality — in particular, it holds equally well for gas and condensed phases, while BOLTZMANN's results apply only to dilute gases — it nevertheless fails to give us many things that BOLTZMANN'S "collision equation" does yield, however imperfectly. For BOLTZMANN'S equation can be applied to irreversible processes; and it gives definite theoretical expressions for transport coefficients (viscosity, diffusion, heat conductivity), while GIBBS' rules refer only to thermal equilibrium, and one has not seen how to extend them beyond that domain. Furthermore, in spite of all my carping about the imprecision of BOLTZMANN'S equation, the fact remains that it has been very successful in giving good numerical values for these transport coefficients; and it does so even for fairly dense gases, where we really have no right to expect such success. So, my adulation of Gibbs must not be carried to the point of rejecting BOLTZMANN'S work; it appears that we need both approaches!

All right. I have now posed the problem as it appeared to me a number of years ago. Can't we learn how to combine the best features of both approaches, into a new theory that retains the unity and formal simplicity of GIBBS' work with the ability to describe irreversible processes (hopefully, a *better* ability) of BOLTZMANN'S work? This question must have occurred to almost every physicist who has made a serious study of statistical mechanics, for the past sixty years. And yet, it has seemed to many a hopelessly difficult task; or even an impossible one. For example, at the 1956 International Congress on Theoretical Physics, L. VAN HOVE [11] remarked, "In contrast to the case of thermodynamical equilibrium, no general set of equations is known to describe the behavior of many-particle systems whenever their state is different from the equilibrium state and, in view of the unlimited diversity of possible nonequilibrium situations, the existence of such a set of equations seems rather doubtful".

Now, while I hesitate to say so at a symposium devoted to Philosophy of Science, the injection of philosophical considerations into science has usually proved fruitless, in the sense that it does not, of itself, lead to any advances in the science. But there is one extremely important exception to this; and it is in exactly the situation now before us. At the stage in development of a theory where we already have a formalism successful in one domain, and we are trying to extend it to a wider one, some kind of philosophy about what the formalism "means" is absolutely essential to provide us with a sense of direction. And it need not even be a "true" philosophy — whatever that may mean — for its real justification will not lie in whether it is "true", but in whether it does point the way to a successful extension of the theory.

In the construction of theories, a philosophy plays somewhat the same role as scaffolding does in the construction of buildings; you need it desperately at a certain phase of the operation, but when the construction is completed you can remove if it you wish; and the structure will still stand of its own accord. This analogy is imperfect, however, because in the case of theories, the scaffolding is rarely ugly, and many will wish to retain it as an integral part of the final structure. At the opposite extreme to this conservative attitude stands the radical positivist, who in his zeal to remove every trace of scaffolding, also tears down part of the building. Almost always, the wisest course will lie somewhere between these extremes.

The point which I am trying to make, in this rather cryptic way, is just the one which we have already noted in the attempt to evaluate and extend BOLTZMANN's collision equation. Different philosophies of what that equation means carry different implications as to its range of validity, and the correct way of extending it. And we are now at just the same impasse with regard to GIBBS' equations; *because their conceptual basis has not been precisely defined, the theory dissolves in ambiguities* which have precented us, for sixty years, from extending to new domains.

2.3. Conceptual Problems of the Ensemble

The fact that two different camps exist, with diametrically opposed views as to the justification of GIBBS' methods, is simply the reflection of two diametrically opposed philosophies about the real meaning of the GIBBS ensemble; and this in turn arises from two different philosophies about the meaning of *any* probability distribution. Thus, the foundations of probability theory itself are involved in the problem of extending GIBBS' methods.

Statistical mechanics has always been troubled with questions concerning the relation between the ensemble and the individual system, even apart from possible extensions to nonequilibrium cases. In the theory, we calculate numbers to compare with experiment by taking ensemble *averages*; that is what we are doing in Eqs. (6) and (7). And yet, our experiments to check these predictions are not performed on ensembles; they are performed on the one *individual* system that exists in the laboratory. Nevertheless, we find that the predictions are verified accurately; a rather astonishing result, but one without which we would have little interest in ensembles. For if it were necessary to repeat a thermodynamic measurement 1,000 times and average the results before any regularities (laws of thermodynamics) began to appear, both thermodynamics and statistical mechanics would be virtually useless to us; and they would not appear in our physics curriculum. Thus, it appears that a major problem is to explain why GIBBS' rules work in practice; and not only why they work so well, but why they work at all!

We can make this dilemma appear still worse by noting that the relation between the ensemble and the individual system is usually described by supposing that the individual system can be regarded as having been drawn "at random" from the ensemble. I personally have never been able to comprehend what "at random" means; for I ask myself: What is the criterion, what is the test, by which we could decide whether it was or was not really "random"? Does it make sense to ask whether it was *exactly* random, or *approximately* random? — and neither the literature nor my introspection give me any answer. However, even without understanding this point, the real difficulty is obvious; for the *same* individual system may surely, and with equal justice, be regarded as having been drawn "at random" from any one of an infinite number of *different* ensembles! But the measured properties of an individual system depend on the *state of the system*; and not on which ensemble you or I regard it as having been "drawn from". How, then is it possible that ensemble averages coincide with experimental values?

The two different philosophical camps try to extricate themselves from this dilemma in two entirely different ways. The "ergodic" camp, of course, is composed of those who believe that a probability distribution describes an objectively real physical situation; that it stands for an assertion about experimentally measurable *frequencies*; that it is therefore either correct or incorrect; and that this can, in principle, be decided by performing "random experiments". They note that what we measure in any experiment is necessarily a time average over a time that is long on the atomic scale of things; and so the success of GIBBS' methods will be accounted for if we can prove, from the microscopic equations of motion, that the *time average* for an individual system is equal to the *ensemble average* over the particular ensembles given by GIBBS.

This viewpoint has much to recommend it. In the first place, physicists have a natural tendency to believe that, since the observed properties of matter "in the large" are simply the resultant of its properties "in the small" multiplied many times over, it ought to be possible to obtain the macroscopic behavior by strict logical deduction from the microscopic laws of physics; and the "ergodic" approach gives promise of being able to do this. Secondly, while the necessary theorems have not been established rigorously and universally, the work done on this problem thus far has made it highly plausible that, in a system interacting with a large heat bath, the *frequencies* with which various microscopic conditions are realized in the long run are indeed given correctly by the GIBBS canonical ensemble. This has been rendered so extremely plausible that I think no reasonable person can seriously doubt that it is true, although we cannot rule out the possibility of occasional "pathological" exceptions. Thus the "ergodic" school of thought has, in my opinion, very nearly succeeded in its aim of establishing equality of time averages and ensemble averages *for the particular case of Gibbs' canonical ensemble*; and in the following I am simply going to grant, for the sake of the argument, that this program has succeeded entirely.

Nevertheless, the "ergodic" school of thought still faces a fundamental difficulty; and one that was first pointed out by BOLTZMANN himself, and stressed in the EHRENFEST review article. Curiously, there exists to this day a group of workers in Europe who refuse to recognize the seriousness of this difficulty, and deny that it invalidates their approach. The difficulty is that, even if one had succeeded in proving these ergodic theorems rigorously and universally, the result would have been established only for time averages *over infinite times*; whereas the experiments which verify GIBBS' rules measure time averages only over finite times. Thus, a further mathematical demonstration would in any event be necessary, to show that these finite time averages have sufficiently approximated their limits for infinite times.

Now we can give simple and general counter-examples proving that such an additional demonstration *cannot* be given; and indeed that any macroscopic system, given a time millions of times the age of the universe, still could not "sample" more than an infinitesimal fraction of all the microscopic states which have high probability in the canonical emsemble; and thus any assertion about the *frequencies* with which different microscopic states are realized in an individual system, is completely devoid of operational meaning.

The easiest way of seeing this is just to note that, if a macroscopic system could sample all microscopic states in the time in which measurements are made, so that the measured time averages would be equal to ensemble averages, then the measured values would necessarily always be the *equilibrium* values; we would not even know about irreversible processes! *The fact that we can measure the rate of an irreversible process already proves that the time required for a representative sampling of microstates must be much longer than the time required to make our measurements.* Thus, any purported proof that time averages over the finite times involved in actual measurements are equal to canonical ensemble averages would, far from justifying statistical mechanics, stand in clear conflict with the very experimental facts about irreversibility that we are trying to account for by extending GIBBS' methods!

The thing which has to be explained is, not that ensemble averages are equal to time averages; but the much stronger statement that ensemble averages are equal to experimental values. The most that ergodic theorems could possibly establish is that ensemble averages are equal to time averages over infinite time, and so the "ergodic" approach cannot even justify equilibrium statistical mechanics without contradicting experimental facts. Obviously, such an approach cannot be extended to irreversible processes where, in order for ensemble theory to be of any use, the ensemble averages must still be equal to experimental values; but the very phenomena to be explained consist of the fact that these are *not* equal to time averages.

The above line of reasoning convinced me, ten years ago, that further advances in the basic formulation of statistical mechanics cannot be made within the framework of the "ergodic" viewpoint; and, rightly or wrongly, it seemed equally clear to me that the really fundamental trouble which was preventing further advances, both in statistical mechanics and in the field of statistics in general, was this dogmatic, single-minded insistence on the frequency theory of probability which had dominated the field for so many years. At that time, virtually every writer on probability theory felt impelled to insert an introductory paragraph or two, expressing his denunciation and total rejection of the so-called "subjective" interpretation of probability, as advocated by LAPLACE, DE MORGAN, POINCARÉ, KEYNES, and JEFFREYS; and this was done, invariably, without any attempt to understand the arguments and results which these people — particularly LAPLACE and JEFFREYS — had advanced. The situation was, psychologically, exactly like the one which has dominated American Politics since about 1930; the Republicans continually analyze the statements of Democrats and issue counter-arguments, which the Democrats contemptuously dismiss without any attempt to understand them or answer them.

On the other hand, I had taken the trouble to read all of JEFFREYS' work, and much of LAPLACE'S, on probability theory; and was unable to find any of the terrible things about which the "frequentist" writers had warned us. On the philosophical side I found their arguments to be, far from irresponsible and useless, so eminently sound and reasonable that I could not imagine any sane person disputing them. On the mathematical side, I found that in problems of statistical estimation and hypothesis testing, any problem for which the "frequentist" offered any solution at all was also solved with ease by the methods of LAPLACE and JEFFREYS; and their results were either the same or demonstrably superior to the ones found by the frequentists. Furthermore, the methods of LAPLACE and JEFFREYS (which were, of course, based on BAYES' theorem as the fundamental tool of statistics) were applied with equal ease to many problems which, according to the frequentist, did not belong to the field of probability theory at all; and they still yielded perfectly reasonable, and scientifically useful, results!

I don't want to dwell at length on the situation in probability theory, because time is running short and a rather large exposition of this, with full mathematical details, is being readied for publication elsewhere. But let me just mention one example of what one finds if he takes the trouble to go beyond polemics and study the mathematical facts of the matter.

In problems of interval estimation of unknown parameters, the frequentist has rejected the method of LAPLACE and JEFFREYS, on grounds that I can only describe as ideological, and has advocated vigorously the method of confidence intervals. Now it is a matter of straightforward mathematics to show that, whenever the frequentist's "estimator" is not a sufficient statistic (in the terminology of FISHER), there is always a class of possible samples for which the method of confidence intervals leads to absurd or dangerously misleading results, in the sense that it yields a wrong answer far more frequently (or, if one prefers, with far higher probability) than one would suppose from the stated confidence level. The confidence interval can, in some cases, contradict what can be proved on strict deductive reasoning from the observed sample. One can even invent problems, which are not at all unrealistic, in which the probability of this happening is greater than the stated confidence level!

This is something which, to the best of my knowledge, you cannot find mentioned in any of the "orthodox" statistical literature; and I shudder to think of some of the possible consequences, if important decisions are being made on the basis of confidence interval analyses. The method of LAPLACE and JEFFREYS is demonstrably free from this defect; it cannot contradict deductive reasoning and, in the case of the aforementioned "bad" class of samples, it automatically detects them and yields a wider interval, so that the probability of a correct decision remains equal to the stated value. Once one is aware of such facts, the arguments advanced against the method of LAPLACE and JEFFREYS and in favor of confidence intervals (i.e. that it is meaningless to speak of the probability that θ lies in a certain interval, because θ is not a "random variable," but only an unknown constant) appear very much like those of the 17th century scholar who claimed his theology had proved there could be *no* moons on Jupiter, and steadfastly refused to look through GALILEO's telescope.

Since the reasoning by which the "frequentist" has rejected LAPLACE's methods is so patently unsound, and since attempts to extend, or even justify, GIBBS' methods in terms of the frequency theory of probability have met with an impasse, it would appear that we ought to explore the possibilities of applying LAPLACE's "subjective" theory of probability to this problem. At any rate, to reject this procedure without bothering to explore its potentialities, is hardly what we mean by a "scientific" attitude! So, I undertook to think through statistical mechanics all over again, using the concept of "subjective" probability.

It became clear, very quickly, that to do this makes all the unsolved problems of the theory appear in a very different light; and possibilities for extension of GIBBS' methods are seen in entirely different directions. Once we clearly and explicitly free ourselves from the delusion that an ensemble describes an "objectively real" physical situation, and recognize that it describes only a certain *state of knowledge*, then it is clear that, in the case of irreversible processes, the knowledge which we have is of a different nature than in the case of equilibrium. We can then see the problem as one which cannot even be formulated in terms of the frequency theory of probability. It is simply this: *What probability assignment to microstates correctly describes the state of knowledge which we have, in practice, about a nonequilibrium state?* Such a question just doesn't make sense in terms of the frequency theory; but, thanks to the work of Gibbs and Shannon, I believe that it makes extremely good sense, and in fact has a very general and mathematically unambiguous solution in terms of subjective probabilities.

3. The General Maximum-Entropy Formalism

If we accept Shannon's interpretation (which can be justified by other mathematical arguments entirely independent of the ones given by Shannon) that the quantity

$$H = -\sum_i p_i \log p_i \qquad (8)$$

is an "information measure" for any probability distribution p_i; i.e. that it measures the "amount of uncertainty" as to the true value of i, then an ancient principle of wisdom — that one ought to acknowledge frankly the full extent of his ignorance — tells us that the distribution that maximizes H subject to constraints which represent whatever information we have, provides the most honest description of what we know. The probability is, by this process, "spread out" as widely as possible without contradicting the available information.

But recognition of this simple principle suddenly makes all the maximum-minimum properties given by Gibbs in his Chapter XI — what I believe to be the climax of Gibbs' work, and just the place where time and energy ran out on him — acquire a much deeper meaning. If we specify the expectation value of the energy, this principle uniquely determines Gibbs' canonical ensemble. If we specify the expectations of energy and mole numbers, it uniquely determines Gibbs' grand canonical ensemble [8]. If we specify the expectations of energy and angular momentum, it uniquely determines Gibbs' rotational ensemble [7]. Thus, all the results of Gibbs on statistical mechanics follow immediately from the principle of maximum entropy; and their derivation is astonishingly short and simple compared to the arguments usually found in textbooks.

But the generalization of Gibbs' formalism to nonequilibrium problems also follows immediately (although I have to confess that I spent six years trying to do this by introducing new and more complicated principles, before I finally saw how simple the problem was). For this principle in no way depends on the physical meaning of the quantities we specify; there is nothing unique about energy, mole numbers, or angular momentum. If we grant that it represents a valid method of reasoning at all, then we must also grant that it applies equally well to *any physical quantity whatsoever*. So, let us jump immediately, in view of the time, to the most sweeping generalization of GIBBS' formalism.

We have a number of physical quantities about which we have some experimental information. Let them be represented by the Heisenberg operators $F_1(x, t)$, $F_2(x, t)$, ... $F_m(x, t)$. In general they will depend on the position x and, through the equations of motion, on the time t. For example, F_1 might be the particle density, F_2 the density of kinetic energy, F_3 the "mass velocity" of the fluid, F_4 the (yz)-component of the stress tensor, F_5 the intensity of magnetization, ..., and so on; whatever information of this type is available, represents our definition of the nonequilibrium state.

Now we wish to construct a density matrix ϱ which incorporates all this information. When I say that a density matrix "contains" certain information, I mean by this simply that, if we apply the usual rule for prediction; i.e. calculate the expectation values

$$\langle F_k(x, t) \rangle = \text{Tr}[\varrho \, F_k(x, t)] \tag{9}$$

we must be able to recover this information from the density matrix. Thus, the mathematical constraints on the problem are that the expectation values (9) must agree with the experimental information:

$$f_k(x, t) = \text{Tr}[\varrho \, F_k(x, t)], \qquad x, t \text{ in } R_k \tag{10}$$

where $f_k(x, t)$ represent the experimental values, and R_k is the space-time region in which we have information about f_k; in general it may be different for different k. Subject to these constraints, we are to maximize the "information entropy"

$$S_I = -\text{Tr}(\varrho \log \varrho) \tag{11}$$

which is the appropriate generalization of (8), as found many years ago by VON NEUMANN. The solution of this variational problem is:

$$\varrho = \frac{1}{Z} \exp\left\{ \sum_{k=1}^{m} \int_{R_k} d^3x \, dt \, \lambda_k(x, t) \, F_k(x, t) \right\} \tag{12}$$

where the $\lambda_k(x, t)$ are a set of real functions to be determined presently (they arise mathematically as Lagrange multipliers in solving the variational problem with constraints), and for normalization the partition function of GIBBS has been generalized to the *partition functional:*

$$Z[\lambda_k(x,t)] \equiv \operatorname{Tr} \exp\left\{\sum_{k=1}^{m} \int_{R_k} d^3x\, dt\, \lambda_k(x,t)\, F_k(x,t)\right\}. \tag{13}$$

The $\lambda_k(x,t)$ are now to be found from the conditions (10), which reduce to

$$f_k(x,t) = \frac{\delta}{\delta \lambda_k(x,t)} \log Z \tag{14}$$

which is a generalization of GIBBS' equation (6); where δ denotes the functional derivative. Mathematical analysis shows that (14) is just sufficient to determine uniquely the integrals in the exponent of (12); it does not necessarily determine the functions $\lambda_k(x,t)$, but it does determine the only property of those functions which is needed in the theory; a very interesting example of mathematical economy.

The density matrix having been thus found, prediction of any other quantity $J(x,t)$ in its space-time dependence is then found by applying the usual rule:

$$\langle J(x,t)\rangle = \operatorname{Tr}[\varrho J(x,t)]. \tag{15}$$

In Eqs. (12) to (15) we have the generalization of GIBBS' algorithm to arbitrary nonequilibrium problems. From this point on, it is simply a question of mathematics to apply the theory to any problem you wish.

Of course, it requires a great deal of nontrivial mathematics to carry out these steps explicitly for any nontrivial problem! If GIBBS' original formalism was somewhat deceptive, in that its formal simplicity conceals an enormous amount of intricate detail, the same is true with a vengeance for this generalization. Nevertheless, it is still only mathematics; and if it were important enough to get a certain result, one could always hire a building full of mathematicians and computers to grind it out; there are no further questions of principle to worry about.

For the past three years, my students and I have been exploring these mathematical problems, and we have a large mass of results that will be reported in due course. Without going into further details, let me just say that all the previously known results in theory of irreversible processes can be derived easily from this algorithm. Dissipative effects such as viscosity, diffusion, heat conductivity are obtained by direct quadratures using (15), with no need for the forward integration and coarse-graining operations characteristic of previous treatments. For static transport coefficients we obtain formulas essentially equivalent to those of KUBO; we can exhibit certain ensembles for which KUBO's results, originally obtained by perturbation theory, are in fact exact.

Because we are freed from the need for time-smoothing and other coarse-graining operations, the theory is no longer restricted to the quasi-stationary, long-wavelength limit. It gives, with equal ease, general formulas for such things as ultrasonic attenuation and for nonlinear effects, such as those due to extremely large temperature or concentration gradients, for which previously no unambiguous theory existed. Because of these results, I feel quite confident that we are on the right track, and that this generalization will prove to be the final form of nonequilibrium statistical mechanics.

Let me close with a couple of philosophical remarks, relating this development to things I mentioned earlier in this talk. In seeking to extend a theory to new domains, some kind of philosophy about what the theory "means" is absolutely essential. The philosophy which led me to this generalization was, as already indicated, my conviction that the "subjective" theory of probability has been subjected to grossly unfair attacks from people who have never made the slightest attempt to examine its potentialities; and that if one does take the trouble to rise above ideology and study the facts, he will find that "subjective" probability is not only perfectly sound philosophically; it is a far more powerful tool for solving practical problems than the frequency theory. I am, moreover, not alone in thinking this, as those familiar with the rise of the "neo-Bayesian" school of thought in statistics are well aware.

Nevertheless, that philosophy of mine was only scaffolding, which served the purpose of telling me in what *specific* way the formalism of GIBBS was to be generalized. Once a philosophy has led to a definite, unambiguous mathematical formalism by which practical calculations may be carried out, then the issue is no longer one of philosophy; but of fact. The formalism either will or will not prove adequate in practice; and it will be judged, quite properly, not by the philosophy which led to it, but by the results which its gives. If you do not like my philosophy, but you find that the formalism, nevertheless, does give useful results, then I am quite sure that you will be able to invent some *other* philosophy by which that formalism can be justified! And, perhaps, that other philosophy will lead to still further generalizations and extensions, to which my own philosophy makes me blind. That is, after all, just the process by which all progress in theoretical physics has been made.

REFERENCES

[1] EPSTEIN, P. S.: Textbook of thermodynamics, p. 27—34. New York: John Wiley & Sons, Inc. 1937.
[2] BELL, E. T.: Men of mathematics, p. 546. New York: Dover Publ. Inc. 1937.
[3] See, for example: Niels Bohr and the development of physics (W. PAULI, ed.), p. 17—28, and footnote, p. 76. New York: Pergamon Press 1955; Observation and interpretation (S. KÖRNER, ed.), p. 41—45. New York: Academic

Press, Inc. 1957; W. HEISENBERG, Physics and philosophy, p. 128—146. New York: Harper & Brothers, Publ. 1958; N. R. HANSON, Am. J. Phys. **27**, 1 (1959); Quanta and reality (D. EDGE, ed.) p. 85—93. Larchmont (New York): Am. Research Council 1962.

[4] DRESDEN, M.: Revs. Mod. Phys. **33**, 265 (1961).

[5] EHRENFEST, P. and T.: Encykl. Math. Wiss. 1912. English translation by M. J. MORAVCSIK, The conceptual foundations of the statistical approach in mechanics. Ithaca (N.Y.): Cornell University Press 1959.

[6] GIBBS, J. W.: Collected works and commentary, vol. II (A. HAAS, ed.), p. 461—488. Yale University Press New Haven (Conn.): 1936.

[7] A successful application of GIBBS' rotationally canonical ensemble to the theory of gyromagnetic effects has since been given: S. P. HEIMS and E. T. JAYNES. Revs. Mod. Phys. **34**, 143 (1962).

[8] JAYNES, E. T.: Phys. Rev. **106**, 620; **108**, 171 (1957).

[9] — Chapter 4 of Statistical physics (1962 Brandeis Lectures) (K. W. FORD, ed.). New York: W. A. Benjamin, Inc. 1963; Gibbs vs Boltzmann entropies, Am. J. Phys. **33**, 391 (1965).

[10] SHANNON, C. E., and W. WEAVER: The mathematical theory of communication. Urbana (Ill.): University of Illinois Press 1949.

[11] HOVE, L. VAN: Revs. Mod. Phys. **29**, 200 (1957).

7. PRIOR PROBABILITIES (1968)

The fundamentally different problems of doing maximum entropy with continuous variables, noted already in a footnote in ITSM I, continued to plague the theory after the recognition of the proper continuous entropy expression in *Brandeis*. Introducing the 'measure' $m(x)$ achieved the needed invariance of the results with respect to parameter changes, but left the practical problem untouched. Given $m(x)$ in one parameter space, the mathematics tells us how to transform it to any other; but first we need a principle to find it in one space. If there is no obvious limiting process from some discrete set, what defines this measure?

It was clear that $m(x)$ was, essentially, what had been called previously 'the prior distribution expressing complete ignorance'. But this took us back to just what Jacob Bernoulli and Laplace had struggled with but had not solved except on finite discrete sets: what do we mean by 'complete ignorance' and what distribution represents it? It seemed that we were back to Square One, from whence it had all started.

But before the mathematical problem could be treated, we had to deal with the semantic one. The phrase 'complete ignorance' is too vague to define any particular mathematical problem; can we state what we really want here in terms that do make mathematical sense?

In 1965 it occurred to me that one very reasonable interpretation of 'complete ignorance' was group invariance. It is in retrospect incredible that it could have required so long to see this, since my thesis advisor had been Eugene Wigner, 'Mr. Group Theory' of theoretical physics. I attended his course and that of Hermann Weyl in Princeton, and had for fifteen years been an enthusiastic teacher of group theory methods in problems of mathematical physics. But better late than never.

Applying this idea in just the way Wigner and Weyl had taught me, I found immediately a much deeper understanding of the Jeffreys prior $d\mu \, d\sigma/\sigma$ in the location-scale parameter problem. This rule had been rejected in my Socony–Mobil Lectures (1958) because Jeffreys' argument in favor of it seemed ad hoc and arbitrary. But now it was clear that the point was not merely that σ was positive, the rationale that Jeffreys had given. The point was that σ was a scale parameter, complete ignorance of which meant invariance under the group of scale changes. I immediately became an advocate, rather than a critic, of the Jeffreys rule, and the nice numerical results that it gives (noted wistfully in the Socony–Mobil lectures as something forbidden to me) were now mine to use after all, with the sanction of a clear rational justification.

This work, which was for me a major advance in thinking, suffered the standard fate. It was submitted to a well-known statistical journal in 1966, and was indignantly rejected. The editor (whom I had thought to be a Bayesian) took the trouble to write me a letter requesting that I never again send him anything like it. But quickly I received an invitation to contribute an article to this IEEE journal's special issue on Decision Theory, and so with the addition of a few introductory pages it became the article which follows.

Although, as noted, group invariance arguments had appeared before this article and have appeared many times in the literature since 1968, to the best of my knowledge no other writer since Poincaré has recognized the 'Desideratum of Consistency' as providing the basic justification for imposing group invariance. Most writers simply dive into the mathematics without saying why they are doing it. I therefore stress this desideratum as something that appears to me not only a necessary axiom for any rational theory of inference, but also of much greater generality than the application to groups given here.

Group invariance arguments are carried further in *The Well-Posed Problem* and in the Appendix to *Marginalization*.

Prior Probabilities

EDWIN T. JAYNES

Abstract—In decision theory, mathematical analysis shows that once the sampling distribution, loss function, and sample are specified, the only remaining basis for a choice among different admissible decisions lies in the prior probabilities. Therefore, the logical foundations of decision theory cannot be put in fully satisfactory form until the old problem of arbitrariness (sometimes called "subjectiveness") in assigning prior probabilities is resolved.

The principle of maximum entropy represents one step in this direction. Its use is illustrated, and a correspondence property between maximum-entropy probabilities and frequencies is demonstrated. The consistency of this principle with the principles of conventional "direct probability" analysis is illustrated by showing that many known results may be derived by either method. However, an ambiguity remains in setting up a prior on a continuous parameter space because the results lack invariance under a change of parameters; thus a further principle is needed.

It is shown that in many problems, including some of the most important in practice, this ambiguity can be removed by applying methods of group theoretical reasoning which have long been used in theoretical physics. By finding the group of transformations on the parameter space which convert the problem into an equivalent one, a basic desideratum of consistency can be stated in the form of functional equations which impose conditions on, and in some cases fully determine, an "invariant measure" on the parameter space. The method is illustrated for the case of location and scale parameters, rate constants, and in Bernoulli trials with unknown probability of success.

In realistic problems, both the transformation group analysis and the principle of maximum entropy are needed to determine the prior. The distributions thus found are uniquely determined by the prior information, independently of the choice of parameters. In a certain class of problems, therefore, the prior distributions may now be claimed to be fully as "objective" as the sampling distributions.

I. BACKGROUND OF THE PROBLEM

SINCE THE time of Laplace, applications of probability theory have been hampered by difficulties in the treatment of prior information. In realistic problems of decision or inference, we often have prior information which is highly relevant to the question being asked; to fail to take it into account is to commit the most obvious inconsistency of reasoning and may lead to absurd or dangerously misleading results.

As an extreme example, we might know in advance that a certain parameter $\theta \leq 6$. If we fail to incorporate that fact into the equations, a conventional statistical analysis might easily lead to the conclusion that the "best" estimate of θ is $\theta^* = 8$, and a shortest 90-percent confidence interval is $(7 \leq \theta \leq 9)$.

Few people will accept an estimate of a parameter which lies outside the parameter space, and so "orthodox" statistical principles such as efficient estimators or shortest

Manuscript received March 1, 1968. This work was supported in part by the National Science Foundation under Grant GP-6210.
The author is with the Department of Physics, Washington University, St Louis. Mo 63130

confidence intervals can break down and leave no definite procedure for inference in the presence of this kind of prior information. Further examples of this phenomenon are given by Kendall and Stuart [1]

With more "gentle" kinds of prior information, which do not absolutely exclude any interval for θ but only render certain intervals highly unlikely, the difficulty is less drastic but still present. Such cases are even more dangerous in practice because the shortcomings of orthodox principles, while just as real, are no longer obvious.

The Bayesian approach to statistics offers some hope of overcoming such difficulties since, of course, both the prior and posterior distributions of θ will vanish outside the parameter space, and so the results cannot conflict with deductive reasoning. However, what determines the prior within the parameter space? After nearly two centuries of discussion and debate, we still do not seem to have the principles needed to translate prior information into a definite prior probability assignment.

For many years the orthodox school of thought, represented by most statisticians, has sought to avoid this problem by rejecting the use of prior probabilities altogether, except in the case where the prior information consists of frequency data However, as the preceding example shows, this places a great restriction on the class of problems which can be treated Usually the prior information does not consist of frequency data, but is nonetheless cogent. As Kendall and Stuart [1] point out, this is a major weakness of the principle of confidence intervals

With the rise of decision theory, this problem has assumed new importance. As we know, this development was started by Wald [2] with the express purpose of finding a new foundation for statistics which would have the generality, but avoid the supposed mistakes, of the work of Bayes and Laplace. But after monumental labors, the mathematical situation uncovered by Wald finally led to a realization that the only consistent procedure for digesting information into the decision process is identical with application of Bayes' theorem, and that, once the loss function, sampling distribution, and sample are given, the only rational basis for choice among different admissible decisions lies in the prior probabilities.

Thus in modern decision theory, it appears that statistical practice has reached a level where the problem of prior probabilities can no longer be ignored or belittled. In current problems of engineering design, quality control, operations research, and irreversible statistical mechanics, we cannot translate the full problem into mathematical terms until we learn how to find the prior probability assignment which describes the prior information. In fact, as shown later, in some of the most important problems the prior information is the only information available, and so decisions must be based entirely on it. In the absence of any principle for setting up prior distributions, such problems cannot be treated mathematically at all.

The "personalistic" school of thought (Savage [3], [4]) recognizes this deficiency, but proceeds to overcompensate it by offering us many different priors for a given state of prior knowledge. Surely, the most elementary requirement of consistency demands that two persons with the same relevant prior information should assign the same prior probabilities. Personalistic doctrine makes no attempt to meet this requirement, but instead attacks it as representing a naive "necessary" view of probability, and even proclaims as one of its fundamental tenets ([3], p. 3) that we are free to violate it without being unreasonable Consequently, the theory of personalistic probability has come under severe criticism from orthodox statisticians who have seen in it an attempt to destroy the "objectivity" of statistical inference by injecting the user's personal opinions into it

Of course, no one denies that personal opinions are entitled to consideration and respect if they are based on factual evidence. For example, the judgment of a competent engineer as to the reliability of a machine, based on calculation of stresses, rate of wear, etc., is fully as cogent as anything we can learn from a random experiment, and methods of reliability testing which fail to take such information into account are not only logically inconsistent, but economically wasteful Nevertheless, the author must agree with the conclusions of orthodox statisticians, that the notion of personalistic probability belongs to the field of psychology and has no place in applied statistics Or, to state this more constructively, objectivity requires that a statistical analysis should make use, not of anybody's personal opinions, but rather the specific factual data on which those opinions are based.

An unfortunate impression has been created that rejection of personalistic probability automatically means the rejection of Bayesian methods in general It will hopefully be shown here that this is not the case; the problem of achieving objectivity for prior probability assignments is not one of psychology or philosophy, but one of proper definitions and mathematical techniques, which is capable of rational analysis. Furthermore, results already obtained from this analysis are sufficient for many important problems of practice, and encourage the belief that with further theoretical development prior probabilities can be made fully as "objective" as direct probabilities

It is sometimes held that this evident difference in the nature of direct and prior probabilities arises from the fact that the former have a clear frequency interpretation usually lacking in the latter. However, there is almost no situation of practice in which the direct probabilities are actually verified experimentally in the frequency sense. In such cases it is hard to see how the mere possibility of *thinking* about direct probabilities as frequencies in a nonexistent experiment can really be essential, or even relevant, to the problem.

Perhaps the real difference between the manifestly "public" nature of direct probabilities and the "private" nature of prior probabilities lies in the fact that in one case there is an established theory, accepted by all (i.e., Bernoulli trials, etc.), which tells how to calculate them, while in the case of prior probabilities, no universally accepted theory exists as yet. If this view is correct, we would expect that with further development of probability theory, the distinction will tend to disappear. The two principles—maximum entropy and transformation groups—discussed in the following sections represent methods for calculating probabilities which apply indifferently to either

II THE BASIC DESIDERATUM

To elaborate the point just made, a prior probability assignment not based on frequencies is necessarily "subjective" in the sense that it describes a state of knowledge, rather than anything which could be measured in an experiment. But if the methods are to have any relevance to science, the prior distribution must be completely "objective" in the sense that it is independent of the personality of the user. On this point, it is believed that even the most ardent Bayesian must agree with orthodox statisticians The measure of success in producing an objective theory of decision or inference is just the extent to which we are able to eliminate all personalistic elements and create a completely "impersonalistic" theory

Evidently, then, we need to find a middle ground between the orthodox and personalistic approaches, which will give us just one prior distribution for a given state of prior knowledge. Historically, orthodox rejection of Bayesian methods was not based at first on any ideological dogma about the "meaning of probability" and certainly not on any failure to recognize the importance of prior information; this has been noted by Kendall and Stuart [1], Lehmann [5], and many other orthodox writers. The really fundamental objection (stressed particularly in the remarks of Pearson in Savage [4]) was the lack of any principle by which the prior probabilities could be made objective in the aforementioned sense Bayesian methods, for all their advantages, will not be entirely satisfactory until we face the problem squarely and show how this requirement may be met.

For later purposes it will be convenient to state this basic desideratum as follows. *in two problems where we have the same prior information, we should assign the same prior probabilities* This is stated in such a way that it seems psychologically impossible to quarrel with it, indeed, it may appear so trivial as to be without useful content A major purpose of the present paper is to show that in many cases, in spite of first appearances, this desideratum may be formulated mathematically in a way which has nontrivial consequences

Some kinds of prior information seem too vague to be translatable into mathematical terms. If we are told that, "Jones was very pleased at the suggestion that θ might be greater than 100," we have to concede that this does constitute prior information about θ; if we have great respect for Jones' sagacity, it might be relevant for inferences about θ. But how can this be incorporated into a mathematical theory of inference? There is a rather definite minimum requirement which the prior information must satisfy before it can be used by any presently known methods.

Definition 1: A piece of information I concerning a parameter θ will be called *testable* if, given any proposed prior probability assignment $f(\theta)\,d\theta$, there is a procedure which will determine unambiguously whether $f(\theta)$ does or does not agree with the information I.

As examples, consider the following statements.

I_1: "$\theta < 6$."
I_2: "The mean value of $\tanh^{-1}(1 - \theta^2)$ in previous measurements was 1.37."
I_3: "In the eighteenth century, Laplace summarized his analysis of the mass of Saturn by writing, 'It is a bet of 11 000:1 that the error of this estimate is not 1/100 of its value.' He estimated this mass as 1/3512 of the sun's mass."
I_4: "There is at least a 90-percent probability that $\theta > 10$."

Statements I_1 and I_2 clearly constitute testable information; they can be used immediately to restrict the form of a prior probability assignment. Statement I_3 becomes testable if we understand the exact meaning of Laplace's words, and very easily so if we know the additional historical fact that Laplace's calculations were based on the incomplete beta distribution. I_4 is also clearly testable, but it is perhaps less clear how it could lead to any unique prior probability assignment.

Perhaps in the future others will discover new principles by which nontestable prior information could be used in a mathematical theory of inference. For the present, however, we restrict ourselves to a search for formal principles by which testable information can be converted into a unique prior probability assignment.

Fortunately, we are not without clues as to how this uniqueness problem might be solved. The principle of maximum entropy (i.e., the prior probability assignment should be the one with the maximum entropy consistent with the prior knowledge) gives a definite rule for setting up priors. The rule is impersonal and has an evident intuitive appeal [6]–[11] as the distribution which "assumes the least" about the unknown parameter. In applications it has a number of advantages, but also some shortcomings which prevent its being regarded as a complete solution to the problem.

We now survey these briefly and aim to supplement the principle in a way that retains the advantages, while correcting the shortcomings

III. Maximum Entropy

We illustrate this method by a simple example which occurred in a physical problem (distribution of impurities in a crystal lattice), and for simplicity consider only a one-dimensional version. An impurity atom may occupy any of n different positions $\{x_1 \cdots x_n\}$, where $x_j = jL$ and L is a fixed length. From experiments on scattering of X rays, it has been determined that there is a moderate tendency to prefer sites at which $\cos(kx_j) > 0$, the specific datum being that in many previous instances the average value of $\cos kx_j$ was

$$\langle \cos kx_j \rangle = 0.3. \qquad (1)$$

This is clearly testable information, and it is desired to find a probability assignment p_j for occupation of the jth site which incorporates this information, but assumes nothing further, from which statistical predictions about future instances can be made.

The mathematical problem is then to find the p_j which will maximize the entropy

$$H = -\sum_{j=1}^{n} p_j \log p_j \qquad (2)$$

subject to the constraints $p_j \geq 0$, and

$$\sum_j p_j = 1 \qquad (3)$$

$$\sum_j p_j \cos(kx_j) = 0.3. \qquad (4)$$

The solution is well known, and in this case takes the form

$$p_j = \frac{1}{Z(\lambda)} \exp[\lambda \cos kx_j] \qquad (5)$$

where $Z(\lambda)$ is the partition function

$$Z(\lambda) \equiv \sum_{j=1}^{n} \exp[\lambda \cos kx_j] \qquad (6)$$

and the value of λ is to be determined from (4)

$$\langle \cos kx_j \rangle = \frac{\partial}{\partial \lambda} \log Z(\lambda) = 0.3. \qquad (7)$$

In the case where $ka \ll 1$, $nka \gg 1$, we may approximate the discrete sums sufficiently well by integrals, leading to

$$Z(\lambda) \simeq nI_0(\lambda) \qquad (8)$$

$$\langle \cos mkx \rangle \simeq \frac{I_m(\lambda)}{I_0(\lambda)} \qquad (9)$$

where $I_m(\lambda)$ are the modified Bessel functions From (1), and (9) in the case $m = 1$, we find $\lambda = 0.63$.

Having found the distribution p_j, we can now use it as the prior from which further information about the impurity location can be incorporated via Bayes' theorem. For example, suppose that if the impurity is at site j, the probability that a neutron incident on the crystal will be reflected is proportional to $\sin^2 kx$, We acquire the new data. "n neutrons incident, r reflected." The posterior probability for the impurity to be at site j would then be

$$p(x_j|n, r) = Ap_j p(r|n, j)$$
$$= B \exp\{\lambda \cos kx_j\} \{\sin^2 kx_j\}^r \{\cos^2 kx_j\}^{n-r} \quad (10)$$

where A, B are normalizing constants.

Alternatively, and representative of a large class of important problems which includes statistical mechanics, the prior distribution p_j may be used directly for certain kinds of decision or inference For example, suppose that before the neutron reflection experiment, we wish to estimate the probability of reflection of r neutrons from n incident Conditional only on the prior information (1), this probability is

$$p(r|n) = \sum_{j=1}^{n} p(r|n, j) p_j$$
$$= \binom{n}{r} \langle \{\sin^2 kx\}^r \{\cos^2 kx\}^{n-r} \rangle \quad (11)$$

the expectation value being taken over the prior distribution (5). In the case $n = r = 1$, it reduces to the probability of reflection at a single trial; using (9), we find

$$\langle \sin^2 kx \rangle = \frac{I_0 - I_2}{2I_0} = \lambda^{-1} \langle \cos kx \rangle = 0.48 \quad (12)$$

which is only slightly below the value 0.50 corresponding to a uniform prior distribution p_j; thus, in agreement with our intuition, the moderate constraint (1) is by no means sufficient to inhibit appreciably the occupation of sites for which $|\sin kx| \ll 1$. On the other hand, if the prior information had given $\langle \cos kx \rangle = 0.95$, repetition of the argument would yield $\langle \sin^2 kx \rangle = 0.09$, indicating now a very appreciable inhibition.

The values of $\langle \sin^2 kx \rangle$ thus calculated represent estimates of $\sin^2 kx$ which are "optimal" in the sense that 1) they are "maximally noncommittal" with regard to all information except the specific datum given, and 2) they minimize the expected square of the error. Of course, in a problem as rudimentary as this, one does not expect that these estimates can be highly reliable; the information available is far too meager to permit such a thing But this fact, too, is automatically incorporated into the maximum-entropy formalism; a measure of the reliability of the estimate is given by the expected "loss function," which in this case is just the variance of $\sin^2 kx$ over the maximum-entropy distribution

$$\sigma^2 = \langle \sin^4 kx \rangle - \langle \sin^2 kx \rangle^2 = \frac{I_0^2 - 2I_2^2 + I_0 I_4}{8 I_0^2} \quad (13)$$

from which we find, in the cases $\langle \cos kx \rangle = 0.3, 0.95$, the values $\sigma = 0.35$, $\sigma = 0.12$, respectively. Thus, if $\langle \cos kx \rangle = 0.3$, no accurate estimate of $\sin^2 kx$ is possible; we can say only that it is reasonably likely to lie in the interval $(0.13, 0.83)$ With the prior datum $\langle \cos kx \rangle = 0.95$, we are in a somewhat better position, and can say that $\sin^2 kx$ is reasonably likely to be less than 0.21.

Evidently the principle of maximum entropy can yield reliable predictions only of those quantities for which it leads to a sharply peaked distribution. If, for example, we find that a maximum-entropy distribution concentrates 99.99 percent of the probability on those values of x for which $6.72 < f(x) < 6.73$, we shall feel justified in predicting that $f(x)$ lies in that interval, and in attributing a very high (but not necessarily 99.99 percent) reliability to our prediction Mathematically, both equilibrium and nonequilibrium statistical mechanics are equivalent to applying the principle of maximum entropy in just this way; and their success derives from the enormous number of possible microstates, which leads to very sharply peaked distributions (typically of relative width 10^{-12}) for the quantities of interest.

Let us now try to understand some conceptual problems arising from the principle of maximum entropy. A common objection to it is that the probabilities thus obtained have no frequency interpretation, and therefore cannot be relevant to physical applications; there is no reason to believe that distributions observed experimentally would agree with the ones found by maximum entropy. We wish to show that the situation is a great deal more subtle than that by demonstrating that 1) there is a sense in which maximum-entropy distributions do have a precise correspondence with frequencies; 2) in most realistic problems, however, this frequency connection is unnecessary for the usefulness of the principle; and 3) in fact, the principle is most useful in just those cases where the empirical distribution fails to agree with the one predicted by maximum entropy

IV. The Correspondence Property

Application of the principle of maximum entropy does not require that the distribution sought be the result of any random experiment (in fact, its main purpose was to extend the range of application of Bayesian methods to problems where the prior probabilities have no reasonable frequency interpretation, such problems being by far the most often encountered in practice) Nevertheless, nothing prevents us from applying it also in cases where the prior distribution is the result of some random experiment, and one would hope that there is some close correspondence between the maximum-entropy distribution and observable frequencies in such cases; indeed, any principle for assigning priors which lacked this correspondence property would surely contain logical inconsistencies.

We give a general proof for the discrete case. The quantity x can take on the values $\{x_1 \cdots x_n\}$ where n may be finite or countably infinite, and the x_i may be specified arbitrarily. The available information about x places a number of constraints on the probability distribution $p_i = p(x_i)$ We assume for convenience, although it is in no way necessary for our argument, that these take the form of mean values of several functions $\{f_1(x), \cdots, f_m(x)\}$, where $m < n$ The probability distribution p_i which incorporates this information, but is free from all other assumptions, is then the one which maximizes

$$H = -\sum_{i=1}^{n} p_i \log p_i \qquad (14)$$

subject to the constraints

$$\sum_i p_i = 1 \qquad (15)$$

$$\sum_i p_i f_k(x_i) = F_k, \quad k = 1, 2, \cdots, m \qquad (16)$$

where the F_k are the prescribed mean values. Again, the well-known solution is

$$p_i = \frac{1}{Z(\lambda_1 \cdots \lambda_m)} \exp[\lambda_1 f_1(x_i) + \cdots + \lambda_m f_m(x_i)] \qquad (17)$$

with the partition function

$$Z(\lambda_1 \cdots \lambda_m) \equiv \sum_{i=1}^{n} \exp[\lambda_1 f_1(x_i) + \cdots + \lambda_m f_m(x_i)] \qquad (18)$$

in which the real constants λ_k are to be determined from the constraints (16), which reduce to the relations

$$F_k = \frac{\partial}{\partial \lambda_k} \log Z(\lambda_1 \cdots \lambda_m). \qquad (19)$$

The distribution (17) is the one which is, in a certain sense, spread out as uniformly as possible without contradicting the given information, i.e., it gives free rein to all possible variability of x allowed by the constraints. Thus it accomplishes, in at least one sense, the intuitive purpose of assigning a prior distribution; it agrees with what is known, but expresses a "maximum uncertainty" with respect to all other matters, and thus leaves a maximum possible freedom for our final decisions to be influenced by the subsequent sample data.

Suppose now that the value of x is determined by some random experiment; at each repetition of the experiment the final result is one of the values x_i. On the basis of the given information, what can we say about the frequencies with which the various x_i will occur? Let the experiment be repeated M times (we are particularly interested in the limit $M \to \infty$, because that is the situation referred to in the usual frequency theory of probability), and let every conceivable sequence of results be analyzed. Each trial could give, independently, any one of the results $\{x_1 \cdots x_n\}$, and so there are a priori n^M conceivable detailed outcomes. However, many of these will be incompatible with the given information about mean values of the $f_k(x)$. We will, of course, assume that the result of the random experiment agrees with this information (if it did not, then the given information was false and we are doing the wrong problem). In the M repetitions of the experiment, the result x_1 will be obtained m_1 times, x_2 will be obtained m_2 times, etc. Of course,

$$\sum_{i=1}^{n} m_i = M \qquad (20)$$

and if the specified mean values are in fact verified, we have the additional relations

$$\sum_{i=1}^{n} m_i f_k(x_i) = MF_k, \quad k = 1, \cdots, m. \qquad (21)$$

If $m < n - 1$, the constraints (20) and (21) are insufficient to determine the relative frequencies $f_i = m_i/M$. Nevertheless, we have strong grounds for preferring some choices of the f_i to others. For out of the original n^M conceivable results, how many would lead to a given set of sample numbers $\{m_1 \cdots m_n\}$? The answer is, of course, the multinomial coefficient

$$W = \frac{M!}{m_1! \cdots m_n!} = \frac{M!}{(Mf_1)! \cdots (Mf_n)!} \qquad (22)$$

and so the set of frequencies $\{f_1 \cdots f_n\}$ which can be realized in the greatest number of ways is the one which maximizes (22) subject to the constraints (20) and (21). We may, equally well, maximize any monotonic increasing function of W, in particular $M^{-1} \log W$, but as $M \to \infty$ we have immediately from the Stirling approximation,

$$M^{-1} \log W \to -\sum_{i=1}^{n} f_i \log f_i = H_f \qquad (23)$$

It is now evident that, in (20)–(23) we have formulated exactly the same mathematical problem as in (14)–(16), and that this identity will persist whether or not the constraints take the form of mean values. Given any testable prior information, the *probability* distribution which maximizes the entropy is numerically identical with the *frequency* distribution which can be realized in the greatest number of ways.

The maximum in W is, furthermore, enormously sharp; to investigate this, let $\{f_i\}$ be the set of frequencies which maximizes W and has entropy H_f, and $\{f_i'\}$ be any other set of frequencies which agrees with the constraints (20) and (21) and has entropy $H_f' < H_f$. The ratio [(number of ways in which $\{f_i\}$ could be realized)/(number of ways in which $\{f_i'\}$ could be realized)] grows asymptotically as

$$\frac{W}{W'} \sim e^{M(H_f - H_f')} \qquad (24)$$

and passes all bounds as $M \to \infty$. Therefore, the distribution predicted by maximum entropy can be realized experimentally in overwhelmingly more ways than can any other. This is the precise connection between maximum-entropy distributions and frequencies promised in Section III.

Now, does this property justify a prediction that the maximum-entropy distribution will, in fact, be observed in a real experiment? Clearly not, in the sense of deductive proof, for different people may have different amounts of information, which will lead them to set up different maximum-entropy distributions. Consider a specific case: Mr. A knows the mean values $\langle f_1(x) \rangle$, $\langle f_2(x) \rangle$; but Mr. B knows in addition $\langle f_3(x) \rangle$. Each sets up a maximum-entropy distribution conditional on his information, and since Mr. B's entropy H_B is maximized subject to one further constraint, we will have

$$H_B \leq H_A. \qquad (25)$$

We note two properties, easily verified from the foregoing equations If Mr. B's additional information is redundant (in the sense that it is only what Mr. A would have predicted from his distribution), then $\lambda_k = 0$, and the distribution is unchanged. In this case, and only in this case, we have equality in (25). Because of this property (which holds generally), it is never necessary when setting up a maximum-entropy problem to determine whether the different pieces of information used are independent, any redundant information will drop out of the equations automatically.

On the other hand, if the given pieces of information are logically contradictory (for example, if it turns out that $f_3(x) = f_1(x) + 2 f_2(x)$, but the given mean values fail to satisfy $\langle f_3 \rangle = \langle f_1 \rangle + 2 \langle f_2 \rangle$), then it will be found that (19) has no simultaneous solution with real λ_k In this case, the method of maximum entropy breaks down, as it should, giving us no distribution at all.

In general, Mr. B's extra information will be neither redundant nor contradictory, and so he will find a maximum-entropy distribution different from that of Mr. A. The inequality will then hold in (25), indicating that Mr. B's extra information was "useful" in further narrowing down the range of possibilities Suppose now that we start performing the random experiment with Mr. A and Mr. B watching. Since Mr. A predicts a mean value $\langle f_k \rangle$ different from the correct one known to Mr. B, it is clear that the experimental distribution cannot agree in all respects with Mr A's prediction. We cannot be sure in advance that it will agree with Mr. B's prediction either, for there may be still further constraints $f_4(x), f_5(x), \cdots,$ etc., operating in the experiment but unknown to Mr B.

However, the property demonstrated above does justify the following weaker statement of frequency correspondence. If the information incorporated into the maximum-entropy analysis includes all the constraints actually operative in the random experiment, then the distribution predicted by maximum entropy is overwhelmingly the most likely to be observed experimentally, because it can be realized in overwhelmingly the greatest number of ways.

Conversely, if the experiment fails to confirm the maximum-entropy prediction, and this disagreement persists on indefinite repetition of the experiment, then we will conclude that the physical mechanism of the experiment must contain additional constraints which were not taken into account in the maximum-entropy calculation. The observed deviations then provide a clue as to the nature of these new constraints. In this way, Mr. A can discover empirically that his information was incomplete.

Now the little scenario just described is an accurate model of just what did happen in one of the most important applications of statistical analysis, carried out by Gibbs. By the year 1900 it was known that in classical statistical mechanics, use of the canonical ensemble (which Gibbs derived as the maximum-entropy distribution over classical phase volume, based on a specified mean value of the energy) failed to predict thermodynamic properties (heat capacities, equations of state, equilibrium constants, etc.) correctly. Analysis of the data showed that the entropy of a real physical system was always less than the value predicted At that time, therefore, Gibbs was in just the position of Mr. A. in the scenario, and the conclusion was drawn that the microscopic laws of physics must involve an additional constraint not contained in the laws of classical mechanics.

In due course, the nature of this constraint was found; first by Planck in the case of radiation, then by Einstein and Debye for solids, and finally by Bohr for isolated atoms The constraint consisted in the discreteness of the possible energy values, thenceforth called energy levels. By 1927, the mathematical theory by which these could be calculated was developed nearly to its present form.

Thus it is an historical fact that the first clues indicating the need for the quantum theory, and indicating some necessary features of the new theory, were uncovered by a seemingly "unsuccessful" application of the principle of maximum entropy. We may expect that such things will happen again in the future, and this is the basis of the remark that the principle of maximum entropy is most useful to us in just those cases where it fails to predict the correct experimental facts.

Since the history of this development is not well known (a fuller account is given elsewhere [12]), the following brief remarks seem appropriate here. Gibbs [13] wrote his probability density in phase space in the form

$$w(q_1 \cdots q_n; p_1 \cdots p_n) = \exp[\eta(q_1 \cdots p_n)] \quad (26)$$

and called his function η the "index of probability of phase " He derived his canonical and grand canonical ensembles ([13], ch. 11) from constraints on average energy, and average energy and particle numbers, respectively, as ([13], p. 143) "the distribution in phase which without violating this condition gives the least value of the average index of probability of phase $\bar{\eta}$ . ." This is, of course, just what we would describe today as maximizing the entropy subject to constraints.

Unfortunately, Gibbs did not give any clear explanation, and we can only conjecture whether he possessed one, as to why this particular function is to be minimized on the average, in preference to all others. Consequently, his procedure appeared arbitrary to many, and for sixty years there was controversy over the validity and justification of Gibbs' method In spite of its enormous practical success when adapted to quantum statistics, few attempts were made to extend it beyond problems of thermal equilibrium.

It was not until the work of Shannon in our own time that the full significance and generality of Gibbs' method could be appreciated. Once we had Shannon's theorem establishing the uniqueness of entropy as an "information measure," it was clear that Gibbs' procedure was an example of a general method for inductive inference, whose applicability is in no way restricted to equilibrium thermodynamics or to physics.

V. Connection with Direct Probability Models

Another important conceptual point is brought out by comparing the frequency correspondence property of maximum-entropy distributions with those obtained from other theoretical models, for example, the standard model of Bernoulli trials We wish to show that this difference is far less than is often supposed

As noted previously, we are not entitled to assert that the distribution predicted by maximum entropy must be observed in a real experiment, we can say only that this distribution is by far the most likely to be observed, provided that the information used includes all the constraints actually operating in the experiment. This requirement, while sufficient, is not always necessary; from the fact that the predicted distribution has been observed, we cannot conclude that no further constraints exist beyond those taken into account We can conclude only that further constraints, if present, must be of such a nature that they do not affect the relative frequencies (although they might affect other observable things such as correlations)

Now what are we entitled to assert about frequency correspondence of probabilities calculated from the theory of Bernoulli trials? Clearly no probability calculation, whether based on maximum entropy or any other principle, can predict with certainty what the result of a real experiment must be, if the information available were sufficient to permit such a thing, we would have no need of probability theory at all.

In the theory of Bernoulli trials, we calculate the probability that we shall obtain r successes in n trials as

$$p(r|n) = \binom{n}{r} p^r (1-p)^{n-r} \qquad (27)$$

in which p is regarded as a given number in $0 < p < 1$. For finite n, there is no r in $0 \leq r \leq n$ which is absolutely excluded by this, and so the observed frequency of success $f \equiv r/n$ cannot be predicted with certainty Nevertheless, we infer from (27) that, as n becomes very large, the frequency $f = p$ becomes overwhelmingly the most likely to be observed, provided that the assumptions which went into the derivation of (27) (numerical value of p, independence of different trials) correctly describe the conditions operative in the real experiment.

Conversely, if the observed frequency fails to agree with the predictions (and this tendency persists on indefinite repetitions of the experiment), we will conclude that the physical mechanism of the experiment is different from that assumed in the calculation, and the nature of the observed deviation gives a clue as to what is wrong in our assumptions

On comparing these statements of probability–frequency correspondence, we see that there is virtually no difference in the logical situation between the principles of maximum entropy and of Bernoulli trials. In both cases, and in every other application of probability theory, the onus is on the user to make sure that all the information, which his common sense tells him is relevant to the problem, is actually incorporated into the equations. There is nothing in the mathematical theory which can determine whether this has been, in fact, accomplished; success can be known only a posteriori from agreement with experiment But in both cases, failure to confirm the predictions gives us an opportunity to learn more about the physical mechanism of the experiment

For these reasons, we are entitled to claim that probabilities calculated by maximum entropy have just as much and just as little correspondence with frequencies as those calculated from any other principle of probability theory

We can make this point still more strongly by exhibiting a mathematical connection between these two methods of calculation, showing that in many cases we can obtain identical results from use of either method. For this purpose, it is convenient to introduce some more of the vocabulary usually associated with information theory. Any random experiment may be regarded as a "message" transmitted to us by nature The "alphabet" consists of the set of all possible outcomes of a single trial; on each repetition of the experiment, nature transmits to us one more letter of the message In the case of Bernoulli trials, we are concerned with a message on a binary alphabet Define the "random variables"

$$y_i \equiv \begin{cases} 1, & \text{if the } i\text{th trial yields success} \\ 0, & \text{if the } i\text{th trial yields failure} \end{cases} \qquad (28)$$

On n repetitions of the experiment, we receive the message

$$M \equiv \{y_1 y_2 \cdots y_n\} \qquad (29)$$

and the total number of successes obtained is

$$r(M) \equiv \sum_{i=1}^{n} y_i \qquad (30)$$

From (27) we find that, for any n, the expected number of successes is

$$\langle r \rangle = np. \qquad (31)$$

Suppose now that we reverse our viewpoint, regard (31) as the primary given datum, and seek the probability of obtaining r successes in n trials by maximum entropy A full probability analysis of the experiment requires that we consider, not just the probabilities on the 2-point sample space of a single trial, but rather the probabilities

$$P_M \equiv p\{y_0 \cdots y_n\} \qquad (32)$$

on the 2^n-point sample space of all possible messages. The problem is then to find the distribution P_M which maximizes the entropy

$$H = -\sum_M P_M \log P_M \qquad (33)$$

subject to the constraint (31). The result is

$$P_M = \frac{1}{Z(\lambda)} e^{\lambda r(M)} \qquad (34)$$

with the partition function

$$Z(\lambda) = \sum_M e^{\lambda r(M)} = (e^\lambda + 1)^n. \quad (35)$$

The value of λ is determined, as always, by (19):

$$\langle r \rangle = \frac{\partial}{\partial \lambda} \log Z = n(e^{-\lambda} + 1)^{-1}$$

or

$$\lambda = \log \frac{\langle r \rangle}{n - \langle r \rangle} = \log \frac{p}{1-p}. \quad (36)$$

Using (35) and (36), the maximum-entropy distribution (34) reduces to

$$P_M = p^r(1-p)^{n-r}. \quad (37)$$

This is the probability of obtaining a specific message, with successes at specified trials. The probability of obtaining r successes regardless of the order then requires the additional binomial coefficient, and so we obtain precisely the result (27) of the Bernoulli model.

From a mathematical standpoint, therefore, it is immaterial whether we approach the theory of Bernoulli trials in the conventional way, or whether we regard it is as an example of maximum-entropy inference on a "higher manifold" than the sample space of a single trial, in which the only information available is the mean value (31).

In a similar way, many other of the so-called "direct probability" calculations may be regarded equally well as the result of applying the principle of maximum entropy on a higher manifold. If we had considered a random experiment with m possible outcomes at a single trial, we would be concerned with messages on an alphabet of m symbols $\{A_1 \cdots A_m\}$, and repetition of the preceding argument leads immediately to the usual multinomial distribution.

We may, perhaps, feel that this result gives us a new insight into the nature of Bernoulli trials. The "independence of different trials" evident already from (34) arises here from the fact that the given information consisted only of statements about individual trials and said nothing about mutual properties of different trials. The principle of maximum entropy thus tells us that, if no information is available concerning correlations between different trials, then we should not assume any such correlations to exist. To do so would reduce the entropy of the distribution P_M and thus reduce the range of variability of different messages below that permitted by the data, i.e., it would amount to introducing new arbitrary assumptions not warranted by the given information. The precise nature of this reduction is described by the asymptotic equipartition theorem [14]. The principle of maximum entropy is just the formal device which ensures that no such hidden arbitrary assumptions have been introduced, and so we are taking into account the full range of possibilities permitted by the information at hand.

If definite information concerning correlations is available, the maximum-entropy method readily digests this information. The usual theory of discrete stochastic processes can be derived by this same application of maximum entropy on a higher manifold, for particular kinds of information about correlations. To give only the simplest example, suppose that in our random experiment with m possible outcomes per trial, we are given information fixing the mean values not only of the "single-letter frequencies" $\langle f_i \rangle$, but also the "digram frequencies" $\langle f_{ij} \rangle$. The maximum-entropy distribution over messages will then take the form

$$P_M = \frac{1}{Z} \exp \left[\sum_i \lambda_i f_i(M) + \sum_{ij} \lambda_{ij} f_{ij}(M) \right] \quad (38)$$

where $n f_i(M)$ is the number of times the letter A_i occurs in the message M, and $(n-1) f_{ij}(M)$ is the number of times the digram $A_i A_j$ occurs in M. The partition function Z is determined by normalization of (38). Calculation of the λ_i and the λ_{ij} from (19) is no longer trivial, however, we find the problem to be exactly solvable [15] For messages of finite length, there are small "end effects," but in the limit of long messages the maximum-entropy distribution (38) reduces to the distribution of a Markov chain with transition probabilities $p_{ij} = \langle f_{ij} \rangle / \langle f_i \rangle$, in agreement with the results of conventional methods.

In a similar way, if the given information includes expectations of trigram frequencies $\langle f_{ijk} \rangle$, we obtain the distribution of a higher type stochastic process, in which the probability of the outcome A_i at any trial depends on the results of the previous two trials, etc.

To point out the possibility of deriving so much of conventional "direct probability" analysis from maximum entropy on a higher manifold is, of course, in no way to suggest that conventional methods of analysis be abandoned in favor of maximum entropy (although this would bring a higher degree of unity into the field), because in these applications the conventional methods usually lead to shorter calculations. The pragmatic usefulness of maximum entropy lies rather in the fact that it is readily applied in many problems (in particular, setting up prior probability assignments) where conventional methods do not apply.

It is, however, important to realize the possibility of deriving much of conventional probability theory from the principle of maximum entropy, firstly, because it shows that this principle fits in neatly and consistently with the other principles of probability theory. Secondly, we still see from time to time some doubts expressed as to the uniqueness of the expression $(- p \log p)$, it has even been asserted that the results of maximizing this quantity have no more significance than those obtained by maximizing any other convex function. In pointing out the correspondence with frequencies and the fact that many other standard results of probability theory follow from the maximum-entropy principle, we have given a constructive answer to such objections. Any alternative expression to $(- p \log p)$ must surely reproduce all these desirable properties before it could be taken seriously. It seems to the author impossible that any alternative quantity could do so, and likely that a rigorous proof of this could now be given.

VI. Continuous Distributions

Thus far we have considered the principle of maximum entropy only for the discrete case and have seen that if the distribution sought can be regarded as produced by a random experiment, there is a correspondence property between probability and frequency, and the results are consistent with other principles of probability theory. However, nothing in the mathematics requires that any random experiment be in fact performed or conceivable; and so we interpret the principle in the broadest sense which gives it the widest range of applicability, i.e., whether or not any random experiment is involved, the maximum-entropy distribution still represents the most "honest" description of our state of knowledge.

In such applications, the principle is easy to apply and leads to the kind of results we should want and expect. For example, in Jaynes [16] a sequence of problems of decision making under uncertainty (essentially, of inventory control) of a type which arises constantly in practice was analyzed. Here the state of nature was not the result of any random experiment; there was no sampling distribution and no sample. Thus it might be thought to be a "no data" decision problem, in the sense of Chernoff and Moses [17]. However, in successive stages of the sequence, there were available more and more pieces of prior information, and digesting them by maximum entropy led to a sequence of prior distributions in which the range of possibilities was successively narrowed down. They led to a sequence of decisions, each representing the rational one on the basis of the information available at that stage, which corresponded to intuitive common-sense judgments in the early stages where intuition was able to see the answer. It is difficult to see how this problem could have been treated at all without use of the principle of maximum entropy, or some other device that turns out in the end to be equivalent to it.

In several years of routine application of this principle in problems of physics and engineering, we have yet to find a case involving a discrete prior where it fails to produce a useful and intuitively reasonable result. To the best of the author's knowledge, no other general method for setting up discrete priors has been proposed. It appears, then, that the principle of maximum entropy may prove to be the final solution to the problem of assigning discrete priors.

Use of this principle in setting up continuous prior distributions, however, requires considerably more analysis because at first glance the results appear to depend on the choice of parameters. We do not refer here to the well-known fact that the quantity

$$H' = -\int p(x) \log p(x)\, dx \quad (39)$$

lacks invariance under a change of variables $x \to y(x)$, for (39) is not the result of any derivation, and it turns out not to be the correct information measure for a continuous distribution. Shannon's theorem establishing (14) as an information measure goes through only for discrete distributions; but to find the corresponding expression in the continuous case we can (in the absence of any more direct argument) pass to the limit from a discrete distribution. As shown previously [7], this leads instead to the quantity

$$H_c = -\int p(x) \log[p(x)/m(x)]\, dx \quad (40)$$

where $m(x)$ is an "invariant measure" function, proportional to the limiting density of discrete points. (In all applications so far studied, $m(x)$ is a well-behaved continuous function, and so we continue to use the notation of Riemann integrals; we call $m(x)$ a "measure" only to suggest the appropriate generalization, readily supplied if a practical problem should ever require it.) Since $p(x)$ and $m(x)$ transform in the same way under a change of variables, H_c is invariant. We examine the form of maximum-entropy inference based on this information measure, in which we may regard x as being either a one-dimensional or multidimensional parameter.

We seek a probability density $p(x)$ which is to be normalized:

$$\int p(x)\, dx = 1 \quad (41)$$

(we understand the range of integration to be the full parameter space); and we have information fixing the mean values of m different functions $f_k(x)$:

$$F_k = \int p(x) f_k(x)\, dx, \quad k = 1, 2, \cdots, m \quad (42)$$

where the F_k are the given numerical values. Subject to these constraints, we are to maximize (40). The solution is again elementary:

$$p(x) = Z^{-1} m(x) \exp[\lambda_1 f_1(x) + \cdots + \lambda_m f_m(x)] \quad (43)$$

with the partition function

$$Z(\lambda_1, \cdots, \lambda_m) \equiv \int m(x) \exp[\lambda_1 f_1(x) + \cdots + \lambda_m f_m(x)]\, dx \quad (44)$$

and the Lagrange multipliers λ_k are determined once again by (19). Our "best" estimate (by quadratic loss function) of any other quantity $q(x)$ is then

$$\langle q \rangle = \int q(x)\, p(x)\, dx. \quad (45)$$

It is evident from these equations that when we use (40) rather than (39) as our information measure not only our final conclusions (45), but also the partition function and Lagrange multipliers are all invariant under a change of parameters $x \to y(x)$. In applications, these quantities acquire definite physical meanings.

There remains, however, a practical difficulty. If the parameter space is not the result of any obvious limiting process, what determines the proper measure $m(x)$? The conclusions, evidently, will depend on which measure we adopt. This is the shortcoming from which the maximum-entropy principle has suffered heretofore, and which must be cleared up before we can regard it as a full solution to the prior probability problem.

Let us note the intuitive meaning of this measure. Consider the one-dimensional case, and suppose it is known that $a < x < b$ but we have no other prior information. Then there are no Lagrange multipliers λ_k, and (43) reduces to

$$p(x) = \left[\int_a^b m(x)\, dx \right]^{-1} m(x), \quad a < x < b. \quad (46)$$

Except for a constant factor, the measure $m(x)$ is also the prior distribution describing "complete ignorance" of x. The ambiguity is, therefore, just the ancient one which has always plagued Bayesian statistics; how do we find the prior representing "complete ignorance?" Once this problem is solved, the maximum-entropy principle will lead to a definite, parameter-independent method for setting up prior distributions based on any testable prior information. Since this problem has been the subject of so much discussion and controversy for 200 years, we wish to state what appears to us a constructive attitude toward it.

To reject the question, as some have done, on the grounds that the state of complete ignorance does not "exist" would be just as absurd as to reject Euclidean geometry on the grounds that a physical point does not exist. In the study of inductive inference, the notion of complete ignorance intrudes itself into the theory just as naturally and inevitably as the concept of zero in arithmetic.

If one rejects the consideration of complete ignorance on the grounds that the notion is vague and ill-defined, the reply is that the notion cannot be evaded in any full theory of inference. So if it is still ill-defined, then a major and immediate objective must be to find a precise definition which will agree with intuitive requirements and be of constructive use in a mathematical theory.

With this in mind, let us survey some previous thought on the problem Bayes suggested, in one particular case, that we express complete ignorance by assigning a uniform prior probability density; and the domain of useful application of this rule is certainly not zero, for Laplace was led to some of the most important discoveries in celestial mechanics by using it in analysis of astronomical data. However, Bayes' rule has the obvious difficulty that it is not invariant under a change of parameters, and there seems to be no criterion telling us which parameterization to use. (We note in passing that the notions of an unbiased estimator, an efficient estimator, and a shortest confidence interval are all subject to just the same ambiguity with equally serious consequences, and so orthodox statistics cannot claim to have solved this problem any better than Bayes did.)

Jeffreys [18], [19] suggested that we assign a prior $d\sigma/\sigma$ to a continuous parameter σ known to be positive, on the grounds that we are then saying the same thing whether we use the parameter σ or σ^m. Such a desideratum is surely a step in the right direction; however, it cannot be extended to more general parameter changes. We do not want (and obviously cannot have) invariance of the form of the prior under all parameter changes; what we want is invariance of content, but the rules of probability theory already determine how the prior must transform, under any parameter changes, so as to achieve this.

The real problem, therefore, must be stated rather differently; we suggest that the proper question to ask is: "For which choice of parameters does a given form such as that of Bayes or Jeffreys apply?" Our parameter spaces seem to have a mollusk-like quality that prevents us from answering this, unless we can find a new principle that gives them a property of "rigidity."

Stated in this way, we recognize that problems of just this type have already appeared and have been solved in other branches of mathematics In Riemannian geometry and general relativity theory, we allow arbitrary continuous coordinate transformations; yet the property of rigidity is maintained by the concept of the invariant line element, which enables us to make statements of definite geometrical and physical meaning independently of the choice of coordinates. In the theory of continuous groups, the group parameter space had just this mollusk-like quality until the introduction of invariant group measure by Hurwitz [20] and Haar [21], [22]. We seek to do something very similar to this for the parameter spaces of statistics.

The idea of utilizing groups of transformations in problems related to this was discussed by Poincaré [23] and more recently by Fraser [24], Hartigan [25], and Stone [26]. In the following section we give three examples of a different group theoretical method of reasoning developed largely by Weyl and Wigner [20], which has met with great success in physical problems and seems uniquely adapted to our problem.

VII TRANSFORMATION GROUPS—EXAMPLES

The method of reasoning is best illustrated by a simple example, which also happens to be one of the most important in practice We sample from a continuous two-parameter distribution

$$p(dx|\,\mu,\,\sigma) = h\left(\frac{x-\mu}{\sigma}\right)\frac{dx}{\sigma} \quad (47)$$

where $h(y)$ is a non-negative and normalized function, and consider the following problem.

Problem 1: Given a sample $\{x_1 \cdots x_n\}$, estimate μ and σ. The problem is indeterminate, both mathematically and conceptually, until we introduce a definite prior distribution

$$f(\mu,\,\sigma)\,d\mu\,d\sigma \quad (48)$$

but if we merely specify "complete initial ignorance," this does not seem to tell us which function $f(\mu,\,\sigma)$ to use.

Now what do we mean by the statement that we are completely ignorant of μ and σ, except for the knowledge that μ is a location parameter and σ a scale parameter? If we know the sampling distribution (47), we can hardly be ignorant of at least that much To answer this, we might reason as follows. If a change of scale can make the problem appear in any way different to us, then we were not completely ignorant; we must have had some kind of prior knowledge about the absolute scale of the problem. Likewise, if a shift of location can make the problem appear in any way different, then it must be that we had some kind of prior knowledge about location. In other words, complete ignorance of a location and scale parameter is a state of knowledge such that a change of scale and a shift of location does not change that state of knowledge. Suppose, therefore, that we carry out a change of variables $(x, \mu, \sigma) \rightarrow (x', \mu', \sigma')$ according to

$$\mu' = \mu + b$$
$$\sigma' = a\sigma \qquad (49)$$
$$x' - \mu' = a(x - \mu)$$

where $0 < a < \infty$, $-\infty < b < \infty$. The distribution (47) expressed in the new variables is unchanged:

$$p(dx'|\mu', \sigma') = h\left(\frac{x' - \mu'}{\sigma'}\right) \frac{dx'}{\sigma'} \qquad (50)$$

but the prior distribution is changed to $g(\mu', \sigma')\, d\mu'\, d\sigma'$ where from the Jacobian of the transformation (49)

$$g(\mu', \sigma') = a^{-1} f(\mu, \sigma). \qquad (51)$$

Now let us consider a second problem.

Problem 2: Given a sample $\{x_1' \cdots x_n'\}$, estimate μ' and σ'. If we are completely ignorant in the preceding sense, then we must consider Problems 1 and 2 as entirely equivalent, for they have identical sampling distributions and our state of prior knowledge about μ' and σ' in Problem 2 is exactly the same as for μ and σ in Problem 1. But our basic desideratum of consistency demands that in two problems where we have the same prior information, we should assign the same prior probabilities. Therefore, f and g must be the same function:

$$f(\mu, \sigma) = g(\mu, \sigma) \qquad (52)$$

whatever the values of (a,b). But the form of the prior is now uniquely determined, for combining (49), (51), and (52), we see that $f(\mu, \sigma)$ must satisfy the functional equation

$$f(\mu, \sigma) = a\, f(\mu + b, a\sigma) \qquad (53)$$

whose general solution is

$$f(\mu, \sigma) = \frac{\text{(const)}}{\sigma} \qquad (54)$$

which is the Jeffreys rule.

As another example, not very different mathematically but differently verbalized, consider a Poisson process. The probability that exactly n events will occur in a time interval t is

$$p(n|\lambda, t) = e^{-\lambda t} \frac{(\lambda t)^n}{n!} \qquad (55)$$

and by observing the number of events we wish to estimate the rate constant λ. We are initially completely ignorant of λ except for the knowledge that it is a rate constant of physical dimensions (seconds)$^{-1}$, i.e., we are completely ignorant of the absolute time scale of the process.

Suppose, then, that two observers, Mr. X and Mr. X', whose watches run at different rates so their measurements of a given interval are related by $t = qt'$, conduct this experiment. Since they are observing the same physical experiment, their rate constants must be related by $\lambda' t' = \lambda t$, or $\lambda' = q\lambda$. They assign prior distributions

$$p(d\lambda | X) = f(\lambda)\, d\lambda \qquad (56)$$
$$p(d\lambda' | X') = g(\lambda')\, d\lambda' \qquad (57)$$

and if these are mutually consistent (i.e., they have the same content), it must be that $f(\lambda)\, d\lambda = g(\lambda')\, d\lambda'$, or $f(\lambda) = q\, g(\lambda')$. But if Mr. X and Mr. X' are both completely ignorant, then they are in the same state of knowledge, and so f and g must be the same function: $f(\lambda) = g(\lambda)$. Combining these relations gives the functional equation $f(\lambda) = q f(q\lambda)$ or

$$p(d\lambda | X) \sim \lambda^{-1}\, d\lambda. \qquad (58)$$

To use any other prior than this will have the consequence that a change in the time scale will lead to a change in the form of the prior, which would imply a different state of prior knowledge, but if we are completely ignorant of the time scale, then all time scales should appear equivalent.

As a third and less trivial example, where intuition did not anticipate the result, consider Bernoulli trials with an unknown probability of success. Here the probability of success is itself the parameter θ to be estimated. Given θ, the probability that we shall observe r successes in n trials is

$$p(r|n, \theta) = \binom{n}{r} \theta^r (1 - \theta)^{n-r} \qquad (59)$$

and again the question is: What prior distribution $f(\theta)\, d\theta$ describes "complete initial ignorance" of θ?

In discussing this problem, Laplace followed the example of Bayes and answered the question with that famous sentence: "When the probability of a simple event is unknown, we may suppose all values between 0 and 1 as equally likely." In other words, Bayes and Laplace used the uniform prior $f_B(\theta) = 1$. However, Jeffreys [18] and Carnap [27] have noted that the resulting rule of succession does not seem to correspond well with the inductive reasoning which we all carry out intuitively. Jeffreys suggested that $f(\theta)$ ought to give greater weight to the endpoints $\theta = 0,1$ if the theory is to account for the kind of inferences made by a scientist.

For example, in a chemical laboratory we find a jar containing an unknown and unlabeled compound. We are at first completely ignorant as to whether a small sample of this compound will dissolve in water or not. But having observed that one small sample does dissolve, we infer immediately that all samples of this compound are water soluble, and although this conclusion does not carry quite the force of deductive proof, we feel strongly that the inference was justified. Yet the Bayes–Laplace rule leads to a negligibly small probability of this being true, and yields only a probability of 2/3 that the next sample tested will dissolve.

Now let us examine this problem from the standpoint of transformation groups. There is a conceptual difficulty here, since $f(\theta)\, d\theta$ is a "probability of a probability." However, it can be removed by carrying the notion of a split personality to extremes; instead of supposing that $f(\theta)$ describes the state of knowledge of any one person, imagine that we have a large population of individuals who hold varying beliefs about the probability of success, and that $f(\theta)$ describes the distribution of their beliefs. Is it possible that, although each individual holds a definite opinion, the population as a whole is completely ignorant of θ? What distribution $f(\theta)$ describes a population in a state of total confusion on the issue?

Since we are concerned with a consistent extension of probability theory, we must suppose that each individual reasons according to the mathematical rules (Bayes' theorem, etc.) of probability theory. The reason they hold different beliefs is, therefore, that they have been given different and conflicting information; one man has read the editorials of the St. Louis Post-Dispatch, another the Los Angeles Times, one has read the Daily Worker, another the National Review, etc., and nothing in probability theory tells one to doubt the truth of what he has been told in the statement of a problem.

Now suppose that, before the experiment is performed, one more definite piece of evidence E is given simultaneously to all of them. Each individual will change his state of belief according to Bayes' theorem. Mr. X, who had previously held the probability of success to be

$$\theta = p(S|X) \qquad (60)$$

will change it to

$$\theta' = p(S|E, X) = \frac{p(S|X)\, p(E|SX)}{p(E|SX)\, p(S|X) + p(E|FX)\, p(F|X)} \qquad (61)$$

where $p(F|X) = 1 - p(S|X)$ is his prior belief in probability of failure. This new evidence thus generates a mapping of the parameter space $0 \leq \theta \leq 1$ onto itself, given from (61) by

$$\theta' = \frac{a\theta}{1 - \theta + a\theta} \qquad (62)$$

where

$$a = \frac{p(E|SX)}{p(E|FX)}. \qquad (63)$$

If the population as a whole can learn nothing from this new evidence, then it would seem reasonable to say that the population has been reduced, by conflicting propaganda, to a state of total confusion on the issue. We therefore define the state of "total confusion" or "complete ignorance" by the condition that after the transformation (62), the number of individuals who hold beliefs in any given range $\theta_1 < \theta < \theta_2$ is the same as before.

The mathematical problem is again straightforward. The original distribution of beliefs $f(\theta)$ is shifted by the transformation (62) to a new distribution $g(\theta')$ with

$$f(\theta)\, d\theta = g(\theta')\, d\theta' \qquad (64)$$

and, if the population as a whole learned nothing, then f and g must be the same function:

$$f(\theta) = g(\theta). \qquad (65)$$

Combining (62), (64), and (65), we find that $f(\theta)$ must satisfy the functional equation

$$a f\left(\frac{a\theta}{1 - \theta + a\theta}\right) = (1 - \theta + a\theta)^2 f(\theta). \qquad (66)$$

This may be solved directly by eliminating the a between (62) and (66) or, in the more usual manner, by differentiating with respect to a and setting $a = 1$. This leads to the differential equation

$$\theta(1 - \theta)\, f'(\theta) = (2\theta - 1)\, f(\theta) \qquad (67)$$

whose solution is

$$f(\theta) = \frac{(\text{const})}{\theta(1 - \theta)} \qquad (68)$$

which has the qualitative property anticipated by Jeffreys. Now that the imaginary population of individuals has served its purpose of revealing the transformation group (62) of the problem, let them coalesce again into a single mind (that of the statistician who wishes to estimate θ), and let us examine the consequences of using (68) as our prior distribution.

If we have observed r successes in n trials, then from (59) and (68) the posterior distribution of θ is (provided that $r \geq 1$, $n - r \geq 1$)

$$p(d\theta|r, n) = \frac{(n - 1)!}{(r - 1)!\,(n - r - 1)!}\, \theta^{r-1}(1 - \theta)^{n-r-1}\, d\theta. \qquad (69)$$

This distribution has expectation value and variance

$$\langle \theta \rangle = \frac{r}{n} = f \qquad (70)$$

$$\sigma^2 = \frac{f(1 - f)}{n + 1}. \qquad (71)$$

Thus the "best" estimate of the *probability* of success, by the criterion of quadratic loss function, is just equal to the observed *frequency* of success f; and this is also equal to the probability of success at the next trial, in agreement with the intuition of everybody who has studied Bernoulli trials On the other hand, the Bayes–Laplace uniform prior would lead instead to the mean value $\langle\theta\rangle_B = (r + 1)/(n + 2)$ of the rule of succession, which has always seemed a bit peculiar.

For interval estimation, numerical analysis shows that the conclusions drawn from (69) are for all practical purposes the same as those based on confidence intervals [i e., the shortest 90-percent confidence interval for θ is nearly equal to the shortest 90-percent posterior probability interval determined from (69)]. If $r \gg 1$ and $(n - r) \gg 1$, the normal approximation to (71) will be valid, and the $100P$ percent posterior probability interval is simply $(f \pm q\sigma)$, where q is the $(1 + P)/2$ percentile of the normal distribution, for the 90-, 95-, and 99-percent levels, $q = 1.645$, $1\,960$, and 2.576, respectively. Under conditions where this normal approximation is valid, differences between this result and the exact confidence intervals are generally less than the differences between various published confidence interval tables, which have been calculated from different approximation schemes.

If $r = (n - r) = 1$, (69) reduces to $p(d\theta|r, n) = d\theta$, the uniform distribution which Bayes and Laplace took as their prior Therefore, we can now interpret the Bayes–Laplace prior as describing not a state of complete ignorance, but the state of knowledge in which we have observed one success and one failure. It thus appears that the Bayes–Laplace choice will be the appropriate prior if the prior information assures us that it is physically possible for the experiment to yield either a success or a failure, while the distribution of complete ignorance (68) describes a "pre-prior" state of knowledge in which we are not even sure of that.

If $r = 0$ or $r = n$, the derivation of (69) breaks down and the posterior distribution remains unnormalizable, proportional to $\theta^{-1}(1 - \theta)^{n-1}$ or $\theta^{n-1}(1 - \theta)^{-1}$, respectively. The weight is concentrated overwhelmingly on the values $\theta = 0$ or $\theta = 1$ The prior (68) thus accounts for the kind of inductive inferences noted in the case of the chemical, which we all make intuitively However, once we have seen at least one success and one failure, then we know that the experiment is a true binary one, in the sense of physical possibility, and from that point on all posterior distributions (69) remain normalized, permitting definite inferences about θ.

The transformation group method therefore yields a prior which appears to meet the common objections raised against the Laplace rule of succession, but we also see that whether (68) or the Bayes–Laplace prior is appropriate depends on the exact prior information available.

To summarize the above results· if we merely specify complete initial ignorance, we cannot hope to obtain any definite prior distribution, because such a statement is too vague to define any mathematically well-posed problem We are defining what we mean by complete ignorance far more precisely if we can specify a set of operations which we recognize as transforming the problem into an equivalent one, and the desideratum of consistency then places nontrivial restrictions on the form of the prior.

VIII TRANSFORMATION GROUPS—DISCUSSION

Further analysis shows that, if the number of independent parameters in the transformation group is equal to the number of parameters in the statistical problem, the "fundamental domain" of the group [20] reduces to a point, and the form of the prior is uniquely determined; thus specification of such a transformation group is an exhaustive description of a state of knowledge.

If the number of parameters in the transformation group is less than the number of statistical parameters, the fundamental domain is of higher dimensionality, and the prior will be only partially determined For example, if in the group (49) we had specified only the change of scale operation and not the shift of location, repetition of the argument would lead to the prior $f(\mu, \sigma) = \sigma^{-1} k(\mu)$, where $k(\mu)$ is an arbitrary function.

It is also readily verified that the transformation group method is consistent with the desideratum of invariance under parameter changes mentioned in Section VI, i.e., that while the form of the prior cannot be invariant under all parameter changes, its content should be. If the transformation group (49) had been specified in terms of some other choice of parameters (α, β), the form of the transformation equations and functional equations would, of course, be different, but the prior to which they would lead in the (α, β) space would be just the one that we obtain by solving the problem in the (μ, σ) space and transforming the result to the (α, β) space by the usual Jacobian rule

The method of reasoning illustrated here is somewhat reminiscent of Laplace's "principle of indifference." However, we are concerned here with indifference between problems, rather than indifference between events. The distinction is essential, for indifference between events is a matter of intuitive judgment on which our intuition often fails even when there is some obvious geometrical symmetry (as Bertrand's paradox shows). However, if a problem is formulated in a sufficiently careful way, indifference between problems is a matter that is determined by the statement of a problem, independently of our intuition, none of the preceding transformation groups corresponded to any particularly obvious geometrical symmetry.

More generally, if we approach a problem with the charitable presumption that it has a definite solution, then every circumstance left unspecified in the statement of the problem defines an invariance property (i e , a transformation to an equivalent problem) which that solution must have. Recognition of this leads tó a resolution of the Bertrand paradox, here we draw straight lines "at random" intersecting a circle and ask for the distribution of chord lengths. But the statement of the problem does not specify the exact position of the circle, therefore, if there is any definite solution, it must not depend on this circumstance The condition that the solution be invariant under infinitesimal displacements of the circle relative to the random straight lines uniquely determines the solution

In such problems, furthermore, the transformation group method is found to have a frequency correspondence property rather like that of the maximum-entropy principle. If (as in the Bertrand problem) the distribution sought can be regarded as the result of a random experiment, then the distribution predicted by invariance under the transformation group is by far the most likely to be observed experimentally, because it requires by far the least "skill," consistently to produce any other would require a "microscopic" degree of control over the exact conditions of the experiment Proof of the statements in the last two paragraphs will be deferred to a later article

The transformation group derivation enables us to see the Jeffreys prior probability rule in a new light It has, perhaps, always been obvious that the real justification of the Jeffreys rule cannot lie merely in the fact that the parameter is positive As a simple example, suppose that μ is known to be a location parameter, then both intuition and the preceding analysis agree that a uniform prior density is the proper way to express complete ignorance of μ. The relation $\mu = \theta - \theta^{-1}$ defines a 1.1 mapping of the region $(-\infty < \mu < \infty)$ onto the region $(0 < \theta < \infty)$, but the Jeffreys rule cannot apply to the parameter θ, consistency demanding that its prior density be taken proportional to $d\mu = (1 + \theta^{-2}) d\theta$ It appears that the fundamental justification of the Jeffreys rule is not merely that a parameter is positive, but that it is a *scale parameter*.

The fact that the distributions representing complete ignorance found by transformation groups cannot be normalized may be interpreted in two ways One can say that it arises simply from the fact that our formulation of the notion of complete ignorance was an idealization that does not strictly apply in any realistic problem A shift of location from a point in St Louis to a point in the Andromeda nebula, or a change of scale from the size of an atom to the size of our galaxy, does not transform any problem of earthly concern into a completely equivalent one. In practice we will always have some kind of prior knowledge about location and scale, and in consequence the group parameters (a, b) cannot vary over a truly infinite range Therefore, the transformations (49) do not, strictly speaking, form a group However, over the range which does express our prior ignorance, the above kind of arguments still apply Within this range, the functional equations and the resulting form of the prior must still hold

However, our discussion of maximum entropy shows a more constructive way of looking at this. Finding the distribution representing complete ignorance is only the first step in finding the prior for any realistic problem The pre-prior distribution resulting from a transformation group does not strictly represent any realistic state of knowledge, but it does define the invariant measure for our parameter space, without which the problem of finding a realistic prior by maximum entropy is mathematically indeterminate.

IX. Conclusion

The analysis given here provides no reason to think that specifying a transformation group is the only way in which complete ignorance may be precisely defined, or that the principle of maximum entropy is the only way of converting testable information into a prior distribution. Furthermore, the procedures described here are not necessarily applicable in all problems, and so it remains an open question whether other approaches may be as good or better. However, before we would be in a position to make any comparative judgments, it would be necessary that some definite alternative procedure be suggested.

At present, lacking this, one can only point out some properties of the methods here suggested. The class of problems in which they can be applied is that in which 1) the prior information is testable, and 2) in the case of a continuous parameter space, the statement of the problem suggests some definite transformation group which establishes the invariant measure We note that satisfying these conditions is, to a large extent, simply a matter of formulating the problem more completely than is usually done.

If these conditions are met, then we have the means for incorporating prior information into our problem, which is independent of our choice of parameters and is completely impersonal, allowing no arbitrary choice on the part of the user. Few orthodox procedures and, to the best of the author's knowledge, no other Bayesian procedures, enjoy this complete objectivity. Thus while the above criticisms are undoubtedly valid, it seems apparent that this analysis does constitute an advance in the precision with which we are able to formulate statistical problems, as well as an extension of the class of problems in which statistical methods can be used. The fact that this has proved possible gives hope that further work along these lines—in particular, directed toward learning how to formulate problems so that condition 2) is satisfied—may yet lead to the final solution of this ancient but vital puzzle, and thus achieve full objectivity for Bayesian methods.

Bibliography

[1] M G Kendall and A Stuart, *The Advanced Theory of Statistics*, vol 2. New York Hafner, 1961.
[2] A Wald, *Statistical Decision Functions*. New York: Wiley, 1950
[3] L J Savage, *The Foundations of Statistics* New York. Wiley, 1954
[4] ———, *The Foundations of Statistical Inference* London. Methuen, 1962.
[5] E L. Lehmann, *Testing Statistical Hypotheses*. New York Wiley, 1959, p 62
[6] E T Jaynes, "Information theory and statistical mechanics," pt I, *Phys Rev.*, vol 106, pp 620–630, March 1957; pt. II, *ibid*, vol 108, pp 171–191, October 1957.
[7] ———, "Information theory and statistical mechanics," in *Statistical Physics*, vol 3, K W. Ford, Ed New York. W. A. Benjamin, Inc., 1963, ch 4, pp 182–218
[8] I J. Good, "Maximum entropy for hypothesis formulation," *Ann. Math Statist*, vol. 34, pp 911–930, 1963.
[9] S Kullback, *Information Theory and Statistics* New York: Wiley, 1959.
[10] E H Wichmann, "Density matrices arising from incomplete measurements," *J. Math. Phys*, vol 4, pp 884–897, July 1963.
[11] M. Dutta, "On maximum entropy estimation," *Sankhya*, ser. A, vol 28, pt 4, pp 319–328, July 1966.

[12] E T Jaynes, "Foundations of probability theory and statistical mechanics," in *Delaware Seminar in the Foundations of Physics*, M. Bunge, Ed. Berlin: Springer, 1967, ch. 6
[13] J. W Gibbs, *Elementary Principles in Statistical Mechanics*. New Haven, Conn. Yale University Press, 1902.
[14] A. Feinstein, *Foundations of Information Theory* New York. McGraw-Hill, 1958
[15] E T Jaynes, unpublished, 1960 Copies available on request
[16] ——, "New engineering applications of information theory," in *Proc. 1st Symp on Engineering Applications of Random Function Theory and Probability*, J L Bogdanoff and F Kozin, Eds. New York Wiley, 1963, pp 163–203
[17] H Chernoff and L. E Moses, *Elementary Decision Theory* New York Wiley, 1959
[18] H Jeffreys, *Theory of Probability* Oxford Clarendon Press, 1939.
[19] ——, *Scientific Inference* London Cambridge University Press, 1957.

[20] E. P. Wigner, *Group Theory*. New York Academic Press, 1959
[21] A Haar, "Der Massbegriff in der Theorie der kontinuierlichen Gruppen," *Ann. Math.*, vol 34, pp 147–169, 1933
[22] L. Pontryagin, *Topological Groups* Princeton, N J Princeton University Press, 1946, ch 4.
[23] H Poincaré, *Calcul des Probabilités* Paris Gauthier-Villars, 1912
[24] D A S Fraser, "On fiducial inference," *Ann. Math. Statist.*, vol 32, pp. 661–676, 1966.
[25] J Hartigan, "Invariant prior distributions," *Ann Math Statist*, vol 35, pp 836–845, 1964.
[26] M Stone, "Right Haar measure for convergence in probability to quasi-posterior distributions," *Ann Math Statist*, vol 36, pp 449–453, 1965
[27] R Carnap, *The Continuum of Inductive Methods* Chicago. University of Chicago Press, 1952

8. THE WELL-POSED PROBLEM (1973)

The idea of assigning probabilities by the principle of group invariance developed in a way parallel to that of the principle of maximum entropy. In both cases the original motivation (at least for me) was that the principle expressed in mathematical terms what seemed intuitively the 'most honest' description of a state of knowledge. In neither case was there any connection with frequencies – or indeed any reference to a repetitive 'random experiment'.

But in the case of maximum entropy a frequency connection appeared at once. *Maxent* on the space S of any experiment corresponds closely to a combinatorial theorem on the extension space S^n of n repetitions of the experiment, the probabilities on S which have maximum entropy subject to any constraints being numerically equal to the frequencies on S^n which could be realized in the greatest number of ways subject to the same constraints.

We could state the situation as follows. The *Maxent* probability distribution is at the same time the frequency distribution most likely to be realized, in the sense that to produce appreciably different frequencies would require additional physical constraints in the experiment, beyond those taken into account in the *Maxent* calculation.

On further meditation it was realized that probabilities determined by group invariance on a space S also have a frequency connection on the extension space S^n. They are numerically equal to the frequencies that require the least 'skill' to produce. That is, to produce frequencies appreciably different from the invariant ones would require some degree of control over the initial conditions of the experiment.

This statement, almost a trivial tautology on first reading, becomes nontrivial when we realize that in many cases an extremely large amount of skill, or control, would be required; and so in those cases group invariance already suffices to make quite reliable predictions of frequencies. Of course, those who spin roulette wheels or drive golf balls are well aware of this. But from the standpoint of principle it seemed important to show that, even though probabilities are not defined as frequencies, they often acquire frequency connections, which may be of several different kinds.

At the purely pragmatic level, there also appears to be something of value here. From the way in which group invariance arguments were able to deal with the Bertrand problem, one may expect that there are many problems in which group invariance can lead us to useful predictions of observable facts.

For readers who may wish to try their hand at this kind of reasoning, here is the String Problem, calling out for solution: A perfectly flexible string of length L is tossed, very unskillfully, onto the floor. What is the probability distribution of the distance between its ends? Please do not cheat by doing the experiment first.

The Well-Posed Problem

E. T. Jaynes

Department of Physics, Washington University, St. Louis, Missouri

Received June 1, 1973

Many statistical problems, including some of the most important for physical applications, have long been regarded as underdetermined from the standpoint of a strict frequency definition of probability; yet they may appear wellposed or even overdetermined by the principles of maximum entropy and transformation groups. Furthermore, the distributions found by these methods turn out to have a definite frequency correspondence; the distribution obtained by invariance under a transformation group is by far the most likely to be observed experimentally, in the sense that it requires by far the least "skill." These properties are illustrated by analyzing the famous Bertrand paradox. On the viewpoint advocated here, Bertrand's problem turns out to be well posed after all, and the unique solution has been verified experimentally. We conclude that probability theory has a wider range of useful applications than would be supposed from the standpoint of the usual frequency definitions.

1. BACKGROUND

In a previous article[1] we discussed two formal principles—maximum entropy and transformation groups—that are available for setting up probability distributions in the absence of frequency data. The resulting distributions may be used as prior distributions in Bayesian inference; or they may be used directly for certain physical predictions. The exact sense in which distributions found by maximum entropy correspond to observable frequencies was given in the previous article; here we demonstrate a similar correspondence property for distributions obtained from transformation groups, using as our main example the famous paradox of Bertrand.

Bertrand's problem[2] was stated originally in terms of drawing a straight line "at random" intersecting a circle. It will be helpful to think of this in a more concrete way; presumably, we do no violence to the problem (i.e., it is still just as "random") if we suppose that we are tossing straws onto the circle, without specifying how they are tossed. We therefore formulate the problem as follows.

A long straw is tossed at random onto a circle; given that it falls so that it intersects the circle, what is the probability that the chord thus defined is longer than a side of the inscribed equilateral triangle? Since Bertrand proposed it in 1889 this problem has been cited to generations of students to demonstrate that Laplace's "principle of indifference" contains logical inconsistencies. For, there appear to be many ways of defining "equally possible" situations, and they lead to different results. Three of these are: Assign uniform probability density to (A) the linear distance between centers of chord and circle, (B) angles of intersections of the chord on the circumference, (C) the center of the chord over the interior area of the circle. These assignments lead to the results $p_A = 1/2$, $p_B = 1/3$, and $p_C = 1/4$, respectively.

Which solution is correct? Of the ten authors cited,[2-12] with short quotations, in the appendix only Borel is willing to express a definite preference, although he does not support it by any proof. Von Mises takes the opposite extreme, declaring that such problems (including the similar Buffon needle problem) do not belong to the field of probability theory at all. The others, including Bertrand, take the intermediate position of saying simply that the problem has no definite solution because it is ill posed, the phrase "at random" being undefined.

In works on probability theory this state of affairs has been interpreted, almost universally, as showing that the principle of indifference must be totally rejected. Usually, there is the further conclusion that the only valid basis for assigning probabilities is frequency in some random experiment. It would appear, then, that the only way of answering Bertrand's question is to perform the experiment.

But do we really believe that it is beyond our power to predict by "pure thought" the result of such a simple experiment? The point at issue is far more important than merely resolving a geometric puzzle; for, as discussed further in Section 7, applications of probability theory to physical experiments usually lead to problems of just this type; i.e., they appear at first to be underdetermined, allowing many different solutions with nothing to choose among them. For example, given the average particle density and total energy of a gas, predict its viscosity. The answer, evidently, depends on the exact spatial and velocity distributions of the molecules (in fact, it depends critically on position–velocity correlations), and nothing in the given data seems to tell us which distribution to assume. Yet physicists *have* made definite choices, guided by the principle of indifference, and they *have* led us to correct and nontrivial predictions of viscosity and many other physical phenomena.

Thus, while in some problems the principle of indifference has led us to paradoxes, in others it has produced some of the most important and successful applications of probability theory. To reject the principle without having anything better to put in its place would lead to consequences so unacceptable that for many years even those who profess the most faithful adherence to the strict frequency definition of probability have managed to overlook these logical difficulties in order to preserve some very useful solutions.

Evidently, we ought to examine the apparent paradoxes such as Bertrand's more closely; there is an important point to be learned about the application of probability theory to real physical situations.

It is evident that if the circle becomes sufficiently large, and the tosser sufficiently skilled, various results could be obtained at will. However, in the limit where the skill of the tosser must be described by a "region of uncertainty" large compared to the circle, the distribution of chord lengths must surely go into one unique function obtainable by "pure thought." A viewpoint toward probability theory which cannot show us how to calculate this function from first principles, or even denies the possibility of doing this, would imply severe—and, to a physicist, intolerable—restrictions on the range of useful applications of probability theory.

An invariance argument was applied to problems of this type by Poincaré,[4] and cited more recently by Kendall and Moran.[8] In this treatment we consider straight lines drawn "at random" in the xy plane. Each line is located by specifying two parameters (u, v) such that the equation of the line is $ux + vy = 1$, and one can ask: Which probability density $p(u, v)$ $du\, dv$ has the property that it is invariant in *form* under the group of Euclidean transformations (rotations and translations) of the plane? This is a readily solvable problem,[8] with the answer $p(u, v) = (u^2 + v^2)^{-3/2}$.

Yet evidently this has not seemed convincing; for later authors have ignored Poincaré's invariance argument, and adhered to Bertrand's original judgment that the problem has no definite solution. This is understandable, for the statement of the problem does not specify that the distribution of straight lines is to have this invariance property, and we do not see any compelling reason to expect that a rain of straws produced in a real experiment would have it. To assume this would seem to be an intuitive judgment resting on no stronger grounds than the ones which led to the three different solutions above. All of these amount to trying to guess what properties a "random" rain of straws should have, by specifying the intuitively "equally possible" events; and the fact remains that different intuitive judgments lead to different results.

The viewpoint just expressed, which is by far the most common in the literature, clearly represents one valid way of interpreting the problem. If we can find another viewpoint according to which such problems *do* have definite solutions, *and define the conditions under which these solutions are experimentally verifiable*, then while it would perhaps be overstating the case to say that this new viewpoint is more "correct" in principle than the conventional one, it will surely be more useful in practice.

We now suggest such a viewpoint, and we understand from the start that we are not concerned at this stage with *frequencies* of various events. We ask rather: Which probability distribution describes our *state of knowledge* when the only information available is that given in the above statement of the problem? Such a distribution must conform to the desideratum of consistency formulated previously[1]: In two problems where we have the same state of knowledge we must assign the same subjective probabilities. The essential point is this: If we start with the presumption that Bertrand's problem has a definite solution *in spite of the many things left unspecified*, then the statement of the problem automatically implies certain invariance properties, which in no way depend on our intuitive judgments. After the subjective solution is found, it may be used as a prior for Bayesian inference whether or not it has any correspondence with frequencies; any frequency connections that may emerge will be regarded as an additional bonus, which justify its use also for direct physical prediction.

Bertrand's problem has an obvious element of rotational symmetry, recognized in all the proposed solutions; however, this symmetry is irrelevant to the distribution of chord lengths. There are two other "symmetries" which are highly relevant: Neither Bertrand's original statement nor our restatement in terms of straws specifies the exact size of the circle, or its exact location. If, therefore, the problem is to have any definite solution at all, it must be "indifferent" to these circumstances; i.e., it must be unchanged by a small change in the size or position of the circle. This seemingly trivial statement, as we will see, fully determines the solution.

It would be possible to consider all these invariance requirements simultaneously by defining a four-parameter transformation group, whereupon the complete solution would appear suddenly, as if by magic. However, it will be more instructive to analyze the effects of these invariances separately, and see how each places its own restriction on the form of the solution.

2. ROTATIONAL INVARIANCE

Let the circle have radius R. The position of the chord is determined by giving the polar coordinates (r, θ) of its center. We seek to answer a more detailed question than Bertrand's: What probability density $f(r, \theta)\,dA = f(r, \theta)\,r\,dr\,d\theta$ should we assign over the interior area of the circle? The dependence on θ is actually irrelevant to Bertrand's question, since the distribution of chord lengths depends only on the radial distribution

$$g(r) = \int_0^{2\pi} f(r, \theta)\,d\theta$$

However, intuition suggests that $f(r, \theta)$ should be independent of θ, and the formal transformation group argument deals with the rotational symmetry as follows.

The starting point is the observation that the statement of the problem does not specify whether the observer is facing north or east; therefore if there is a definite solution, it must not depend on the direction of the observer's line of sight. Suppose, therefore, that two different observers, Mr. X and Mr. Y, are watching this experiment. They view the experiment from different directions, their lines of sight making an angle α. Each uses a coordinate system oriented along his line of sight. Mr. X assigns the probability density $f(r, \theta)$ in his coordinate system S; and Mr. Y assigns $g(r, \theta)$ in his system S_α. Evidently, if they are describing the same situation, then it must be true that

$$f(r, \theta) = g(r, \theta - \alpha) \tag{1}$$

which expresses a simple change of variables, transforming a fixed distribution f to a new coordinate system; this relation will hold whether or not the problem has rotational symmetry.

But now we recognize that, because of the rotational symmetry, the problem appears exactly the same to Mr. X in his coordinate system as it does to Mr. Y in his. Since they are in the same state of knowledge, our desideratum of consistency demands that they assign the same probability distribution; and so f and g must be the same function:

$$f(r, \theta) = g(r, \theta) \tag{2}$$

These relations must hold for all α in $0 \leqslant \alpha \leqslant 2\pi$; and so the only possibility is $f(r, \theta) = f(r)$.

This formal argument may appear cumbersome when compared to our obvious flash of intuition; and of course it is, when applied to such a trivial problem. However, as Wigner[13] and Weyl[14] have shown in other physical problems, it is this cumbersome argument that generalizes at once to nontrivial cases where our intuition fails us. It always consists of two steps: We first find a transformation equation like (1) which shows how two problems are related to each other, irrespective of symmetry; then a symmetry relation like (2) which states that we have formulated two equivalent *problems*. Combining them leads in most cases to a functional equation which imposes some restriction on the form of the distribution.

3. SCALE INVARIANCE

The problem is reduced, by rotational symmetry, to determining a function $f(r)$, normalized according to

$$\int_0^{2\pi} \int_0^R f(r)\, r\, dr\, d\theta = 1 \qquad (3)$$

Again, we consider two different problems; concentric with a circle of radius R, there is a circle of radius aR, $0 < a \leqslant 1$. Within the smaller circle there is a probability $h(r)\, r\, dr\, d\theta$ which answers the question: Given that a straw intersects the smaller circle, what is the probability that the center of its chord lies in the area $dA = r\, dr\, d\theta$?

Any straw that intersects the small circle will also define a chord on the large one; and so, within the small circle $f(r)$ must be proportional to $h(r)$. This proportionality is, of course, given by the standard formula for a conditional probability, which in this case takes the form

$$f(r) = 2\pi h(r) \int_0^{aR} f(r)\, r\, dr, \qquad 0 < a \leqslant 1, \quad 0 \leqslant r \leqslant aR \qquad (4)$$

This transformation equation will hold whether or not the problem has scale invariance.

But we now invoke scale invariance; to two different observers with different size eyeballs, the problems of the large and small circles would appear exactly the same. If there is any unique solution independent of the size of the circle, there must be another relation between $f(r)$ and $h(r)$, which expresses the fact that one problem is merely a scaled-down version of the other. Two elements of area $r\, dr\, d\theta$ and $(ar)\, d(ar)\, d\theta$ are related to the large and small circles respectively in the same way; and so they must be assigned the same probabilities by the distributions $f(r)$ and $h(r)$, respectively:

$$h(ar)(ar)\, d(ar)\, d\theta = f(r)\, r\, dr\, d\theta$$

or
$$a^2 h(ar) = f(r) \tag{5}$$

which is the symmetry equation. Combining (4) and (5), we see that invariance under change of scale requires that the probability density satisfy the functional equation

$$a^2 f(ar) = 2\pi f(r) \int_0^{aR} f(u)\, u\, du, \quad 0 < a \leqslant 1,\ 0 \leqslant r \leqslant R \tag{6}$$

Differentiating with respect to a, setting $a = 1$, and solving the resulting differential equation, we find that the most general solution of (6) satisfying the normalization condition (3) is

$$f(r) = qr^{q-2}/2\pi R^q \tag{7}$$

where q is a constant in the range $0 < q < \infty$, not further determined by scale invariance.

We note that the proposed solution B in the introduction has now been eliminated, for it corresponds to the choice $f(r) \sim (R^2 - r^2)^{-1/2}$, which is not of the form (7). This means that if the intersections of chords on the circumference were distributed in angle uniformly and independently on one circle, this would not be true for a smaller circle inscribed in it; i.e., the probability assignment of B could be true for, at most, only one size of circle. However, solutions A and C are still compatible with scale invariance, corresponding to the choices $q = 1$ and $q = 2$ respectively.

4. TRANSLATIONAL INVARIANCE

We now investigate the consequences of the fact that a given straw S can intersect two circles C, C' of the same radius R, but with a relative displacement b. Referring to Fig. 1, the midpoint of the chord with respect to circle C is the point P, with coordinates (r, θ); while the same straw defines a midpoint of the chord with respect to C' at the point P' whose coordinates are (r', θ). From Fig. 1 the coordinate transformation $(r, \theta) \to (r', \theta')$ is given by

$$r' = |r - b \cos \theta| \tag{8}$$

$$\theta' = \begin{cases} \theta, & r > b \cos \theta \\ \theta + \pi, & r < b \cos \theta \end{cases} \tag{9}$$

As P varies over the region Γ, P' varies over Γ', and vice versa; thus the straws define a 1:1 mapping of Γ onto Γ'.

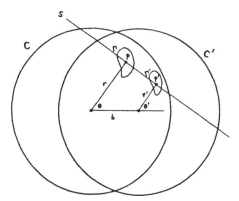

Fig. 1. A straw S intersects two slightly displaced circles C and C'.

Now we note the translational symmetry; since the statement of the problem gave no information about the location of the circle, the problems of C and C' appear exactly the same to two slightly displaced observers O and O'. Our desideratum of consistency then demands that they assign probability densities in C and C' respectively which have the same form (7) with the same value of q.

It is further necessary that these two observers assign equal probabilities to the regions Γ and Γ', respectively, since (a) they are probabilities of the same event, and (b) the probability that a straw which intersects one circle will also intersect the other, thus setting up this correspondence, is also the same in the two problems. Let us see whether these two requirements are compatible.

The probability that a chord intersecting C will have its midpoint in Γ is

$$\int_\Gamma f(r)\, r\, dr\, d\theta = (q/2\pi R^q) \int_\Gamma r^{q-1}\, dr\, d\theta \qquad (10)$$

The probability that a chord intersecting C' will have its midpoint in Γ' is

$$\frac{q}{2\pi R^q} \int_{\Gamma'} (r')^{q-1}\, dr'\, d\theta' = \frac{q}{2\pi R^q} \int_\Gamma |r - b\cos\theta|^{q-1}\, dr\, d\theta \qquad (11)$$

where we have transformed the integral back to the variables (r, θ) by use of (8) and (9), noting that the Jacobian is unity. Evidently, (10) and (11) will be equal for arbitrary Γ if and only if $q = 1$; and so our distribution $f(r)$ is now uniquely determined.

The proposed solution C in the introduction is thus eliminated for lack of translational invariance; a rain of straws which had the property assumed with respect to one circle, could not have the same property with respect to a slightly displaced one.

5. FINAL RESULTS

We have found that invariance requirements determine the probability density

$$f(r, \theta) = 1/2\pi Rr, \quad 0 \leqslant r \leqslant R, \quad 0 \leqslant \theta \leqslant 2\pi \tag{12}$$

corresponding to solution A in the introduction. It is interesting that this has a singularity at the center, the need for which can be understood as follows. The condition that the midpoint (r, θ) falls within a small region Δ imposes restrictions on the possible directions of the chord. But as Δ moves inward, as soon as it includes the center of the circle all angles are suddenly allowed. Thus there is an infinitely rapid change in the "manifold of possibilities."

Further analysis (almost obvious from contemplation of Fig. 1) shows that the requirement of translational invariance is so stringent that it already determines the result (12) uniquely; thus the proposed solution B is incompatible with either scale or translational invariance, and in order to find (12), it was not really necessary to consider scale invariance. However, the solution (12) would in any event have to be tested for scale invariance, and if it failed to pass that test, we would conclude that the problem as stated has *no* solution; i.e., although at first glance it appears underdetermined, it would have to be regarded, from the standpoint of transformation groups, as overdetermined. As luck would have it, these requirements *are* compatible; and so the problem has one unique solution.

The distribution of chord lengths follows at once from (12). A chord whose midpoint is at (r, θ) has a length $L = 2(R^2 - r^2)^{1/2}$. In terms of the reduced chord lengths, $x \equiv L/2R$, we obtain the universal distribution law

$$p(x)\, dx = \frac{x\, dx}{(1 - x^2)^{1/2}}, \quad 0 \leqslant x < 1 \tag{13}$$

in agreement with Borel's conjecture.[3]

6. FREQUENCY CORRESPONDENCE

From the manner of its derivation, the distribution (13) would appear to have only a subjective meaning; while it describes the only possible state of knowledge corresponding to a unique solution in view of the many things left unspecified in the statement of Bertrand's problem, we have as yet given no reason to suppose that it has any relation to frequencies observed in the actual experiment. In general, of course, no such claim can be made; the mere fact that my state of knowledge gives me no reason to prefer one event over another is not enough to make them occur equally often! Indeed, it is clear that no "pure thought" argument, whether based on transformation groups or any other principle, can predict with certainty what must happen in a real experiment. And we can easily imagine a very precise machine which tosses straws in such a way as to produce any distribution of chord lengths we please on a given circle.

Nevertheless, we are entitled to claim a definite frequency correspondence for the result (13). For there is one "objective fact" which *has* been proved by the above derivation: Any rain of straws which does *not* produce a frequency distribution agreeing with (13) will necessarily produce different distributions on different circles.

But this is all we need in order to predict with confidence that the distribution (13) *will* be observed in any experiment where the "region of uncertainty" is large compared to the circle. For, if we lack the skill to toss straws so that, with certainty, they intersect a given circle, then surely we lack *a fortiori* the skill consistently to produce different distributions on different circles *within* this region of uncertainty!

It is for this reason that distributions predicted by the method of transformation groups turn out to have a frequency correspondence after all. Strictly speaking, this result holds only in the limiting case of "zero skill," but as a moment's thought will show, the skill required to produce any appreciable deviation from (13) is so great that in practice it would be difficult to achieve even with a machine.

Of course, the above arguments have demonstrated this frequency correspondence in only one case. In the following section we adduce arguments indicating that it is a general property of the transformation group method.

These conclusions seem to be in direct contradiction to those of von Mises,[10,11] who denied that such problems belong to the field of probability theory at all. It appears to us that if we were to adopt von Mises' philosophy of probability theory strictly and consistently, the range of legitimate physical applications of probability theory would be reduced almost to the vanishing point. Since we have made a definite, unequivocal prediction, this issue has now been removed from the realm of philosophy into that of verifiable fact. The predictive power of the transformation group method can be put to the test quite easily in this and other problems by performing the experiments.

The Bertrand experiment has, in fact, been performed by the writer and Dr. Charles E. Tyler, tossing broom straws from a standing position onto a 5-in.-diameter circle drawn on the floor. Grouping the range of chord lengths into ten categories, 128 successful tosses confirmed Eq. (13) with an embarrassingly low value of chi-squared. However, experimental results will no doubt be more convincing if reported by others.

7. DISCUSSION

Bertrand's paradox has a greater importance than appears at first glance, because it is a simple crystallization of a deeper paradox which has permeated much of probability theory from its beginning. In "real" physical applications when we try to formulate the problem of interest in probability terms we find almost always that a statement emerges which, like Bertrand's, appears too vague to determine any definite solution, because apparently essential things are left unspecified.

We elaborate the example noted in the introduction: Given a gas of N molecules in a volume V, with known intermolecular forces, total energy E, predict from this its molecular velocity distribution, pressure, distribution of pressure fluctuations, viscosity, thermal conductivity, and diffusion constant. Here again the viewpoint expressed by most writers on probability theory would lead one to conclude that the problem has no definite solution because it is ill posed; the things specified are grossly inadequate to determine any unique probability distribution over microstates. If we reject the principle of indifference, and insist that the only valid basis for assigning probabilities is frequency in some random experiment, it would again appear that the only way of determining these quantities is to perform the experiments.

It is, however, a matter of record that over a century ago, without benefit of any frequency data on positions and velocities of molecules, James Clerk Maxwell was able to predict all these quantities correctly by a "pure thought" probability analysis which amounted to recognizing the "equally possible" cases. In the case of viscosity the predicted dependence on density appeared at first to contradict common sense, casting doubt on Maxwell's analysis. But when the experiments were performed they confirmed Maxwell's predictions, leading to the first great triumph of kinetic theory. These are solid, positive accomplishments; and they cannot be made to appear otherwise merely by deploring his use of the principle of indifference.

Likewise, we calculate the probability of obtaining various hands at poker; and we are so confident of the results that we are willing to risk money on bets which the calculations indicate are favorable to us. But underlying these calculations is the intuitive judgment that all distributions of cards are equally likely; and with a different judgment our calculations would give different results. Once again we are predicting definite, verifiable facts by "pure thought" arguments based ultimately on recognizing the "equally possible" cases; and yet present statistical doctrine, both orthodox and personalistic, denies that this is a valid basis for assigning probabilities!

The dilemma is thus apparent; on the one hand, one cannot deny the force of arguments which, by pointing to such things as Bertrand's paradox, demonstrate the ambiguities and dangers in the principle of indifference. But on the other hand, it is equally undeniable that use of this principle has, over and over again, led to correct, nontrivial, and useful predictions. Thus it appears that while we cannot wholly accept the principle of indifference, we cannot wholly reject it either; to do so would be to cast out some of the most important and successful applications of probability theory.

The transformation group method grew out of the writer's conviction, based on pondering this situation, that the principle of indifference has been unjustly maligned in the past; what it has needed was not blanket condemnation, but recognition of the proper way to apply it. We agree with most other writers on probability theory that it is dangerous to apply this principle at the level of indifference between *events*, because our intuition is a very unreliable guide in such matters, as Bertrand's paradox illustrates.

However, the principle of indifference may, in our view, be applied legitimately at the more abstract level of indifference between *problems*; because that is a matter that is definitely determined by the statement of a problem, independently of our intuition. Every circumstance left unspecified in the statement of a problem defines an invariance property which the solution must have if there is to be any definite solution at all. The transformation group, which expresses these invariances mathematically, imposes definite restrictions on the form of the solution, and in many cases fully determines it.

Of course, not all invariances are useful. For example, the statement of Bertrand's problem does not specify the time of day at which the straws are tossed, the color of the circle, the luminosity of Betelgeuse, or the number of oysters in Chesapeake Bay; from which we infer, correctly, that if the problem as stated is to have a unique solution, it must not depend on these circumstances. But this would not help us unless we had previously thought that these things might be germane.

Study of a number of cases makes it appear that the aforementioned dilemma can now be resolved as follows. We suggest that the cases in which the principle of indifference has been applied successfully in the past are just the ones in which the solution can be "reverbalized" so that the actual calculations used are seen as an application of indifference between problems, rather than events.

For example, in the case of poker hands the statement of the problem does not specify the order of cards in the deck before shuffling; therefore if the problem is to have any definite solution, it must not depend on this circumstance; i.e., it must be invariant under the group of 52! permutations or cards, each of which transforms the problem into an equivalent one. Whether we verbalize the solution by asserting that all distributions of cards in the final hands are "equally likely" or by saying that the solution shall have this invariance property, we shall evidently do just the same calculation and obtain the same final results.

There remains, however, a difference in the logical situation. After having applied the transformation group argument in this way we are not entitled to assert that the predicted distribution of poker hands *must* be observed in practice. The only thing that can be proved by transformation groups is that if this distribution is *not* forthcoming then the probability of obtaining a given hand will necessarily be different for different initial orders of the cards; or, as we would state it colloquially, the cards are not being "properly" shuffled. This is, of course, just the conclusion we do draw in practice, whatever our philosophy about the "meaning of probability."

Once again it is clear that the invariant solution is overwhelmingly the most likely one to be produced by a person of ordinary skill; to shuffle cards in such a way that one particular aspect of the initial order is retained consistently in the final order requires a "microscopic" degree of control over the exact details of shuffling (in this case, however, the possession of such skill is generally regarded as dishonest, rather than impossible).

We have not found any general proof that the method of transformation groups will always lead to solutions with this frequency correspondence property; however, analysis of some dozen problems like the above has failed to produce any counterexample, and its general validity is rendered plausible as follows.

In the first place, we recognize that every circumstance which our common sense tells us may exert some influence on the result of an experiment ought to be given explicitly in the statement of a problem. If we fail to do that, then of course we have no right to expect agreement between prediction and observation; but this is not a failure of probability theory, but rather a failure on our part to state the full problem. If the statement of a problem *does* properly include all such information, then it would appear that any circumstances that are still left unspecified must correspond to some lack of control over the conditions of the experiment, which makes it impossible for us to state them. But invariance under the corresponding transformation group is just the formal expression of this lack of control, or lack of skill.

One has the feeling that this situation can be formalized more completely; perhaps one can define some "space" corresponding to all possible degrees of skill and define a measure in this space, which proves to be concentrated overwhelmingly on those regions leading to the invariant solution. Up to the present, however, we have not seen how to carry out such a program; perhaps others will.

8. CONJECTURES

There remains the interesting, and still unanswered, question of how to define precisely the class of problems which can be solved by the method illustrated here. There are many problems in which we do not see how to apply it unambiguously; von Mises' water-and-wine problem is a good example. Here we are told that a mixture of water and wine contains at least half wine, and are asked: What is the probability that it contains at least three-quarters wine? On the usual viewpoint this problem is underdetermined; nothing tells us which quantity should be regarded as uniformly distributed. However, from the standpoint of the invariance group, it may be more useful to regard such problems as *overdetermined*; so many things are left unspecified that the invariance group is too large, and no solution can conform to it.

It thus appears that the "higher-level problem" of how to formulate statistical problems in such a way that they are neither underdetermined nor overdetermined may itself be capable of mathematical analysis. In the writer's opinion it is one of the major weaknesses of present statistical practice that we do not seem to know how to formulate statistical problems in this way, or even how to judge whether a given problem is well posed. Again, the Bertrand paradox is a good illustration of this difficulty, for it was long thought that not enough was specified to determine any unique solution, but from the viewpoint which recognizes the full invariance group implied by the above statement of the problem, it now appears that it was well posed after all.

In many cases, evidently, the difficulty has been simply that we have not been reading out all that is implied by the statement of a problem; the things left unspecified must be taken into account just as carefully as the ones that are specified. Presumably, a person would not seriously propose a problem unless he supposed that it had a definite solution. Therefore, as a matter of courtesy and in keeping with a worthy principle of law, we might take the view that *a problem shall be presumed to have a definite solution until the contrary has been proved.* If we accept this as a reasonable attitude, then we must recognize that we are not in a position to judge whether a problem is well posed until we have carried out a transformation group analysis of all the invariances implied by its statement.

The question whether a problem is well posed is thus more subtle in probability theory than in other branches of mathematics, and any results which could be obtained by study of the "higher-level problem" might be of immediate use in applied statistics.

APPENDIX: COMMENTS ON BERTRAND'S PROBLEM

Bertrand (Ref. 2, pp. 4–5): "Aucune de trois n'est fausse, aucune n'est exacte, la question est mal posée."

Borel (Ref. 3, pp. 110–113): "...il est aisé de voir que la plupart des procédés naturels que l'on peut imaginer conduiser à la première."

Poincaré (Ref. 4, pp. 118–130): "... nous avons definie la probabilité de deux manières différentes."

Uspensky (Ref. 5, p. 251): "... we are really dealing with two different problems."

Northrup (Ref. 6, pp. 181–183): "One guess is as good as another."

Gnedenko (Ref. 7, pp. 40–41): The three results "would be appropriate" in three different experiments.

Kendall and Moran (Ref. 8, p. 10): "All three solutions are correct, but they really refer to different problems."

Weaver (Ref. 9, pp. 356–357): "... you have to watch your step."

Von Mises (Ref. 10, pp. 160–166): "Which one of these or many other assumptions should be made is a question of fact and depends on how the needles are thrown. It is not a problem of probability calculus to decide which distribution prevails...." Von Mises, in the preface to Ref. 11, also charges that, "Neither Laplace nor any of his followers, including Poincaré, ever reveals how, starting with *a priori* premises concerning equally possible cases, the sudden transition to the description of real statistical events is to be made." It appears to us that this had already been accomplished in large part by James Bernoulli (1703) in his demonstration of the weak law of large numbers, the first theorem establishing a connection between probability and frequency. Reference 1 and the present article may be regarded as further contributions toward answering von Mises' objections.

Mosteller (Ref. 12, p. 40): "Until the expression 'at random' is made more specific, the question does not have a definite answer.... We cannot guarantee that any of these results would agree with those obtained from some physical process...."

REFERENCES

1. E. T. Jaynes, Prior probabilities, *IEEE Trans. Systems Sci. Cybernetics* **SSC-4** (3), 227–241 (1968).
2. J. Bertrand, *Calcul des probabilités* (Gauthier-Villars, Paris, 1889), pp. 4–5.

3. E. Borel, *Éléments de la théorie des probabilités* (Hermann et Fils, Paris, 1909), pp. 110–113.
4. H. Poincaré, *Calcul des probabilités* (Paris, 1912), pp. 118–130.
5. J. V. Uspensky, *Introduction to Mathematical Probability* (McGraw-Hill, New York, 1937), p. 251.
6. E. P. Northrup, *Riddles in Mathematics* (van Nostrand, New York, 1944), pp. 181–183.
7. B. V. Gnedenko, *The Theory of Probability* (Chelsea Publ. Co., New York, 1962), pp. 40–41.
8. M. G. Kendall and P. A. P. Moran, *Geometrical Probability* (Hafner Publ. Co., New York, 1963), p. 10.
9. W. Weaver, *Lady Luck: the Theory of Probability* (Doubleday-Anchor, Garden City, New York, 1963), pp. 356–357.
10. R. von Mises, in *Mathematical Theory of Probability and Statistics*, H. Geiringer, ed. (Academic Press, New York, 1964), pp. 160–166.
11. R. von Mises, *Probability, Statistics and Truth* (Macmillan, New York, 1957).
12. F. Mosteller, *Fifty Challenging Problems in Probability* (Addison-Wesley, Reading, Massachusetts, 1965), p. 40.
13. E. P. Wigner, *Gruppentheorie und ihre Anwendung auf die Quantenmechanik der Atomspektren* (Fr. Vieweg, Braunschweig, 1931).
14. H. Weyl, *The Classical Groups* (Princeton University Press, Princeton, New Jersey, 1946).

9. CONFIDENCE INTERVALS vs BAYESIAN INTERVALS (1976)

We come now to the most polemical of all my articles. There are several reasons for this heated style. Firstly, most of it was written in 1963, as a reply to the 'astonishing article' of Bross referred to, whose polemics make mine seem unimaginative. Indeed, his anti-Bayesian tirade contained nothing but polemics, unsupported by a single technical fact. But had he taken the trouble to read Jeffreys, he would have found demonstrations, on the level of technical fact with no polemics, of the falsity of his charges.

But this distortion of known facts — naturally infuriating to a Bayesian — was hardly limited to Bross. Almost every 'orthodox' textbook written for decades had charged Bayesian methods with nonexistent defects, while ignoring the demonstrable defects in orthodox methods.

It seemed appropriate that a Bayesian point out these things, and so I collected a number of case histories, with mathematical details, from the same areas that Bross had alluded to and showed that, contrary to his assertions, Bayesian methods correct the shortcomings of orthodox methods.

My attempts to get the work published met with rebuff twice from those who had so quickly accepted the Bross article; clearly, different standards of acceptance existed for works differently slanted. After ten years of waiting, the opportunity arrived in an invitation to present a paper at the 1973 London Symposium. But three more years passed before the Proceedings Volume appeared, and so it required thirteen years to get this reply before the public. Still, none of it was obsolete — an interesting commentary on the rate of progress of orthodox thinking.

In the Proceedings Volume, there appear also comments on my presentation by Margaret Maxfield and Oscar Kempthorne, and my replies. Portions of the latter, which extend the main message, are included here.

The reply to Kempthorne also becomes polemical in places, but for an entirely different reason. Just as it was being written, a student of mine went forth into the world with a fresh Ph.D. degree, seeking a teaching position. Since he knew some statistical theory as well as some theoretical physics, one well-known institution suggested that he teach a statistics course, which he was well qualified to do. But on learning that he had been exposed to Bayesian thinking, he was taken aside and told that it was a condition of his employment that he agree to teach straight out of Hoel, expounding no Bayesian ideas.

My revulsion at such thought control — by persons who would rightly denounce it anywhere else — resulted in my coming down on poor old Oscar a bit harder than I would have otherwise (still, he will be the first to admit that he delivers fully as much as he receives, and neither of us takes it personally).

The manuscript was sent to the publisher with some trepidations over the polemical style, but those fears were groundless; in fact, I have received more favorable fan mail over this article than on any two others. Many quite well-known figures have told me in confidence: 'Bravo! These things needed to be said, but I cannot say them because my position requires me to maintain diplomatic relations with both sides in the Great Debate. You are just enough of an outsider so you can get away with it.'

# CONFIDENCE INTERVALS VS BAYESIAN INTERVALS

ABSTRACT. For many years, statistics textbooks have followed this 'canonical' procedure: (1) the reader is warned not to use the discredited methods of Bayes and Laplace, (2) an orthodox method is extolled as superior and applied to a few simple problems, (3) the corresponding Bayesian solutions are *not* worked out or described in any way. The net result is that no evidence whatsoever is offered to substantiate the claim of superiority of the orthodox method.

To correct this situation we exhibit the Bayesian and orthodox solutions to six common statistical problems involving confidence intervals (including significance tests based on the same reasoning). In every case, we find that the situation is exactly the opposite; i.e., the Bayesian method is easier to apply and yields the same or better results. Indeed, the orthodox results are satisfactory only when they agree closely (or exactly) with the Bayesian results. No contrary example has yet been produced.

By a refinement of the orthodox statistician's own criterion of performance, the best confidence interval for any location or scale parameter is proved to be the Bayesian posterior probability interval. In the cases of point estimation and hypothesis testing, similar proofs have long been known. We conclude that orthodox claims of superiority are totally unjustified; today, the original statistical methods of Bayes and Laplace stand in a position of proven superiority in actual performance, that places them beyond the reach of mere ideological or philosophical attacks. It is the continued teaching and use of orthodox methods that is in need of justification and defense.

I. INTRODUCTION[1]

The theme of our meeting has been stated in rather innocuous terms: how should probability theory be (1) formulated, (2) applied to statistical inference; and (3) to statistical physics? Lurking behind these bland generalities, many of us will see more specific controversial issues: (1) frequency vs. nonfrequency definitions of probability, (2) 'orthodox' vs. Bayesian methods of inference, and (3) ergodic theorems vs. the principle of maximum entropy as the basis for statistical mechanics.

When invited to participate here, I reflected that I have already held forth on issue (3) at many places, for many years, and at great length. At the moment, the maximum entropy cause seems to be in good hands and advancing well, with no need for any more benedictions from me; in any event, I have little more to say beyond what is already in print.[2] So it seemed time to widen the front, and enter the arena on issue (2).

Why a physicist should have the temerity to do this, when no statistician has been guilty of invading physics to tell us how we ought to do our jobs, will become clear only gradually; but the main points are: (A) we were here first, and (B) because of our past experiences, physicists may be in a position to help statistics in its present troubles, well described by Kempthorne (1971). More specifically:

(A) Historically, the development of probability theory in the 18'th and early 19'th centuries from a gambler's amusement to a powerful research tool in science and many other areas, was the work of people – Daniel Bernoulli, Laplace, Poisson, Legendre, Gauss, and several others – whom we would describe today as mathematical physicists. In the 19'th century, a knowledge of their work was considered an essential part of the training of any scientist, and it was taught largely as a part of physics.

A radical change took place early in this century when a new group of workers, not physicists, entered the field. They proceeded to reject virtually everything done by Laplace and sought to develop statistics anew, based on entirely different principles. Simultaneously with this development, the physicists – with Sir Harold Jeffreys as almost the sole exception – quietly retired from the field, and statistics disappeared from the physics curriculum.

This departure of physicists from the field they had created was not, of course, due to the new competition; rather, it was just at this time that relativity theory burst upon us, X-rays and radioactivity were discovered, and quantum theory started to develop. The result was that for fifty years physicists had more than enough to do unravelling a host of new experimental facts, digesting these new revolutions of thought, and putting our house back into some kind of order. But the result of our departure was that this extremely aggressive new school in statistics soon dominated the field so completely that its methods are now known as 'orthodox statistics'. For these historical reasons, I ask you to think with me, that for a physicist to turn his attention now to statistics, is more of a homecoming than an invasion.

(B) Today, a physicist revisiting statistics to see how it has fared in our absence, sees quickly that something has gone wrong. For over fifteen years now, statistics has been in a state of growing ideological crisis – literally a crisis of conflicting ideas – that shows no signs of resolving itself, but yearly grows more acute; but it is one that physicists can recognize as basically the same thing that physics has been through several times (Jaynes, 1967). Having seen how these crises work themselves out, I think physicists may be in a position to prescribe a physic that will speed up the process in statistics.

The point we have to recognize is that issues of the kind facing us are never resolved by mere philosophical or ideological debate. At that level of discussion, people will persist in disagreeing, and nobody will be able to prove his case. In physics, we have our own ideological disputes, just as deeply felt by the protagonists as any in statistics; and at the moment I happen to be involved in one that strikes at the foundations of quantum theory (Jaynes, 1973). But in physics we have been perhaps more fortunate in that we have a universally recognized Supreme Court, to which all disputes are taken eventually, and from whose verdict there is no appeal. I refer, of course, to direct experimental observation of the facts.

This is an exciting time in physics, because recent advances in technology (lasers, fast computers, etc.) have brought us to the point where issues which have been debated fruitlessly on the philosophical level for 45 years, are at last reduced to issues of fact, and experiments are now underway testing controversial aspects of quantum theory that have never before been accessible to direct check. We have the feeling that, very soon now, we are going to know the real truth, the long debate can end at last, one way or the other; and we will be able to turn a great deal of energy to more constructive things. Is there any hope that the same can be done for statistics?

I think there is, and history points the way. It is to Galileo that we owe the first demonstration that ideological conflicts are resolved, not by debate, but by observation of fact. But we also recall that he ran into some difficulties in selling this idea to his contemporaries. Perhaps the most striking thing about his troubles was not his eventual physical persecution, which was hardly uncommon in those days; but rather the quality of logic that was used by his adversaries. For example, having turned his new telescope to the skies, Galileo announced discovery of the moons of Jupiter. A contemporary scholar ridiculed the idea, asserted that his theology had proved there could be *no* moons about Jupiter; and steadfastly refused to look through Galileo's telescope. But to everyone who did take a look, the evidence of his own eyes somehow carried more convincing power than did any amount of theology.

Galileo's telescope was able to reveal the truth, in a way that transcended all theology, because it could *magnify* what was too small to be perceived by our unaided senses, up into the range where it could be seen directly by all. And that, I suggest, is exactly what we need in statistics if this conflict is ever to be resolved. Statistics cannot take its dispute to the Supreme Court of the physicist; but there is another. It was recognized by Laplace in that famous remark, "Probability theory is nothing but common sense reduced to calculation".

Let me make what, I fear, will seem to some a radical, shocking suggestion: *the merits of any statistical method are not determined by the ideology which led to it.* For, many different, violently opposed ideologies may all lead to the same final 'working equations' for dealing with real problems. Apparently, this phenomenon is something new in statistics; but it is so commonplace in physics that we have long since learned how to live with it. Today, when a physicist says, "Theory A is better than theory B", he does not have in mind any ideological considerations; he means simply, "There is at least one specific application where theory A leads to a better result than theory B".

I suggest that we apply the same criterion in statistics: *the merits of any statistical method are determined by the results it gives when applied to specific problems.* The Court of Last Resort in statistics is simply our commonsense judgment of those results. But our common sense, like our unaided vision, has a limited resolving power. Given two different statistical methods (e.g., an orthodox and a Bayesian one), in many cases they lead to final numerical results which are so nearly alike that our common sense is unable to make a clear decision between them. What we need, then, is a kind of Galileo telescope for statistics; let us try to invent an extreme case where a small difference is magnified to a large one, or if possible to a qualitative difference in the conclusions. Our common sense will then tell us which method is preferable, in a way that transcends all ideological quibbling over 'subjectivity', 'objectivity', the 'true meaning of probability', etc.

I have been carrying out just this program, as a hobby, for many years, and have quite a mass of results covering most areas of statistical practice. They all lead to the same conclusion, and I have yet to find one exception to it. So let me give you just a few samples from my collection.

CONFIDENCE INTERVALS VS BAYESIAN INTERVALS

(a) INTERVAL ESTIMATION

Time not permitting even a hurried glimpse at the entire field of statistical inference, it is better to pick out a small piece of it for close examination. Now we have already a considerable Underground Literature on the relation of orthodox and Bayesian methods in the areas of point estimation and hypothesis testing, the topics most readily subsumed under the general heading of Decision Theory. [I say underground, because the orthodox literature makes almost no mention of it. Not only in textbooks, but even in such a comprehensive treatise as that of Kendall and Stuart (1961), the reader can find no hint of the existence of the books of Good (1950), Savage (1954), Jeffreys (1957), or Schlaifer (1959), all of which are landmarks in the modern development of Bayesian statistics].

It appears that much less has been written about this comparison in the case of interval estimation; so I would like to examine here the orthodox principle of confidence intervals (including significance tests based on the same kind of reasoning), as well as the orthodox criteria of performance and method of reporting results; and to compare these with the corresponding Bayesian reasoning and results, with magnification.

The basic ideas of interval estimation must be ancient, since they occur inevitably to anyone involved in making measurements, as soon as he ponders how he can most honestly communicate what he has learned to others, short of giving the entire mass of raw data. For, if you merely give your final best number, some troublesome fellow will demand to know how accurate the number is. And you will not appease him merely by answering his question; for if you reply, "It is within a tenth of a percent", he will only ask, "How sure are you of that? Will you make a 10:1 bet on it?"

It is not enough, then, to give a number or even an interval of possible error; at the very minimum, one must give both an interval and some indication of the reliability with which one can assert that the true value lies within it. But even this is not really enough; ideally (although this goes beyond current practice) one ought to give many different intervals – or even a continuum of all possible intervals – with some kind of statement about the reliability of each, before he has fully described his state of knowledge. This was noted by D. R. Cox (1958), in producing a nested sequence of confidence intervals; evidently, a Bayesian posterior probability accomplishes the same thing in a simpler way.

Perhaps the earliest formal quantitative treatment of interval estimation was Laplace's analysis of the accuracy with which the mass of Saturn was known at the end of the 18'th century. His method was to apply Bayes' theorem with uniform prior density; relevant data consist of the mutual perturbations of Jupiter and Saturn, and the motion of their moons, but the data are imperfect because of the finite accuracy with which angles and time intervals can be measured. From the posterior distribution $P(M)\,dM$ conditional on the available data, one can determine the shortest interval which contains a specified amount of posterior probability, or equally well the amount of posterior probability contained in a specified interval. Laplace chose the latter course, and announced his result as follows: "... it is a bet of 11 000 against 1 that the error of this result is not 1/100 of its value". In the light of present knowledge, Laplace would have won his bet; another 150 years' accumulation of data has increased the estimate by 0.63 percent.

Today, orthodox teaching holds that Laplace's method was, in Fisher's words, "founded upon an error". While there are some differences of opinion within the orthodox school, most would hold that the proper method for this problem is the confidence interval. It would seem to me that, in order to substantiate this claim, the orthodox writers would have to (1) produce the confidence interval for Laplace's problem, (2) show that it leads us to numerically different conclusions, and (3) demonstrate that the confidence interval conclusions are more statisfactory than Laplace's. But, in some twenty years of searching the orthodox literature, I have yet to find one case where such a program is carried out, on any statistical problem.

Invariably, the superiority of the orthodox method is asserted, not by presenting evidence of superior performance, but by a kind of ideological invective about 'objectivity' which perhaps reached its purple climax in an astonishing article of Bross (1963), whose logic recalls that of Galileo's colleague. In his denunciation of everything Bayesian, Bross specifically brings up the matter of confidence intervals and orthodox significance tests (which are based on essentially the same reasoning, and often amount to one-sided confidence intervals). So we will do likewise; in the following, we will examine these same methods and try to supply what Bross omitted; the demonstrable facts concerning them.

We first consider three significance tests appearing in the recent literature of reliability theory. The first two, which turn out to be so clear that no magnification is needed, will also bring out an important point concerning orthodox methods of reporting results.

II. SIGNIFICANCE TESTS

> Significance tests, in their usual form, are not compatible with a Bayesian attitude.
>
> C. A. B. Smith (1962)

> At any rate, what I feel quite sure at the moment to be needed is simple illustration of the new [i.e., Bayesian] notions on real, everyday statistical problems.
>
> E. S. Pearson (1962)

(a) EXAMPLE 1. DIFFERENCE OF MEANS

One of the most common of the 'everyday statistical problems' concerns the difference of the means of two normal distributions. A good example, with a detailed account of how current orthodox practice deals with such problems, appears in a recent book on reliability engineering (Roberts, 1964).

Two manufacturers, A and B, are suppliers for a certain component, and we want to choose the one which affords the longer mean life. Manufacturer A supplies 9 units for test, which turn out to have a (mean $\pm$ standard deviation) lifetime of (42 ± 7.48) hours. B supplies 4 units, which yield (50 ± 6.48) hours.

I think our common sense tells us immediately, without any calculation, that this constitutes fairly substantial (but not overwhelming) evidence in favor of B. While we should certainly prefer a larger sample, B's units did give a longer mean life, the difference being appreciably greater than the sample standard deviation; and so if a decision between them must be made on this basis, we should have no hesitation in choosing B. However, the author warns against drawing any such conclusion, and says that, if you are tempted to reason this way, then "perhaps statistics is not for you!" In any event, when we have so little evidence, it is imperative that we analyze the data in a way that does not throw any of it away.

The author then offers us the following analysis of the problem. He first asks whether the two variances are the same. Applying the F-test, the hypothesis that they are equal is not rejected at the 95 percent significance level, so without further ado he assumes that they *are* equal, and pools the data for an estimate of the variance. Applying the *t*-test, he then finds that, at the 90 percent level, the sample affords no significant evidence in favor of either manufacturer over the other.

Now, any statistical procedure which fails to extract evidence that is already clear to our unaided common sense, is certainly *not* for me! So, I carried out a Bayesian analysis. Let the unknown mean lifetimes of A's and B's components be a, b respectively. If the question at issue is whether $b > a$, the way to answer it is to calculate the *probability* that $b > a$, conditional on all the available data. This is

$$(1) \quad \text{Prob}(b > a) = \int_{-\infty}^{\infty} da \int_{a}^{\infty} db \, P_n(a) \, P_m(b)$$

where $P_n(a)$ is the posterior distribution of a, based on the sample of $n = 9$ items supplied by A, etc. When the variance is unknown, we find that these are of the form of the 'Student' *t*-distribution:

$$(2) \quad P_n(a) \sim [s_A^2 + (a - \bar{\imath}_A)^2]^{-n/2}$$

where $\bar{\imath}_A$, $s_A^2 = \overline{\imath_A^2} - \bar{\imath}_A^2$ are the mean and variance of sample A. Carrying out the integration (1), I find that the given data yield a probability of 0.920, or odds of 11.5 to 1, that B's components *do* have a greater mean life – a conclusion which, I submit, conforms nicely to the indications of common sense.[3]

But this is far from the end of the story; for one feels intuitively that if the variances are assumed equal, this ought to result in a more selective test than one in which this is not assumed; yet we find the Bayesian test without assumption of equal variance yielding an apparently sharper result than the orthodox one with that assumption. This suggests that we repeat the Bayesian calculation, using the author's assumption of equal variances. We have again an integral like (1), but a and b are no longer independent, their joint posterior distribution being proportional to

$$(3) \quad P(a, b) \sim \{n [s_A^2 + (a - \bar{\imath}_A)^2] + m [s_B^2 + (b - \bar{\imath}_B)^2]\}^{-1/2 (n+m)}$$

Integrating this over the same range as in (1) – which can be done simply by consulting the *t*-tables after carrying out one integration analytically –

I find that the Bayesian analysis now yields a probability of 0.948, or odds of 18:1, in favor of B.

How, then, could the author have failed to find significance at the 90 percent level? Checking the tables used we discover that, without having stated so, he has applied the *equal tails* t-test at the 90 percent level. But this is surely absurd; it was clear from the start that there is no question of the data supporting A; the only purpose which can be served by a statistical analysis is to tell us *how strongly* it supports B.

The way to answer this is to test the null hypothesis $b=a$ against the one-sided alternative $b>a$ already indicated by inspection of the data; using the 90 percent equal-tails test throws away half the 'resolution' and gives what amounts to a one-sided test at the 95 percent level, where it just barely fails to achieve significance.

In summary, the data yield clear significance at the 90 percent level; but the above orthodox procedure (which is presumably now being taught to many students) is a compounding of two errors. Assuming the variances equal makes the difference $(\bar{t}_B - \bar{t}_A)$ appear, unjustifiedly, even more significant; but then use of the equal tails criterion throws away more than was thus gained, and we still fail to find significance at the 90 percent level.

Of course, the fact that orthodox methods are capable of being misused in this way does not invalidate them; and Bayesian methods can also be misused, as we know only too well. However, there must be something in orthodox teaching which predisposes one toward this particular kind of misuse, since it is very common in the literature and in everyday practice. It would be interesting to know why most orthodox writers will not use – or even mention – the Behrens-Fisher distribution, which is clearly the correct solution to the problem, has been available for over forty years (Fisher, 1956; p. 95), and follows immediately from Bayes' theorem with the Jeffreys prior (Jeffreys, 1939; p. 115).

(b) EXAMPLE 2. SCALE PARAMETERS

A recent Statistics Manual (Crow *et al.*, 1960) proposes the following problem: 31 rockets of type 1 yield a dispersion in angle of 2237 mils^2, and 61 of type 2 give instead 1347 mils^2. Does this constitute significant evidence for a difference in standard deviation of the two types?

I think our common sense now tells us even more forcefully that, in view of the large samples and the large observed difference in dispersion, this constitutes absolutely unmistakable evidence for the superiority of type 2 rockets. Yet the authors, applying the equal-tails F-test at the 95 percent level, find it not significant, and conclude: "We need not, as far as this experiment indicates, differentiate between the two rockets with respect to their dispersion".

Suppose you were a military commander faced with the problem of deciding which type of rocket to adopt. You provide your statistician with the above data, obtained at great trouble and expense, and receive the quoted report. What would be your reaction? I think that you would fire the statistician on the spot; and henceforth make decisions on the basis of your own common sense, which is evidently a more powerful tool than the equal-tails F-test.

However, if your statistician happened to be a Bayesian, he would report[4] instead: "These data yield a probability of 0.9574, or odds of 22.47:1, in favor of type 2 rockets". I think you would decide to keep this fellow on your staff, because his report not only agrees with common sense; it is stated in a far more useful form. For, you have little interest in being told merely whether the data constitute 'significant evidence for a difference'. It is already obvious without any calculation that they *do* constitute highly significant evidence in favor of type 2; the only purpose that can be served by a statistical analysis is, again, to tell us quantitatively *how significant* that evidence is. Traditional orthodox practice fails utterly to do this, although the point has been noted recently by some.

What we have found in these two examples is true more generally. The orthodox statistician conveys little useful information when he merely reports that the null hypothesis is or is not rejected at some arbitrary preassigned significance level. If he reports that it is rejected at the 90 percent level, we cannot tell from this whether it would have been rejected at the 92 percent, or 95 percent level. If he reports that it is not rejected at the 95 percent level, we cannot tell whether it would have been rejected at the 50 percent, or 90 percent level. If he uses an equal-tails test, he in effect throws away half the 'resolving power' of the test, and we are faced with still more uncertainty as to the real import of the data.

Evidently, the orthodox statistician would tell us far more about what the sample really indicates if he would report instead *the critical significance level at which the null hypothesis is just rejected in favor of the one-*

sided alternative indicated by the data; for we then know what the verdict would be at all levels, and no resolution has been lost to a superfluous tail. Now two possible cases can arise: (I) the number thus reported is identical with the Bayesian posterior probability that the alternative is true; (II) these numbers are different.

If case (I) arises (and it does more often than is generally realized), the Bayesian and orthodox tests are going to lead us to exactly the same numerical results and the same conclusions, with only a verbal disagreement as to whether we should use the word 'probability' or 'significance' to describe them. In particular, the orthodox t-test and F-test against one-sided alternatives would, if their results were reported in the manner just advocated, be precisely equivalent to the Bayesian tests based on the Jeffreys prior $d\mu d\sigma/\sigma$. Thus, if we assume the variances equal in the above problem of two means, the observed difference is just significant by the one-sided t-test at the 94.8 percent level; and in the rocket problem a one-sided F-test just achieves significance at the 95.74 percent level.

It is only when case (II) is found that one could possibly justify any 'objective' claim for superiority of either approach. Now it is just these cases where we have the opportunity to carry out our 'magnification' process; and if we can find a problem for which this difference is magnified sufficiently, the issue cannot really be in doubt. We find this situation, and a number of other interesting points of comparison, in one of the most common examples of acceptance tests.

(c) EXAMPLE 3. AN ACCEPTANCE TEST

The probability that a certain machine will operate without failure for a time t is, by hypothesis, $\exp(-\lambda t)$, $0 < t < \infty$. We test n units for a time t, and observe r failures; what assurance do we then have that the mean life $\theta = \lambda^{-1}$ exceeds a preassigned value θ_0?

Sobel and Tischendorf (1959) (hereafter denoted ST) give an orthodox solution with tables that are reproduced in Roberts (1964). The test is to have a critical number C (i.e., we accept only if $r \leq C$). On the hypothesis that we have the maximum tolerable failure rate, $\lambda_0 = \theta_0^{-1}$, the probability that we shall see r or fewer failures is the binomial sum

(4) $$W(n, r) = \sum_{k=0}^{r} \binom{n}{k} e^{-(n-k)\lambda_0 t} (1 - e^{-\lambda_0 t})^k$$

and so, setting $W(n, C) \leqslant 1 - P$ gives us the sample size n required in order that this test will assure $\theta \geqslant \theta_0$ at the 100 P percent significance level. From the ST tables we find, for example, that if we wish to test only for a time $t = 0.01\ \theta_0$ with $C = 3$, then at the 90 percent significance level we shall require a test sample of $n = 668$ units; while if we are willing to test for a time $t = \theta_0$ with $C = 1$, we need test only 5 units.

The amount of testing called for is appalling if $t \ll \theta_0$; and out of the question if the units are complete systems. For example, if we want to have 95 percent confidence (synonymous with significance) that a space vehicle has $\theta_0 \geqslant 10$ years, but the test must be made in six months, then with $C = 1$, the ST tables say that we must build and test 97 vehicles! Suppose that, nevertheless, it had been decreed on the highest policy level that this degree of confidence *must* be attained, and you were in charge of the testing program. If a more careful analysis of the statistical problem, requiring a few man-years of statisticians' time, could reduce the test sample by only one or two units, it would be well justified economically. Scrutinizing the test more closely, we note four points:

(1) We know from the experiment not only the total number r of failures, but also the particular times $\{t_1 \ldots t_r\}$ at which failure occurred. This informaion is clearly relevant to the question being asked; but the ST test makes no use of it.

(2) The test has a 'quasi-sequential' feature; if we adopt an acceptance number $C = 3$, then as soon as the fourth failure occurs, we know that the units are going to be rejected. If no failures occur, the required degree of confidence will be built up long before the time t specified in the ST tables. In fact, t is the *maximum possible* testing time, which is actually required only in the marginal case where we observe exactly C failures. A test which is 'quasi-sequential' in the sense that it terminates when a clear rejection or the required confidence is attained, will have an expected length less than t; conversely, such a test with the expected length set at t will require fewer units tested.

(3) We have relevant prior information; after all, the engineers who designed the space vehicle knew in advance what degree of reliability was needed. They have chosen the quality of materials and components, and the construction methods, with this in mind. Each sub-unit has had its own tests. The vehicles would never have reached the final testing stage unless the engineers knew that they were operating satisfactorily. In

other words, we are not testing a completely unknown entity. The ST test (like most orthodox procedures) ignores all prior information, except perhaps when deciding which hypotheses to consider, or which significance level to adopt.

(4) In practice, we are usually concerned with a different question than the one the ST test answers. An astronaut starting a five-year flight to Mars would not be particularly comforted to be told, "We are 95 percent confident that the average life of an imaginary population of space vehicles like yours, is at least ten years". He would much rather hear, "There is 95 percent probability that *this* vehicle will operate without breakdown for ten years". Such a statement might appear meaningless to an orthodox statistician who holds that (probability)≡(frequency). But such a statement would be very meaningful indeed to the astronaut.

This is hardly a trivial point; for if it were *known* that $\lambda^{-1} = 10$ yr, the probability that a particular vehicle will actually run for 10 yrs would be only $1/e = 0.368$; and the period for which we are 95 percent sure of success would be only $-10 \ln(0.95)$ years, or 6.2 months. Reports which concern only the 'mean life' can be rather misleading.

Let us first compare the ST test with a Bayesian test which makes use of exactly the same information; i.e., we are allowed to use only the total number of failures, not the actual failure times. On the hypothesis that the failure rate is λ, the probability that exactly r units fail in time t is

$$(5) \qquad p(r \mid n, \lambda, t) = \binom{n}{r} e^{-(n-r)\lambda t} (1 - e^{-\lambda t})^r.$$

I want to defer discussion of nonuniform priors; for the time being suppose we assign a uniform prior density to λ. This amounts to saying that, before the test, we consider it extremely unlikely that our space vehicles have a mean life as long as a microsecond; nevertheless it will be of interest to see the result of using this prior. The posterior distribution of λ is then

$$(6) \qquad p(d\lambda \mid n, r, t) = \frac{n!}{(n-r-1)! \, r!} e^{-(n-r)\lambda t} (1 - e^{-\lambda t})^r \, d(\lambda t).$$

The Bayesian acceptance criterion, which ensures $\theta \geq \lambda_0^{-1}$ with 100 P

percent probability, is then

(7) $$\int_{\lambda_0}^{\infty} p(d\lambda \mid n, r, t) \leq 1 - P.$$

But the left-hand side of (7) is identical with $W(n, r)$ given by (4); this is just the well-known identity of the incomplete Beta function and the incomplete binomial sum, given already in the original memoir of Bayes (1763).

In this first comparison we therefore find that the ST test is mathematically identical with a Bayesian test in which (1) we are denied use of the actual failure times; (2) because of this it is not possible to take advantage of the quasi-sequential feature; (3) we assign a ridiculously pessimistic prior to λ; (4) we still are not answering the question of real interest for most applications.

Of these shortcomings, (2) is readily corrected, and (1) undoubtedly could be corrected, without departing from orthodox principles. On the hypothesis that the failure rate is λ, the probability that r specified units fail in the time intervals $\{dt_1 \ldots dt_r\}$ respectively, and the remaining $(n-r)$ units do not fail in time t, is

(8) $$p(dt_1 \ldots dt_r \mid n, \lambda, t) = [\lambda^r e^{-\lambda r \bar{t}} dt_1 \ldots dt_r] [e^{-(n-r)\lambda t}]$$

where $\bar{t} \equiv r^{-1} \sum t_i$ is the mean life of the units which failed. There is no single 'statistic' which conveys all the relevant information; but r and $\bar{t}$ are jointly sufficient, and so an optimal orthodox test must somehow make use of both. When we seek their joint sampling distribution $p(r, d\bar{t} \mid n, \lambda, t)$ we find, to our dismay, that for given r the interval $0 < \bar{t} < t$ is broken up into r equal intervals, with a different analytical expression for each. Evidently a decrease in r, or an increase in $\bar{t}$, should incline us in the direction of acceptance; but at what rate should we trade off one against the other? To specify a definite critical region in both variables would seem to imply some postulate as to their relative importance. The problem does not appear simple, either mathematically or conceptually; and I would not presume to guess how an orthodox statistician would solve it.

The relative simplicity of the Bayesian analysis is particularly striking in this problem; for all four of the above shortcomings are corrected effortlessly. For the time being, we again assign the pessimistic uniform prior to λ; from (8), the posterior distribution of λ is then

(9) $\quad p(d\lambda \mid n, t, t_1 \ldots t_r) = \dfrac{(\lambda T)^r}{r!} e^{-\lambda T} d(\lambda T)$

where

(10) $\quad T \equiv r\bar{t} + (n - r) t$

is the total unit-hours of failure-free operation observed. The posterior probability that $\lambda \geqslant \theta_0$ is now

(11) $\quad B(n, r) = \dfrac{1}{r!} \displaystyle\int_{\lambda_0 T}^{\infty} x^r e^{-x} dx = e^{-\lambda_0 T} \sum_{k=0}^{r} \dfrac{(\lambda_0 T)^k}{k!}$

and so, $B(n, r) \leqslant 1 - P$ is the new Bayesian acceptance criterion at the 100 P percent level; the test can terminate with acceptance as soon as this inequality is satisfied.

Numerical analysis shows little difference between this test and the ST test in the usual range of practical interest where we test for a time short compared to θ_0 and observe only a very few failures. For, if $\lambda_0 t \ll 1$, and $r \ll n$, then the Poisson approximation to (4) will be valid; but this is just the expression (11) except for the replacement of T by nt, which is itself a good approximation. In this region the Bayesian test (11) with maximum possible duration t generally calls for a test sample one or two units smaller than the ST test. Our common sense readily assents to this; for if we see only a few failures, then information about the actual failure time adds little to our state of knowledge.

Now let us magnify. The big differences between (4) and (11) will occur when we find many failures; if all n units fail, the ST test tells us to reject at all confidence levels, even though the observed mean life may have been thousands of times our preassigned θ_0. The Bayesian test (11) does not break down in this way; thus if we test 9 units and all fail, it tells us to accept at the 90 percent level if the observed mean life $\bar{t} \geqslant 1.58\, \theta_0$. If we test 10 units and 9 fail, the ST test says we can assert with 90 percent confidence that $\theta \geqslant 0.22\, t$; the Bayesian test (11) says there is 90 percent probability that $\theta \geqslant 0.63\, \bar{t} + 0.07\, t$. Our common sense has no difficulty in deciding which result we should prefer; thus taking the actual failure times into account leads to a clear, although usually not spectacular, improvement in the test. The person who rejects the use of Bayes' theorem in the manner of Equation (9) will be able to obtain a comparable improvement only with far greater difficulty.

But the Bayesian test (11) can be further improved in two respects. To correct shortcoming (4), and give a test which refers to the reliability of the individual unit instead of the mean life of an imaginary 'population' of them, we note that if λ were known, then by our original hypothesis the probability that the lifetime θ of a given unit is at least θ_0, is

$$(12) \quad p(\theta \geq \theta_0 \mid \lambda) = e^{-\lambda \theta_0}.$$

The probability that $\theta \geq \theta_0$, conditional on the evidence of the test, is therefore

$$(13) \quad p(\theta \geq \theta_0 \mid n, t_1 \ldots t_r) =$$

$$= \int_0^\infty e^{-\lambda \theta_0} p(d\lambda \mid n, t_1 \ldots t_r) = \left(\frac{T}{T+\theta_0}\right)^{r+1}.$$

Thus, the Bayesian test which ensures, with 100 P percent probability, that the life of an *individual unit* is at least θ_0, has an acceptance criterion that the expression (13) is $\geq P$; a result which is simple, sensible, and as far as I can see, utterly beyond the reach of orthodox statistics.

The Bayesian tests (11) and (13) are, however, still based on a ridiculous prior for λ; another improvement, even further beyond the reach of orthodox statistics, is found as a result of using a reasonable prior. In 'real life' we usually have excellent grounds based on previous experience and theoretical analyses, for predicting the general order of magnitude of the lifetime in advance of the test. It would be inconsistent from the standpoint of inductive logic, and wasteful economically, for us to fail to take this prior knowledge into account.

Suppose that initially, we have grounds for expecting a mean life of the order of t_i; or a failure rate of about $\lambda_i \cong t_i^{-1}$. However, the prior information does not justify our being too dogmatic about it; to assign a prior centered sharply about λ_i would be to assert so much prior knowledge that we scarcely need any test. Thus, we should assign a prior that, while incorporating the number t_i, is still as 'spread out' as possible, in some sense.

Using the criterion of maximum entropy, we choose that prior density $p_i(\lambda)$ which, while yielding an expectation equal to λ_i, maximizes the 'measure of ignorance' $H = -\int p_i(\lambda) \log p_i(\lambda) \, d\lambda$. The solution is: $p_i(\lambda) = t_i \exp(-\lambda t_i)$. Repeating the above derivation with this prior, we find that the posterior distribution (9) and its consequences (11)–(13) still hold, but that Equation (11) is now to be replaced by

(14) $\quad T = r\bar{t} + (n-r)\,t + t_i$.

Subjecting the resulting solution to various extreme conditions now shows an excellent correspondence with the indications of common sense. For example, if the total unit-hours of the test is small compared to t_i, then our state of knowledge about λ can hardly be changed by the test, unless an unexpectedly large number of failures occurs. But if the total unit-hours of the test is large compared to t_i, then for all practical purposes our final conclusions depend only on what we observed in the test, and are almost independent of what we thought previously. In intermediate cases, our prior knowledge has a weight comparable to that of the test; and if $t_i \gtrsim \theta_0$, the amount of testing required is appreciably reduced. For, if we were already quite sure the units *are* satisfactory, then we require less additional evidence before accepting them. On the other hand, if $t_i \ll \theta_0$, the test approaches the one based on a uniform prior; if we are initially very doubtful about the units, then we demand that the test itself provide compelling evidence in favor of them.

These common-sense conclusions have, of course, been recognized qualitatively by orthodox statisticians; but only the Bayesian approach leads automatically to a means of expressing all of them explicitly and quantitatively in our equations. As noted by Lehmann (1959), the orthodox statistician can and does take his prior information into account, in some degree, by moving his significance level up and down in a way suggested by the prior information. But, having no formal principle like maximum entropy that tells him how much to move it, the resulting procedure is far more 'subjective' (in the sense of varying with the taste of the individual) than anything in the Bayesian approach which recognizes the role of maximum entropy and transformation groups in determining priors.

No doubt, the completely indoctrinated orthodoxian will continue to reject priors based even on the completely impersonal (and parameter-independent) principles of maximum entropy and transformation groups, on the grounds that they are still 'subjective' because they are not frequencies [although I believe I have shown (Jaynes, 1968, 1971) that if a random experiment is involved, the probabilities calculated from maximum entropy and transformation groups have just as definite a connection with frequencies as probabilities calculated from any other principle of probability theory]. In particular, he would claim that the prior just introduced into the ST test represents a dangerous loss of 'objectivity' of that test.

To this I would reply that the judgment of a competent engineer, based on data of past experience in the field, represents information fully as 'objective' and reliable as anything we can possibly learn from a random experiment. Indeed, most engineers would make a stronger statement; since a random experiment is, by definition, one in which the outcome – and therefore the conclusion we draw from it – is subject to uncontrollable variations, it follows that the only fully 'objective' means of judging the reliability of a system is through analysis of stresses, rate of wear, etc., which avoids random experiments altogether.

In practice, the real function of a reliability test is to check against the possibility of completely unexpected modes of failure; once a given failure mode is recognized and its mechanism understood, no sane engineer would dream of judging its chances of occurring merely from a random experiment.

(d) SUMMARY

In the article of Bross (1963) – and in other places throughout the orthodox literature – one finds the claim that orthodox significance tests are 'objective' and 'scientific', while the Bayesian approach to these problems is erroneous and/or incapable of being applied in practice. The above comparisons have covered some important types of tests arising in everyday practice, and in no case have we found any evidence for the alleged superiority, or greater applicability, of orthodox tests. In every case, we have found clear evidence of the opposite.

The mathematical situation, as found in these comparisons and in many others, is just this: some orthodox tests are equivalent to the Bayesian ones based on non-informative priors, and some others, when sufficiently improved both in procedure and in manner of reporting the results, can be made Bayes-equivalent. We have found this situation when the orthodox test was (A) based on a sufficient statistic, and (B) free of nuisance parameters. In this case, we always have asymptotic equivalence for tests of a simple hypothesis against a one-sided alternative. But we often find exact equivalence for all sample sizes, for simple mathematical reasons; and this is true of almost all tests which the orthodox statistician himself considers fully satisfactory.

The orthodox t-test of the hypothesis $\mu=\mu_0$ against the alternative $\mu>\mu_0$ is exactly equivalent to the Bayesian test for reasons of symmetry; and there are several cases of exact equivalence even when the distribution is not symmetrical in parameter and estimator. Thus, for the Poisson distribution the orthodox test for $\lambda=\lambda_0$ against $\lambda>\lambda_0$ is exactly equivalent to the Bayesian test because of the identity

$$\frac{1}{n!}\int_\lambda^\infty x^n e^{-x}\,dx = \sum_{k=0}^n \frac{e^{-\lambda}\lambda^k}{k!}$$

and the orthodox F-test for $\sigma_1=\sigma_2$ against $\sigma_1>\sigma_2$ is exactly Bayes-equivalent because of the identity

$$\frac{(n+m+1)!}{n!\,m!}\int_0^P x^n(1-x)^m\,dx = \sum_{k=0}^m \frac{(n+k)!}{n!\,k!}P^{n+1}(1-P)^k.$$

In these cases, two opposed ideologies lead to just the same final working equations.

If there is no single sufficient statistic (as in the ST test) the orthodox approach can become extremely complicated. If there are nuisance parameters (as in the problem of two means), the orthodox approach is faced with serious difficulties of principle; it has not yet produced any unambiguous and fully satisfactory way of dealing with such problems.

In the Bayesian approach, neither of these circumstances caused any difficulty; we proceeded in a few lines to a definite and useful solution. Furthermore, Bayesian significance tests are readily extended to permit us to draw inferences about the specific case at hand, rather than about some purely imaginary 'population' of cases. In most real applications, it is just the specific case at hand that is of concern to us; and it is hard to see how frequency statements about a mythical population or an imaginary experiment can be considered any more 'objective' than the Bayesian statements. Finally, no statistical method which fails to provide any way of taking prior information into account can be considered a full treatment of the problem; it will be evident from our previous work (Jaynes, 1968) and the above example, that Bayesian significance tests are extended just as readily to incorporate any testable prior information.

III. TWO-SIDED CONFIDENCE INTERVALS

> The merit of the estimator is judged by the distribution of estimates to which it gives rise, i.e., by the properties of its sampling distribution.
>
> We must content ourselves with formulating a rule which will give good results 'in the long run' or 'on the average' ....
>
> Kendall and Stuart (1961)

The above examples involved some one-sided confidence intervals, and they revealed some cogent evidence concerning the role of sufficiency and nuisance parameters; but they were not well adapted to studying the principle of reasoning behind them. When we turn to the general principle of two-sided confidence intervals some interesting new features appear.

(a) EXAMPLE 4. BINOMIAL DISTRIBUTION

Consider Bernoulli trials B_2 (i.e., two possible outcomes at each trial, independence of different trials). We observe r successes in n trials, and asked to estimate the limiting frequency of success f, and give a statement about the accuracy of the estimate. In the Bayesian approach, this is a very elementary problem; in the case of a uniform prior density for f [the basis of which we have indicated elsewhere (Jaynes, 1968) in terms of transformation groups; it corresponds to prior knowledge that it is *possible* for the experiment to yield either success or failure], the posterior distribution is proportional to $f^r(1-f)^{n-r}$ as found in Bayes' original memoir, with mean value $\hat{f} = (r+1)/(n+2)$ as given by Laplace (1774), and variance $\sigma^2 = \hat{f}(1-\hat{f})/(N+3)$.

The $(\hat{f} \pm \sigma)$ thus found provide a good statement of the 'best' estimate of f, and if $\hat{f}$ is not too close to 0 or 1, an interval within which the true value is reasonably likely to be. The full posterior distribution of f yields more detailed statements; if $r \gg 1$ and $(n-r) \gg 1$, it goes into a normal distribution $(\bar{f}, \sigma)$. The $100\,P$ percent interval (i.e., the interval which contains $100\,P$ percent of the posterior probability) is then simply $(\bar{f} \pm q\sigma)$, where q is the $(1+P)/2$ percentile of the normal distribution; for the 90, 95, and 99% levels, $q = 1.645, 1.960, 2.576$ respectively.

When we treat this same problem by confidence intervals, we find that it is no longer an undergraduate-level homework problem, but a research project. The final results are so complicated that they can hardly be expressed analytically at all, and we require a new series of tables and charts.

In all of probability theory there is no calculation which has been subjected to more sneering abuse from orthodox writers than the Bayesian one just described, which contains Laplace's rule of succession. But suppose we take a glimpse at the final numerical results, comparing, say, the 90% confidence belts with the Bayesian 90% posterior probability belts.

This must be done with caution, because published confidence intervals all appear to have been calculated from approximate formulas which yield wider intervals than is needed for the stated confidence level. We use a recently published (Crow et al., 1960) recalculated table which, for the case $n = 10$, gives intervals about 0.06 units smaller than the older Pearson-Clopper values.

If we have observed 10 successes in 20 trials, the upper 90% confidence limit is given as 0.675; the above Bayesian formula gives 0.671. For 13 successes in 26 trials, the tabulated upper confidence limit is 0.658; the Bayesian result is 0.652.

Continued spot-checking of this kind leads one to conclude that, quite generally, the Bayesian belts lie just inside the confidence belts; the difference is visible graphically only for wide belts for which, in any event, no accurate statement about f was possible. The inaccuracy of published tables and charts is often greater than the difference between the Bayesian interval and the correct confidence interval. Evidently, then, claims for the superiority of the confidence interval must be based on something other than actual performance. The differences are so small that I could not magnify them into the region where common sense is able to judge the issue.

Once aware of these things the orthodox statistician might well decide to throw away his tables and charts, and obtain his confidence intervals from the Bayesian solution. Of course, if one demands very accurate intervals for very small samples, it would be necessary to go to the incomplete Beta-function tables; but it is hard to imagine any real problem where one would care about the exact width of a very wide belt. When $r \gg 1$ and $(n-r) \gg 1$, then to all the accuracy one can ordinarily use, the required interval is simply the above ($\bar{f} \pm q\sigma$). Since, as noted, published confidence intervals are 'conservative' – a common euphemism – he can even improve his results by this procedure.

Let us now seek another problem, where differences can be magnified to the point where the equations speak very clearly to our common sense.

(b) EXAMPLE 5. TRUNCATED EXPONENTIAL DISTRIBUTION

The following problem has occurred in several industrial quality control situations. A device will operate without failure for a time θ because of a protective chemical inhibitor injected into it; but at time θ the supply of this chemical is exhausted, and failures then commence, following the exponential failure law. It is not feasible to observe the depletion of this inhibitor directly; one can observe only the resulting failures. From data on actual failure times, estimate the time θ of guaranteed safe operation by a confidence interval. Here we have a continuous sample space, and we are to estimate a location parameter θ, from the sample values $\{x_1 ... x_N\}$, distributed according to the law

$$(15) \quad p(dx \mid \theta) = \begin{cases} \exp(\theta - x)\, dx, & x > \theta \\ 0, & x < \theta \end{cases}.$$

Let us compare the confidence intervals obtained from two different estimators with the Bayesian intervals. The population mean is $E(x) = \theta + 1$, and so

$$(16) \quad \theta^*(x_1 ... x_N) \equiv \frac{1}{N} \sum_{i=1}^{N} (x_i - 1)$$

is an unbiased estimator of θ. By a well-known theorem, it has variance $\sigma^2 = N^{-1}$, as we are accustomed to find. We must first find the sampling distribution of θ^*; by the method of characteristic functions we find that

CONFIDENCE INTERVALS VS BAYESIAN INTERVALS

it is proportional to $y^{N-1} \exp(-Ny)$ for $y>0$, where $y \equiv (\theta^* - \theta + 1)$. Evidently, it will not be feasible to find the shortest confidence interval in closed analytical form, so in order to prevent this example from growing into another research project, we specialize to the case $N=3$, suppose that the observed sample values were $\{x_1, x_2, x_3\} = \{12, 14, 16\}$; and ask only for the shortest 90% confidence interval.

A further integration then yields the cumulative distribution function $F(y) = [1 - (1 + 3y + 9y^2/2) \exp(-3y)]$, $y>0$. Any numbers y_1, y_2 satisfying $F(y_2) - F(y_1) = 0.9$ determine a 90% confidence interval. To find the shortest one, we impose in addition the constraint $F'(y_1) = F'(y_2)$. By computer, this yields the interval

(17) $\qquad \theta^* - 0.8529 < \theta < \theta^* + 0.8264$

or, with the above sample values, the shortest 90% confidence interval is

(18) $\qquad 12.1471 < \theta < 13.8264.$

The Bayesian solution is obtained from inspection of (15); with a constant prior density [which, as we have argued elsewhere (Jaynes, 1968) is the proper way to express complete ignorance of location parameter], the posterior density of θ will be

(19) $\qquad p(\theta | x_1 \ldots x_N) = \begin{Bmatrix} N \exp N(\theta - x_1), & \theta < x_1 \\ 0 & , \theta > x_1 \end{Bmatrix}$

where we have ordered the sample values so that x_1 denotes the least one observed. The shortest posterior probability belt that contains 100 P percent of the posterior probability is thus $(x_1 - q) < \theta < x_1$, where $q = -N^{-1} \log(1-P)$. For the above sample values we conclude (by slide-rule) that, with 90% probability, the true value of θ is contained in the interval

(20) $\qquad 11.23 < \theta < 12.0.$

Now what is the verdict of our common sense? The Bayesian interval corresponds quite nicely to our common sense; the confidence interval (18) is over twice as wide, and *it lies entirely in the region $\theta > x_1$ where it is obviously impossible for θ to be!*.

I first presented this result to a recent convention of reliability and quality control statisticians working in the computer and aerospace industries; and at this point the meeting was thrown into an uproar, about a dozen people trying to shout me down at once. They told me, "This is complete nonsense. A method as firmly established and thoroughly worked over as confidence intervals couldn't possibly do such a thing. You are maligning a very great man; Neyman would never have advocated a method that breaks down on such a simple problem. If you can't do your arithmetic right, you have no business running around giving talks like this".

After partial calm was restored, I went a second time, very slowly and carefully, through the numerical work leading to (18), with all of them leering at me, eager to see who would be the first to catch my mistake [it is easy to show the correctness of (18), at least to two figures, merely by applying parallel rulers to a graph of $F(y)$]. In the end they had to concede that my result was correct after all.

To make a long story short, my talk was extended to four hours (all afternoon), and their reaction finally changed to: "My God – why didn't somebody tell me about these things before? My professors and textbooks never said anything about this. Now I have to go back home and recheck everything I've done for years".

This incident makes an interesting commentary on the kind of indoctrination that teachers of orthodox statistics have been giving their students for two generations now.

(c) WHAT WENT WRONG?

Let us try to understand what is happening here. It is perfectly true that, *if* the distribution (15) is indeed identical with the limiting frequencies of various sample values, and *if* we could repeat all this an indefinitely large number of times, then use of the confidence interval (17) *would* lead us, in the long run, to a correct statement 90% of the time. But it would lead us to a wrong answer 100% of the time in the subclass of cases where $\theta^* > x_1 + 0.85$; and *we know from the sample whether we are in that subclass.*

That there must be a very basic fallacy in the reasoning underlying the principle of confidence intervals, is obvious from this example. The difficulty just exhibited is generally present in a weaker form, where it escapes detection. The trouble can be traced to two different causes.

Firstly, it has never been a part of 'official' doctrine that confidence intervals must be based on sufficient statistics; indeed, it is usually held to be a particular advantage of the confidence interval method that it leads to exact frequency-interpretable intervals without the need for this. Kendall and Stuart (1961), however, noting some of the difficulties that may arise, adopt a more cautious attitude and conclude (loc. cit., p. 153): "... confidence interval theory is possibly not so free from the need for sufficiency as might appear".

We suggest that the general situation, illustrated by the above example, is the following: whenever the confidence interval is not based on a sufficient statistic, it is possible to find a 'bad' subclass of samples, *recognizable from the sample,* in which use of the confidence interval would lead us to an incorrect statement more frequently than is indicated by the confidence level; and also a recognizable 'good' subclass in which the confidence interval is wider than it needs to be for the stated confidence level. The point is not that confidence intervals fail to do what is claimed for them; the point is that, if the confidence interval is not based on a sufficient statistic, it is possible to do better in the individual case by taking into account evidence from the sample that the confidence interval method throws away.

The Bayesian literature contains a multitude of arguments showing that it is precisely the original method of Bayes and Laplace which does take into account all the relevant information in the sample; and which will therefore always yield a superior result to any orthodox method not based on sufficient statistics. That the Bayesian method does have this property (i.e., the 'likelihood principle') is, in my opinion, now as firmly established as any proposition in statistics. Unfortunately, many orthodox textbook writers and teachers continue to ignore these arguments; for over a decade hardly a month has gone by without the appearance of some new textbook which carries on the indoctrination by failing to present both sides of the story.

If the confidence interval *is* based on a sufficient statistic, then as we saw in Example 4, it turns out to be so nearly equal to the Bayesian interval that it is difficult to produce any appreciable difference in the numerical results; in an astonishing number of cases, they are identical. That is the case in the example just given, where x_1 is a sufficient statistic, and it yields a confidence interval identical with the Bayesian one (20).

Similarly, the shortest confidence interval for the mean of a normal distribution, whether the variance is known or unknown; and for the variance of a normal distribution, whether the mean is known or unknown; and for the width of a rectangular distribution, all turn out to be identical with the shortest Bayesian intervals at the same level (based on a uniform prior density for location parameters and the Jeffreys prior $d\sigma/\sigma$ for scale parameters). Curiously, these are just the cases cited most often by textbook writers, after warning us not to use those erroneous Bayesian methods, as an illustration of their more 'objective' orthodox methods.

The second difficulty in the reasoning underlying confidence intervals concerns their criteria of performance. In both point and interval estimation, orthodox teaching holds that the reliability of an estimator is measured by its performance 'in the long run', i.e., by its sampling distribution. Now there are some cases (e.g., fixing insurance rates) in which long-run performance *is* the sole, all-important consideration; and in such cases one can have no real quarrel with the orthodox reasoning (although the same conclusions are found just as readily by Bayesian methods). However, in the great majority of real applications, long-run performance is of no concern to us, because it will never be realized.

Our job is not to follow blindly a rule which would prove correct 90% of the time in the long run; there are an infinite number of radically different rules, all with this property. Our job is to draw the conclusions that are most likely to be right in the specific case at hand; indeed, the problems in which it is most important that we get this theory right are just the ones (such as arise in geophysics, econometrics, or antimissile defense) where we know from the start that the experiment can *never* be repeated.

To put it differently, the sampling distribution of an estimator is not a measure of its reliability in the individual case, because considerations about samples that have *not* been observed, are simply not relevant to the problem of how we should reason from the one that *has* been observed. A doctor trying to diagnose the cause of Mr. Smith's stomachache would not be helped by statistics about the number of patients who complain instead of a sore arm or stiff neck.

This does not mean that there are no connections at all between individual case and long-run performance; for if we have found the procedure which is 'best' in each individual case, it is hard to see how it could fail to be 'best' also in the long run.

CONFIDENCE INTERVALS VS BAYESIAN INTERVALS

The point is that the converse does not hold; having found a rule whose long-run performance is proved to be as good as can be obtained, it does not follow that this rule is necessarily the best in any particular individual case. One can trade off increased reliability for one class of samples against decreased reliability for another, in a way that has no effect on long-run performance; but has a very large effect on performance in the individual case.

Now, if I closed the discussion of confidence intervals at this point, I know what would happen; because I have seen it happen several times. Many persons, victims of the aforementioned indoctrination, would deny and ridicule what was stated in the last five paragraphs, claim that I am making wild, irresponsible statements; and make some reference like that of Bross (1963) to the 'first-rate mathematicians' who have already looked into these matters.

So, let us turn to another example, in which the above assertions are demonstrated explicitly, and so simple that all calculations can be carried through analytically.

(d) EXAMPLE 6. THE CAUCHY DISTRIBUTION

We sample two members $\{x_1, x_2\}$ from the Cauchy population

$$(21) \quad p(dx \mid \theta) = \frac{1}{\pi} \frac{dx}{1 + (x - \theta)^2}$$

and from them we are to estimate the location parameter θ. The translational and permutation symmetry of this problem suggests that we use the estimator

$$(22) \quad \theta^*(x_1, x_2) = \tfrac{1}{2}(x_1 + x_2)$$

which has a sampling distribution $p(d\theta^* \mid \theta)$ identical with the original distribution (21); an interesting feature of the Cauchy law.

It is just this feature which betrays a slight difficulty with orthodox criteria of performance. For x_1, x_2, and θ^* have identical sampling distributions; and so according to orthodox teaching it cannot make any difference which we choose as our estimator, for either point or interval estimation. They will all give confidence intervals of the same length, and in the long run they will all yield correct statements equally often.

But now, suppose you are confronted with a *specific* problem; the first measurement gave $x_1 = 3$, the second $x_2 = 5$. You are not concerned in the slightest with the 'long run', because you know that, if your estimate of θ *in this specific case* is in error by more than one unit, the missile will be upon you, and you will not live to repeat the measurement. Are you now going to choose $x_1 = 3$ as your estimate when the evidence of that $x_2 = 5$ stares you in the face? I hardly think so! Our common sense thus forces us to recognize that, contrary to orthodox teaching, the reliability of an estimator is not determined merely by its sampling distribution.

The Bayesian analysis tells, us, in agreement with common sense, that for this sample, by the criterion of any loss function which is a monotonic increasing function of $|\theta^* - \theta|$ (and, of course, for which the expected loss converges), the estimator (22) is uniquely determined as the optimal one. By the quadratic loss criterion, $L(\theta^*, \theta) = (\theta^* - \theta)^2$, it is the unique optimal estimator whatever the sample values.

The confidence interval for this problem is easily found. The cumulative distribution of the estimator (22) is

$$(23) \qquad p(\theta^* < \theta' \mid \theta) = \tfrac{1}{2} + \frac{1}{\pi} \tan^{-1}(\theta' - \theta)$$

and so the shortest $100\,P$ percent confidence interval is

$$(24) \qquad (\theta^* - q) < \theta < (\theta^* + q)$$

where

$$(25) \qquad q = \tan(\pi P/2).$$

At the 90% level, $P = 0.9$, we find $q = \tan(81°) = 6.31$. Let us call this the 90% CI.

Now, does the CI make use of all the information in the sample that is relevant to the question being asked? Well, we have made use of $(x_1 + x_2)$; but we also know $(x_1 - x_2)$. Let us see whether this extra information from the individual sample can help us. Denote the sample half-range by

$$(26) \qquad y = \tfrac{1}{2}(x_1 - x_2).$$

The sampling distribution $p(\mathrm{d}y \mid \theta)$ is again a Cauchy distribution with the same width as (21) but with zero median.

CONFIDENCE INTERVALS VS BAYESIAN INTERVALS

Next, we transform the distribution of samples, $p(dx_1, dx_2 | \theta) = p(dx_1 | \theta) p(dx_2 | \theta)$ to the new variables (θ^*, y). The jacobian of the transformation is just 2, and so the joint distribution is

$$(27) \quad p(d\theta^*, dy | \theta) = \frac{2}{\pi^2} \frac{d\theta^* \, dy}{[1 + (\theta^* - \theta + y)^2][1 + (\theta^* - \theta - y)^2]}.$$

While (x_1, x_2) are independent, (θ^*, y) are not. The conditional cumulative distribution of θ^*, when y is known, is therefore not (23), but

$$(28) \quad p(\theta^* < \theta' | \theta, y) = \tfrac{1}{2} + \frac{1}{2\pi} [\tan^{-1}(\theta' - \theta + y) + \tan^{-1} \times \\ \times (\theta' - \theta - y)] + \frac{1}{4\pi y} \log \left[\frac{1 + (\theta' - \theta + y)^2}{1 + (\theta' - \theta - y)^2} \right]$$

and so, in the subclass of samples with given $(x_1 - x_2)$, the probability that the confidence interval (24) will yield a correct statement is not $P = (2/\pi) \tan^{-1} q$, but

$$(29) \quad w(y, q) = \frac{1}{\pi} [\tan^{-1}(q + y) + \tan^{-1}(q - y)] + \\ + \frac{1}{2\pi y} \log \left[\frac{1 + (q + y)^2}{1 + (q - y)^2} \right].$$

Numerical values computed from this equation are given in Table I,

TABLE I

Performance of the 90% confidence interval for various sample half-ranges y

y	$w(y, 6.31)$	$F(y)$
0	0.998	1.000
2	0.991	0.296
4	0.952	0.156
6	0.702	0.105
8	0.227	0.079
10	0.111	0.064
12	0.069	0.053
14	0.047	0.046
>14	$\dfrac{4q}{\pi(1+y^2)}$	$\dfrac{2}{\pi y}$

in which we give the actual frequency $w(y, 6.31)$ of correct statements obtained by use of the 90% confidence interval, for various half-ranges y. In the third column we give the fraction of all samples, $F(y) = (2/\pi) \tan^{-1}(1/y)$ which have half-range greater than y.

It appears that information about $(x_1 - x_2)$ was indeed relevant to the question being asked. In the long run, the 90% CI will deliver a right answer 90% of the time; however, its merits appear very different in the individual case. In the subclass of samples with reasonably small range, the 90% CI is too conservative; we can choose a considerably smaller interval and still make a correct statement 90% of the time. If we are so unfortunate as to get a sample with very wide range, then it is just too bad; but the above confidence interval would have given us a totally false idea of the reliability of our result. In the 6% of samples of widest range, the supposedly '90%' confidence interval actually yields a correct statement less than 10% of the time – a situation that ought to alarm us if confidence intervals are being used to help make important decisions.

The orthodox statistician can avoid this dangerous shortcoming of the confidence interval (24), without departing from his principles, by using instead a confidence interval based on the conditional distribution (28). For every sample he would choose a different interval located from (29) so as to be the shortest one which *in that subclass* will yield a correct statement 90% of the time. For small-range samples this will give a narrower interval, and for wide-range samples a correct statement more often, than will the confidence interval (24). Let us call this the 90% 'uniformly reliable' (UR) estimation rule.

Now let us see some numerical analysis of (29), showing how much improvement has been found. The 90% UR rule will also yield a correct statement 90% of the time; but for 87% of all samples (those with range less than 9.7) the UR interval is shorter than the confidence interval (24). For samples of very small range, it is 4.5 times shorter, and for half of all samples, the UR interval is less than a third of the confidence interval (24). In the 13% of samples of widest range, the confidence interval (24) yields correct statements less than 90% of the time, and so in order actually to achieve the claimed reliability, the UR interval must be wider, if we demand that it be simply connected. But we can find a UR region of two disconnected parts, whose total length remains less than a third of the CI (24) as $y \to \infty$.

The situation, therefore, is the following. For the few 'bad' samples of very wide range, no accurate estimate of θ is possible, and the confidence interval (24), being of fixed width, cannot deliver the presumed 90% reliability. In order to make up for this and hold the average success for all samples at 90%, it is then forced to cheat us for the great majority of 'good' samples by giving us an interval far wider than is needed. The UR rule never misleads us as to its reliability, neither underestimating it nor overestimating it for any sample; and for most samples it gives us a much shorter interval.

Finally, we note the Bayesian solution to this problem. The posterior distribution of θ is, from (21) in the case of a uniform prior density,

$$(30) \quad p(d\theta \mid x_1, x_2) = \frac{2}{\pi} \frac{(1 + y^2) \, d\theta}{[1 + (\theta - x_1)^2][1 + (\theta - x_2)^2]}$$

and, to find the shortest 90% posterior probability interval, we compute the cumulative distribution:

$$(31) \quad p(\theta < \theta' \mid x_1, x_2) = \tfrac{1}{2} + \frac{1}{2\pi}\left[\tan^{-1}(\theta' - x_1) + \tan^{-1}(\theta' - x_2)\right] + \frac{1}{4\pi y}\log\left[\frac{1 + (\theta' - x_2)^2}{1 + (\theta' - x_1)^2}\right]$$

and so, – but there is no need to go further. At this point, simply by comparing (31) with (28), the horrible truth appears: the uniformly reliable rule is precisely the Bayesian one! And yet, if I had simply introduced the Bayesian solution *ab initio*, the orthodox statistician would have rejected it instantly on grounds that have nothing to do with its performance.

(e) GENERAL PROOF

The phenomenon just illustrated is not peculiar to the Cauchy distribution or to small samples; it holds for any distribution with a location parameter. For, let the sampling distribution be

$$(32) \quad p(dx_1 \ldots dx_n \mid \theta) = f(x_1 \ldots x_n; \theta) \, dx_1 \ldots dx_n.$$

The statement that θ is a location parameter means that

$$(33) \quad f(x_1 + a, x_2 + a, \ldots x_n + a; \theta + a) = f(x_1 \ldots x_n; \theta),$$
$$-\infty < a < \infty.$$

Now transform the sample variables $\{x_1 \ldots x_n\}$ to a new set $\{y_1 \ldots y_n\}$:

(34) $\quad y_1 \equiv \bar{x} = n^{-1} \sum x_i$

(35) $\quad y_i = x_i - x_1, \quad i = 2, 3, \ldots n.$

From (33), (34), (35), the sampling distribution of the $\{y_1 \ldots y_n\}$ has the form

(36) $\quad p(dy_1 \ldots dy_n \mid \theta) = g(y_1 - \theta; y_2 \ldots y_n) \, dy_1 \ldots dy_n.$

If y_1 is not a sufficient statistic, a confidence interval based on the sampling distribution $p(dy_1 \mid \theta)$ will be subject to the same objection as was (24); i.e., knowledge of $\{y_2 \ldots y_n\}$ will enable us to define 'good' and 'bad' subclasses of samples, in which the reliability of the confidence interval is better or worse than indicated by the stated confidence level. To obtain the Uniformly Reliable interval, we must use instead the distribution conditional on all the 'ancillary statistics' $\{y_2 \ldots y_n\}$. This is

(37) $\quad p(dy_1 \mid y_2 \ldots y_n; \theta) = Kg(y_1 - \theta; y_2 \ldots y_n) \, dy_1$

where K is a normalizing constant. But the Bayesian posterior distribution of θ based on uniform prior is:

$$p(d\theta \mid x_1 \ldots x_n) = p(d\theta \mid y_1 \ldots y_n) =$$
(38) $\quad = Kg(y_1 - \theta; y_2 \ldots y_n) \, d\theta$

which has exactly the same density function as (37). Therefore, by a refined orthodox criterion of performance, the 'best', (i.e., Uniformly Reliable) confidence interval for any location parameter is identical with the Bayesian posterior probability interval (based on a uniform prior) at the same level.

With a scale parameter σ, data $\{q_1 \ldots q_n\}$, set $\theta = \log \sigma$, $x_i = \log q_i$, and the above argument still holds; the UR confidence interval for any scale parameter is identical with the Bayesian interval based on the Jeffreys prior $d\sigma/\sigma$.

IV. POLEMICS

Seeing the above comparisons, one naturally asks: on what grounds was it ever supposed that confidence intervals represent an advance over the

CONFIDENCE INTERVALS VS BAYESIAN INTERVALS   183

original treatment of Laplace? On this point the record is clear and abundant; orthodox arguments against Laplace's use of Bayes' theorem, and in favor of confidence intervals, have never considered such mundane things as demonstrable facts concerning performance. They consist of ideological slogans, such as "Probability statements can be made only about random variables. It is meaningless to speak of the probability that θ lies in a certain interval, because θ is not a random variable, but only an unknown constant".

On such grounds we are to be denied the derivation via Equations (1), (6), (9), (19), (30), (38) which in each case leads us in a few lines to a result that is either the same as the best orthodox result or demonstrably superior to it. On such grounds it is held to be very important that we use the words, "the probability that the interval covers the true value of θ" and we must *never, never* say, "the probability that the true value of θ lies in the interval". Whenever I hear someone belabor this distinction, I feel like the little boy in the fable of the Emperor's New Clothes.

Suppose someone proposes to you a new method for carrying out the operations of elementary arithmetic. He offers scathing denunciations of previous methods, in which he never examines the results they give, but attacks their underlying philosophy. But you discover that application of the new method leads to the conclusion that $2+2=5$. I think all protestations to the effect that, "Well, the case of $2+2$ is a peculiar pathological one, and I didn't intend the method to be used there", will fall on deaf ears. A method of reasoning which leads to an absurd result in *one* problem is thereby proved to contain a fallacy. At least, that is a rule of evidence universally accepted by scientists and mathematicians.

Orthodox statisticians appear to use different rules of evidence. It is clear from the foregoing that one can produce any number of examples, at first sight quite innocent-looking, in which use of confidence intervals or orthodox significance tests leads to absurd or dangerously misleading results. And, note that the above examples are not pathological freaks; every one of them is an important case that arises repeatedly in current practice. To the best of my knowledge, nobody has ever produced an example where the Bayesian method fails to yield a reasonable result; indeed, in the above examples, and in those noted by Kendall and Stuart (1961), the only cases where confidence intervals appear satisfactory at all are just the ones where they agree closely (or often exactly) with the Bayesian intervals. From our general proof, we understand why. And, year after year, the printing presses continue to pour out textbooks whose authors extoll the virtues of confidence intervals and warn the student against the thoroughly discredited method of Bayes and Laplace.

A physicist viewing this situation finds it quite beyond human understanding. I don't think the history of science can offer any other example in which a method which has always succeeded was rejected on doctrinaire grounds in favor of one which often fails.

Proponents of the orthodox view often describe themselves, as did Bross (1963), as 'objective', and 'fact-oriented', thereby implying that Bayesians are not. But the foundation-stone of the orthodox school of thought is this dogmatic insistence that the word 'probability' *must* be interpreted as 'frequency in some random experiment'; and that any other meaning is metaphysical nonsense. Now, assertions about the 'true meaning of probability', whether made by the orthodox or the Bayesian, are not statements of demonstrable fact. They are statements of ideological belief about a matter that cannot be settled by logical demonstration, or by taking votes. The only fully objective, fact-oriented criterion we have for deciding issues of this type, is just the one scientists use to test any theory: sweeping aside all philosophical clutter, which approach leads us to the more reasonable and useful results? I propose that we make some use of this criterion in future discussions.

Mathematically, or conceptually, there is absolutely nothing to prevent us from using probability theory in the broader Laplace interpretation, as the 'calculus of inductive reasoning'. Evidence of the type given above indicates that to do so greatly increases both the power and the simplicity of statistical methods; in almost every case, the Bayesian result required far less calculation. The main reason for this is that both the *ad hoc* step of 'choosing a statistic' and the ensuing mathematical problem of finding its sampling distribution, are eliminated. In particular, the F-test and the t-test, which require considerable mathematical demonstration in the orthodox theory, can each be derived from Bayesian principles in a few lines of the most elementary mathematics; the evidence of the sample is already fully displayed in the likelihood function, which can be written down immediately.

Now, I understand that there are some who are not only frightened to death by a prior probability, they do not even believe this last statement, the so-called 'likelihood principle', although a proof has been given (Birnbaum, 1962). However, I don't think we need a separate formal proof if we look at it this way. Nobody questions the validity of applying Bayes' theorem in the case where the parameter θ is itself a 'random variable'. But in this case the entire evidence provided by the sample *is* contained in the likelihood function; independently of the prior distribution, different intervals $d\theta$ are indicated by the sample to an extent precisely proportional to $L(\theta) d\theta$. It is already conceded by all that the likelihood function has this property when θ is a random variable with an arbitrary frequency distribution; is it then going to lose this property in the special case where θ is a constant? Indeed, isn't it a matter of the most elementary common sense to recognize that, in the specific problem at hand, θ is always just an unknown constant? Whether it would or would not be different in some other case that we are not reasoning about, is just not relevant to our problem; to adopt different methods on such grounds is to commit the most obvious inconsistency.

I am unable to see why 'objectivity' requires us to interpret every probability as a frequency in some random experiment; particularly when we note that in virtually every problem of real life, the direct probabilities are not determined by any real random experiment; they are calculated from a theoretical model whose choice involves 'subjective' judgment. The most 'objective' probabilities appearing in most problems are, therefore, frequencies only in an *ad hoc,* imaginary universe invented just for the purpose of allowing a frequency interpretation. The Bayesian could also, with equal ease and equal justification, conjure up an imaginary universe in which all his probabilities are frequencies; but it is idle to pretend that a mere act of the imagination can confer any greater objectivity on our methods.

According to Bayes' theorem, the posterior probability is found by multiplying the prior probability by a numerical factor, which is determined by the data and the model. The posterior probabilities therefore partake of whatever 'qualities' the priors have:

(A) If the prior probabilities are real frequencies, then the posterior probabilities are also real frequencies.

(B) If the prior probabilities are frequencies in an imaginary universe, then the posterior probabilities are frequencies in that same universe.

(C) If the prior probabilities represent what it is reasonable to believe before the experiment, by any criterion of 'reasonable', then the posterior probabilities will represent what it is equally reasonable to believe after the experiment, by the same criterion.

In no case are there any grounds for questioning the use of Bayes' theorem, which after all is just the condition for consistency of the product rule of probability theory; i.e., $p(AB \mid C)$ is symmetric in the propositions A and B, and so it can be expanded two different ways: $p(AB \mid C) = p(A \mid BC) p(B \mid C) = p(B \mid AC) p(A \mid C)$. If $p(B \mid C) \neq 0$, the last equality is just Bayes' theorem:

$$P(A \mid BC) = p(A \mid C) \frac{P(B \mid AC)}{P(B \mid C)}.$$

To recognize these things in no way forces us to accept the 'personalistic' view of probability (Savage, 1954, 1962). 'Objectivity' clearly does demand at least this much: the results of a statistical analysis ought to be independent of the personality of the user. In particular, our prior probabilities should describe the prior information; and not anybody's vague personal feelings.

At present, this is an ideal that is fully achieved only in particularly simple cases where all the prior information is testable in the sense defined previously (Jaynes, 1968). In the case of the aforementioned 'competent engineer' the determination of the exact prior is, of course, not yet completely formalized. But, as stressed before, the measure of our success in achieving 'objectivity' is just the extent to which we are able to eliminate all personalistic elements, and approach a completely 'impersonalistic' theory of inference or decision; on this point I must agree whole-heartedly with orthodox statisticians.

The real issue facing us is not an absolute value judgment but a relative one; it is not whether Bayesian methods are 100% perfect, or whether their underlying philosophy is opprobrious; but simply whether, at the present time, they are better or worse than orthodox methods in the results they give in practice. Comparisons of the type given here and in the aforementioned Underground Literature – and the failure of orthodoxy to produce any counter-examples – show that the original statistical methods of Laplace stand today in a position of proven superiority, that places them beyond the reach of attacks on the philosophical level, and *a fortiori* beyond any need for defense on that level.

Presumably, the future will bring us still better statistical methods; I predict that these will be found through further refinement and generalization of our present Bayesian principles. After all, the unsolved problems of Bayesian statistics are ones (such as treatment of nontestable prior information) that, for the most part, go so far beyond the domain of orthodox methods that they cannot even be formulated in orthodox terms.

It would seem to me, therefore, that instead of attacking Bayesian methods because we still have unsolved problems, a rational person would want to be constructive and recognize the unsolved problems as the areas where it is important that further research be done. My work on maximum entropy and transformation groups is an attempt to contribute to, and not to tear down, the beautiful and powerful intellectual achievement that the world owes to Bayes and Laplace.

Dept. of Physics, Washington University,
St. Louis, Missouri 63130

REFERENCES

Note: Two recent objections to the principle of maximum entropy (Rowlinson, 1970; Friedman and Shimony, 1971) appear to be based on misunderstandings of work done seventeen years ago (Jaynes, 1957). In the meantime, these objections had been anticipated and answered in other articles (particularly Jaynes, 1965, 1967, 1968), of which these authors take no note. To help avoid further misunderstandings of this kind, the following references include a complete list of my publications in which maximum entropy is discussed, although not all are relevant to the present topic of Bayesian interval estimation.

Bayes, Rev. Thomas, 'An Essay Toward Solving a Problem in the Doctrine of Chances', *Phil. Trans. Roy. Soc.* 330–418 (1763). Reprint, with biographical note by G. A. Barnard, in *Biometrika* 45, 293–315 (1958) and in *Studies in the History of Statistics and Probability*, E. S. Pearson and M. G. Kendall, (eds), C. Griffin and Co. Ltd., London, (1970). Also reprinted in *Two Papers by Bayes with Commentaries*, (W. E. Deming, ed.), Hafner Publishing Co., New York, (1963).

Birnbaum, Allen, 'On the Foundations of Statistical Inference', *J. Am. Stat. Ass'n* 57 269 (1962).

Bross, Irwin D. J., 'Linguistic Analysis of a Statistical Controversy', *The Am. Statist.* 17, 18 (1963).

Cox, D. R., 'Some Problems Connected with Statistical Inference', *Ann. Math. Stat.* 29, 357 (1958).

Crow, E. L., Davis, F. A., and Maxfield, M. W., *Statistics Manual*, Dover Publications, Inc., New York (1960).

Fisher, R. A., *Statistical Methods and Scientific Inference*, Hafner Publishing Co., New York (1956).
Friedman, K. and Shimony, A., 'Jaynes' Maximum Entropy Prescription and Probability Theory', *J. Stat. Phys.* 3, 381–384 (1971).
Good, I. J., *Probability and The Weighing of Evidence*, C. Griffin and Co. Ltd., London (1950).
Good, I. J., *The Estimation of Probabilities*, Research Monograph #30, The MIT Press, Cambridge, Mass. (1965); paperback edition, 1968.
Jaynes, E. T., 'Information Theory and Statistical Mechanics, I, II', *Phys. Rev.* **106**, 620–630; **108**, 171–190 (1957).
Jaynes, E. T., *Probability Theory in Science and Engineering*, No. 4 of *Colloquium Lectures on Pure and Applied Science*, Socony-Mobil Oil Co., Dallas, Texas (1958).
Jaynes, E. T., 'Note on Unique Decipherability', IRE Trans. on Information Theory, p. 98 (September 1959).
Jaynes, E. T., 'New Engineering Applications of Information Theory', in *Engineering Uses of Random Function Theory and Probability*, J. L. Bogdanoff and F. Kozin, (eds.), J. Wiley & Sons, Inc., N.Y. (1963); pp. 163–203.
Jaynes, E. T., 'Information Theory and Statistical Mechanics', in *Statistical Physics*, K. W. Ford, (ed.), W. A. Benjamin, Inc., (1963); pp. 181–218.
Jaynes, E. T., 'Gibbs vs. Boltzmann Entropies', *Am. J. Phys.* **33**, 391 (1965).
Jaynes, E. T., 'Foundations of Probability Theory and Statistical Mechanics', Chap. 6 in *Delaware Seminar in Foundations of Physics*, M. Bunge, (ed.), Springer-Verlag, Berlin (1967); Spanish translation in *Modern Physics*, David Webber, (ed.), Alianza Editorial s/a, Madrid 33 (1973).
Jaynes, E. T., 'Prior Probabilities', IEEE Trans. on System Science and Cybernetics, SSC-4, (September 1968), pp. 227–241.
Jaynes, E. T., 'The Well-Posed Problem', in *Foundations of Statistical Inference*, V. P. Godambe and D. A. Sprott, (eds.), Holt, Rinehart and Winston of Canada, Toronto (1971).
Jaynes, E. T., 'Survey of the Present Status of Neoclassical Radiation Theory', in *Coherence and Quantum Optics*, L. Mandel and E. Wolf, (eds.), Plenum Publishing Corp., New York (1973), pp. 35–81.
Jeffreys, H., *Theory of Probability*, Oxford University Press (1939).
Jeffreys, H., *Scientific Inference*, Cambridge University Press (1957).
Kempthorne, O., 'Probability, Statistics, and the Knowledge Business', in *Foundations of Statistical Inference*, V. P. Godambe and D. A. Sprott, (eds.), Holt, Rinehart and Winston of Canada, Toronto (1971).
Kendall, M. G. and Stuart, A., *The Advanced Theory of Statistics*, Volume 2, C. Griffin and Co., Ltd., London (1961).
Lehmann, E. L., *Testing Statistical Hypotheses*, J. Wiley & Sons, Inc., New York (1959), p. 62.
Pearson, E. S., Discussion in Savage (1962); p. 57.
Roberts, Norman A., *Mathematical Methods in Reliability Engineering*, McGraw-Hill Book Co., Inc., New York (1964) pp. 86–88.
Rowlinson, J. S., 'Probability, Information and Entropy', *Nature* **225**, 1196–1198 (1970).
Savage, L. J., *The Foundations of Statistics*, John Wiley, & Sons, Inc., New York (1954).

Savage, L. J., *The Foundations of Statistical Inference*, John Wiley & Sons, Inc., New York (1962).
Schlaifer, R., *Probability and Statistics for Business Decisions*, McGraw-Hill Book Co., Inc., New York (1959).
Sobel, M. and Tischendorf, J. A., Proc. Fifth Nat'l Symposium on Reliability and Quality Control, I.R.E., pp. 108-118 (1959).
Smith, C. A. B., Discussion in Savage (1962); p. 60.

NOTES

[1] Supported by the Air Force Office of Scientific Research, Contract No. F44620-60-0121.

[2] For those who had hoped, or at least expected, to hear instead a summary of the present status of maximum entropy, see the Note at the beginning of the References.

[3] This analysis is mathematically equivalent to use of the Behrens-Fisher distribution; however, the numerical work was done directly from Equation (1) rather than relying on tables which have been so little used and which would require a risky kind of interpolation. The first integration can be done analytically, and the second is easily done numerically to all the accuracy needed. Tail areas for $a<0$ need not be truncated, since they contribute to (1) only in the sixth decimal place.

[4] IBM 7092 calculation by Mr. Robert Schainker. Using the Jeffreys prior, $d\sigma/\sigma$, the posterior distributions have the form $p(d\sigma \mid s) = x^r e^{-x} dx/r!$, where $x \equiv ns^2/2\sigma^2$, $2r = n-3$, and $s_1^2 = 2,237$, etc. The required probability is then an integral like (1), which can be expressed as a finite sum for numerical work. Alternatively, it can be expressed in terms of the incomplete Beta function, so that in principle the F-tables could be used; however, these tables use too widely separated values of the significance level for accurate interpolation.

JAYNES' REPLY TO KEMPTHORNE'S COMMENTS

Such a magnificent confirmation of my main thesis could hardly have been hoped for; he has surely silenced those critics who thought that my account of the orthodox position was exaggerated.

Before venturing into areas where we presently differ I want to say that, during our five days acquaintance at this Conference, I have developed a warm personal affection for Oscar Kempthorne, and came to seek him out for many between-sessions and after-dinner discussions, all pleasant and valuable to me for reasons ranging from his interesting comments to the aroma of his cigars. Although it may not be apparent to the casual reader, there is a very wide area of agreement between us; on most of the issues discussed at this Conference, we would stand together.

For example, we both see at a glance the sterility of efforts to refine the mathematics without refining the concepts; or to axiomatize old ideas without any creative development of new ones. We are, I think, equally appalled at the prospect of changing the principles of logic to accommodate an illogical theory of physics.

We both tend to place more emphasis on the practical working rules and less on highflown mathematical and philosophical aspects of statistics than some of our younger colleagues, because we have seen enough ambitious but short-lived efforts with the generic title: 'A New Foundation for Statistics' to become a bit weary of them. And we have seen enough putative 'foundations' develop a fluid character unlike real foundations and adapt themselves to the unyielding practical realities, to become a bit wary of them.

It is clear to me that, on a much deeper level than the superficial differences being aired here, Oscar Kempthorne and I are kindred souls, with the same basic outlook and value judgments. On studying his comments, I am convinced that our differences arise almost entirely from misapprehensions concerning the nature of Bayesian methods *as they exist today,* which could have been cleared up if only we had more time to thresh matters out. Surely, there is no difference in our real aims to improve the power and scope of statistical methods at the practical, working level.

But granting all this, the differences between us do involve issues of crucial importance to statistics, and it would be a disservice to minimize them. This 120-year-old hangup over prior probabilities, started by Boole, must come to an end, because it is the direct cause of the troubles that today prevent orthodox statistics from giving any useful solutions to many important, real statistical problems.

Thus, linear regression with both variables subject to error is one of the most common statistical problems faced by experimenters; yet orthodox theory is helpless to deal with it because with n data points we have $(n+2)$ nuisance parameters. In irreversible statistical mechanics, and in some mathematically similar problems of communication theory and business decisions, the only probabilities involved are prior probabilities. The possibility of any useful solutions at all depends on principles such as maximum entropy, for translating prior information into prior distributions.

This debate has gone on for over 100 years, with the same old arguments and counter-arguments repeated back and forth for generations, without ever getting anywhere. Philosophical disputation may be great fun; but through recorded history its score for actually solving problems is, I believe, precisely zero. Anybody who genuinely wants to see these issues resolved must recognize the need for a better method.

Now the present condition of statistics is just the condition physics was in until the late 16th century, when Galileo showed us a better method – the direct cause of the advances that physics has made since. Instead of arguing about how objects 'ought' to move according to some philosophical or theological preconceptions, or by quoting ancient authorities such as Aristotle, why don't we just use the evidence of our own eyes? We are surrounded daily by moving objects; so any proposed theory about how they move can be tested by direct observation of the facts.

But, as this Conference showed very dramatically, 400 years of 'enlightenment' have not changed basic human nature. Today, statisticians regard themselves as the guardians of 'scientific objectivity' in drawing conclusions from data. Yet when I suggested that their own methods be judged, not by the philosophical preconceptions underlying them, but by examination of the facts of their actual performance, this appeared to many – as I knew it would – just as radical and shocking at as it did to Galileo's contemporaries. After my talk, a half-dozen people remonstrated with me, trying to inform me about the terrible defects of Bayesian methods by repeating the same tired old Boole-Venn clichés that we all learned as children. Not one of these individuals took the slightest note of the contrary facts (the mathematically demonstrable relations between actual performance of Bayesian and orthodox methods) that I had just pointed out. So we had an exact 20'th century repetition of Galileo's experience with the colleague who refused to look through his telescope.

To answer fully every point raised by Kempthorne would require a document much longer than my original presentation. Therefore, this reply must be confined to a brief summary of the situation, followed by specific comments only on those points of fact which are of general interest, and which would propagate confusion if they were allowed to go unanswered.

SUMMARY

My presentation was concerned with examining the relative merits of orthodox and Bayesian statistical methods by considering specific real problems, giving for each *an* orthodox solution which has been advocated in the recent literature, and adding what cannot be found in that literature, namely *the* Bayesian solution *which makes use of the same information* (i.e., is based on a noninformative prior). In Example 3, we also examined the further improvement obtainable when definite prior information is put in by maximum entropy. From these comparisons, several substantive conclusions emerge, which can be summed up as follows: Orthodox methods, when improved to the maximum possible extent (by using one-sided tests, reporting critical significance levels, using sufficient statistics or conditioning on all ancillary information, etc.) become mathematically equivalent to the Bayesian methods based on noninformative priors, provided that no nuisance parameters are present, and a sufficient statistic or complete set of ancillary statistics exists. Otherwise, mathematical equivalence cannot be achieved, and magnification then shows the Bayesian result to be superior.

This conclusion is supported in part by general theorems, in part by examination of specific cases. By now, we have a multitude of specific worked-out examples supporting it; and anyone who has understood my analysis can see that we are prepared to mass-produce any number of additional examples. Orthodox statistics has yet to produce *one* counter-example. The reason for this is clear to one who has studied the theorems of R. T. Cox (1946, 1961). He shows that any method of plausible reasoning in which we represent degrees of plausibility by real numbers, is necessarily either equivalent to Laplace's, or inconsistent.

Even though an orthodox statistician may, in the words between his equations, vociferously denounce the use of Bayes' theorem, it is nevertheless a matter of straightforward mathematics to see if his actual conclusions can be derived from Bayes' theorem. Either they can or they cannot. If they can, then it is obvious that his rejection of the Bayesian method is not based on its actual performance. If the conclusions are different, then we have the opportunity to judge that difference by Galileo's method. If we can magnify the difference sufficiently, it will become quite obvious which method is giving sensible results, and which is not.

Let me stress this point. Doubtless, some readers will jump to the conclusion that I deliberately chose examples to support my prejudices; and that one can just as easily produce examples on the other side. In fact, I hope that every reader of the orthodox persuasion will come to exactly that conclusion, and set about immediately to produce six examples where an orthodox method yields a result that simple common sense can see is preferable to the Bayesian result. For it is not in the passive reading of my words, but in the active attempt to produce these counter-examples, that one's eyes will be opened.

(2) My topic was the relative merits of orthodox and Bayesian methods, and not how they correlate with intelligence. Not having studied the latter topic, I have nothing more to add to the conclusions already reported by professional statisticians, viz:

I believe, for example, that it would be very difficult to persuade an intelligent physicist that current statistical practice was sensible, but that there would be much less difficulty with an approach via likelihood and Bayes' theorem.

G. E. P. Box (1962)

A student of statistical methods tends to be one of two types; either he accepts the technique in its entirety and applies it to every conceivable situation, or he is more intelligent and questions the applicability at all.

O. Kempthorne (1952)

With regard to the other remark, I think an historical study would show that the reasons for the interest of both Laplace and Jeffreys in probability theory arose from the problem of extracting 'signals' (i.e., new systematic effects) from the 'noise' of imperfect observations, in astronomy and geophysics respectively. The procedures would today be called 'significance tests', and I wish every one who has not already done so, would read Jeffreys' (1939) beautiful and comprehensive chapters on significance tests, then compare them from the standpoint of solid content and usefulness in real problems, with any work ever written on the subject from the orthodox point of view.

Likewise, my own interest in statistics arose from problems of extracting signals from noise in several applications ranging from optimum design of radar receivers and magnetic resonance probes, to land mine detectors. I am on record (Jaynes, 1963) as claiming that there is no area of physics, from elementary particle theory to cyclotron design, in which the phenomenon of noise does not present itself.

In view of all this, one can imagine my consternation at the suggestion that "there has been little attention to unavoidable noise" in physics. Physicists were actively studying noise and, thanks to Laplace, knew the proper way to deal with it, long before there was any such thing as a Statistician.

(3) (a) Of course, by 'the orthodox solution' I mean the particular one *which I am describing*; and likewise for 'the Bayesian solution'. Of course, there are many different orthodox solutions to a given problem – but I think that is the last thing a defender of orthodoxy would wish to bring to our attention.

Dirac did not in any way suggest that "working statisticians would estimate a probability to be a negative number", as a reading of his lecture will show. On the other hand, it *is* a matter of documentable fact that some orthodox statisticians suggest estimating a parameter known to be positive by an estimator which can become negative for some samples [KF, p. 203, Equation (7.42)].

(b) It is really discouraging to find – 25 years after the birth of information theory (Shannon, 1948), 17 years after its bearing on the prior probability problem was shown (Jaynes, 1957), ten years after the generalization to continuous distributions (Jaynes, 1963), six years after the resulting functional analysis generalization of Gibb's work to irreversible statistical mechanics was given (Jaynes, 1967), five years after it was shown that the theory becomes parameter-independent if one uses the entropy relative to the invariant measure on the parameter space (Jaynes, 1968), and two years after the frequency interpretation of that invariant measure was demonstrated (Jaynes, 1971) – that an eminent worker in statistics is still writing that attempts to produce prior distributions by logical analysis have 'failed'.

It is true that the principles of maximum entropy and transformation groups have not yet led to the solutions of every conceivable statistical problem; and I know that there are some who reject the entire program just for that reason. Presumably these same critics do not condemn the use of insulin on the grounds that it will not cure all diseases. The point is that we have solved *some* problems, in a way which I believe will be recognized by history as the final answer; and in fact we have succeeded in a wide enough class of problems to cover perhaps 90% of current applications. Criticisms of Bayesian methods on the grounds that we still have unsolved problems, come with particularly ill grace from those who have in the past, by their discouraging negative attitude, done everything in their power to prevent these problems from being solved.

I would think that anyone might recognize that a meaningful comparison of Bayesian and orthodox solutions must use the Bayesian solution which makes use of the *same* information as does the orthodox solution. A Bayesian solution which makes use of extra prior information that the orthodox method cannot use at all, will of course be superior for that reason alone; it is more instructive – and in a sense fairer – to make comparisons using a Bayesian solution based on a noninformative prior. Now, a noninformative prior is one which is uniform, not necessarily with respect to Lebesgue measure for any particular choice of the parameter, but with respect to the invariant measure defined by the transformation group on the parameter space. As explained in my work referred to, this is just the mathematical statement of the basic desideratum of consistency: in two problems where we have the same prior information, we should assign the same prior probabilities.

My previous work (1968) shows how to construct priors for location and scale parameters, the rate constant of a Poisson process, and the parameter of a binomial distribution, by logical analysis. Evidently, the point needs to be made repeatedly and with more examples; so let me show briefly how to find the prior in the parameter space (α, β) of the standard regression problem $y = \alpha + \beta x$, by logical analysis, for the case that x, y are variables of the same kind (for example, the departure from average barometric pressure at New York and Boston), so that it is as natural to consider regression of $(x$ on $y)$ as $(y$ on $x)$. Given any proposed element of prior probability $f(\alpha, \beta)\,d\alpha\,d\beta$, interchange x and y. The estimated line becomes $x = \alpha' + \beta' y$, with a prior probability element $g(\alpha', \beta')\,d\alpha'\,d\beta'$. From the Jacobian of the transformation $\alpha' = -\beta^{-1}\alpha$, $\beta' = \beta^{-1}$, we find $g(\alpha'\beta') = \beta^3 f(\alpha, \beta)$. This transformation equation holds whatever the function f.

Now if we are 'completely ignorant' of (α, β), the interchange of (x, y) shouldn't matter; we are also 'completely ignorant' of (α', β'). But consistency demands that in two problems where we have the same state of knowledge, we must assign the same probabilities. Therefore f and g must be the same function; i.e., the prior density representing 'complete ignorance' must satisfy the functional equation $\beta^3 f(\alpha, \beta) = f(-\beta^{-1}\alpha, \beta^{-1})$, which has the solution $f(\alpha, \beta) = (1+\beta^2)^{-3/2}$. Thus, setting $\beta = \tan\theta$, the invariant measure of the parameter space is

$$d\mu = d\alpha\,d\sin\theta.$$

Why is this not uniformly distributed in θ rather than in $\sin\theta$? Answer: it is uniform in $\sin\theta$ only for fixed α; but under rotations of the (x, y) plane α also varies [indeed, under any Euclidean transformation $(x, y) \to (x', y')$, where $x = x'\cos\phi - y'\sin\phi + x_0$, $y = y'\cos\phi + x'\sin\phi + y_0$, the estimated line $y = \alpha + \beta x$ goes into $y' = \alpha' + \beta' x'$, where $\alpha' = (\alpha - y_0 + \beta x_0)/(\cos\phi + \beta\sin\phi)$, $\beta' = (\beta\cos\phi - \sin\phi)/(\cos\phi + \beta\sin\phi) = \tan\theta'$; and we readily verify the invariance: $d\alpha'\,d\sin\theta' = d\alpha\,d\sin\theta$, while $d\alpha\,d\theta$ is not invariant].

This invariance of the measure $d\mu$ means that, however we draw the x and y axes, the prior $d\mu = d\alpha\,d\sin\theta$ expresses exactly the same state of prior knowledge about the position of the regression line. It thus leaves the entire decision to the subsequent evidence of the sample – which, of course, is exactly what Fisher insisted that a method of inference ought to do. But as we see, if this is the property we want to have, the goal is not achieved by closing our eyes to the very existence of a prior. It can be achieved only by logical analysis showing us *which* prior has the desired property. If we do have relevant prior information, it can now be incorporated into the problem by finding the probability measure dp that maximizes the entropy relative to $d\mu$: $H = -\int dp\,\log(dp/d\mu)$, subject to whatever constraint the prior information imposes on dp; if the constraints take the form of mean values, this reduces to the canonical ensemble formalism of statistical mechanics of J. Willard Gibbs.

Now the simple facts, made understandable by Cox's theorems, illustrated in my presentation and in many other examples throughout the Bayesian literature, explain what we have observed throughout the history of orthodox statistics; every advance in orthodox practice has brought the actual procedures back closer and closer to the original methods of Laplace. The rise of decision theory was, in fact, the main spark that touched off the present 'Bayesian Revolution'. Other examples are Fisher's introduction of conditioning, discussed below, and his introduction of the notion of sufficiency.

The discovery of sufficiency was, of course, a great advance *in orthodox statistics*; because in an important class of problems it removed the ambiguity in deciding which statistic should be used; if a sufficient statistic for θ exists, it is rather hard to justify using any other for inference about θ, for reasons illustrated in my Example 5 and explained under 'What Went Wrong?' But in Bayesian statistics there never was any ambiguity of this type to resolve. Fisher's definition of sufficiency can be stated more succinctly (and in my view, more meaningfully) as: If the posterior distribution of θ depends on the sample $(x_1 \ldots x_n)$ only through the value of a certain function $\theta^*(x_1 \ldots x_n)$, then θ^* is a sufficient statistic for θ. Evidently, if a sufficient statistic exists, application of Bayes' theorem will lead us to it automatically without our having to take any special note of the idea. But Bayes' theorem will lead us to the optimum inference whether or not any sufficient statistic exists; i.e., sufficiency is a convenience affecting the amount of calculation but not the quality of the inference.

I am afraid that to castigate Bayesian methods, but not orthodox ones, on grounds of lack of uniqueness, is to get it exactly backwards. It is orthodox statistics that offers us many different solutions to a single problem, (i.e., given prior information, sampling distribution, and sample), depending on whose school of thought, whose textbook within that school, and even which chapter of that textbook, you read. An estimator ought to be unbiased, efficient, consistent, etc.; but in general orthodoxy gives us no criterion as to the relative importance of these, nor any method by which a 'best' estimator can be constructed. The use of an unbiased estimator or a shortest confidence interval will lead us to different conclusions with different choices of parameters. KF (p. 316) cannot make up their minds about whether to accept the principle of conditioning, and advocate significance tests in which the conclusions depend on the arbitrary ordering you or I might assign to data sets *which were not observed!* Indeed, there is scarcely any problem of inference for which KF offer any definite preferred solution; in most cases there is an inconclusive discussion that terminates abruptly with the remark that 'it is all very difficult', leaving the reader in utter confusion as to which method should be used. But with all this ambiguity, orthodox methods provide no means for taking prior information into account.

In sharp contrast to this, for a given sampling distribution and sample, different Bayesian results correspond, as rational inferences should, to and *only* to, differences in the prior information. When priors are determined by the principles of maximum entropy and transformation groups, Bayesian methods achieve complete invariance under parameter changes (Jaynes, 1968).

(4) We are now told that even to utter the words 'Fisher-Neyman-Pearson theory' is a calumny on Fisher's views (but apparently not on Neyman's or Pearson's); and again for the 'benefit' (precious little) of readers not present at the Conference, may I state that I first heard this phrase from the lips of Professor Oscar Kempthorne, shortly before my talk was given. I repeated it only to say that I would follow common practice by using the word 'orthodox' as an approximate synonym.

However, since the issue has been raised, I would like to state that the term 'Fisher-Neyman-Pearson approach' appears to me as an entirely accurate and appropriate term for a certain area of statistical thought. To use it is in no way to ignore, much less deny, the fact that there were differences between Fisher on the one hand, and Neyman-Pearson on the other. However, this should not blind us to the fact that there is a very much larger area of agreement; i.e., a corpus of ideas which are not in Bayesian statistics, but are common to the Fisher and Neyman-Pearson points of view and which therefore characterize their union. I refer to the ideas that (1) the word 'probability' must be used only in the sense of 'frequency in a random experiment', (2) inference requires that we find sampling distributions of some 'statistics' in addition to the direct sample distribution $p(dx \mid \theta)$, (3) the conclusions we draw from an experiment can depend on the probabilities of data sets which were not observed, or the psychological state of mind of the experimenter (optional stopping),

(4) we can improve the precision of our results by throwing away relevant information instead of taking it into account (the procedure euphemistically called 'randomization'), (5) the attempt to dispense with prior probabilities.

Recalling the difference between the Fisher and Neyman-Pearson camps over confidence intervals vs fiducial probabilities, let's just see how great this calumny is. Given a basic sample distribution $p(\mathrm{d}x \mid \theta)$, choose two 'statistics' $\theta_1(x_1...x_n)$, $\theta_2(x_1...x_n)$ such that $\mathrm{prob}(\theta_1 < \theta < \theta_2) = P$; this defines a $100\,P$ percent confidence interval. Letting $\theta_1 \to \theta_{\min}$, the lower bound of the parameter space, we have $\mathrm{prob}(\theta < \theta_2) = P$, which is Fisher's definition (Collected works, 27.253) of the fiducial distribution of θ, based on the statistic θ_2. As we see, the deep, profound difference in basic approach is fully as great as that between Tweedledee and Tweedledum.

The difference is not in the approach, but in the perception with which it was used. Fisher, with his vastly greater intuitive understanding, saw at once something which still does not seem generally recognized by others; that all this is valid only when we are using sufficient statistics. Even in the Fisher obituary notice, Kendall (1963, p. 4) questions the need for sufficiency. My Example 5 was intended to make Fisher's point by demonstrating just what can happen when we use a confidence interval not based on a sufficient statistic. Obviously, anyone who rejects fiducial probability, but endorses the use of confidence intervals, is not doing so on grounds of their actual performance.

(7) We apparently agree that a statistical method should be judged by the results it gives in practice. Well and good. However, I categorically deny that "the Bayesian idea was rejected by Boole, Venn, Fisher and Neyman" on these grounds. It is just the weakness of their work that they rejected Bayesian methods on purely philosophical or ideological grounds, *without* examining their actual performance.

Since the case of Boole and Venn has been brought up, let us examine the work of these gentlemen and see for ourselves the validity of their actual criticisms, and the accuracy with which their work is reported today in the orthodox literature. I believe that Boole, like most other critics of Laplace, failed to comprehend fully his definition of probability. Since Laplace has been quoted out of context so many times in this and other matters, let us take the trouble to quote his definition in full. The first volume of his *Théorie Analytique* is concerned with mathematical preliminaries, and the actual development of probability theory begins in Volume 2. The first sentence of Volume 2 is: "The probability of an event is the ratio of the number of cases favorable to it, to the number of all cases possible when nothing leads us to expect that any one of these cases should occur more than any other, which renders them, for us, equally possible".

This definition has stated only the finite discrete case, but we know how to generalize it. The point is that Laplace defined probability in a way which clearly represents *a state of knowledge*; and not a frequency. Of course, as Laplace demonstrates over and over again, connections between probability and frequency appear later, as mathematical consequences of the theory. I claim that these derivable connections (the limit theorems of Jacob Bernoulli and de Moivre-Laplace, Laplace's rule of succession, the de Finetti exchangeability theorem, etc.) include all the ones actually used in applications.

If one has no prior knowledge other than enumeration of the possibilities (i.e., specification of the sample space), then to assign equal probabilities is clearly the only honest way one can describe that state of knowledge. This can be formalized more completely than Laplace did, by the aforementioned desideratum of consistency: if we were to assign any distribution other than the uniform one it would be possible, by a mere permutation of labels, to exhibit a second problem in which our state of knowledge is exactly the same, but in which we are assigning different probabilities. But in this case Laplace surely considered the argument and result so obvious that he would insult the reader's intelligence by mentioning them. The only serious error Laplace made was overestimating the intelligence of his readers.

Boole (1854), not perceiving this, rejected Laplace's work on the ground that the prior was 'arbitrary', i.e., not determined by the data. He did *not* reject it in the ground of the actual performance of Laplace's results in the case of uniform prior because he, like Laplace's other critics, never bothered to examine the actual performance under these conditions, much less to compare it with alternative methods. Had he done so, he might have discovered the real facts about performance, presented 85 years later by Jeffreys. Curiously, Boole, after criticizing Laplace's prior distribution based on the principle of indifference, then invokes that principle to defend his own methods against the criticisms of Wilbraham (see several articles in *Phil. Mag*, Vols. vii and viii. 1854).

This brings up another matter that needs to be mentioned. Boole's unjust criticism of Laplace has been quoted approvingly, over and over again, in the orthodox literature, Fisher (1956) being a very generous contributor. But in that same literature, a conspiracy of silence hides the fact that Boole's own work on probability theory (Boole, 1854, Chapters 16–21) contains ludicrous errors, far worse than any committed by Laplace. Some were noted by Wilbraham (1854), McColl (1897) and Keynes (1921). See his Example 6, page 286, where by a confusion of propositions [taking the probability of the proposition: 'If X is true, Y is true' as the conditional probability $p(Y \mid X)$] he arrives at the conclusion that two propositions with the same truth value can have different probabilities. He not only fails to see the absurdity of this, but even calls it to the reader's attention as something which 'deserves to be specially noticed'. Or his solution to another problem, page 324, Equation (10), which reduces to an absurdity in the special cases $c_1 = c_2 = 1$ and $c_1 = p_1 = 1$. While Laplace considered real problems and got scientifically useful answers, Boole invented artificial school-room type problems, and often gave absurd answers. Finally, it is mathematically trivial to show that all of 'Boolean algebra' was contained already in the rules of probability theory given by Laplace – in the limit as all probabilities go to zero or unity, any equation of Laplace's 'Calculus of Inductive Reasoning' reduces to one of Boolean algebra.

Now let's turn to the case of Venn (1866), who expresses his disdain for mathematical demonstration very clearly throughout his book and its preface. Venn's Chapter 6 is an attack on Laplace's rule of succession, so viciously unfair that even Fisher (1956) was impelled to come to Laplace's defense on this issue. Fisher questions whether Venn was even aware of the fact that Laplace's rule had a mathematical basis, and like other mathematical theorems has 'stipulations specific for its validity'. He proceeds to give examples in which, unlike those of the 'great thinker' Venn, the stipulations are satisfied, and Laplace's rule is the correct one to use.

How is it possible for one human mind to reject Laplace's rule of succession; and then advocate a frequency definition of probability? Anybody who assigns a probability to an event equal to its observed frequency in many trials, is doing just what Laplace's rule tells him to do. In my Example 4, we examined Laplace's calculation underlying this rule, and learned that anybody who rejects Laplace's methods in favor of confidence intervals for the binomial, is certainly not doing so on grounds of actual performance.

I would like to plead here for a greater concern for historical accuracy, in writing on these matters. For over a century, there has been a conspiracy in the statistical literature to rewrite history and denigrate Laplace, first in the Boole-Venn manner, then by denying him credit when his principles were rediscovered (examples below). An *ad hominem* attack on Laplace (as 'a consummate politician') has even befouled the air of this Conference. I have long since learned never to accept the word of a biased source (Boole, Venn, Von Mises, Fisher, E. T. Bell, Cramér, Feller, etc.) on *any* question of what Laplace did or did not do. When working in my study, Laplace's *Théorie Analytique* is always at my elbow; and when any question about him comes up, I go straight to the original source. It is for this reason that my judgment of Laplace differs so radically from that presented in the literature from Boole on.

Not only those who are ignorant of history, but also those who will not profit by its lessons, are doomed to repeat it. Starting with Condorcet and his omelette, those who scorned Laplace's outlook and methods – whether in science or politics – and tried to do things differently, have shared a common experience.

(A) In George Gamow's book, *The Biography of the Earth* (1941), Laplace's theory of the origin of the solar system is torn to shreds. But in 1944, Weiszäcker pointed out a few things that Laplace's critics had overlooked; and the 1948 edition of Gamow's book had a new 15 page section entitled, '*Laplace was right after all!*'

(B) Abraham Wald, in his mimeographed course notes of 1941, rejected Laplace's methods of parameter estimation and hypothesis testing and asserted that such problems cannot be solved by the principles of probability theory. During the 1940's Wald sought a new foundation for statistics based on the idea of rational decisions, which had the aim of avoiding the mistakes of Laplace; but in Wald's final 1950 book, *Statistical Decision Functions,* the fundamental place of 'Bayes strategies' is finally recognized. As it turned out, Wald's life work was to prove, very much against his will, that the original methods developed by

Laplace in the 18'th century, which he and many other statisticians had scorned for years, were in fact the unique solution to the problem of rational decisions. *Laplace was right after all.*

(C) I had the same experience. In 1951, I somehow came to the conclusion that Bayes' theorem did not adequately represent the full variety of inductive reasoning, and sought to develop a two-valued theory of probability, very much like the one presented here by Shafer, except that my numbers corresponded to the sum and difference of his. I even expounded this in a Round Table Discussion at one of the Berkeley Statistical Symposiums. However, I then made the tactical error of trying to apply this theory to some real problems. At about the third attempt, the scales fell from my eyes and I saw that a two-valued theory contains nothing that is not already given by Laplace's original one-valued theory, by going to a deeper sample space. In other words, the defects that I thought I saw in Laplace's theory were my own defects, in not having the ingenuity to invent an adequate model. *Laplace was right after all.*

Now, I don't know how many other people are doomed to follow this path – already far more man-years of potentially useful talent have been wasted on futile attempts to evade Laplace's principles, than were ever invested in circle-squaring and perpetual motion machines. But just as Lindemann's proof put an end to circle-squaring for all who could see its implications, so Cox's theorems (1946) ought to have put an end, twenty-five years ago, to these unceasing efforts to evade what cannot be evaded. The situation is described in more detail in my review of Cox (1961). This is why I can say the following to latter-day Don Quixotes:

Many of us have already explored the road you are following, and we know what you will find at the end of it. It doesn't matter how many new words you drag into this discussion to avoid having to utter the word 'probability' in a sense different from frequency: likelihood, confidence, significance, propensity, support, credibility, acceptability, indifference, consonance, tenability, – and so on, until the resources of the good Dr Roget are exhausted. All of these are attempts to represent degrees of plausibility by real numbers, and they are covered automatically by Cox's theorems. It doesn't matter which approach you happen to like philosophically – by the time you have made your methods fully consistent, you will be forced, kicking and screaming, back to the ones given by Laplace. Until you have achieved mathematical equivalence with Laplace's methods, it will be possible, by looking at specific problems with Galileo's magnification, to exhibit the defects in your methods.

Here are two typical examples of the kind of factual distortion that we find in the literature. KF (p. 314) quote approvingly a statement of Fisher (1956, p. 4) that: "So early as Darwin's experiments on growth rate the need was felt for some sort of a test of whether an apparent effect might reasonably be due to chance". More specifically, Fisher (p. 81) then states that the 'Student' t-test was "the first exact test of significance." Neither book makes any mention of the historical fact that Laplace developed many significance tests to determine whether discrepancies between prediction and observation 'might reasonably be due to chance' and used them to decide which astronomical problems were worth working on: a bit of wisdom that might well be noted by scientists today. Laplace also illustrates the use of these tests, including two-way classifications, in many other problems of geodesy, meteorology, population statistics, etc. As I hope to show in detail elsewhere, Laplace's significance tests were in no way inferior – and were in some cases demonstrably superior – to tests advocated in the orthodox literature today.

Likewise, both KF and Fisher denounce the use of Bayes' theorem and uphold the 'student' t-test as a great advance in statistical practice; but of course neither mentions the fact that precisely the same result follows in two lines *from* Bayes' theorem; given the data $D=\{x_1 \ldots x_n\}$, the likelihood function is $L(\mu, \sigma)=\sigma^{-n} \exp(-nQ/2\sigma^2)$, where $Q=s^2+(\bar{x}-\mu)^2$. Integrating out σ with respect to Jeffreys' prior, the posterior density of μ is $\sim Q^{-n/2}$, which but for notation is just the t-distribution. Students reading these works obtain a completely false picture of both the historical and mathematical facts about significance tests.

THE WEATHERMAN'S JOB

In a certain city, the joint frequencies of the actual weather and the weatherman's predictions are given by:

		Actual	
		Rain	Shine
Predicted	Rain	¼	½
	Shine	0	¼

An enterprising fellow trained in orthodox statistics (but not in meteorology) notices that, while the weatherman is right only 50% of the time, a prediction of 'shine' everyday would be right 75% of the time, and applies for the weatherman's job. Should he get it? Which would you rather have in your city?

The weatherman is delivering useful information at a rate I = (entropy of distribution of predictions) + (entropy of actual weather distribution) − (entropy of joint distribution) = (0.562 + 0.562 − 1.040)/ln 2 = 0.123 bits/day. As explained previously (Jaynes, 1968) this means that in the course of a year the weatherman's information has reduced the number of reasonably probable weather sequences by a factor of $W = \exp(0.123 \times \times 365 \times \ln 2) = 2.92 \times 10^{13}$. With the weatherman on the job, you will never be caught out in an unpredicted rain; with the orthodox statistician this would happen to you one day out of four.

As this example once more forces one to recognize, the value of an inference lies in its usefulness *in the individual case*, and not in its long-run frequency of success; they are not necessarily even positively correlated. The question of how often a given situation would arise is utterly irrelevant to the question how we should reason when it *does* arise. I don't know how many times this simple fact will have to be pointed out before statisticians of 'frequentist' persuasions will take note of it; but I think it is important that we keep trying.

(15) *'Improper' Priors.* Let me try to explain the situation. 'Complete initial ignorance' of a scale parameter σ corresponds formally to use of the Jeffreys prior $d\sigma/\sigma = d \log \sigma$. But as noted before (Jaynes, 1968), to apply this within infinite limits $(-\infty < \log \sigma < \infty)$ would not represent any realistic state of prior information. For example, if x is a measured length of some material object on the earth, we surely know that the standard error σ_x of the measurement cannot be less than the size of one atom, $\sim 10^{-8}$ cm; or greater than the size of the earth, $\sim 10^9$ cm. So we know in advance that $(-8 < \log_{10} \sigma_x < +9)$. Outside this range, the prior density must be zero.

Similarly, if x is the measured breaking stress of some structural material, we know in advance that σ_x surely cannot be less than the pressure of sound waves, ~ 1 dyne cm^{-2}, due to people talking in the room; nor greater than 10^{14} dynes cm^{-2}, which is 1000 times the tensile strength of any known material. So the prior density must be all contained in (0 <

$< \ln \sigma_x < 33$). If x is a time interval measured in seconds, we can be pretty sure in advance that $(-12 < \log_{10} \sigma_x < 18)$.

Generally, thinking about any problem in this way will lead one to specify prior limits σ_{min}, σ_{max} within which the unknown value surely lies; within this interval the invariance arguments leading to the form $d\sigma/\sigma$ still apply if there is no other prior information (Jaynes, 1968). Therefore, the prior is normalizable, and we have a well-behaved mathematical problem.

Now if our final conclusions depend appreciably on the exact prior limits chosen, then obviously we should analyze our prior information more carefully than I did above, to get more reliable numerical values for σ_{min}, σ_{max}. But it just wouldn't be very intelligent to go to all that work, only to discover that σ_{min}, σ_{max} cancel out of the expressions representing our final conclusions (which might be the first few moments, or the quartiles, of a posterior distribution). So it will be good strategy to work through the solution first for general limits, whereupon the mathematics will tell us under just what conditions the prior limits matter; and when they don't.

Having thus formulated the problem, the conclusion is fairly obvious: if the likelihood function is sufficiently concentrated (i.e., if the experiment is a sufficiently informative one), then the prior limits cannot matter appreciably as long as they are outside the region of appreciable likelihood. To put it in a way somewhat crude, but not really wrong: if the amount of likelihood [integral of $L(\sigma)$] lying outside the limits ($\sigma_1 < \sigma < \sigma_2$) is less than 10^{-6} of the total likelihood, then as long as our prior limits are still wider ($\sigma_{min} < \sigma_1 < \sigma_2 < \sigma_{max}$), the exact values of σ_{min}, σ_{max} can't make more than about one part in 10^6 difference in our conclusions. If, then, we don't worry about them, and just take the limiting form of the solution as $\sigma_{min} \to 0$, $\sigma_{max} \to \infty$ for mathematical convenience, we are committing no worse a sin than does the person who laboriously determines the proper values of σ_{min}, σ_{max}, works out the exact solution based on them – and then rounds off his final result to six significant figures. We are only getting that result with an order of magnitude less labor.

If, on the other hand, we should encounter a non-normalizable posterior distribution in this limit, the theory is telling us that the experiment is so uninformative that our exact state of prior information is still important, and must be taken into account explicitly. This phenomenon, far from being a defect of Bayesian methods, is a valuable safety device that warns us when an experiment is too uninformative to justify, by itself, any definite inferences. If someone ignores the warning, and gets into trouble with 'improper priors', what we are witnessing is not a failure, but only a misapplication, of Bayesian methods.

Finally, let us keep in mind that we are really concerned here with relative value judgments; and so if anyone attacks Bayesian methods because of the possible situation just described, fairness demands that he also takes note of what happens to orthodox methods in the same problems. Now one of the substantive factual issues illustrated in my presentation, is this: orthodox methods, when improved to the maximum possible degree, reduce ultimately to procedures that are mathematically identical with applying Bayes' theorem *with just the noninformative improper prior* about which Professor Kempthorne expresses such alarm! We saw this phenomenon in Examples 2, 3, 5 and 6. As we have just seen, this causes difficulty only when the experiment is so uninformative that our final conclusions must, necessarily, still depend strongly on our prior knowledge. The Bayesian can correct this at once by using a realistic prior, leading to the inferences that *are* justified by the total information at hand; but the orthodoxian cannot, because his ideology forbids him to recognize the existence of any prior which is not also a known frequency.

In fact, we had just this situation in the first part of my Example 3, where we took no note of the actual failure times. If all units tested fail, the test provides no evidence against the hypothesis of arbitrarily large λ. The Bayesian test (6) based on a uniform prior then yields a non-normalizable posterior distribution $p(\mathrm{d}\lambda \mid n, r, t) \sim (1 - e^{-\lambda t})^n \, \mathrm{d}\lambda$, which tells us that λ is almost certainly greater than $(t^{-1} \log n)$, but gives no upper limit. In this way, the safety device warns us that our prior information concerning the possibility of very large λ, remains relevant; by taking it explicitly into account, rational inferences about λ are still possible, as I showed by the maximum entropy prior.

But we saw that the orthodox ST test was, in the absence of such pathology, mathematically identical with this Bayesian test; so what happens to it? Well, this is just the case already noted where the ST test breaks down entirely, telling us to reject at all significance levels. In problems where the Bayesian cannot use the approximation of an improper prior, orthodox methods give no warning, but simply yield absurd results; and only the alertness and common sense of the user can save him from the consequences. As we see, it is the orthodoxian, and not the Bayesian, who is going to be in trouble in cases where 'improper priors' cannot be used.

CONCLUSION

I suppose it is possible, without actual logical contradiction, to maintain that Bayesian methods are utterly wrong, but that through a series of fortuitous accidents they always happen to give the right answer in every particular problem. However, I cannot believe that anybody will want to take that position. Now the person who, after studying the evidence given here and in the rest of the Bayesian literature, still wishes to claim that orthodox methods are superior, must realize that, if he is to avoid being forced into exactly that position, mere linguistics and ideological slogans will no longer suffice. The burden of proof is squarely on him to show us specific problems, with mathematical details, in which orthodox methods give a satisfactory result and Bayesian methods do not. My own studies have convinced me that such a problem does not exist.

Whether I am right or wrong in this belief, we now have a large mass of factual evidence showing that (a) orthodox methods contain dangerous fallacies, and must in any event be revised; and (b) Bayesian methods are easier to apply and give better results. As a teacher, I therefore feel that to continue the time honored practice – still in effect in many schools – of teaching pure orthodox statistics to students, with only a passing sneer at Bayes and Laplace, is to perpetuate a tragic error which has already wasted thousands of man-years of our finest mathematical talent in pursuit of false goals. If this talent had been directed toward understanding Laplace's contributions and learning how to use them properly, statistical practice would be far more advanced today than it is.

REFERENCES

Note: The following list includes only those works not already cited in my main presentation or Kempthorne's reply.

Barnard, G. A., 'Comments on Stein's "A Remark on the Likelihood Principle"', *J. Roy. Stat. Soc.* (A) **125**, 569 (1962).

Cox, R. T., *Am. J. Phys.* **17**, 1 (1946).

Cox, R. T., *The Algebra of Probable Inference*, Johns Hopkins University Press, 1961; Reviewed by E. T. Jaynes, *Am. J. Phys.* **31**, 66 (1963).

Deming, W. E., *Statistical Adjustment of Data*, J. Wiley, New York (1943).

Fisher, R. A., *Contributions to Mathematical Statistics*, W. A. Shewhart, (ed.), J. Wiley and Sons, Inc. New York (1950); Referred to above as 'Collected Works'.

Fisher, R. A., *Statistical Methods and Scientific Inference*, Hafner Publishing Co., New York (1956).
Fisher, R. A., *Statistical Methods for Research Workers*, Hafner Publishing Co., New York: Thirteenth Edition (1958).
Forney, G. D., *Information Theory*, (EE376 Course Notes, Stanford University, 1972); p. 26.
Hoel, P. G., *Introduction to Mathematical Statistics*, Fourth Edition, J. Wiley and Sons, Inc., New York (1971).
Jaynes, E. T., 'Review of *Noise and Fluctuations*', by D. K. C. MacDonald, *Am. J. Phys.* **31**, 946 (1963).
Kendall, M. G., 'Ronald Aylmer Fisher, 1890–1962', *Biometrika* **50**, 1–15 (1963); reprinted in *Studies in the History of Statistics and Probability*, E. S. Pearson and M. G. Kendall, (eds)., Hafner Publishing Co., Darien, Conn. (1970).
Mandel, J., *The Statistical Analysis of Experimental Data*, Interscience Publishers, New York (1964); p. 290.
McColl, H., 'The Calculus of Equivalent Statements', *Proc. Lond. Math. Soc.* **28**, p. 556 (1897).
Pearson, Karl, 'Method of Moments and Method of Maximum Likelihood', *Biometrika* **28**, 34 (1936).
Pratt, John W., 'Review of *Testing Statistical Hypothesis*' (Lehmann, 1959); *J. Am. Stat. Assoc.* Vol. **56**, pp. 163–166 (1961).
Roberts, Harry V., 'Statistical Dogma: One Response to a Challenge', Multilithed, University of Chicago (1965).
Thornber, Hodson, 'An Autoregressive Model: Bayesian Versus Sampling Theory Analysis', Multilithed, Dept. of Economics, University of Chicago, Chicago, Illinois (1965).
Wilbraham, H., *Phil. Mag. Series*, 4, Vol. vii, (1854).
Zellner, Arnold, 'Bayesian Inference and Simultaneous Equation Models', Multilithed, University of Chicago, Chicago, Illinois (1965).

10. WHERE DO WE STAND ON MAXIMUM ENTROPY? (1978)

In May 1978 a three-day Symposium on 'The Maximum-Entropy Formalism' held at M.I.T. provided a good opportunity to put into the record a more comprehensive summing up than had been attempted until then. The following lengthy article was written hastily in a few weeks, to meet a publication deadline; and the rough unpolished edges show everywhere. Still, the Proceedings Volume did appear in December 1978, probably a speed record.

For some years, in dealing with both physicists and statisticians, I had been struck by the fact that neither group was aware of the history or current problems of the other's field, even though they were often doing nearly the same thing. So I wrote two parallel histories, in the hope of giving at least a sense of a common background. It is hard to understand why the two fields have developed in almost complete isolation from each other since the time of Laplace; surely this has been detrimental to both.

Also, for some time I had tried to follow Galton's wise advice that one should neither resent nor reply to criticisms of his own work, leaving that judgment to others. But an increasing number of correspondents urged me to change this policy, on the grounds that my continued failure to answer two attacks published several years previously, was giving the impression that they were unanswerable. I was finally persuaded, but found the Shimony criticism so vague and confused that there was really nothing specific to reply to; one could only point out that *Maxent* and Bayes' theorem are different things, and different problems have different solutions.

But replying to Rowlinson became a genuine pleasure, because he was so clear that one could see exactly where the difficulty lay, and answer with definite, interesting calculations. Many have since told me that the calculation of that combinatorial ratio (10^{86}) and the analysis of Wolf's dice data are the high points of this work.

However, in the haste of writing a technical error crept in, when I argued for five degrees of freedom on the grounds that a still lower value of Chi-squared could have been achieved had the parameters been chosen by that criterion instead of by the *Maxent* one. In fact, the criteria were numerically indistinguishable; for the next Chapter of that story, see the 'Concentration of Distributions at Entropy Maxima' article in this volume. Rowlinson's criticism has now led to an advance in the technique of hypothesis testing, using numerical values of entropy rather than of Chi-squared.

WHERE DO WE STAND ON MAXIMUM ENTROPY?

Edwin T. Jaynes

A. Historical Background

B. Present Features

C. Speculations for the Future

D. An Application: Irreversible Statistical Mechanics

Summary. In Part A we place the Principle of Maximum Entropy in its historical perspective as a natural extension and unification of two separate lines of development, both of which had long used special cases of it. The first line is identified with the names Bernoulli, Laplace, Jeffreys, Cox; the second with Maxwell, Boltzmann, Gibbs, Shannon.

Part B considers some general properties of the present maximum entropy formalism, stressing its consistency and interderivability with the other principles of probability theory. In this connection we answer some published criticisms of the principle.

In part C we try to view the principle in the wider context of Statistical Decision Theory in general, and speculate on possible future applications and further theoretical developments. The Principle of Maximum Entropy, together with the seemingly disparate principles of Group Invariance and Marginalization, may in time be seen as special cases of a still more general principle for translating information into a probability assignment.

Part D, which should logically precede C, is relegated to the end because it is of a more technical nature, requiring also the full formalism of quantum mechanics. Readers not familiar with this will find the first three Sections a self-contained exposition.

In Part D we present some of the details and results of what is at present the most highly developed application of the Principle of Maximum Entropy; the extension of Gibbs' formalism to irreversible processes. Here we consider the most general application of the principle, without taking advantage of any special features (such as interest in only a subspace of states, or a subset of operators) that might be found in particular problems. An alternative formulation, which does take such advantage--and is thus closer to the spirit of previous "kinetic equation" approaches at the cost of some generality, appears in the presentation of Dr. Baldwin Robertson.

A. Historical Background

The ideas to be discussed at this Symposium are found clearly expressed already in ancient sources, particularly the Old Testament, Herodotus, and Ovennus. All note the virtue of making wise decisions by taking into account all possibilities, i.e., by not presuming more information than we possess. But probability theory, in the form which goes beyond these moral exhortations and considers actual numerical values of probabilities and expectations, begins with the Ludo aleae of Gerolamo Cardano, some time in the mid-sixteenth century. Wilks (1961) places this "around 1520," although Cardano's Section "On Luck in Play" contains internal evidence that shows the date of its writing to be 1564, still 90 years before the Pascal-Fermat correspondence.

Already in these earliest works, special cases of the Principle of Maximum Entropy are recognized intuitively and, of necessity, used. For there is no application of probability theory in which one can evade that all-important first step: assigning some initial numerical values of probabilities so that the calculation can get started. Even in the most elementary homework problems, such as "Find the probability of getting at least two heads in four tosses of a coin," we have no basis for the calculation until we make some initial judgment, usually that "heads" shall have the probability 1/2 independently at each toss. But by what reasoning does one arrive at this initial assignment? If it is questioned, how shall we defend it?

The basis underlying such initial assignments was stated as an explicit formal principle in the Ars Conjectandi of James (= Jacob) Bernoulli (1713). Unfortunately, it was given the curious name: Principle of Insufficient Reason which has had, ever since, a psychologically repellant quality that prevents many from seeing the positive merit of the idea itself. Keynes (1921) helped somewhat by renaming it the Principle of Indifference; but by then the damage had been done. Had Bernoulli called his principle, more appropriately, the Desideratum of Consistency, nobody would have ventured to deprecate it. and today statistical theory would be in considerably better shape than it is.

The essence of the principle is just: (1) we recognize that a probability assignment is a means of describing a certain state of knowledge. (2) if the available evidence gives us no reason to consider proposition A_1 either more or

WHERE DO WE STAND ON MAXIMUM ENTROPY? 213

less likely than A_2, then the only honest way we can describe that state of knowledge is to assign them equal probabilities: $p_1 = p_2$. Any other procedure would be inconsistent in the sense that, by a mere interchange of the labels (1, 2) we could then generate a new problem in which our state of knowledge is the same but in which we are assigning different probabilities. (3) Extending this reasoning, one arrives at the rule

$$p(A) = \frac{M}{N} = \frac{\text{(Number of cases favorable to A)}}{\text{(Total number of equally possible cases)}} \quad (A1)$$

which served as the basic definition of probability for the next 150 years.

The only valid criticism of this principle, it seems to me, is that in the original form (enumeration of the "equally possible" cases) it cannot be applied to all problems. Indeed, nobody could have emphasized this more strongly than Bernoulli himself. After noting its use where applicable, he adds, "But here, finally, we seem to have met our problem, since this may be done only in a very few cases and almost nowhere other than in games of chance the inventors of which, in order to provide equal chances for the players, took pains to set up so that the numbers of cases would be known and --- so that all these cases could happen with equal ease." After citing some examples, Bernoulli continues in the next paragraph, "But what mortal will ever determine, for example, the number of diseases --- these and other such things depend upon causes completely hidden from us ---."

It was for the explicitly stated purpose of finding probabilities when the number of "equally possible" cases is infinite or beyond our powers to determine, that Bernoulli turns next to his celebrated theorem, today called the weak law of large numbers. His idea was that, if a probability p cannot be calculated in the manner p = M/N by direct application of the Principle of Insufficient Reason, then in some cases we may still reason backwards and estimate the ratio M/N approximately by observing frequencies in many trials.

That there ought to be some kind of connection between a theoretical probability and an observable frequency was a vaguely seen intuition in the earlier works; but Bernoulli, seeing clearly the distinction between the concepts, recognized that the existence of a connection between them cannot be merely postulated; it requires mathematical demonstration. If in a binary experiment we assign a constant probability of success p, independently at each trial, then we find for the probability of seeing m successes in n trials the binomial distribution

$$P(m|n,p) = \binom{n}{m} p^m (1-p)^{n-m} \quad . \tag{A2}$$

Bernoulli then shows that as $n \to \infty$, the observed frequency $f = m/n$ of successes tends to the probability p in the sense that for all $\varepsilon > 0$,

$$P(p - \varepsilon < f < p + \varepsilon | p, n) \to 1 \tag{A3}$$

and thus (in a sense made precise only in the later work of Bayes and Laplace) for sufficiently large n, the observed frequency is practically certain to be close to the number p sought.

But Bernoulli's result does not tell us how large n must be for a given accuracy. For this, one needs the more detailed limit theorem; as n increases, f may be considered a continuous variable, and the probability that $(f < m/n < f + df)$ goes into a gaussian, or normal, distribution:

$$P(df|n,p) \sim \left[\frac{n}{2\pi p(1-p)}\right]^{\frac{1}{2}} \exp\left[-\frac{n(f-p)^2}{2p(1-p)}\right] df \tag{A4}$$

in the sense of the leading term of an asymptotic expansion. For example, if $p = 2/3$, then from (A4), in $n = 1000$ trials, there is a 99% probability that the observed f will lie in the interval 0.667 ± 0.038, and an even chance that it will fall in 0.667 ± 0.010. The result (A4) was first given in this generality by Laplace; it had been found earlier by de Moivre for the case $p = \frac{1}{2}$. And in turn, the de Moivre-Laplace theorem (A4) became the ancestor of our present Central Limit Theorem.

Since these limit theorems are sometimes held to be the most important and sophisticated fruits of probability theory, we note that they depend crucially on the assumption of independence of different trials. The slightest positive correlation between trials i and j, if it persists for arbitrarily large $|i - j|$, will render these theorems qualitatively incorrect.

Laplace's contributions to probability theory go rather far beyond mere analytical refinements of other peoples' results. Most important for statistical theory today, he saw the general principle needed to solve problems of the type formulated by Bernoulli, but left unfinished by the Bernoulli and de Moivre-Laplace limit theorems. These results concern only the so-called "sampling distribution." That is, given $p = M/N$, what is the probability that we shall see particular sample numbers (m,n)? The results (A1)-(A4) describe a state of knowledge in which the "population numbers" (M,N) are known, the sample number unknown. But in the problem Bernoulli tried to solve, the sample is known and the population is not only unknown--

its very existence is only a tentative hypothesis (what mortal will ever determine the number of diseases, etc.).

We have, therefore, an inversion problem. The above theorems show that, given (M,N) and the correctness of the whole conceptual model, then it is likely that in many trials the observed frequency f will be close to the probability p. Presumably, then, given the observed f in many trials, it is likely that p is close to f. But can this be made into a precise theorem like (A4)? The binomial law (A2) gives the probability of m, given (M,N,n). Can we turn this around and find a formula for the probability of M, given (m,N,n)? This is the problem of inverse probabilities.

A particular inversion of the binomial distribution was offered by a British clergyman and amateur mathematician, Thomas Bayes (1763) in what has become perhaps the most famous and controversial work in probability theory. His reasoning was obscure and hard to describe; but his actual result is easy to state. Given the data (m,n), he finds for the probability that M/N lies in the interval $p < (M/N) < p + dp$,

$$P(dp|m,n) = \frac{(n+1)!}{m!(n-m)!} p^m (1-p)^{n-m} dp , \qquad (A5)$$

today called a Beta distribution. It is not a binomial distribution because the variable is p rather than m and the numerical coefficient is different, but it is a trivial mathematical exercise [expand the logarithm of (A5) in a power series about its peak] to show that, for large n, (A5) goes asymptotically into just (A4) with f and p everywhere interchanged. Thus, if in n = 1000 trials we observe m = 667 successes, then on this evidence there would be a 99% probability that p lies in (0.667 ± 0.038), etc.

In the gaussian approximation, according to Bayes' solution, there is complete mathematical symmetry between the probability of f given p, and of p given f. This would certainly seem to be the neatest and simplest imaginable solution to Bernoulli's inversion problem.

Laplace, in his famous memoir of 1774 on the "probabilities of causes," perceived the principle underlying inverse probabilities in far greater generality. Let E stand for some observable event and $\{C_1 \ldots C_N\}$ the set of its conceivable causes. Suppose that we have found, according to some conceptual model, the "sampling distribution" or "direct" probabilities of E for each cause: $P(E|C_i)$, $i = 1,2,\ldots,N$. Then, says Laplace, if initially the causes C_i are considered equally likely, then having seen the event E, the different causes are indicated with probability proportional to $P(E|C_i)$. That is, with uniform prior probabilities, the posterior probabilities of the C_i are

$$P(C_i|E) = \left[\sum_{j=1}^{N} P(E|C_j)\right]^{-1} P(E|C_i) . \tag{A6}$$

This is a tremendous generalization of the Bernoulli-Bayes results (A2), (A5). If the event E consists in finding m successes in n trials, and the causes C_j correspond to the possible values of M in the Bernoulli model, then $P(E|C_j)$ is the binomial distribution (A2); and in the limit $N \to \infty$ (A6) goes into Bayes' result (A5).

Later, Laplace generalized (A6) further by noting that, if initially the C_i are not considered equally likely, but have prior probabilities $P(C_i|I)$, where I stands for the prior information, then the terms in (A6) should be weighted according to $P(C_i|I)$:

$$P(C_i|E,I) = \frac{P(E|C_i)P(C_i|I)}{\sum_j P(E|C_j)P(C_j|I)} \tag{A7}$$

but, following long-established custom, it is Laplace's result (A7) that is always called, in the modern literature, "Bayes' theorem."

Laplace proceeded to apply (A6) to a variety of problems that arise in astronomy, meteorology, geodesy, population statistics, etc. He would use it typically as follows. Comparing experimental observations with some existing theory, or calculation, one will never find perfect agreement. Are the discrepancies so small that they might reasonably be attributed to measurement errors, or are they so large that they indicate, with high probability, the existence of some new systematic cause? If so, Laplace would undertake to find that cause. Such uses of inverse probability--what would be called today "significance tests" by statisticians, and "detection of signals in noise" by electrical engineers--led him to some of the most important discoveries in celestial mechanics.

Yet there were difficulties that prevented others from following Laplace's path, in spite of its demonstrated usefulness. In the first place, Laplace simply stated the results (A6),(A7) as intuitive, ad hoc recipes without any derivation from compelling desiderata; and this left room for much agonizing over their logical justification and uniqueness. For an account of this, see Keynes (1921). However, we now know that Laplace's result (A7) is, in fact, the entirely correct and unique solution to the inversion problem.

More importantly, it became apparent that, in spite of first appearances, the results of Bayes and Laplace did not, after all, solve the problem that Bernoulli had set out to deal with. Recall, Bernoulli's original motivation was that the Principle of Insufficient Reason is inapplicable in so many real problems, because we are unable to break things down into an enumeration of "equally possible" cases. His hope--left unrealized at his death in 1705--had been that, by inversion of his theorem one could avoid having to use Insufficient Reason. Yet when the inversion problem was finally solved by Bayes and Laplace, the prior probabilities $P(C_i|I)$ that Bernoulli had sought to avoid, intruded themselves inevitably right back into the picture!

The only useful results Laplace got came from (A6), based on the uniform prior probabilities $P(C_i|I) = 1/N$ from the Principle of Insufficient Reason. That is, of course, not because Laplace failed to understand the generalization (A7) as some have charged--it was Laplace who, in his Essai Philosophique, pointed out the need for that generalization. Rather, Laplace did not have any principle for finding prior probabilities in cases where the prior information fails to render the possibilities "equally likely."

At this point, the history of statistical theory takes a sharp 90° turn away from the original goal, and we are only slowly straightening out again today. One might have thought, particularly in view of the great pragmatic success achieved by Laplace with (A6), that the next workers would try to build constructively on the foundations laid down by him. The next order of business should have been seeking new and more general principles for determining prior probabilities, thus extending the range of problems where probability theory is useful to (A7). Instead, only fifteen years after Laplace's death, there started a series of increasingly violent attacks on his work. Totally ignoring the successful results they had yielded, Laplace's methods based on (A6) were rejected and ridiculed, along with the whole conception of probability theory expounded by Bernoulli and Laplace. The main early references to this counter-stream of thought are Ellis (1842), Boole (1854), Venn (1866), and von Mises (1928).

As already emphasized, Bernoulli's definition of probability (A1) was developed for the purpose of representing mathematically a particular state of knowledge; and the equations of probability theory then represent the process of plausible, or inductive, reasoning in cases where there is not enough information at hand to permit deductive reasoning. In particular, Laplace's result (A7) represents the process of "learning by experience," the prior probability $P(C|I)$ changing to the posterior probability $P(C|E,I)$ as a result of obtaining new evidence E.

This counter-stream of thought, however, rejected the notion of probability as describing a state of knowledge, and insisted that by "probability" one must mean only "frequency in a random experiment." For a time this viewpoint dominated the field so completely that those who were students in the period 1930-1960 were hardly aware that any other conception had ever existed.

If anyone wishes to study the properties of frequencies in random experiments he is, of course, perfectly free to do so; and we wish him every success. But if he wants to talk about frequencies, why can't he just use the word "frequency?" Why does he insist on appropriating the word "probability," which had already a long-established and very different technical meaning?

Most of the debate that has been in progress for over a century on "frequency vs. non-frequency definitions of probability" seems to me not concerned with any substantive issue at all; but merely arguing over who has the right to use a word. Now the historical priority belongs clearly to Bernoulli and Laplace. Therefore, in the interests not only of responsible scholarship, but also of clear exposition and to avoid becoming entangled in semantic irrelevancies, we ought to use the word "probability" in the original sense of Bernoulli and Laplace; and if we mean something else, call it something else.

With the usage just recommended, the term "frequency theory of probability" is a pure incongruity; just as much so as "theory of square circles." One might speak properly of a "frequency theory of inference," or the better term "sampling theory," now in general use among statisticians (because the only distributions admitted are the ones we have called sampling distributions). This stands in contrast to the "Bayesian theory" developed by Laplace, which admits the notion of probability of an hypothesis.

Having two opposed schools of thought about how to handle problems of inference, the stage is set for an interesting contest. The sampling theorists, forbidden by their ideology to use Bayes' theorem as Laplace did in the form (A6), must seek other methods for dealing with Laplace's problems. What methods, then, did they invent? How do their procedures and results compare with Laplace's?

The sampling theory developed slowly over the first half of this Century by the labors of many, prominent names being Fisher, "Student," Pearson, Neyman, Kendall, Cramér, Wald. They proceeded through a variety of ad hoc intuitive principles, each appearing reasonable at first glance, but for which defects or limitations on generality always appeared. For

example, the Chi-squared test, maximum likelihood, unbiased
and/or efficient estimators, confidence intervals, fiducial
distributions, conditioning on ancillary statistics, power
functions and sequential methods for hypothesis testing.
Certain technical difficulties ("nuisance" parameters, nonexistence of sufficient or ancillary statistics, inability
to take prior information into account) remained behind as
isolated pockets of resistance which sampling theory has never
been able to overcome. Nevertheless, there was discernible
progress over the years, accompanied by an unending stream of
attacks on Laplace's ideas and methods, sometimes degenerating
into personal attacks on Laplace himself [see, for example,
the biographical sketch by E. T. Bell (1937), entitled "From
Peasant to Snob"].

Enter Jeffreys. After 1939, the sampling theorists had another
target for their scorn. Sir Harold Jeffreys, finding in geophysics some problems of "extracting signals from noise" very
much like those treated by Laplace, found himself unconvinced
by Fisher's arguments, and produced a book in which the methods
of Laplace were reinstated and applied, in the precise, compact
modern notation that did not exist in the time of Laplace, to
a mass of current scientific problems. The result was a vastly
more comprehensive treatment of inference than Laplace's, but
with two points in common: (A) the applications worked out
beautifully, encountering no such technical difficulties as
the "nuisance parameters" noted above; and yielding the same
or demonstrably better results than those found by sampling
theory methods. For many specific examples, see Jaynes (1976).
(B) Unfortunately, like Laplace, Jeffreys did not derive his
principles as necessary consequences of any compelling desiderata; and thus left room to continue the same old arguments
over their justification.

The sampling theorists, seizing eagerly upon point (B) while
again totalling ignoring point (A), proceeded to give Jeffreys
the same treatment as Laplace, which he had to endure for some
thirty years before the tide began to turn.

As a student in the mid-1940's, I discovered the book of
Jeffreys (1939) and was enormously impressed by the smooth,
effortless way he was able to derive the useful results of
the theory, as well as the sensible philosophy he expressed.
But I too felt that something was missing in the exposition
of fundamentals in the first Chapter and, learning about the
attacks on Jeffreys' methods by virtually every other writer
on statistics, felt some mental reservations.

But just at the right moment there appeared a work that
removed all doubts and set the direction of my own life's

work. An unpretentious little article by Professor R. T. Cox (1946) turned the problem under debate around and, for the first time, looked at it in a constructive way. Instead of making dogmatic <u>assertions</u> that it is or is not legitimate to use probability in the sense of degree of plausibility rather than frequency, he had the good sense to ask a <u>question</u>: Is it possible to construct a consistent set of mathematical rules for carrying out plausible, rather than deductive, reasoning? He found that, if we try to represent degrees of plausibility by real numbers, then the conditions of consistency can be stated in the form of functional equations, whose general solutions can be found. The results were: out of all possible monotonic functions which might in principle serve our purpose, there exists a particular scale on which to measure degrees of plausibility which we henceforth call <u>probability</u>, with particularly simple properties. Denoting various propositions by A, B, etc., and using the notation, $AB \equiv$ "Both A and B are true," $\bar{A} \equiv$ "A is false," $p(A|B) \equiv$ probability of A given B, the consistent rules of combination take the form of the familiar product rule and sum rule:

$$p(AB|C) = p(A|BC) \, p(B|C) \, , \tag{A8}$$

$$p(A|B) + p(\bar{A}|B) = 1 \, . \tag{A9}$$

By mathematical transformations we can, of course, alter the <u>form</u> of these rules; but what Cox proved was that any alteration of their <u>content</u> will enable us to exhibit inconsistencies (in the sense that two methods of calculation, each permitted by the rules, will yield different results). But (A8), (A9) are, in fact, the basic rules of probability theory; all other equations needed for applications can be derived from them. Thus, Cox proved that any method of inference in which we represent degrees of plausibility by real numbers, is necessarily either equivalent to Laplace's, or inconsistent.

For me, this was exactly the argument needed to clinch matters; for Cox's analysis makes no reference whatsoever to frequencies or random experiments. From the day I first read Cox's article I have never for a moment doubted the basic soundness and inevitability of the Laplace-Jeffreys methods, while recognizing that the theory needs further development to extend its range of applicability.

Indeed, such further development was started by Jeffreys. Recall, in our narrative we left Laplace (or rather, Laplace left us) at Eq. (A6), seeing the need but not the means to make the transition to (A7), which would open up an enormously wider range of applications for Bayesian inference. Since the

WHERE DO WE STAND ON MAXIMUM ENTROPY?   221

function of the prior probabilities is to describe the prior information, we need to develop new or more general principles for determination of those priors by logical analysis of prior information when it does not consist of frequencies; just what should have been the next order of business after Laplace.

Recognizing this, Jeffreys resumed the constructive development of this theory at the point where Laplace had left off. If we need to convert prior information into a prior probability assignment, perhaps we should start at the beginning and learn first how to express "complete ignorance" of a continuously variable parameter, where Bernoulli's principle will not apply.

Bayes and Laplace had used uniform prior densities, as the most obvious analog of the Bernoulli uniform discrete assignment. But it was clear, even in the time of Laplace, that this rule is ambiguous because it is not invariant under a change of parameters. A uniform density for θ does not correspond to a uniform density for $\alpha = \theta^3$; or $\beta = \log \theta$; so for which choice of parameters should the uniform density apply?

In the first (1939) Edition of his book, Jeffreys made a tentative start on this problem, in which he found his now famous rule: to express ignorance of a scale parameter σ, whose possible domain is $0 < \sigma < \infty$, assign uniform prior density to its logarithm: $P(d\sigma|I) = d\sigma/\sigma$. The first arguments advanced in support of this rule were not particularly clear or convincing to others (including this writer). But other desiderata were found; and we have now succeeded in proving via the integral equations of marginalization theory (Jaynes, 1979) that Jeffreys' prior $d\sigma/\sigma$ is, in fact, uniquely determined as the only prior for a scale parameter that is "completely uninformative" in the sense that it leads us to the same conclusions about other parameters θ as if the parameter σ had been removed from the model [see Eq. (C33) below].

In the second (1948) Edition, Jeffreys gave a much more general "Invariance Theory" for determining ignorance priors, which showed amazing prevision by coming within a hair's breadth of discovering both the principles of Maximum Entropy and Transformation Groups. He wrote down the actual entropy expression (note the date!), but then used it only to generate a quadratic form by expansion about its peak. Jeffreys' invariance theory is still of great importance today, and the question of its relation to other methods that have been proposed is still under study.

In the meantime, what had been happening in the sampling theory camp? The culmination of this approach came in the late 1940's when for the first time, Abraham Wald succeeded in removing all <u>ad hockeries</u> and presenting general rules of conduct for making decisions in the face of uncertainty, that he proved to be uniquely optimal by certain very simple and compelling desiderata of reasonable behavior. But quickly a number of people--including I. J. Good (1950), L. J. Savage (1954), and the present writer--realized independently that, if we just ignore Wald's entirely different vocabulary and diametrically opposed philosophy, and look only at the specific mathematical steps that were now to be used in solving specific problems, <u>they were identical with the rules given by Laplace in the eighteenth century</u>, which generations of statisticians had rejected as metaphysical nonsense!

It is one of those ironies that make the history of science so interesting, that the missing Bayes-optimality proofs, which Laplace and Jeffreys had failed to supply, were at last found inadvertently, while trying to prove the opposite, by an early ardent disciple of the von Mises "collective" approach. It is also a tribute to Wald's intellectual honesty that he was able to recognize this, and in his final work (Wald, 1950) he called these optimal rules, "Bayes strategies."

Thus came the "Bayesian Revolution" in statistics, which is now all but over. This writer's recent polemics (Jaynes, 1976) will probably be one of the last battles waged. Today, most active research in statistics is Bayesian, a good deal of it directed to the above problem of determining priors by logical analysis; and the parts of sampling theory which do not lie in ruins are just the ones (such as sufficient statistics and sequential analysis) that can be justified in Bayesian terms.

This history of basic statistical theory, showing how developments over more than two centuries set the stage naturally for the Principle of Maximum Entropy, has been recounted at some length because it is unfamiliar to most scientists and engineers. Although the second line converging on this principle is much better known to this audience, our account can be no briefer because there is so much to be unlearned.

<u>The Second Line: Maxwell, Boltzmann, Gibbs, Shannon</u>. Over the past 120 years another line of development was taking place, which had astonishingly little contact with the "statistical inference" line just described. In the 1850's James Clerk Maxwell started the first serious work on the application of probability analysis to the kinetic theory of gases. He was confronted immediately with the problem of assigning initial probabilities to various positions and velocities of molecules. To see how he dealt with it, we quote his first (1859) words on the problem of finding the probability distribution for velocity direction of a spherical molecules after an impact: "In order that a collision may take place, the line of motion of one of the balls must pass the center of the other at a

distance less than the sum of their radii; that is, it must pass through a circle whose centre is that of the other ball, and radius the sum of the radii of the balls. Within this circle every position is equally probable, and therefore --- ."

Here again, as that necessary first step in a probability analysis, Maxwell had to apply the Principle of Indifference; in this case to a two-dimensional continuous variable. But already at this point we see a new feature. As long as we talk about some abstract quantity θ without specifying its physical meaning, we see no reason why we could not as well work with $\alpha = \theta^3$, or $\beta = \log \theta$; and there is an unresolved ambiguity. But as soon as we learn that our quantity has the physical meaning of position within the circular collision cross-section, our intuition takes over with a compelling force and tells us that the probability of impinging on any particular region should be taken proportional to the <u>area</u> of that region; and not to the cube of the area, or the logarithm of the area. If we toss pennies onto a wooden floor, something inside us convinces us that the probability of landing on any one plank should be taken proportional to the <u>width</u> of the plank; and not to the cube of the width, or the logarithm of the width.

In other words, merely knowing the physical meaning of our parameters, <u>already constitutes highly relevant prior information</u> which our intuition is able to use at once; in favorable cases its effect is to give us an inner conviction that there is no ambiguity after all in applying the Principle of Indifference. Can we analyze how our intuition does this, extract the essence, and express it as a formal mathematical principle that might apply in cases where our intuition fails us? This problem is not completely solved today, although I believe we have made a good start on it in the principle of transformation groups (Jaynes, 1968, 1973, 1979). Perhaps these remarks will encourage others to try their hand at resolving these puzzles; this is an area where important new results might turn up with comparatively little effort, given the right inspiration on how to approach them.

Maxwell built a lengthy, highly non-trivial, and needless to say, successful analysis on the foundation just quoted. He was able to predict such things as the equation of state, velocity distribution law, diffusion coefficient, viscosity, and thermal conductivity of the gas. The case of viscosity was particularly interesting because Maxwell's theory led to the prediction that viscosity is independent of density, which seemed to contradict common sense. But when the experiments were performed, they confirmed Maxwell's prediction; and what had seemed a difficulty with his theory became its greatest triumph.

Enter Boltzmann. So far we have considered only the problem
of expressing initial ignorance by a probability assignment.
This is the first fundamental problem, since "complete initial
ignorance" is the natural and inevitable starting point from
which to measure our positive knowledge; just as zero is the
natural and inevitable starting point when we add a column of
numbers. But in most real problems we do not have initial
ignorance about the questions to be answered. Indeed, unless
we had some definite prior knowledge about the parameters to
be measured or the hypotheses to be tested, we would seldom
have either the means or the motivation to plan an experiment
to get more knowledge. But to express positive initial
knowledge by a probability assignment is just the problem of
getting from (A6) to (A7), bequeathed to us by Laplace.

The first step toward finding an explicit solution to this
problem was made by Boltzmann, although it was stated in very
different terms at the time. He wanted to find how molecules
will distribute themselves in a conservative force field (say,
a gravitational or centrifugal field; or an electric field
acting on ions). The force acting on a molecule at position
x is then $F = -\text{grad } \phi$, where $\phi(x)$ is its potential energy. A
molecule with mass m, position x, velocity v thus has energy
$E = \frac{1}{2} mv^2 + \phi(x)$. We neglect the interaction energy of molecules
with each other and suppose they are enclosed in a container
of volume V, whose walls are rigid and impermeable to both
molecules and heat. But Boltzmann was not completely ignorant
about how the molecules are distributed, because he knew that
however they move, the total number N of molecules present cannot change, and the total energy

$$E = \sum_{i=1}^{N} \left[\frac{1}{2} m v_i^2 + \phi(x_i) \right] \tag{A10}$$

must remain constant. Because of the energy constraint,
evidently, all positions and velocities are not equally likely.

At this point, Boltzmann found it easier to think about
discrete distributions than continuous ones (a kind of prevision of quantum theory); and so he divided the phase space
(position-momentum space) available to the molecules into
discrete cells. In principle, these could be defined in any
way; but let us think of the k'th cell as being a region R_k
so small that the energy E_k of a molecule does not vary appreciably within it; but also so large that it can accommodate
a large number, $N_k \gg 1$, of molecules. The cells $\{R_k, 1 \leq k \leq s\}$
are to fill up the accessible phase space (which because of the
energy constraint has a finite volume) without overlapping.

The problem is then: given N, E, and $\phi(x)$, what is the
best prediction we can make of the number of N_k of molecules

WHERE DO WE STAND ON MAXIMUM ENTROPY? 225

in R_k? In Boltzmann's reasoning at this point, we have the beginning of the Principle of Maximum Entropy. He asked first: In how many ways could a given set of occupation numbers N_k be realized? The answer is the multinomial coefficient

$$W(N_k) = \frac{N!}{N_1! \, N_2! \, \ldots \, N_s!} \; . \tag{A11}$$

This particular distribution will have total energy

$$E = \sum_{k=1}^{s} N_k E_k \tag{A12}$$

and of course, the N_k are also constrained by

$$N = \sum_{k=1}^{s} N_k \; . \tag{A13}$$

Now any set $\{N_k\}$ of occupation numbers for which E, N agree with the given information, represents a <u>possible</u> distribution, compatible with all that is specified. Out of the millions of such possible distributions, which is most likely to be realized? Boltzmann's answer was that the "most probable" distribution is the one that can be realized in the greatest number of ways; i.e., the one that maximizes (A11) subject to the constraints (A12), (A13), if the cells are equally large (phase volume).

Since the N_k are large, we may use the Stirling approximation for the factorials, whereupon (A11) can be written

$$\log W = -N \sum_{k=1}^{s} \left(\frac{N_k}{N}\right) \log\left(\frac{N_k}{N}\right) \; . \tag{A14}$$

The mathematical solution by Lagrange multipliers is straightforward, and the result is: the "most probable" value of N_k is

$$\hat{N}_k = \frac{N}{Z(\beta)} \exp\left(-\beta E_k\right) \tag{A15}$$

where

$$Z(\beta) \equiv \sum_{k=1}^{s} \exp\left(-\beta E_k\right) \tag{A16}$$

and the parameter β is to be chosen so that the energy constraint (A12) is satisfied.

This simple result contains a great deal of physical information. Let us choose a particular set of cells R_k as follows. Divide up the coordinate space V and the velocity space into cells X_a, Y_b respectively, such that the potential and kinetic energies $\phi(x)$, $\frac{1}{2} m v^2$ do not vary appreciably within them, and take $R_k = X_a \otimes Y_b$. Then, writing $N_k = N_{ab}$, Boltzmann's prediction of the number of molecules in X_a irrespective of their velocity, is from (A15)

$$\hat{N}_a = \sum_b \hat{N}_{ab} = A(\beta) \exp(-\beta \phi_a) \qquad (A17)$$

where the normalization constant $A(\beta)$ is determined from $\sum N_a = N$. This is the famous Boltzmann distribution law. In a gravitational field, $\phi(x) = mgz$, it gives the usual "barometric formula" for decrease of the atmospheric density with height:

$$\rho(z) = \rho(0) \exp(-\beta mgz) \qquad . \qquad (A18)$$

Now this can be deduced also from the macroscopic equation of state: for one mole, $PV = RT$, or $P(z) = (RT/mN_0)\rho(z)$, where N_0 is Avogadro's number. But hydrostatic equilibrium requires $-dP/dz = g\rho(z)$, which gives on integration, for uniform temperature, $\rho(z) = \rho(0) \exp(-N_0 mgz/RT)$. Comparing with (A18), we find the meaning of the parameter: $\beta = (kT)^{-1}$, where T is the Kelvin temperature and $k \equiv R/N_0$ is Boltzmann's constant.

We can, equally well, sum (A15) over the space cells X_a and find the predicted number of molecules with velocity in the cell Y_b, irrespective of their position in space; but a far more interesting result is contained already in (A15) without this summation. Let us ask, instead; What fraction of the molecules in the space cell X_a are predicted to have velocity in the cell Y_b? This is, from (A15) and (A17),

$$f_b = \hat{N}_{ab}/\hat{N}_a = B(\beta) \exp(-\beta m v_b^2/2) \qquad (A20)$$

This is, of course, just the Maxwellian velocity distribution law; but with the new and at first sight astonishing feature that it is independent of position in space. Even though the force field is accelerating and decelerating molecules as they move from one region to another, when they arrive at their new location they have exactly the same mean square velocity as when they started! If this result is correct (as indeed it proved to be) it means that a Maxwellian velocity distribution, once established, is maintained automatically, without any help from collisions, as the molecules move about in any conservative force field.

From Boltzmann's reasoning, then, we get a very unexpected and nontrivial dynamical prediction by an analysis that, seemingly, ignores the dynamics altogether! This is only the first of many such examples where it appears that we are "getting something for nothing," the answer coming too easily to believe. Poincaré, in his essays on "Science and Method,"

felt this paradox very keenly, and wondered how by exploiting
our ignorance we can make correct predictions in a few lines
of calculation, that would be quite impossible to obtain if we
attempted a detailed calculation of the 10^{23} individual
trajectories.

It requires very deep thought to understand why we are not,
in this argument and others to come, getting something for
nothing. In fact, Boltzmann's argument does take the dynamics
into account, but in a very efficient manner. Information about
the dynamics entered his equations at two places: (1) the conservation of total energy; and (2) the fact that he defined his
cells in terms of phase volume, which is conserved in the
dynamical motion (Liouville's theorem). The fact that this was
enough to predict the correct spatial and velocity distribution
of the molecules shows that the millions of intricate dynamical
details that were not taken into account, were actually irrelevant to the predictions, and would have cancelled out anyway if
he had taken the trouble to calculate them.

Boltzmann's reasoning was super-efficient; far more so than
he ever realized. Whether by luck or inspiration, he put into
his equations only the dynamical information that happened to
be relevant to the questions he was asking. Obviously, it would
be of some importance to discover the secret of how this come
about, and to understand it so well that we can exploit it in
other problems.

If we can learn how to recognize and remove irrelevant information at the beginning of a problem, we shall be spared
having to carry out immense calculations, only to discover at
the end that practically everything we calculated was irrelevant to the question we were asking. And that is exactly what
we are after by applying Information Theory [actually, the
secret was revealed in my second paper (Jaynes, 1957b); but to
the best of my knowledge no other person has yet noticed it
there; so I will explain it again in Section D below. The
point is that Boltzmann was asking only questions about experimentally reproducible equilibrium properties].

In Boltzmann's "method of the most probable distribution,"
we have already the essential mathematical content of the
Principle of Maximum Entropy. But in spite of the conventional
name, it did not really involve probability. Boltzmann was not
trying to calculate a probability distribution; he was estimating
some physically real occupation numbers N_k, by a criterion
(value of W) that counts the number of real physical possibilities; a definite number that has nothing to do with anybody's
state of knowledge. The transition from this to our present
more abstract Principle of Maximum Entropy, although mathematically trivial, was so difficult conceptually that it required

almost another Century to bring about. In fact, this required
three more steps and even today the development of irreversible
Statistical Mechanics is being held up as much by conceptual
difficulties as by mathematical ones.

Enter Gibbs. Curiously, the ideas that we associate today with
the name of Gibbs were stated briefly in an early work of Boltzmann (1871); but were not pursued as Boltzmann became occupied
with his more specialized H-theorem. Further development of
the general theory was therefore left to Gibbs (1902). The
Boltzmann argument just given will not work when the molecules
have appreciable interactions, since then the total energy cannot be written in the additive form (A12). So we go to a much
more abstract picture. Whereas the preceding argument was
applied to an actually existing large collection of molecules,
we now let the entire macroscopic system of interest become, in
effect, a "molecule," and imagine a large collection of copies
of it.

This idea, and even the term "phase" to stand for the collection of all coordinates and momenta, appears also in a work
of Maxwell (1876). Therefore, when Gibbs adopted this notion,
which he called an "ensemble," it was not, as is apparently
thought by those who use the term "Gibbs ensemble," an innovation on his part. He used ensemble language rather as a concession to an already established custom. The idea became associated later with the von Mises "Kollektiv" but was actually
much older, dating back to Venn (1866); and Fechner's book
Kollektivmasslehre appeared in 1897.

It is important for our purposes to appreciate this little
historical fact and to note that, far from having invented the
notion of an ensemble, Gibbs himself (loc cit., p. 17) deemphasized its importance. We can detect a hint of cynicism
in his words when he states: "It is in fact customary in the
discussion of probabilities to describe anything which is imperfectly known as something taken at random from a great number
of things which are completely described." He continues that,
if we prefer to avoid any reference to an ensemble of systems,
we may recognize that we are merely talking about "the probability that the phase of a system falls within certain limits
at a certain time ---."

In other words, even in 1902 it was customary to talk about
a probability as if it were a frequency; even if it is a
frequency only in an imaginary ad hoc collection invented just
for that purpose. Of course, any probability whatsoever can be
thought of in this way if one wishes to; but Gibbs recognized
that in fact we are only describing our imperfect knowledge
about a single system.

The reason it is important to appreciate this is that we then understand Gibbs' later treatment of several topics, one of which had been thought to be a serious omission on his part. If we are describing only a state of knowledge about a single system, then clearly there can be nothing physically real about frequencies in the ensemble; and it makes no sense to ask, "which ensemble is the correct one?" In other words: different ensembles are not in 1:1 correspondence with different physical situations; they correspond only to different states of knowledge about a single physical situation. Gibbs understood this clearly; and that, I suggest, is the reason why he does not say a word about ergodic theorems, or hypotheses, but instead gives a totally different reason for his choice of the canonical ensembles.

Technical details of Gibbs' work will be deferred to Sec. D below, where we generalize his algorithm. Suffice it to say here that Gibbs introduces his canonical ensemble, and works out its properties, without explaining why he chooses that particular distribution. Only in Chap. XII, after its properties--including its maximum entropy property--have been set forth, does he note that the distribution with the minimum expectation of log p (i.e., maximum entropy) for a prescribed distribution of the constants of the motion has certain desirable properties. In fact, this criterion suffices to generate all the ensembles--canonical, grand canonical, microcanonical, and rotational--discussed by Gibbs.

This is, clearly, just a generalized form of the Principle of Indifference. The possibility of a different justification in the frequency sense, via ergodic theorems, had been discussed by Maxwell, Boltzmann, and others for some thirty years; as noted in more detail before (Jaynes, 1967) if Gibbs thought that any such further justification was needed, it is certainly curious that he neglected to mention it.

After Gibbs' work, however, the frequency view of probability took such absolute control over mens' minds that the ensemble became something physically real, to the extent that the following phraseology appears. Thermal equilibrium is defined as the situation where the system is "in a canonical distribution." Assignment of uniform prior probabilities was considered to be not a mere description of a state of knowledge, but a basic postulate of physical fact, justified by the agreement of our predictions with experiment.

In my student days this was the kind of language always used, although it seemed to me absurd; the individual system is not "in a distribution;" it is in a *state*. The experiments, moreover, do not verify "equal *a priori* probabilities" or "random *a priori* phases;" they verify only the predicted macroscopic equation of state, heat capacity, etc., and the predictions for these would have been the same for many ensembles, uniform or nonuniform microscopically. Therefore, the reason for the success of Statistical Mechanics must be altogether different from our having found the "correct" ensemble.

Intuitively, it must be true that use of the canonical ensemble, while sufficient to predict thermal equilibrium properties, is very far from necessary; in some sense, "almost every" member of a very wide class of ensembles would all lead to the same predictions for the particular macroscopic quantities actually observed. But I did not have any hint as to exactly what that class is; and needless to say, had not the faintest success in persuading anyone else of such heretical views.

We stress that, on this matter of the exact status of ensembles, you have to read Gibbs' own words in order to know accurately what his position was. For example, Ter Haar (1954, p. 128) tells us that "Gibbs introduced ensembles in order to use them for statistical considerations rather than to illustrate the behavior of physical systems ---." But Gibbs himself (loc. cit. p. 150) says, "--- our ensembles are chosen to illustrate the probabilities of events in the real world ---."

It might be thought that such questions are only matters of personal taste, and a scientist ought to occupy himself with more serious things. But one's personal taste determines which research problems he believes to be the important ones in need of attention; and the total domination by the frequency view caused all attention to be directed instead to the aforementioned "ergodic" problems; to justify the methods of Statistical Mechanics by proving from the dynamic equations of motion that the canonical ensemble correctly represents the frequencies with which, over a long time, an individual system coupled to a heat bath, finds itself in various states.

This problem metamorphosed from the original conception of Boltzmann and Maxwell that the phase point of an isolated (system + heat bath) ultimately passes through every state compatible with the total energy, to the statement that the time average of any phase function $f(p,q)$ for a single system is equal to the ensemble average of f; and this statement in turn was reduced (by von Neumann and Birkhoff in the 1930's) to the condition of metric transitivity (i.e., the full phase space shall have no subspace of positive measure that is invariant under the motion). But here things become extremely complicated, and there is little further progress. For example, even if one proves that in a certain sense "almost every" continuous flow is metrically transitive, one would still have to prove that the particular flows generated by a Hamiltonian are not exceptions.

WHERE DO WE STAND ON MAXIMUM ENTROPY? 231

Such a proof certainly cannot be given in generality, since counter-examples are known. One such is worth noting: in the writer's "Neoclassical Theory" of electrodynamics (Jaynes, 1973) we write a complete classical Hamiltonian system of equations for an atom (represented as a set of harmonic oscillators) interacting with light. But we find [loc. cit. Eq. (52)] that not only is the total energy a constant of the motion, the quantity $\Sigma_n W_n/\nu_n$ is conserved, where W_n, ν_n are the energy and frequency of the n'th normal mode of oscillation of the atom.

Setting this new constant of the motion equal to Planck's constant h, we have a classical derivation of the $E = h\nu$ law usually associated with quantum theory! Indeed, quantum theory simply takes this as a basic empirically justified postulate; and never makes any attempt to explain why such a relation exists. In Neoclassical Theory it is explained as a consequence of a new uniform integral of the motion, of a type never suspected in classical Statistical Mechanics. Because of it, for example, there is no Liouville theorem in the "action shell" subspace of states actually accessible to the system, and statistical properties of the motion are qualitatively different from those of the usual classical Statistical Mechanics. But all this emerges from a simple, innocent-looking classical Hamiltonian, involving only harmonic oscillators with a particular coupling law (linear in the field oscillators, bilinear in the atom oscillators). Having seen this example, who can be sure that the same thing is not happening more generally?

This was recognized by Truesdell (1960) in a work that I recommend as by far the clearest exposition, carried to the most far-reaching physical results, of any discussion of ergodic theory. He comes up against, "--- an old problem, one of the ugliest which the student of statistical mechanics must face: What can be said about the integrals of a dynamical system?" The answer is, "Practically nothing." In view of such simple counter-examples as that provided by Neoclassical theory, confident statements to the effect that real systems are almost certainly ergodic, seem like so much whistling in the dark.

Nevertheless, ergodic theory considered as a topic in its own right, does contain some important results. Unlike some others, Truesdell does not confuse the issue by trying to mix up probability notions and dynamical ones. Instead, he states unequivocally that his purpose is to calculate time averages. This is a definite, well posed dynamical problem having nothing to do with any probability considerations; and Truesdell proceeds to show, in greater depth than any other writer known to me, exactly what implications the Birkhoff theorem has for this question. Since we cannot prove, and in view of counter-examples have no valid reason to expect, that the flow is metrically transitive over the entire phase space S, the original hopes of Boltzmann and Maxwell must remain unrealized; but in return for this we get something far more valuable, which just misses being noticed.

The flow will be metrically transitive on some (unknown) sub-space S' determined by the (unknown) uniform integrals of the motion; and the time average of any phase function $f(p,q)$ will, by the Birkhoff theorem, be equal to its phase space average over that subspace. Furthermore, the fraction of time that the system spends in any particular region s in S' is equal to the ratio of phase volumes: $\sigma(s)/\sigma(S')$.

These are just the properties that Boltzmann and Maxwell wanted; but they apply only to some subspace S' <u>which cannot be known until we have determined all the uniform integrals of the motion</u>. That is the purely dynamical theorem; and I think that if today we could resurrect Maxwell and tell it to him, his reaction would be: "Of course, that is obviously right and it is just what I was trying to say. The trouble was that I was groping for words, because in my day we did not have the mathematical vocabulary, arising out of measure theory and the theory of transformation groups, that is needed to state it precisely."

That more valuable result is tantalizingly close when Truesdell considers "--- the idea that however many integrals a system has, generally we shall not know the value of any but the energy, so we should assign equal <u>a priori</u> probability to the possible values of the rest, which amounts to disregarding the rest of them. Now an idea of this sort, by itself, is just unsound." It is indeed unsound, in the context of Truesdell's purpose to calculate correct time averages from the dynamics; for those time averages must in general depend on all the integrals of the motion, whether or not we happen to know about them.

The point that he just fails to see is that if, nevertheless, we only have the courage to go ahead and do the calculation he rejects as unsound, we can then compare its results with experimental time averages. If they disagree, then <u>we have obtained experimental evidence of the existence of new integrals of the motion</u>, and the nature of the deviation gives a clue as to what they may be. So, if our calculation should indeed prove to be "unsound," the result would be far more valuable to physics than a "successful" calculation!

To all this, however, one proviso must be added. Even if one could prove transitivity for the entire phase space, this result would not explain the success of equilibrium statistical mechanics, for reasons expounded in great detail before (Jaynes, 1967). These theorems apply only to time averages over enormous

(strictly, infinite) time; and an average over a finite time T
will approach its limiting value for $T \to \infty$ only if T is so long
that the phase point of the system has explored a "representative
sample" of the accessible phase volume. But the very
existence of time-dependent irreversible processes shows that
the "representative sampling time" must be very long compared
to the time in which our measurements are made. So the equality
of phase space averages with infinite time averages fails, on
two counts, to explain the equality of canonical ensemble
averages and experimental values. We can conclude only that
the "ergodic" attempts to justify Gibbs' statistical mechanics
foundered not only on impossibly difficult technical problems
of integrals of the motion; but also on a basic logical defect
arising from the impossibly long averaging times.

Enter Shannon. It was the work of Claude Shannon (1948) on
Information Theory which showed us the way out of this dilemma.
Like all major advances, it had many precursors, whose full
significance could be seen only later. One finds them not only
in the work of Boltzmann and Gibbs just noted, but also in that
of G. N. Lewis, L. Szilard, J. von Neumann, and W. Elsasser, to
mention only the most obvious examples.

Shannon's articles appeared just at the time when I was taking
a course in Statistical Mechanics from Professor Eugene Wigner;
and my mind was occupied with the difficulties, which he always
took care to stress, faced by the theory at that time; the
short sketch above notes only a few of them. Reading Shannon
filled me with the same admiration that all readers felt, for
the beauty and importance of the material; but also with a
growing uneasiness about its meaning. In a communication
process, the message M_i is assigned probability p_i, and the
entropy $H = -\Sigma p_i \log p_i$ is a measure of "information." But
whose information? It seems at first that if information is
being "sent," it must be possessed by the sender. But the
sender knows perfectly well which message he wants to send;
what could it possibly mean to speak of the probability that
he will send message M_i?

We take a step in the direction of making sense out of this
if we suppose that H measures, not the information of the sender,
but the ignorance of the receiver, that is removed by receipt
of the message. Indeed, many subsequent commentators appear
to adopt this interpretation. Shannon, however, proceeds to
use H to determine the channel capacity C required to transmit
the message at a given rate. But whether a channel can or
cannot transmit message M in time T obviously depends only on
properties of the message and the channel--and not at all on

the prior ignorance of the receiver! So this interpretation will not work either.

Agonizing over this, I was driven to conclude that the different messages considered must be the set of all those that will, or might be, sent over the channel during its useful life; and therefore Shannon's H measures the degree of ignorance of the <u>communication engineer</u> when he designs the technical equipment in the channel. Such a viewpoint would, to say the least, seem natural to an engineer employed by the Bell Telephone Laboratories--yet it is curious that nowhere does Shannon see fit to tell the reader explicitly <u>whose</u> state of knowledge he is considering, although the whole content of the theory depends crucially on this.

It is the obvious importance of Shannon's theorems that first commands our attention and respect; but as I realized only later, it was just his vagueness on these conceptual questions--allowing every reader to interpret the work in his own way--that made Shannon's writings, like those of Niels Bohr, so eminently suited to become the Scriptures of a new Religion, as they so quickly did in both cases.

Of course, we do not for a moment suggest that Shannon was deliberately vague; indeed, on other matters few writers have achieved such clarity and precision. Rather, I think, a certain amount of caution was forced on him by a growing paradox that Information Theory generates within the milieu of probability theory as it was then conceived--a paradox only vaguely sensed by those who had been taught only the strict frequency definition of probability, and clearly visible only to those familiar with the work of Jeffreys and Cox. What do the probabilities p_i mean? Do they stand for the <u>frequencies</u> with which the different messages are sent?

Think, for a moment, about the last telegram you sent or received. If the Western Union Company remains in business for another ten thousand years, how many times do you think it will be asked to transmit that identical message?

The situation here is not really different from that in statistical mechanics, where our first job is to assign probabilities to the various possible quantum states of a system. In both cases the number of possibilities is so great that a time millions of times the age of the universe would not suffice to realize all of them. But it seems to be much easier to think clearly about messages than quantum states. Here at last, it seemed to me, was an example where the absurdity of a frequency interpretation is so obvious that no one can fail to see it; but the usefulness of the probability approach was equally clear. The probabilities assigned to individual messages are not measurable frequencies; they are only a means of describing a state of knowledge; just the original sense in which Laplace and Jeffreys interpreted a probability distribution.

The reason for the vagueness is then apparent; to a person who has been trained to think of probability only in the sense of frequency in a random experiment (as was surely the case for anyone educated at M.I.T. in the 1930's!), the idea that a probability distribution represents a mere state of knowledge is strictly taboo. A probability distribution would not be "objective" unless it represents a real physical situation. The question: "Whose information are we describing?" doesn't make sense, because the notion of a probability for a person with a certain state of knowledge just doesn't exist. So Shannon is forced to do the most careful egg-walking, speaking of a probability as if it were a real, measurable frequency, while using it in a way that shows clearly that it is not.

For example, Shannon considers the entropies H_1 calculated from single letter frequencies, H_2 from digram frequencies, H_3 from trigram frequencies, etc., as a sequence of successive approximations to the "true" entropy of the source, which is $H = \lim H_n$ for $n \to \infty$. Application of his theorems presupposes that all this is known. But suppose we try to determine the "true" ten-gram frequencies of English text. The number of different ten-grams is about 1.4×10^{14}; to determine them all to something like five percent accuracy, we should need a sample of English text containing about 10^{17} ten-grams. That is thousands of times greater than all the English text in the Library of Congress, and indeed much greater than all the English text recorded since the invention of printing.

If we had overcome that difficulty, and could measure those ten-gram frequencies (by scanning the entire text) at the rate of 1000 per second, it would require about 4400 years to take the data; and to record it on paper at a rate of 1000 entries per sheet, would require a stack of paper about 7000 miles high. Evidently, then, we are destined never to know the "true" entropy of the English language; and in the application of Shannon's theorems to real communication systems we shall have to accept some compromise.

Now, our story reaches its climax. Shannon discusses the problem of encoding a message, say English text, into binary digits in the most efficient way. The essential step is to assign probabilities to each of the conceivable messages in a way which incorporates the prior knowledge we have about the structure of English. Having this probability assignment, a construction found independently by Shannon and R. M. Fano yields the encoding rules which minimize the expected transmission time of a message.

But, as noted, we shall never know the "true" probabilities of English messages; and so Shannon suggests the principle by which we may construct the distribution p_i actually used for applications: "--- we may choose to use some of our statistical knowledge of English in constructing a code, but not all of it. In such a case we consider the source with the <u>maximum entropy subject to the statistical conditions we wish to retain</u>. The entropy of this source determines the channel capacity which is necessary and sufficient." [emphasis mine].

Shannon does not follow up this suggestion with the equations, but turns at this point to other matters. But if you start to solve this problem of maximizing the entropy subject to certain constraints, you will soon discover that you are writing down some very familiar equations. The probability distribution over messages is just the Gibbs canonical distribution with certain parameters. To find the values of the parameters, you must evaluate a certain partition function, etc.

Here was a problem of statistical inference--or what is the same thing, statistical decision theory--in which we are to decide on the best way of encoding a message, making use of certain partial information about the message. The solution turns out to be mathematically identical with the Gibbs formalism of statistical mechanics, which physicists had been trying, long and unsuccessfully, to justify in an entirely different way.

The conclusion, it seemed to me, was inescapable. We can have our justification for the rules of statistical mechanics, in a way that is incomparably simpler than anyone had thought possible, if we are willing to pay the price. The price is simply that we must loosen the connections between probability and frequency, by returning to the original viewpoint of Bernoulli and Laplace. The only new feature is that their Principle of Insufficient Reason is now generalized to the Principle of Maximum Entropy. Once this is accepted, the general formalism of statistical mechanics--partition functions, grand canonical ensemble, laws of thermodynamics, fluctuation laws--can be derived in a few lines without wasting a minute on ergodic theory. The pedagogical implications are clear.

The price we have paid for this simplification is that we cannot interpret the canonical distribution as giving the <u>frequencies</u> with which a system goes into the various states. But nobody had ever justified or needed that interpretation anyway. In recognizing that the canonical distribution represents only our state of knowledge when we have certain partial information derived from macroscopic measurements, we are not losing anything we had before, but only frankly admitting the situation that has always existed; and indeed, which Gibbs had recognized.

On the other hand, what we have gained by this change in interpretation is far more than we bargained for. Even if one had been completely successful in proving ergodic theorems, and had continued to ignore the difficulty about length of time over which the averages have to be taken, this still would have given a justification for the methods of Gibbs only in the equilibrium case. But the principle of maximum entropy, being entirely independent of the equations of motion, contains no such restriction. If one grants that it represents a valid method of reasoning at all, one must grant that it gives us also the long-hoped-for general formalism for treatment of irreversible processes!

The last statement above breaks into new ground, and claims for statistical mechanics based on Information Theory, a far wider range of validity and applicability than was ever claimed for conventional statistical mechanics. Just for that reason, the issue is no longer one of mere philosophical preference for one viewpoint or another; the issue is now one of definite mathematical fact. For the assertion just made can be put to the test by carrying out specific calculations, and will prove to be either right or wrong.

Some Personal Recollections. All this was clear to me by 1951; nevertheless, no attempt at publication was made for another five years. There were technical problems of extending the formalism to continuous distributions and the density matrix, that were not solved for many years; but the reason for the initial delay was quite different.

In the Summer of 1951, Professor G. Uhlenbeck gave his famous course on Statistical Mechanics at Stanford, and following the lectures I had many conversations with him, over lunch, about the foundations of the theory and current progress on it. I had expected, naively, that he would be enthusiastic about Shannon's work, and as eager as I to exploit these ideas for Statistical Mechanics. Instead, he seemed to think that the basic problems were, in principle, solved by the then recent work of Bogoliubov and van Hove (which seemed to me filling in details, but not touching at all on the real basic problems)--and adamantly rejected all suggestions that there is any connection between entropy and information.

His initial reaction to my remarks was exactly like my initial reaction to Shannon's: "Whose information?" His position, which I never succeeded in shaking one iota, was: "Entropy cannot be a measure of 'amount of ignorance,' because different people have different amounts of ignorance; entropy is a definite physical quantity that can be measured in the laboratory with thermometers and calorimeters." Although the answer to this was clear in my own mind, I was unable, at the time, to convey that answer to him. In trying to explain a new idea I was, like Maxwell, groping for words because the way of thinking and habits of language then current had to be broken before I could express a different way of thinking.

Today, it seems trivially easy to answer Professor Uhlenbeck's objection as follows: "Certainly, different people have different amounts of ignorance. The entropy of a thermodynamic system is a measure of the degree of ignorance of a person whose sole knowledge about its microstate consists of the values of the macroscopic quantities X_i which define its thermodynamic state. This is a completely 'objective' quantity, in the sense that it is a function only of the X_i, and does not depend on anybody's personality. There is then no reason why it cannot be measured in the laboratory."

It was my total inability to communicate this argument to Professor Uhlenbeck that caused me to spend another five years thinking over these matters, trying to write down my thoughts more clearly and explicitly, and making sure in my own mind that I could answer all the objections that Uhlenbeck and others had raised. Finally, in the Summer of 1956 I collected this into two papers, sending the first off to the Physical Review on August 29.

Now another irony takes place; it is left to the Reader to guess to whom the Editor (S. Goudsmit) sent it for refereeing. That Unknown Referee's comments (now framed on my office wall as an encouragement to young men who today have to fight for new ideas against an Establishment that wants only new mathematics) opine that the work is clearly written, but since it expounds only a certain philosophy of interpretation and has no application whatsoever in Physics, it is out of place in a Physics journal. But a second referee thought differently, and so the papers were accepted after all, appearing in 1957. Within a year there were over 2000 requests for reprints.

Needless to say, my own understanding of the technical problems continued to evolve for many years afterward. A schoolboy, having just learned the rules of arithmetic, does not see immediately how to apply them to the extraction of cube roots, although he has in his grasp all the principles needed for this. Similarly, I did not see how to set down the explicit equations for irreversible processes because I simply could not believe that the solution to such a complicated problem could be as simple as the Maximum Entropy Principle was giving; and spent six more years (1956-1962) trying to mutilate the principle by grafting new and more complicated rococo embellishments onto it. In my Brandeis lectures of 1962, tongue and pen somehow managed to state the right rule [Eq. (50)]; but the inner mind did not fully assent; it still seemed like getting something for nothing.

The final breakthrough came in the Christmas vacation period of 1962 when, after all else had failed, I finally had the courage to sit down and work out all the details of the calculations that result from using only the Maximum Entropy Principle; and nothing else. Within three days the new formalism was in hand, masses of the known correct results of Onsager, Wiener, Kirkwood, Callen, Kubo, Mori, MacLennon, were pouring out as special cases, just as fast as I could write them down; and it was clear that this was it. Two months later, my students were the first to have assigned homework problems to predict irreversible processes by solving Wiener-Hopf integral equations.

As it turned out, no more principles were needed beyond those stated in my first paper; one has merely to take them absolutely literally and apply them, putting into the equations the macroscopic information that one does, in fact, have about a nonequilibrium state; and all else follows inevitably.

From this the reader will understand why I have considerable sympathy for those who today have difficulty in accepting the Principle of Maximum Entropy, because (1) the results seem to come too easily to believe; and (2) it seems at first glance as if the dynamics has been ignored. In fact, I struggled for eleven years with exactly the same feeling, before seeing clearly not only why, but also in detail how the formalism is able to function so efficiently.

The point is that we are not ignoring the dynamics, and we are not getting something for nothing, because we are asking of the formalism only some extremely simple questions; we are asking only for predictions of experimentally reproducible things; and for these all circumstances that are not under the experimenter's control must, of necessity, be irrelevant.

If certain macroscopically controlled conditions are found, in the laboratory, to be sufficient to determine a reproducible outcome, then it must follow that information about those macroscopic conditions tells us everything about the microscopic state that is relevant for theoretical prediction of that outcome. It may seem at first "unsound" to assign equal a priori probabilities to all other details, as the Maximum Entropy Principle does; but in fact we are assigning uniform probabilities only to details that are irrelevant for questions about reproducible phenomena.

To assume further information by putting some additional fine-grained structure into our ensembles would, in all probability, not lead to incorrect predictions; it would only force us to calculate intricate details that would, in the end, cancel out of our final predictions. Solution by the Maximum Entropy Principle is so unbelievably simple just because it eliminates those irrelevant details right at the beginning of the calculation by averaging over them.

To discover this argument requires only that one think, very carefully, about why Boltzmann's method of the most probable distribution was able to predict the correct spatial and velocity distribution of the molecules; and this could have been done at any time in the past 100 years. Whether or not one wishes to recognize it, this--and not ergodic properties--is the real reason why all Statistical Mechanics works. But once the argument is understood, it is clear that it applies equally well whether the macroscopic state is equilibrium or non-equilibrium, and whether the observed phenomenon is reversible or irreversible.

I hope that this historical account will also convey to the reader that the Principle of Maximum Entropy, although a powerful tool, is hardly a radical innovation. Its philosophy was clearly foreshadowed by Laplace and Jeffreys; its mathematics by Boltzmann and Gibbs.

B. Present Features and Applications.

Let us set down, for reference, a bit of the basic Maximum Entropy formalism for the finite discrete case, putting off generalizations until they are needed. There are n different possibilities, which would be distinguished adequately by a single index $(i = 1, 2, \ldots, n)$. Nevertheless we find it helpful, both for notation and for the applications we have in mind, to introduce in addition a real variable x, which can take on the discrete values $(x_i, 1 \leq i \leq n)$, defined in any way and not necessarily all distinct. If we have certain information I about x, the problem is to represent this by a probability distribution $\{p_i\}$ which has maximum entropy while agreeing with I.

Clearly, such a problem cannot be well-posed for arbitrary information; I must be such that, given any proposed distribution $\{p_i\}$, we can determine unambiguously whether I does or does not agree with $\{p_i\}$. Such information will be called testable. For example, consider:

$I_1 \equiv$ "It is certain that tanh x < 0.7."

$I_2 \equiv$ "There is at least a 90% probability that tanh x < 0.7."

$I_3 \equiv$ "The mean value of tanh x is 0.675."

WHERE DO WE STAND ON MAXIMUM ENTROPY? 241

$I_4 \equiv$ "The mean value of tanh x is probably less than 0.7."

$I_5 \equiv$ "There is some reason to believe that tanh x = 0.675."

Statements I_1, I_2, I_3 are testable, and may be used as constraints in maximizing the entropy. I_4 and I_5, although clearly relevant to inference about x, are too vague to be testable, and we have at present no formal principle by which such information can be used in a mathematical theory. However, the fact that our intuitive common sense does make use of nontestable information suggests that new principles for this, as yet undiscovered, must exist.

Since n is finite, the entropy has an absolute maximum value log n, and any constraint can only lower this. If we think of the $\{p_i\}$ as cartesian coordinates of a point P in an n-dimensional space, P is constrained by $p_i \geq 0$, $\Sigma p_i = 1$ to lie on a domain D which is a "triangular" segment of an (n-1)-dimensional hyperplane. On D the entropy varies continuously, taking on all values in $0 \leq H \leq \log n$ and reaching its absolute maximum at the center. Any testable information will restrict P to some subregion D' of D, and clearly the entropy has some least upper bound $H \leq \log n$ on D'. So the maximum entropy problem must have a solution if D' is a closed set.

There may be more than one solution: for example, the information $I_6 \equiv$ "The entropy of the distribution $\{p_i\}$ is not greater than log(n-1)" is clearly testable, and if $n > 2$ it yields an infinite number of solutions. Furthermore, strictly speaking, if D' is an open set there may not be any solution, the upper bound being approached but not actually reached on D'. Such a case is generated by $I_7 \equiv$ "$p_1^2 + p_2^2 < n^{-2}$." However, since we are concerned with physical problems where the distinction between open and closed sets cannot matter, we would accept a point on the closure of D' (in this example, on its boundary) as a valid solution, although corresponding strictly only to $I_8 \equiv$ "$p_1^2 + p_2^2 \leq n^{-2}$."

But these considerations are mathematical niceties that one has to mention only because he will be criticized if he does not. In the real applications that matter, we have not yet found a case which does not have a unique solution.

In principle, every different kind of testable information will generate a different kind of mathematical problem. But there is one important class of problems for which the general solution was given once and for all, by Gibbs. If the constraints consist of specifying mean values of certain functions $\{f_1(x), f_2(x), \ldots, f_m(x)\}$:

$$\sum_{i=1}^{n} p_i f_k(x_i) = F_k , \qquad 1 \leq k \leq m \qquad (B1)$$

where $\{F_k\}$ are numbers given in the statement of the problem, then if $m < n$, entropy maximization is a standard variational problem solvable by stationarity using the Lagrange multiplier technique. It has the formal solution:

$$p_i = \frac{1}{Z(\lambda_1 \ldots \lambda_m)} \exp\left[-\lambda_1 f_1(x_i) - \ldots - \lambda_m f_m(x_i)\right] \qquad (B2)$$

where

$$Z(\lambda_1, \ldots, \lambda_m) \equiv \sum_{i=1}^{n} \exp\left[-\lambda_1 f_1(x_i) - \ldots - \lambda_m f_m(x_i)\right] \qquad (B3)$$

is the partition function and $\{\lambda_k\}$ are the Lagrange multipliers, which are chosen so as to satisfy the constraints (B1). This is the case if

$$F_k = -\frac{\partial}{\partial \lambda_k} \log Z , \qquad 1 \leq k \leq m \qquad (B4)$$

a set of m simultaneous equations for m unknowns. The value of the entropy maximum then attained is, as noted in my reminiscences, a function only of the given data:

$$S(F_1 \ldots F_m) = \log Z + \sum_k \lambda_k F_k \qquad (B5)$$

and if this function were known, the explicit solution of (B4) would be

$$\lambda_k = \frac{\partial S}{\partial F_k} , \qquad 1 \leq k \leq m . \qquad (B6)$$

Given this distribution, the best prediction we can make (in the sense of minimizing the expected square of the error) of any quantity $q(x)$, is then

$$\langle q(x) \rangle = \sum_{i=1}^{n} p_i q(x_i)$$

and numerous covariance and reciprocity rules are contained in the identity

$$<qf_k> - <q><f_k> = -\frac{\partial <q>}{\partial \lambda_k} \tag{B7}$$

[note the special cases $q(x) = f_j(x)$, and $j = k$]. The functions $f_k(x)$ may contain also some parameters α_j:

$$f_k = f_k(x;\alpha_1 \ldots \alpha_s)$$

(which in physical applications might have the meaning of volume, magnetic field intensity, angular velocity, etc.); and we have an important variational property; if we make an arbitrary small change in all the data of the problem $\{\delta F_k, \delta\alpha_r\}$, we may compare two slightly different maximum-entropy solutions. The difference in their entropies is found, after some calculation, to be

$$\delta S = \sum_k \lambda_k \, \delta Q_k \tag{B8}$$

where

$$\delta Q_k \equiv \delta<f_k> - <\delta f_k> \quad . \tag{B9}$$

The meaning of this identity has a familiar ring: there is no such function as $Q_k(F_1 \ldots F_m;\alpha_1 \ldots \alpha_s)$ because δQ_k is not an exact differential. However, the Lagrange multiplier λ_k is an integrating factor such that $\Sigma \lambda_k \, \delta Q_k$ is the exact differential of a "state function" $S(F_1 \ldots F_m;\alpha_1 \ldots \alpha_s)$.

I believe that Clausius would recognize here an interesting echo of his work, although we have only stated some general rules for plausible reasoning, making no necessary reference to physics. This is enough of the bare skeleton of the formalism to serve as the basis for some examples and discussion.

The Brandeis Dice Problem. First, we illustrate the formalism by working out the numerical solution to a problem which was used in the Introduction to my 1962 Brandeis lectures merely as a qualitative illustration of the ideas, but has since become a cause célèbre as some papers have been written attacking the Principle of Maximum Entropy on the grounds of this very example. So a close look at it will take us straight to the heart of some of the most common misconceptions and, I hope, give us some appreciation of what the Principle of Maximum Entropy does and does not (indeed, should not) accomplish for us.

When a die is tossed, the number of spots up can have any value i in $1 \le i \le 6$. Suppose a die has been tossed N times and we are told only that the average number of spots up was not 3.5 as we might expect from an "honest" die but 4.5. Given this information, and nothing else, what probability should we assign to i spots on the next toss? The Brandeis lectures started with a qualitative graphical discussion of this problem, which showed (or so I thought) how ordinary common sense forces us to a result with the qualitative properties of the maximum-entropy solution.

Let us see what solution the Principle of Maximum Entropy gives for this problem, if we interpret the data as imposing the mean value constraint

$$\sum_{i=1}^{6} i\, p_i = 4.5 \quad . \tag{B10}$$

The partition function is

$$Z(\lambda) = \sum_i e^{-\lambda i} = x(1-x)^{-1}(1-x^6) \tag{B11}$$

where $x \equiv e^{-\lambda}$. The constraint (B10) then becomes

$$-\frac{\partial}{\partial \lambda} \log Z = \frac{1 - 7x^6 + 6x^7}{(1-x)(1-x^6)} = 4.5$$

or

$$3x^7 - 5x^6 + 9x - 7 = 0 \quad . \tag{B12}$$

By computer, the desired root of this is $x = 1.44925$, which yields $\lambda = -0.37105$, $Z = 26.66365$, $\log Z = 3.28330$. The maximum-entropy probabilities are $p_i = Z^{-1} x^i$, or

$$\{p_1 \ldots p_6\} = \{0.05435, 0.07877, 0.11416, 0.16545, 0.23977, 0.34749\} \tag{B13}$$

From (B5), the entropy of this distribution is

$$S = 1.61358 \text{ natural units} \tag{B14}$$

as compared to the maximum of $\log_e 6 = 1.79176$, corresponding to no constraints and a uniform distribution.

Now, what does this result mean? In the first place, it is a distribution $\{p_r, 1 \le r \le 6\}$ on a space of only six points; the sample space S of a single trial. Therefore, our result as it stands is only a mean of describing a state of knowledge about the outcome of a single trial. It represents a state of

WHERE DO WE STAND ON MAXIMUM ENTROPY? 245

knowledge in which one has only (1) the enumeration of the six possibilities; and (2) the mean value constraint (B10); <u>and no other information</u>. The distribution is "maximally noncommittal" with respect to all other matters; it is as uniform (by the criterion of the Shannon information measure) as it can get without violating the given constraint.

Any probability distribution over some sample space S enables us to make statements about (i.e., assign probabilities to) propositions or events defined within that space. It does not--and by its very nature cannot--make statements about any event lying outside that space. Therefore, our maximum-entropy distribution does not, and cannot, make any statement about <u>frequencies</u>.

Anything one says about a frequency in n tosses is a statement about an event in the n-fold extension space $S^n = S \otimes S \otimes \ldots \otimes S$ of n tosses, containing 6^n points (and of course, in any higher space which has S^n as a subspace).

It may be common practice to jump to the conclusion that a probability in one space is the same as a frequency in a different space; and indeed, the level of many expositions is such that the distinction is not recognized at all. But the first thing one has to learn about using the Principle of Maximum Entropy in real problems is that the mathematical rules of probability theory must be obeyed strictly; all conceptual sloppiness of this sort must be recognized and expunged.

There is, indeed, a connection between a <u>probability</u> p_i in space S and a <u>frequency</u> g_i in S^n; but we are justified in using only those connections <u>which are deducible from the mathematical rules of probability theory</u>. As we shall see in connection with fluctuation theory, some common attempts to identify probability and frequency actually stand in conflict with the rules of probability theory.

<u>Probability and Frequency</u>. To derive the simplest and most general connection, the sample space S^n of n trials may be labeled by $\{r_1, r_2, \ldots, r_n\}$, where $1 \leq r_k \leq 6$, and r_k is the number of spots up on the k'th toss. The most general probability assignment on S^n is a set of non-negative real numbers $P(r_1 \ldots r_n)$ such that

$$\sum_{r_1=1}^{6} \ldots \sum_{r_n=1}^{6} P(r_1 \ldots r_n) = 1 \quad . \tag{B15}$$

In any given sequence $\{r_1 \ldots r_n\}$ of results, the frequency with which i spots occurs is

$$g_i(r_1 \ldots r_n) = n^{-1} \sum_{k=1}^{n} \delta(r_k, i) \qquad (B16)$$

This can take on (n+1) discrete values, and its expectation is

$$\langle g_i \rangle = \frac{1}{n} \sum_{k=1}^{n} \sum_{r_1=1}^{6} \ldots \sum_{r_n=1}^{6} P(r_1 \ldots r_n) \delta(r_k, i)$$

$$= \frac{1}{n} \left[p_1(i) + p_2(i) + \ldots + p_n(i) \right] \qquad (B17)$$

where $p_k(i)$ is the probability of getting i spots on the k'th toss, regardless of what happens in other tosses. The expected frequency of an event is always equal to its <u>average</u> probability over the different trials.

Many experiments fall into the category of <u>exchangeable sequences</u>; i.e., it is clear that the underlying "mechanism" of the experiment, although unknown, is not changing from one trial to another. The probability of any particular sequence of results $\{r_1 \ldots r_n\}$ should then depend only on how many times a particular outcome $r = i$ happened; and not on which particular trials. Then the probability distribution $P\{r_k\}$ is invariant under permutations of the labels k. In this case, the probability of i spots is the same at each trial: $p_1(i) = p_2(i) = \ldots = p_n(i) = p_i$, and (B17) becomes

$$\langle g_i \rangle = p_i \qquad (B18)$$

In an exchangeable sequence, the probability of an event at one trial is not the same as its frequency in many trials; but it is numerically equal to the <u>expectation</u> of that frequency; and this connection holds whatever correlations may exist between different trials.

The probability is therefore the "best" estimate of the frequency, in the sense that it minimizes the expected square of the error. But the result (B18) tells us nothing whatsoever about whether this is a <u>reliable</u> estimate; and indeed nothing <u>in the space</u> S of a single trial can tell us anything about the reliability of (B18).

To investigate this, note that by a similar calculation, the expected product of two frequencies is

$$\langle g_i g_j \rangle = n^{-2} \sum_{k,m=1}^{n} p_k(i) \, p(j,m|i,k) \qquad (B19)$$

where $p(j,m|i,k)$ is the conditional probability that the m'th trial gives the result j, given that the k'th trial had the outcome i. Of course, if $m = k$ we have simply $p(jk|ik) = \delta_{ij}$.

In an exchangeable sequence $p(jm|ik)$ is independent of m,k for $m \neq k$; and so $p_k(i) \, p(jm|ik) = p_{ij}$, the probability of getting the outcomes i,j respectively at any two different tosses. The covariance of g_i, g_j then reduces to

$$\langle g_i g_j \rangle - \langle g_i \rangle \langle g_j \rangle = (p_{ij} - p_i p_j) + \frac{1}{n}(\delta_{ij} p_i - p_{ij}) . \quad (B20)$$

If the probabilities are not independent, $p_{ij} \neq p_i p_j$, this does not go to zero for large n.

Let us examine the case $i = j$ more closely. Writing $p_{ii} = \alpha_i p_i$, α_i is the conditional probability that, having obtained the result i on one toss, we shall get it at some other specified toss. The variance of g_i is, from (B20), dropping the index i,

$$\langle g^2 \rangle - \langle g \rangle^2 = p(\alpha - p) + \frac{1}{n} p(1 - \alpha) . \quad (B21)$$

Two extreme cases of inter-trial correlations are contained in (B21). For complete independence, $\alpha = p$, the variance reduces to $n^{-1} p(1-p)$, just the result of the de Moivre-Laplace limit theorem (A4). But as cautioned before, in any other case the variance does not tend to zero at all; there is no "law of large numbers." For complete dependence, $\alpha = 1$ (i.e., having seen the result of one toss, the die is certain to give the same result at all others), (B21) reduces to $p(1-p)$ which again makes excellent sense; in this case our uncertainty about the frequency in any number of tosses must be just our uncertainty about the first toss.

Note that the variance (B21) becomes zero for a slight negative correlation:

$$\alpha = p - \frac{1-p}{n-1} \quad (B22)$$

Due to the permutation invariance of $P(r_1 \ldots r_n)$ it is not possible to have a negative correlation stronger than this; as $n \to \infty$ it is not possible to have any negative correlation in an exchangeable sequence. This corresponds to the famous de Finetti (1937) representation theorem; in the literature of pure mathematics it is called the Hausdorff moment problem. An almost unbelievably simple proof has just been found by Heath and Sudderth (1976).

To summarize: given any probability assignment $P(r_1 \ldots r_n)$ on the space S^n, we can determine the probability distribution $W_i(t)$ for the frequency g_i to take on any of its possible values $g_i = (t/n)$, $0 \leq t \leq n$. The (mean) ± (standard deviation) over this distribution then provide a reasonable statement of our "best" estimate of g_i and its accuracy. In the case of an exchangeable sequence, this estimate is

$$(g_i)_{est} = p_i \pm \sqrt{p_i(1-p_i)}\left[R_i + \frac{1-R_i}{n}\right]^{1/2} \quad (B23)$$

where $R_i \equiv (\alpha_i - p_i)/(1-p_i)$ is a measure of the inter-trial correlation, ranging from $R = 0$ for complete independence to $R = 1$ for complete dependence.

Evidently, then, to suppose that a probability assignment at a single trial is also an assertion about a frequency in many trials in the sense of the Bernoulli and de Moivre-Laplace limit theorems, is in general unjustified unless (1) the successive trials form an exchangeable sequence, and (2) the correlation of different trials is strictly zero. However, there are other kinds of connections between probability and frequency; and maximum-entropy distributions have an exact and close relation to frequencies after all, as we shall see presently.

<u>Relation to Bayes' Theorem</u>. To prepare us to deal with some objections to the maximum-entropy solution (B13) we turn back to the basic product and sum rules of probability theory (A8), (A9) derived by Cox from requirements of consistency. Just as any argument of deductive logic can be resolved ultimately into many syllogisms, so any calculation of inductive logic (i.e., probability theory) is reducible to many applications of these rules.

We stress that these rules make no reference to frequencies; or to any random experiment. The numbers $p(A|B)$ are simply a convenient numerical scale on which to represent degrees of plausibility. As noted at the beginning of this work, it is the problem of determining initial numerical values by logical analysis of the prior information in more general cases than solved by Bernoulli and Laplace, that underlies our study.

Furthermore, in neither the statement nor the derivation of these rules is there any reference to the notion of a sample space. In a formally qualitative sense, therefore, they may be applied to any propositions A, B, C,... with unambiguous meanings. Their complete qualitative correspondence with ordinary common sense was demonstrated in exhaustive detail by Polya (1954).

But in quantitative applications we find at once that merely defining two propositions, A, B is not sufficient to determine any numerical value for $p(A|B)$. This numerical value depends not only on A, B, but also on which alternative propositions A', A", etc. are to be considered if A should be false; and the problem is mathematically indeterminate until those alternatives are fully specified. In other words, we must define

WHERE DO WE STAND ON MAXIMUM ENTROPY? 249

our "sample space" or "hypothesis space" before we have any mathematically well-posed problem.

In statistical applications (parameter estimation, hypothesis testing), the most important constructive rule is just the statement that the product rule is consistent; i.e., $p(AB|C)$ is symmetric in A and B, so $p(A|BC)p(B|C) = p(B|AC)p(A|C)$. If $p(B|C) \neq 0$, we thus obtain

$$p(A|BC) = p(A|C) \frac{p(B|AC)}{p(B|C)} \qquad (B24)$$

in which we may call C the prior information, B the conditioning information. In typical applications, C represents the general background knowledge or assumptions used to formulate the problem, B is the new data of some experiment, and A is some hypothesis being tested. For example, in the Millikan oil-drop experiment, we might take A as the hypothesis: "the electronic charge lies in the interval $4.802 < e < 4.803$," while C represents the general assumed known laws of electrostatics and viscous hydrodynamics and the results of previous measurements, while B stands for the new data being used to find a revised "best" value of e. Equation (B24) then shows how the prior probability $p(A|C)$ is changed to the posterior probability $p(A|BC)$ as a result of acquiring the new information B.

In this kind of application, $p(B|AC)$ is a "direct" or "sampling" probability, since we reason in the direction of the causal influence, from an assumed cause A to a presumed observable result B: and $p(A|BC)$ is an "inverse" probability, in which we reason from an observed result B to an assumed cause A. On comparing with (A7) we see that (B24) is a more general form of Laplace's rule, in which we need not have an exhaustive set of possible causes. Therefore, since (A7) is always called "Bayes' theorem," we may as well apply the same name to (B24).

At the risk--or rather the certainty--of belaboring it, we stress again that we are concerned here with inductive reasoning of any kind, not necessarily related to random experiments or any repetitive process. On the other hand, nothing prevents us from applying the theory to a repetitive situation (i.e., n tosses of a die); and propositions about frequencies g_i are then just as legitimate pieces of data or objects of inquiry as any other propositions. Various kinds of connection between probability and frequency then appear, as mathematical consequences of (A8), (A9). We have just seen one of them.

But now, could we have solved the Brandeis dice problem by applying Bayes' theorem instead of maximum entropy? If so, how do the results compare? Friedman and Shimony (1971)

(hereafter denoted FS) claimed to exhibit an inconsistency in
the Principle of Maximum Entropy (hereafter denoted PME) by
an argument which introduced a proposition d_ε, so ill-defined
that they tried to use it as (1) a constraint in PME, (2) a
conditioning statement in Bayes' theorem; and (3) an hypothesis
whose posterior probability is calculated. Therefore, let us
note the following.

If a statement d referring to a probability distribution in
space S is testable (for example, if it specifies a mean value
$<f>$ for some function $f(i)$ defined on S), then it can be used
as a constraint in PME; but it cannot be used as a conditioning
statement in Bayes' theorem because it is not a statement about
any event in S or any other space.

Conversely, a statement D about an event in the space S^n
(for example, an observed frequency) can be used as a conditioning statement in applying Bayes' theorem, whereupon it
yields a posterior distribution on S^n which may be contracted
to a marginal distribution on S; but D cannot be used as a
constraint in applying PME in space S, because it is not a
statement about any event in S, or about any probability distribution over S; i.e., it is not testable information in S.

At this point, informed students of statistical mechanics
will be astonished at the suggestion that there is any inconsistency between application of PME in space S and of Bayes'
theorem in S^n, since the former yields a canonical distribution,
while the latter is just the Darwin-Fowler method, originally
introduced as a rigorous way of justifying the canonical distribution! The mathematical fact shown by this well-known
calculation (Schrödinger, 1948) is that, whether we use
maximum entropy in space S with a constraint fixing an average
$<f>$ over a probability distribution, or apply Bayes' theorem
in S^n with a conditioning statement fixing a numerically equal
average $\bar{f}$ over sample values, we obtain for large n identical
distributions in the space S. The result generalizes at once
to the case of several simultaneous mean-value constraints.

This not only illustrates--contrary to the claims of FS--
the consistency of PME with the other principles of probability
theory, but it shows what a powerful tool PME is; i.e., how
much simpler and more convenient mathematically it is to use
PME in statistical calculations if the distribution on S is
what we are seeking. PME leads us directly to the same final
result, without any need to go into a higher space S^n and carry
out passage to the limit $n \to \infty$ by saddle-point integration.

Of course, it is as true in probability theory as in carpentry that introduction of more powerful tools brings with
it the obligation to exercise a higher level of understanding
and judgment in using them. If you give a carpenter a fancy

WHERE DO WE STAND ON MAXIMUM ENTROPY? 251

new power tool, he may use it to turn out more precise work in greater quantity; or he may just cut off his thumb with it. It depends on the carpenter.

The FS article led to considerably more discussion (see the references collected with the FS one) in which severed thumbs proliferated like hydras; but the level of confusion about the points already noted is such that it would be futile to attempt any analysis of the FS arguments.

FS suggest that a possible way of resolving all this is to deny that the probability of d_ε can be well-defined. Of course it cannot be; however, to understand the situation we need no "deep and systematic analysis of the concept of reasonable degree of belief." We need only raise our standards of exposition to the same level that is required in any other application of probability theory; i.e., we must define our propositions and sample spaces with enough precision to make a determinate mathematical problem.

There is a more serious difficulty in trying to reply to these criticisms. If FS dislike the maximum-entropy solution (B13) to this problem strongly enough to write three articles attacking it, then it would seem to follow that they prefer a different solution. But what different solution? One cannot form any clear idea of what is really troubling them, because in all these publications FS give no hint as to how, in their view, a more acceptable solution ought to differ from (B13).

The Rowlinson Criticism. In sharp contrast to the FS criticisms is that of J. S. Rowlinson (1970), who considers the same dice problem but does offer an alternative solution. For this reason, it is easy to give a precise quantitative reply to his criticism.

He starts with the all too familiar line: "Most scientists would say that the probability of an event is (or represents) the frequency with which it occurs in a given situation." Likewise, a critic of Columbus could have written (after he had returned from his first voyage): "Most geographers would say that the earth is flat."

Clarification of the centuries-old confusion about probability and frequency will not be achieved by taking votes; much less by quoting the philosophical writings of Leslie Ellis (1842). Rather, we must examine the mathematical facts concerning the rules of probability theory and the different sample spaces in which probabilities and frequencies are defined. We have seen, in the discussion following (B14) above, that anyone who glibly supposes that a probability in one space can be equated to a frequency in another, is assuming something which is not only not generally deducible from the principles of probability theory; it may stand in conflict with those principles.

There is no stranger experience than seeing printed criticisms which accuse one of saying the exact opposite of what he has said, explicitly and repeatedly. Thus my bewilderment at Rowlinson's statement that I reject "the methods used by Gibbs to establish the rules of statistical mechanics." I believe I can lay some claim to being the foremost current advocate and defender of Gibbs' methods! Anyone who takes the trouble to read Gibbs will see that, far from rejecting Gibbs' methods, I have adopted them enthusiastically and (thanks to the deeper understanding from Shannon) extended their range of application.

One of the major unsolved riddles of probability theory is: how to explain to another person exactly what is the problem being solved? It is well established that merely stating this in words does not suffice; repeatedly, starting with Laplace, writers have given the correct solution to a problem, only to have it attacked on the grounds that it is not the solution to some entirely different problem. This is at least the tenth time it has happened to me. As I tried to stress, the maximum-entropy solution (B13) describes the state of knowledge in which we are given the enumeration of the six possibilities, the mean value $<i> = 4.5$, and nothing else. But Rowlinson proceeds to introduce models with an urn containing seven white and three black balls (or a population of urns with varying contents) from which one makes various numbers of random draws with replacement. One expects that different problems will have different solutions.

In Rowlinson's Urn model, we perform Bernoulli trials five times, with constant probability of success $p = 0.7$. Then the numbers s of successes is in $0 \leq s \leq 5$, and the expected number is $<s> = 5 \times 0.7 = 3.5$. Setting $i \equiv s+1$, we have $1 \leq i \leq 6$, $<i> = 4.5$, the conditions stated in my dice problem. Thus he offers as a counter-proposal the binomial distribution

$$p'_i = \binom{5}{i-1} p^{i-1} (1-p)^{6-i}, \qquad 1 \leq i \leq 6 \ . \tag{B25}$$

These numbers are

$$\{p'_1 \ldots p'_6\} = \{0.00243, 0.02835, 0.1323, 0.3087, 0.36015, 0.16807\}. \tag{B26}$$

and they yield an entropy $S' = 1.413615$, 0.2 unit lower than that of (B13). This lower entropy indicates that the urn model puts further constraints on the solution beyond that used in (B13). We see that these consist in the extreme values ($i = 1, 6$) receiving less probability than before (only one of

WHERE DO WE STAND ON MAXIMUM ENTROPY? 253

$2^5 = 32$ possible outcomes can lead to $i = 1$, while ten of them yield $i = 3$, etc.).

Now if we knew that the experiment consisted of drawing five times from an urn with just the composition specified by Rowlinson, the result (B25) would indeed be the correct solution. But by what right does one assume this elaborate model structure when it is not given in the statement of the problem? One could, with equal right, assume any one of a hundred other specific models, leading to a hundred other counter-proposals. But it is just the point of the maximum-entropy principle that it achieves "objectivity" of our inferences, in the sense that we base our predictions only on the information that we do, in fact, have; and carefully avoid introducing any such gratuitous assumptions not warranted by our data. Any such assumption is far more likely to impose false constraints than to happen, by luck, onto an unknown correct one (which would be like guessing the combination to a safe).

At this point, Rowlinson says, "Those who favour the automatic use of the principle of maximum entropy would observe that the entropy of [our Eq. (B25)], 1.4136, is smaller than that of [B13], and so say that in proposing [B25] as a solution, 'information' has been assumed for which there is no justification." We do indeed say this, although Rowlinson simply rejects it out of hand without giving a reason. So to sustain our claim, let us calculate explicitly just how much Rowlinson's solution assumes without justification.

To clarify what is meant by "assuming information," suppose that an economist, Mr. A, is trying to forecast future price trends for some commodity. The condition of next week's market cannot be known with certainty, because it depends on intentions to buy or sell hidden in the minds of many different individuals. Evidently, a rational method of forecasting must somehow take account of all these unknown possibilities. Suppose that Mr. A's data are found to be equally compatible with 100 different possibilities. If he arbitrarily picked out 10 of these which happened to suit his fancy, and based his forecast only on them, ignoring the other 90, we should certainly consider that Mr. A was guilty of an egregious case of assuming information without justification. Our present problem is similar in concept, but quite different in numerical values.

We have stressed that, fundamentally, the maximum-entropy solution (B13) describes only a state of knowledge about a single trial, and is not an assertion about frequencies. But Rowlinson, as noted, also rejects this distinction and wants to judge the issue on the grounds of frequencies. Very well; let us now bring out the frequency connection that a maximum-entropy distribution does, after all, have (and which, incidentally, was pointed out in my Brandeis lectures, from which Rowlinson got this dice problem).

In N tosses, a set of observed frequencies $\{g_i\} = \{N_i/N\}$ (called g to avoid collision with previous notation) can be realized in

$$W = \frac{N!}{(Ng_1)! \ (Ng_2)! \ \ldots \ (Ng_6)!} \tag{B27}$$

different ways. As we noted from Boltzmann's work, Eq. (A14), the Stirling approximation to the factorials yields an asymptotic formula

$$\log W \sim NS \tag{B28}$$

where

$$S = - \sum_{i=1}^{6} g_i \log g_i \tag{B29}$$

is the entropy of the observed <u>frequency</u> distribution. Given two different sets of frequencies $\{g_i\}$ and $\{g_i'\}$, the ratio: (number of ways g_i can be realized)/(number of ways g_i' can be realized) is given by an asymptotic formula

$$\frac{W}{W'} \sim A \ \exp[N(S - S')] \left\{ 1 + \frac{B}{N} + O(N^{-2}) \right\} \tag{B30}$$

where

$$A \equiv \prod_i (g_i'/g_i)^{\frac{1}{2}} \tag{B31}$$

$$B \equiv \frac{1}{12} \sum_i \left(\frac{1}{g_i'} - \frac{1}{g_i} \right) \tag{B32}$$

are independent of N, and represent corrections from the higher terms in the Stirling approximation. We write them down only to allay any doubts about the accuracy of the numbers to follow. In all cases considered here it is easily seen that they have no effect on our conclusions, and only the exponential factor matters.

Rowlinson mentions an experiment involving 20,000 throws of a die, to which we shall return later; but in the present comparison this leads to numbers beyond human comprehension. To keep the results more modest, let us assume only $N = 1000$ throws. If we take $\{g_i\}$ as the maximum-entropy distribution (B13) and $\{g_i'\}$ as Rowlinson's solution (B26), we find $A = 0.159$, $B = 34$, $S - S' = 0.200$; and thus, with $N = 1000$,

$$\frac{W}{W'} = 1.19 \times 10^{86} \ . \qquad (B34)$$

Both distributions agree with the datum $<i> = 4.5$; but for every way in which Rowlinson's distribution can be realized, there are over 10^{86} ways in which the maximum entropy distribution can be realized (the age of the universe is less than 10^{18} seconds). It appears that information was indeed "assumed for which there is no justification."

This example should help to give us a proper respect for just what we are accomplishing when we maximize entropy. It shows the magnitude of the indiscretion we commit if we accept a distribution whose entropy is 0.2 unit less than the maximum value compatible with our data. In this example, to accept any distribution whose entropy is as much as 0.005 below the maximum value, would be to ignore over 99 percent of all possible ways in which the average $<i> = 4.5$ could be realized.

For reasons unexplained, Rowlinson seizes upon the particular value $p_1 = 0.05435$ from the maximum-entropy solution (B13), and asks: "But what basis is there for trusting in this last number?" but fails to ask the same question about his own very different result $p_1' = 0.00243$. Since it is so seldom that one is able to give a quantitative reply to a rhetorical question, we should not pass up this opportunity.

<u>Answer to the Rhetorical Question</u>. Let us, as before, count up the number of possibilities compatible with the given data. In the original problem we were to find $\{p_1 \ldots p_n\}$ so as to maximize $H = -\Sigma p_i \log p_i$ subject to the constraints $\Sigma p_i = 1$, $<i> = \Sigma i p_i$, a specified numerical value. If now we impose the additional constraint that p_1 is specified, we can define conditional probabilities

$$p_i' = \frac{p_i}{1 - p_1} \ , \qquad i = 2, 3, \ldots n \qquad (B35)$$

with entropy

$$H' = -\sum_{i=2}^{n} p_i' \log p_i' \ . \qquad (B36)$$

These quantities are related by Shannon's basic functional equation

$$H(p_1 \ldots p_n) = H(p_1, 1 - p_1) + (1 - p_1) H'(p_2' \ldots p_n') \qquad (B37)$$

and so, maximizing H with p_1 held fixed is equivalent to maximizing H'. We have the reduced maximum entropy problem:

maximize H' subject to

$$\sum_{i=2}^{n} p_i' = 1 \tag{B38}$$

$$<i>' = \sum_{i=2}^{n} i p_i' = 1 + \frac{<i> - 1}{1 - p_1} \quad . \tag{B39}$$

The solution proceeds as before, but now the maximum attainable entropy is a function $H_{max} = S(p_1, <i>)$ of the specified value of p_1, as well as $<i>$. The maximum of $S(p_1, 4.5)$ is of course the previous value (B14) of 1.61358, attained at the maximum-entropy value $p_1 = 0.05435$. Evaluating this also for Rowlinson's p_1', I find $S(p_1', 4.5) = 1.55716$, lower by 0.05642 units. By (B30) this means that, in 1000 tosses, for every way in which Rowlinson's value could be realized, regardless of all other frequencies except for the constraint $<i> = 4.5$, there are over 10^{24} ways in which the maximum-entropy frequency could be realized.

We may give a more detailed answer: expanding $S(p_1, 4.5)$ about its peak, we find that as we depart from 0.05435, the number of ways in which the frequency g_1 could be realized drops off like

$$\exp[-14,200(g_1 - 0.05435)^2] \tag{B40}$$

and so, for example, for 99% of all possible ways in which the average $<i> = 4.5$ can be realized, g_1 lies in the interval (0.05435 ± 0.0153).

This would seem to be an adequate answer to the question, "But what basis is there for trusting in this number?" I stress that the numerical results just given are theorems, involving only a straightforward counting of the possibilities allowed by the given data. Therefore they stand independently of anybody's personal opinions about either dice or probability theory.

However, it is necessary that we understand very clearly the meaning of these frequency connections. They concern only the number of possible ways in which certain frequencies $\{g_i\}$ could be realized, compatible with our constraints. They do not assert that the maximum-entropy frequencies will be observed in a real experiment; indeed, neither the Principle of Maximum Entropy nor any other principle of probability theory can predict with certainty what will happen in a real experiment. The correct statement is rather: the frequency distribution $\{g_i\}$ with maximum entropy calculated from certain constraints is overwhelmingly the most likely one to be observed in a real experiment, provided that the physical constraints operative in the experiment are the same as those assumed in the calculation.

In our mathematical formalism, a "constraint" is some piece of information that leads us to modify a probability distribution; in the case of a mean value constraint, by inserting an exponential factor $\exp[-\lambda f(x)]$ with an adjustable Lagrange multiplier λ. It is perhaps not yet clear just what we mean by "constraints" in a physical experiment. Of course, by these we do not mean the gross constraining linkages by levers, cables, and gears of a mechanics textbook, but something more subtle. In our applications, a "physical constraint" is any physical influence that exerts a systematic tendency--however slight--on the outcome of an experiment. We give some specific examples of physical constraints in die tossing below.

From the above numbers we can understand the success of the work of J. C. Keck and R. D. Levine reported here. I am sure that their results must seem like pure magic to those who have not understood the maximum-entropy formalism. To find a distribution of populations over 20 molecular energy levels might seem to require 19 independent pieces of data. But if one knows, from approximate rate coefficients or from past experience, which constraints exist (in practice, even if only the one or two most important ones are taken into account), one can make quite confident predictions of distributions over many levels simply by maximizing the entropy.

In fact, most frequency distributions produced in real experiments are maximum-entropy distributions, simply because these can be realized in so many more ways than can any other. As $N \to \infty$, the combinatorial factors become so sharply peaked at the maximum entropy point that to produce any appreciably different distribution would require very effective physical constraints. Any statistically significant departure from a maximum-entropy prediction then constitutes strong--and if it persists, conclusive--evidence of the existence of new constraints that were not taken into account in the calculation. Thus the maximum-entropy formalism has the further "magical" property that it provides the most efficient procedure by which, if unknown constraints exist, they can be discovered. But this is only an updated version of the process noted in Section A by which Laplace discovered new systematic effects.

It is, perhaps, sufficiently clear from this how much a Physical Chemist has to gain by understanding, rather than attacking, maximum entropy methods.

But we still have not dealt with the most fundamental misunderstandings in the Rowlinson article. He turns next to the shape of the maximum-entropy distribution (B13), with another

258   E. T. JAYNES

rhetorical question: "--- is there anything in the mechanics of throwing dice which suggests that if a die is not true the probabilities of scores 1,2,...6, should form the geometrical progression [our Eq. (B13)]?" He then cites some data of Wolf on 20,000 throws of a die which gave an average $\langle i \rangle = 3.5983$, plots the observed frequencies against the maximum-entropy distribution based on that constraint, and concludes that "departures from the random value of 1/6 bear no resemblance to those calculated from the rule of maximum entropy. What is clearly wrong with the indiscriminate use of this rule, and of the older rules from which it stems, is that they ignore the physics of the problem."

We have here a total, absolute misconception about every point I have been trying to explain above. If Wolf's data depart significantly from the maximum-entropy distribution based only on the constraint $\langle i \rangle = 3.5983$, then the proper conclusion is not that maximum entropy methods "ignore the physics" but rather that the maximum entropy method <u>brings out</u> the physics by showing us that another physical constraint exists beyond that used in the calculation. Unable to see the new physical information here revealed, he lashes out blindly against the principle that has revealed it.

Therefore, let us now give an analysis of Wolf's dice data showing just what things maximum entropy can give us here, if we only open our eyes to them.

<u>Wolf's Dice Data</u>. In the period roughly 1850-1890, the Zurich astronomer R. Wolf conducted and reported a mass of "random experiments." An account is given by Czuber (1908). Our present concern is with a particular die (identified as "Weiszer Würfel" in Czuber's two-way table, <u>loc</u>. <u>cit</u> p. 149) that was tossed 20,000 times and yielded the aforementioned mean value $\langle i \rangle = 3.5983$. We shall look at all details of the data presently, but first let us note a few elementary things about that "ignored" physics.

We all feel intuitively that a perfectly symmetrical die, fairly tossed, ought to show all faces equally often (but that statement is really circular, since there is no other way to define a "fair" method of tossing; so, suppose that by experimenting on a die known to be true, we have found such a fair method, and we continue to use it). The uniform frequency distribution $\{g_i = 1/6, 1 < i < 6\}$ then represents the nominal "unconstrained" situation of maximum possible entropy $S = \log 6$. Any imperfection in the die may then give rise to a "physical constraint" as we have defined that term. A little physical common sense can anticipate what these imperfections are likely to be.

WHERE DO WE STAND ON MAXIMUM ENTROPY? 259

The most obvious imperfection is that different faces have different numbers of spots. This affects the center of gravity, because the weight of ivory removed from a spot is obviously not (in any die I have seen) compensated by the paint then applied. Now the numbers of spots on opposite faces add up to seven. Thus the center of gravity is moved toward the "3" face, away from "4", by a small distance ε corresponding to the one spot discrepancy. The effect of this must be a slight frequency difference which is surely, for very small ε, proportional to ε:

$$g_4 - g_3 = \alpha\varepsilon \tag{B41}$$

where the coefficient α would be very difficult to calculate, but could be measured by experiments on dies with known ε. But the (2-5) face direction has a discrepancy of three spots, and (1-6) of five. Therefore, we anticipate the ratios:

$$(g_4 - g_3):(g_5 - g_2):(g_6 - g_1) = 1:3:5 \quad . \tag{B42}$$

But this says only that the spot frequencies vary linearly with i:

$$g_i = \frac{1}{6} + \alpha\varepsilon \, f_1(i) \quad , \qquad 1 \leq i \leq 6 \tag{B43}$$

where

$$f_1(i) \equiv (i - 3.5) \quad . \tag{B44}$$

The spot imperfections should then lead to a small linear skewing favoring the "6." This is the most obvious "physical constraint," and it changes the expected number of spots to

$$<i> = \sum i \, g_i = 3.5 + 17.5 \, \alpha\varepsilon \tag{B45}$$

or, to state it more suggestively, the function $f_1(i)$ acquires a non-zero expectation

$$<f_1> = 17.5 \, \alpha\varepsilon \quad . \tag{B46}$$

Now, what is the next most obvious imperfection to be expected? Evidently, it will involve departure from a perfect cube, the specific kind depending on the manufacturing methods; but let us consider only the highest quality die that a factory would be likely to make. If you were assigned the job of making a perfect cube of ivory, how would you do it with equipment likely to be available in a Physics Department shop or a small factory?

I think you would head for the milling machine, and mount your lump of ivory on a dividing head clamped to the work table with axis vertical. The first cut would be with an end mill, making the "top" face of the die. The construction of the machine guarantees that this will be accurately plane. Then you use side cutters to make the four side faces. For the finish cuts you will move the work table only in the direction of the cut, rotating the dividing head 90° from one face to the next. The accuracy of the equipment guarantees that you now have five of the faces of your cube, all very accurately plane and all angles accurately 90°, the top face accurately square.

But now the trouble begins; to make the final "bottom" face you have to remove the work from its mount, place it upside down on the table, and go over it with the end mill. Again, the construction of the machine guarantees that this final face will be accurately plane and parallel to the "top;" <u>but it will be practically impossible to adjust the work table height so accurately that the final dimension is exactly equal to the other two</u>. Of course, a skilled artisan with a great deal more time and equipment could do better; but this would run up the cost of manufacture for something that would never be detected in use. For factory production, there would be no motivation to do better than we have described.

Thus, the most likely geometrical imperfection in a high quality die is not lack of parallelism or of true 90° angles, but rather that one dimension will be slightly different from the other two.

Again, it is clear what kind of effect this will have on frequencies. Suppose the die comes out slightly "oblate," the (1-6) dimension being shorter than the (2-5) and (3-4) by some small amount δ. If the die were otherwise perfect, this would evidently increase the frequencies g_1, g_6 by some small amount $\beta\delta$, and decrease the other four to keep the sum equal to unity, where β is another coefficient hard to calculate but measurable. The result can be stated thus: the function

$$f_3(i) \equiv \begin{Bmatrix} +2, & i = 1,6 \\ -1, & i = 2,3,4,5 \end{Bmatrix} \tag{B47}$$

defined on the sample space, acquires a non-zero expectation

$$\langle f_3 \rangle = 6\beta\delta \tag{B48}$$

and the frequencies are

$$\begin{aligned} g_i &= \frac{1}{6} + \frac{1}{2} \beta\delta \, f_3(i) \\ &= \frac{1}{6}[1 + 3\beta\delta \, f_3(i)] \quad . \end{aligned} \tag{B49}$$

WHERE DO WE STAND ON MAXIMUM ENTROPY? 261

If now both imperfections are present, since the perturbations are so small we can in first approximation just superpose their effects:

$$g_i \simeq \frac{1}{6}[1 + 6\alpha\epsilon\ f_1(i)][1 + 3\beta\delta\ f_3(i)] \ . \tag{B50}$$

But this is hardly different from

$$g_i = \frac{1}{6} \exp[6\alpha\epsilon\ f_1(i) + 3\beta\delta\ f_3(i)] \tag{B51}$$

and so a few elementary physical common-sense arguments have led us to something which begins to look familiar.

If we had done maximum entropy using the constraints (B46), (B48), we would find a distribution proportional to $\exp[-\lambda_1 f_1(i) - \lambda_3 f_3(i)]$, so that (B51) is a maximum-entropy distribution based on those constraints. We see that the Lagrange multiplier by which any <u>information</u> constraint is coupled into our <u>probability</u> distribution, is just a measure of the strength of the <u>physical</u> constraint required to realize a numerically equal <u>frequency</u> distribution:

$$\lambda_1 = -6\alpha\epsilon \tag{B52}$$

$$\lambda_3 = -3\beta\delta \tag{B53}$$

and if our die has no other imperfections beyond the two noted, then it is overwhelmingly more likely to produce the distribution (B51) than any other.

If the observed frequencies show any statistically significant departure from (B51), then we have extracted from the data evidence of a third imperfection, which probably would have been totally invisible in the raw data; i.e., only when we have used the maximum entropy principle to "subtract off" the effect of the stronger influences, can we hope to detect a weaker one.

Our program for the maximum-entropy analysis of the die—or any other random experiment—is now defined except for the final step; how we decide whether a discrepancy is "statistically significant?"

The reader is cautioned that in all this discussion relating to Rowlinson we are being careless about distinctions between probability and frequency, because Rowlinson himself makes no distinction between them, and trying to correct this at every point quickly became tedious. The following analysis should be restated much more carefully to bring out the fact that it is only a very special case, although to the "frequentist" it appears to be the general case.

We have some "null hypothesis" H_0 about our die, that leads us to assign the probabilities $\{p_1 \ldots p_6\}$. We obtain data from N tosses, in which the observed frequencies are $\{g_i = N_i/N, 1 \le i \le 6\}$. If the numbers $\{g_1 \ldots g_6\}$ are sufficiently close to $\{p_1 \ldots p_6\}$ we shall say the fit is satisfactory; the null hypothesis is consistent with our data, and so there is no need, as far as this experiment indicates, to seek a better hypothesis. But how close is "close?" How do we measure the "distance" between the two distributions; and how large may that distance be before we begin to doubt the null hypothesis?

Early in this Century, Karl Pearson invented an intuitive, ad hoc procedure, called the Chi-squared test, to deal with this problem, which has been since widely adopted. Here we calculate the quantity

$$\chi^2 = N \sum_{i=1}^{6} \frac{(g_i - p_i)^2}{p_i} \tag{B54}$$

and if it is greater than a certain "critical value" given in Tables, we reject the null hypothesis. In the present case (six categories, five "degrees of freedom" after normalization), the critical value at the conventional 5% significance level is

$$\chi_c^2 = 11.07 \tag{B55}$$

which means that, if the null hypothesis is true there is only a 5% chance of seeing a value greater than χ_c^2. The critical value is independent of N, because for a frequentist who believes that p_i is an assertion of a limiting frequency in the sense of the de Moivre-Laplace limit theorem (A4), if H_0 is true, then the deviations should fall off as $|g_i - p_i| = O(N^{-\frac{1}{2}})$. A more careful approach shows that this holds only if our model is an exchangeable sequence with zero correlations; and even in this case the χ^2 criterion of "closeness" has no theoretical justification (i.e., no uniqueness property) in the basic principles of probability theory.

In fact, for the case of independent exchangeable trials, there is a criterion with a direct information-theory justification (Kullback, 1959) in the "minimum discrimination information statistic"

$$\psi \equiv N \sum_{i=1}^{6} g_i \log(g_i/p_i) \tag{B56}$$

and the numerical value of ψ, rather than χ^2, will lead us to inferences directly justifiable by Bayes' theorem. If the deviations $(g_i - p_i)$ are large, these criteria can be very different.

WHERE DO WE STAND ON MAXIMUM ENTROPY? 263

However, by a lucky mathematical accident, if the deviations are small (as we already know them to be for our dice problem) an expansion in powers of $(g_i - p_i)$ [in the logarithm, write $g/p = 1 + (g-p)/g + (g-p)^2/gp$] yields

$$\psi = \frac{1}{2}\chi^2 + O(N^{-\frac{1}{2}}) \tag{B57}$$

the neglected terms falling off as indicated, <u>provided</u> that $|g_i - p_i| = O(N^{-\frac{1}{2}})$. The result is that in our problem, from a pragmatic standpoint it doesn't matter whether we use χ^2 or ψ. So I shall apply the χ^2 test to Wolf's data, because it is so much more familiar to most people.

Wolf's empirical frequencies $\{g_i\}$ are given in the second column of Table 1. As a first orientation, let us test them against the null hypothesis $\{H_o: p_i = 1/6, 1 < i < 6\}$ of a uniform die. We find the result

$$\chi_o^2 = 271 \tag{B58}$$

over twenty times the critical value (B55). The hypothesis H_o is decisively rejected.

Next, let us follow Rowlinson by considering a new hypothesis H_1 which prescribes the maximum-entropy solution based on Wolf's average $<i> = 3.5983$, or,

$$<f_1(i)> = 0.0983 \quad . \tag{B59}$$

This will give us a distribution $p_i \sim \exp[-\lambda f_1(i)]$. From the partition function (B11) with this new datum we find $\lambda = 0.03373$ and the probabilities given in the third column of Table 1. The fourth column gives the differences $\Delta_i = g_i - p_i$, while in the fifth we list the partial contributions to Chi-squared:

$$c_i \equiv 20,000 \frac{(g_i - p_i)^2}{p_i}$$

which add up to the value

$$\chi_1^2 = 199.4 \quad . \tag{B60}$$

The fit is improved only slightly; and H_1 is also decisively rejected.

Table 1. One Constraint

i	g_i	p_i	Δ_i	c_i
1	0.16230	0.15294	+ 0.0094	11.46
2	0.17245	0.15818	+ 0.0143	25.75
3	0.14485	0.16361	- 0.0188	43.02
4	0.14205	0.16922	- 0.0272	87.25
5	0.18175	0.17502	+ 0.0067	5.18
6	0.19660	0.18103	+ 0.0156	26.78
				199.43

At this point, Rowlinson wants to reject not only H₁, but also the whole principle of maximum entropy. But now I stress still another time what the principle is really telling us: a statistically significant deviation is evidence of a new physical constraint; and the nature of the deviation gives us a clue as to what that constraint is. After subtracting off, by maximum entropy, the deviation attributable to the first constraint, the nature of the most important remaining one is revealed. Indeed, from a glance at the deviations $\Delta_i = g_i - p_i$ the answer leaps out at us; Wolf's die was slightly "prolate," the (3-4) dimension being greater than the (2-5) and (1-6) ones. So, instead of (B47), the new constraint is

$$f_2(i) \equiv \begin{cases} +1, & i = 1,2,5,6 \\ -2, & i = 3,4 \end{cases} \tag{B61}$$

and Wolf's data yield the result

$$\langle f_2 \rangle = 0.1393 . \tag{B62}$$

So now let us subtract off, by maximum entropy, the effect of both of these constraints; and thus discover whether Wolf's die had a third imperfection.

With the two constraints (B59), (B62) we have two Lagrange multipliers and a partition function

$$Z(\lambda_1,\lambda_2) = \sum_{i=1}^{6} \exp\left[-\lambda_1 f_1(i) - \lambda_2 f_2(i)\right]$$

$$= x^{-5/2} y(1+x)(1+x^4+x^2 y^{-3}) \tag{B63}$$

where $x \equiv \exp(-\lambda_1)$, $y \equiv \exp(-\lambda_2)$. The maximum-entropy probabilities are then

$$\{p_1 \ldots p_6\} = Z^{-1} x^{-5/2} y \{1, x, x^2 y^{-3}, x^3 y^{-3}, x^4, x^5\} \tag{B64}$$

Writing out the constraint equations (B4) and eliminating y from them, we find that x is determined by

$$(6F_1 - 4F_2 - 11)x^5 + (6F_1 - 4F_2 - 5)x^4 + (6F_1 + 4F_2 + 5)x$$
$$+ (6F_1 + 4F_2 + 11) = 0 \qquad (B65)$$

or, with Wolf's numerical values (B59), (B62),

$$5.4837x^5 + 2.4837x^4 - 3.0735x - 6.0735 = 0 \qquad (B66)$$

This has only one real root, at $x = 1.032233$, from which we have $\lambda_1 = -0.0317244$, $y = 1.074415$, $\lambda_2 = -0.0717764$. The new maximum-entropy probabilities are given in Table 2, which contains the same information as Table 1, but for the new hypothesis H_2.

Table 2. Two Constraints

i	g_i	p_i	Δ_i	c_i
1	0.16230	0.16433	− 0.0020	0.502
2	0.17245	0.16963	+ 0.0028	0.938
3	0.14485	0.14117	+ 0.0037	1.919
4	0.14205	0.14573	− 0.0037	1.859
5	0.18175	0.18656	− 0.0048	2.480
6	0.19660	0.19258	+ 0.0040	1.678
				9.375

We see that the second constraint has greatly improved the fit. Chi-squared has been reduced to

$$\chi_2^2 = 9.375 \qquad (B67)$$

This is less than the critical value 11.07, so there is now no statistically significant evidence for any further imperfections; i.e., if the given p_i were the "exact" values, it is reasonably likely that the distribution g_i would deviate from p_i by the observed amount, by chance alone. Or, to put it in a way perhaps more appropriate to this problem, if the die were tossed another 20,000 times, we would not expect the frequencies g_i to be repeated exactly; the new frequencies g_i', might reasonably be expected to deviate from the first set g_i by about as much as the distributions g_i, p_i differ.

That this is reasonable can be seen directly without calculating Chi-squared. For if the result i is obtained n_i times in N tosses, we might expect this to fluctuate in successive repetitions of the whole experiment by about $\pm\sqrt{n_i}$. Thus the

observed frequencies $g_i = n_i/N$ should fluctuate by about
$\Delta g_i = \pm\sqrt{n_i}/N$; for $g_i = 1/6$, $N = 20,000$, this gives $\Delta g_i \simeq 0.0029$.
But this is just of the order of the observed deviations Δ_i.
Therefore, it would be futile to search the $\{\Delta_i\}$ of Table 2
for a third imperfection. Not only does their distribution
fail to suggest any simple hypothesis; if the die were tossed
another 20,000 times, in all probability the new Δ_i' would be
entirely different. With our two-parameter hypothesis H_2 we
are down "in the noise" of random variations, and any further
systematic influences are too small to be seen unless we go up
to a million tosses, by which time the die will be changed anyway by wear.

A technical point might be raised by Statisticians: "You
have estimated two parameters λ_1, λ_2 from the data; therefore
you should use the test for three degrees of freedom rather
than five." This reduction is appropriate if the parameters
are chosen by the criterion of minimizing χ^2. That is, if we
choose them for the express purpose of making χ^2 small and
still fail to do so, it does not speak well for the hypothesis
and a penalty is in order. But our parameters were chosen by
a criterion that took no note of χ^2; and therefore the proper
question is only; "How well does the result fit the data?"
and not: "How did you find the parameters?" Had we chosen
our parameters to minimize χ^2, we would have found a still
lower value; but one that is not relevant to the point being
made here, which is the performance of the maximum entropy
criterion, as advocated long before this die problem was thought
of.

The maximum entropy method with two Lagrange multipliers
thus successfully determines a distribution with five independent quantities. The "ensemble" canonical with respect to the
constraints $f_1(i)$, $f_2(i)$ describing the two imperfections that
common sense leads us to expect in a die, agrees with Wolf's
data about as well as can be hoped for in a statistical problem.

It was stressed above that in this theory the connections
between probability and frequency are loosened and we noted,
in the discussion following (B40), that the connections remaining are now theorems rather than conjectures. As we now
see, they are not loosened enough to hamper us in dealing with
real random experiments. If we had been given only the two
constraints (B59), (B62) we could have reproduced, by maximum
entropy, all of Wolf's frequency data.

This is an interesting caricature of the results of Keck and
Levine, and shows again how much our critics would gain by
understanding, rather than attacking, this principle. Far from
"ignoring the physics," it leads us to concentrate our attention
on the part of the physics that is relevant. Success in using

WHERE DO WE STAND ON MAXIMUM ENTROPY? 267

it does not require that we take into account all dynamical details; it is enough if we can recognize, whether by commonsense analysis or by inspection of data, what are the systematic influences at work, that represent the "physical constraints?" If by any means we can recognize these, maximum entropy then takes over and supplies the rest of the solution, which does not depend on dynamical details but only on counting the possibilities.

In effect, then, by subtracting off the systematic effects we reduce the problem to Bernoulli's "equally possible" cases; the deviations Δ_i from the canonical distribution that remain in Table 2 are the same as the deviations from $p_i = 1/6$ that we would expect if the die had no imperfections.

Success of our predictions is not guaranteed in advance, as Rowlinson supposed it should be when he wanted to reject the entire principle at the stage of Table 1. But this supposition merely reflects his rejection, at the very outset, of the distinction between probability and frequency that I keep stressing. If one is not moved by theoretical arguments for that distinction, we now see a pragmatic reason for it. The probabilities p_i in Table 1 are an entirely correct description of our state of knowledge about a single toss, when we know about only the constraint $f_1(i)$. It is a theorem that they are also numerically equal to the frequencies which could happen in the greatest number of ways if no other physical constraint existed. But our probabilities will agree with measured frequencies only when we have recognized and put into our equations the constraints representing all the systematic influences at work in the real experiment.

This, I submit, is exactly as it should be in a statistical theory; at no point are we ever justified in claiming that our predictions must be right; only that, in order to make any better ones we should need more information than was given. It is when a theory purports to do more than this (by failing to recognize the distinction between probability and frequency) that it may be charged with promising us something for nothing.

Since the fit is now satisfactory, the above values of λ_1, λ_2 give us the numerical values of the systematic influences in Wolf's experiment: from (B52), (B53) we have

$$\alpha\epsilon = \frac{0.03172}{6} = 0.0053 , \qquad (B68)$$

$$\beta\delta = \frac{0.07178}{3} = 0.024 . \qquad (B69)$$

So, if today some enterprising person at Monte Carlo or Las Vegas will undertake to measure for us the coefficients α, β, then we can determine--100 years after the fact--just how far (in terms of its nominal dimensions) the center of gravity of Wolf's die was displaced (presumably by excavation of the spots), and how much longer it was in the (3-4) direction than in the (2-5) or (1-6). We can also certify that it had no other significant imperfections (at least, none that affected its frequencies). Note, however, that α, β are not, strictly speaking, physical constants only of the die; a little further common-sense reasoning makes it clear that they must depend also on how the die was tossed; for example, tossing it with a large angular momentum about a (3-4) axis will decrease the effect of the $f_1(i)$ constraint, while if it spins about the (1-6) axis the effect of $f_2(i)$ will be less; and with a (2-5) spin axis both constraints will be weakened.

Indeed, as soon as the die is unsymmetrical, all sorts of physical conditions that were irrelevant for a perfectly symmetrical one, become relevant. The frequencies will surely depend not only on its center of gravity but also on all the second moments of its mass distribution, the sharpness of its edges, the smoothness, elasticity, and coefficient of friction of the table, etc.

However, we conjecture that α, β depend very little on these factors within the small range of conditions usually employed (i.e., small angular momentum in tossing, etc.); and suspect that in that range the coefficient α is already well known to those who deal with loaded dice.

I really must thank Rowlinson for giving us (albeit inintentionally) such a magnificent test case by which the nature and power of the Principle of Maximum Entropy can be demonstrated, in a context entirely removed from the conceptual problems of quantum theory. And indeed, all the criticisms he made were richly deserved; for he was not, after all, criticizing the Principle of Maximum Entropy; only a gross misunderstanding of it. Rowlinson's criticisms were, however, taken up and extended by Lindhard (1974); in view of the long commentary above we may leave it as an exercise for the reader to deal with his arguments.

The Constraint Rule. There is a further point of logic about our use of maximum entropy that has troubled some who are able to see the distinction between probability and frequency. In imposing the mean-value constraint (B1) we are simply appropriating a sample average obtained from N measurements that yielded f_j on the j'th observation:

$$F = \bar{f} = \frac{1}{N} \sum_{j=1}^{N} f_j \tag{B70}$$

and equating it to a probability average

$$<f> = \sum_{i=1}^{n} p_i \, f(x_i) \quad . \tag{B71}$$

Is there not an element of arbitrariness about this? A cynic might say that after all these exhortations about the distinction between probability and frequency, we proceed to confuse them after all, by using the word "average" in two quite different senses.

Our rule can be justified in more than one way; in Section D below we argue in terms of what it means to say that certain information is "contained" in a probability distribution. Let us ask now whether the constraint rule (B1) is consistent with, or derivable from, the usual principles of Bayesian inference.

If we decide to use maximum entropy based on expectations of certain specified functions $\{f_1(x)...f_m(x)\}$, then we know in advance that our final distribution will have the mathematical form

$$p(x_i|H) = \frac{1}{Z(\lambda_1...\lambda_m)} \exp\left[-\lambda_1 \, f_1(x_i) \, ... \, - \lambda_m \, f_m(x_i)\right] \tag{B72}$$

and nothing prevents us from thinking of this as defining a class of sampling distributions parameterized by the Lagrange multipliers λ_k, the parameter space consisting of all values of $\{\lambda_1...\lambda_m\}$ which lead to normalizable distributions (B72). Choosing a specific distribution from this class is then equivalent to making an estimate of the parameters λ_k. But parameter estimation is a standard problem of statistical inference.

The class C of hypothesis being considered is thus specified; any particular choice of the $\{\lambda_1...\lambda_m\}$ may be regarded as defining a particular hypothesis H∈C. However, the class C does not determine any particular choice of the functions $\{f_1(x),...,f_m(x)\}$. For, if A is any nonsingular (m x m) matrix, we can carry out a linear transformation

$$\sum_{k=1}^{m} \lambda_k \, f_k(x) = \sum_{j=1}^{m} \lambda_j^* \, f_j^*(x) \tag{B73}$$

where

$$\lambda_j^* \equiv \sum_k \lambda_k \, A_{kj} \tag{B74a}$$

$$f_j^*(x) \equiv \sum_k (A^{-1})_{jk} \, f_k(x) \tag{B74b}$$

and the class of distributions (B72) can be written equally well as

$$P(x_i|H) = \frac{1}{Z(\lambda_1^* \ldots \lambda_m^*)} \exp\left\{-\lambda_1^* f_1^*(x_i) - \ldots - \lambda_m^* f_m^*(x_i)\right\}. \tag{B75}$$

As the $\{\lambda_1^* \ldots \lambda_m^*\}$ vary over their range, we generate exactly the same family of probability distributions as (B72). The class C is therefore characteristic, not of any particular choice of the $\{f_1(x) \ldots f_m(x)\}$, but of the <u>linear manifold</u> M(C) spanned by them.

If the $f_k(x)$ are linearly independent, the manifold M(C) has dimensionality m. Otherwise, M(C) is of some lower dimensionality m' < m; the set of functions $\{f_1(x) \ldots f_m(x)\}$ is then redundant, in the sense that at least one of them could be removed without changing the class C. While the presence of redundant functions $f_k(x)$ proves to be harmless in that it does not affect the actual results of entropy maximization (Jaynes, 1968), it is a nuisance for present purposes [Eq. (B81) below]. In the following we assume that any redundant functions have been removed, so that m' = m.

Suppose now that x_i is the result of some random experiment that has been repeated r times, and we have obtained the data

$$D \equiv \{x_1 \text{ true } r_1 \text{ times, } x_2 \text{ true } r_2 \text{ times, } \ldots, x_n \text{ true } r_n \text{ times}\}. \tag{B76}$$

Of course, $\Sigma r_i = r$. Out of all hypotheses H∈C, which is most strongly supported by the data D according to the Bayesian, or likelihood, criterion? To answer this, choose any particular hypothesis $H_0 \equiv \{\lambda_1^{(0)} \ldots \lambda_m^{(0)}\}$ as the "null hypothesis" and test it against any other hypothesis $H \equiv \{\lambda_1 \ldots \lambda_m\}$ in C by Bayes' theorem. The log-likelihood ratio in favor of H over H_0 is

$$L \equiv \log \frac{P(D|H)}{P(D|H_0)} = \sum_{i=1}^{n} r_i \log\left[p_i/p_i^{(0)}\right]$$

$$= r\left[\log(Z_0/Z) + \sum_{k=1}^{m} \left(\lambda_k^{(0)} - \lambda_k\right)\bar{f}_k\right] \tag{B77}$$

where

$$\bar{f}_k \equiv \frac{1}{r} \sum_{i=1}^{n} r_i f_k(x_i) \tag{B78}$$

is the <u>measured</u> average of $f_k(x)$, as found in the experiment. Out of <u>all</u> hypotheses in class C the one most strongly supported by the data D is the one for which the first variation vanishes:

WHERE DO WE STAND ON MAXIMUM ENTROPY? 271

$$\delta L = -r \sum_{k=1}^{m} \left[\frac{\partial}{\partial \lambda_k} \log Z + \bar{f}_k\right]\delta\lambda_k = 0 \quad . \tag{B79}$$

But from (B4), this yields just our constraint rule (B1):

$$\{<f_k> = \bar{f}_k \quad , \quad 1 \leq k \leq m\} \quad . \tag{B80}$$

To show that this yields a true maximum, form the second variation and note that the covariance matrix

$$\frac{\partial^2 \log Z}{\partial \lambda_j \partial \lambda_k} = <f_j f_k> - <f_j><f_k> \tag{B81}$$

is positive definite almost everywhere on the parameter space if the $f_k(x)$ are linearly independent.

Evidently, this result is invariant under the aforementioned linear transformations (B74); i.e., we shall be led to the same final distribution satisfying (B80) however the $f_k(x)$ are defined. Therefore, we can state our conclusion as follows:

$$\left\{\begin{array}{l}\text{Out of all hypotheses in class C, the data D support}\\ \text{most strongly that one for which the expectation}\\ <f(x)> \text{ is equal to the measured average } \bar{f}(x) \text{ for every}\\ \text{function } f(x) \text{ in the linear manifold M(C)}.\end{array}\right\} \tag{B82}$$

This appears to the writer as a rather complete answer to some objections that have been raised to the constraint rule. We are not, after all, confusing two averages; it is a derivable consequence of probability theory that we should set them equal. Maximizing the entropy subject to the constraints (B80), is equivalent to (i.e., it leads to the same result as) maximizing the likelihood over the manifold of sampling distributions picked out by maximum entropy.

Forney's Question. An interesting question related to this was put to me by G. David Forney in 1963. The procedure (B1) uses only the numerical value of F, and it seems to make no difference whether this was a measured average over 20 observations, or 20,000. Yet there is surely a difference in our state of knowledge--our degree of confidence in the accuracy of F-- that depends on N. The maximum-entropy method seems to ignore this. Shouldn't our final distribution depend on N as well as F?

It is better to answer a question 15 years late than not at all. We can do this on both the philosophical and the technical level. Philosophically, we are back to the question: "What is the specific problem being solved?"

In the problem I am considering F is simply a number given to us in the statement of the problem. Within the context of that problem, F is exact by definition and it makes no difference how it was obtained. It might, for example, be only the guess of an idiot, and not obtained from any measurement at all. Nevertheless, that is the number given to us, and our job is not to question it, but to do the best we can with it.

This may seem like an inflexible, cavalier attitude; I am convinced that nothing short of it can ever remove the ambiguity of "What is the problem?" that has plagued probability theory for two centuries.

Just as Rowlinson was impelled to invent an Urn Model that was not specified in the statement of the problem, you and I might, in some cases, feel the urge to put more structure into this problem than I have used. Indeed, we demand the right to do this. But then, let us recognize that we are considering a different problem than pure "classical" maximum entropy; and it becomes a technical question, not a philosophical one, whether with some new model structure we shall get different results. Clearly, the answer must be sometimes yes, sometimes no, depending on the specific model structure assumed. But it turns out that the answer is "no" far more often than one might have expected.

Perhaps the first thought that comes to one's mind is that any uncertainty as to the value of F ought to be allowed for by averaging the maximum-entropy distribution $p_i(F)$ over the possible values of F. But the maximum-entropy distribution is, by construction, already as "uncertain" as it can get for the stated mean value. Any averaging can only result in a distribution with still higher entropy, which will therefore necessarily violate the mean value number given to us. This hardly seems to take us in the direction wanted; i.e., we are already up against the wall from having maximized the entropy in the first place.

But such averaging was only an ad hoc suggestion; and in fact the Principle of Maximum Entropy already provides the proper means by which any testable information can be built into our probability assignments. If we wish only to incorporate information about the accuracy with which f is known, no new model structure is needed; the way to do this is to impose another constraint. In addition to $<f>$ we may specify $<f^2>$; or indeed, any number of moments $<f^n>$ or more general functions $<h(f)>$. Each such constraint will be accompanied by its Lagrange multiplier λ, and the general maximum-entropy formalism already allows for this.

Of course, whenever information of this kind is available it should in principle be taken into account in this way. I would

"hold it to be self-evident" that for any problem of inference,
the ideal toward which we should aim is that all the relevant
information we have ought to be incorporated explicitly into
our equations; while at the same time, "objectivity" requires
that we carefully avoid assuming any information that we do not
possess. The Principle of Maximum Entropy, like Ockham, tells
us to refrain from inventing Urn Models when we have no Urn.

But in practice, some kinds of information prove to be far
more relevant than others, and this extra information about
the accuracy of F usually affects our actual conclusions so
little that it is hardly worth the effort. This is particularly true in statistical mechanics, due to the enormously
high dimensionality of the phase space. Here the effect of
specifying any reasonable accuracy in F is usually completely
negligible. However, there are occasional exceptions; and
whenever this extra information does make an appreciable difference it would, of course, be wrong to ignore it.

C. Speculations for the Future

The field of statistical Inference--in or out of Physics--is so
wide that there is no hope of guessing every area in which new
advances might be made. But we can indicate a few areas where
progress may be predicted rather safely because it is already
underway, with useful results being found at a rate proportional
to the amount of effort invested.

Current progress is taking place at several different levels:
  I Application of existing techniques to existing problems
  II Extension of present theory to new problems.
  III More powerful mathematical methods.
  IV Further development of the basic theory of inference.

However, I shall concentrate on I and IV, because II is so
enveloped in fog that nothing can be seen clearly, and III seems
to be rather stagnant except for development of new specialized
computer techniques, which I am not competent even to describe,
much less predict.

There are important current areas that seem rather desperately
in need of the same kind of house-cleaning that statistical
mechanics has received. What they all have in common is:
(a) long-standing problems, still unsolved after decades of
mathematical efforts, (b) domination by a mental outlook that
leads one to concentrate all attention on the analogs of ergodic
theory. That is, in the belief that a probability is not respectable unless it is also a frequency, one attempts a direct
calculation of frequencies, or tries to guess the right "statistical assumption" about frequencies, even though the available information does not consist of frequencies, but consists
rather of partial knowledge of certain "macroscopic" parameters

$\{\alpha_i\}$; and the predictions desired are not frequencies, but estimates of certain other parameters $\{\theta_i\}$. It is not yet realized that, by looking at the problems this way one is not making efficient use of probability theory; by restricting its meaning one is denying himself nearly all its real power.

The real problem is not to determine frequencies, but to describe one's <u>state of knowledge</u> by a probability distribution. If one does this correctly, he will find that whatever frequency connections are relevant will appear automatically, not as "statistical assumptions" but as mathematical consequences of probability theory.

Examples are the theory of hydrodynamic turbulence, optical coherence, quantum field theory, and surprisingly, communication theory which after thirty years has hardly progressed beyond the stage of theorems which presuppose all the ten-gram frequencies known in advance.

In early 1978 I attended a Seminar talk by one of the current experts on turbulence theory. He noted that the basic theory is in a quandary because "Nobody knows what statistical assumptions to make." Yet the objectives of turbulence theory are such things as: given the density, compressibility, and viscosity of a fluid, predict the conditions for onset of turbulence, the pressure difference required to maintain turbulent flow, the rate of heat transfer in a turbulent fluid, the distortion and scattering of sound waves in a turbulent medium, the forces exerted on a body in the fluid, etc. Even if one's objective were only to predict some <u>frequencies</u> g_i related to turbulence, statements about the best estimate of g_i and the reliability of that estimate, can only be derived from probabilities that are not themselves frequencies.

We indicated a little of this above [Equations (B15)-(B23)]; now let us see in a more realistic case why the <u>frequencies</u> with which various things happen in a time-dependent process are not the same as their <u>probabilities</u>; but that, nevertheless, there are always definite connections between probability and frequency, derivable as consequences of probability theory.

<u>Fluctuations</u>. Consider some physical quantity $f(t)$. What follows will generalize at once to field quantities $f(x,t)$; but to make the present point it is sufficient to consider only time variations. Therefore, we may think of $f(t)$ as the net force exerted on an area A by some pressure $P(x,t)$:

$$f(t) = \int_A P(x,t) \, dA \qquad (C1)$$

or the net force in the x-direction exerted by an electric field on a charge distributed with density $\rho(x)$: $f(t) = \int E_x(x,t)\rho(x)d^3x$

WHERE DO WE STAND ON MAXIMUM ENTROPY? 275

or as the total magnetic flux passing through an area A, or the number of molecules in an observed volume V; or the difference in magnetic and electrostatic energy stored in V:

$$f(t) = \frac{1}{8\pi} \int_V [H^2(x,t) - E^2(x,t)] d^3x \qquad (C2)$$

and so on! For any such physical meaning, the following considerations will apply.

Given any probability distribution (which we henceforth call, for brevity an ensemble) for $f(t)$, the best prediction of $f(t)$ that we can make from it—"best" in the sense of minimizing the expected square of the error—is the ensemble average

$$<f(t)> = <f> \qquad (C3)$$

which is independent of t if it is an equilibrium ensemble, as we henceforth assume. But this may or may not be a <u>reliable</u> prediction of $f(t)$ at any particular time. The mean square expected deviation from the prediction (C3) is the variance

$$[\Delta f(t)]^2 = <f^2> - <f>^2 \qquad (C4)$$

again independent of t by our assumption. Only if $|\Delta f/<f>| \ll 1$ is the ensemble making a sharp prediction of the measurable value of f.

Basically, the quantity Δf just defined represents only the <u>uncertainty of the prediction</u>; i.e., the degree of ignorance about f expressed by the ensemble. Yet Δf is held, almost universally in the literature of fluctuation theory, to represent also the <u>measurable</u> RMS fluctuations in f. Clearly, this is an additional assumption, which might or might not be true; for, obviously, the mere fact that I know f only to ±1% accuracy, is not enough to make it fluctuate by ±1%! Therefore, we note there is logically no room for any postulate that Δf is the measurable RMS fluctuation; whether this is or is not true is mathematically determined by the probability distribution. To understand this we need a more careful analysis of the relation between $<f>$, Δf, and experimentally measurable quantities.

More generally, we can consider a large class of functionals of $f(t)$ in some time interval $(0 < t < T)$; for example,

$$K[f(t)] \equiv T^{-n} \int_0^T dt_1 \ldots \int_0^T dt_n \, G[f(t_1) \ldots f(t_n)] \qquad (C5)$$

with $G(f_1 \ldots f_n)$ a real function. For any such functional, the ensemble will determine some probability distribution $P(K)dK$, and the best prediction we can make by the mean-square-error criterion is its expectation $<K>$. What is the necessary and sufficient condition that, as $T \to \infty$, the ensemble predicts a sharp value for K? It is, as always, that

$$(\Delta K)^2 \equiv \langle K^2 \rangle - \langle K \rangle^2 \to 0 \quad . \tag{C6}$$

For any such functional, this condition may be written out explicitly; let us give two examples that will surprise some readers.

One of the sources of confusion in this field is that the word "average" is used in several different senses. We try to avoid this by using different notations for different kinds of average. For the single system that exists in the laboratory, the observable average is not the ensemble average $\langle f \rangle$, but a time average, which we denote by a bar (reserving the angular brackets to mean only ensemble averages):

$$\bar{f} \equiv \frac{1}{T} \int_0^T f(t) dt \tag{C7}$$

which corresponds to (C5) with $G \equiv f(t_1)$. The averaging time T is left arbitrary for the time being because the results (C8), (C11), (C18), to be derived next, being exact for any T, then provide a great deal of insight that would be lost if we pass to the limit too soon.

In the state of knowledge represented by the ensemble, the best prediction of $\bar{f}$ by the mean square error criterion, is

$$\langle \bar{f} \rangle = \left\langle \frac{1}{T} \int_0^T f(t) dt \right\rangle = \frac{1}{T} \int_0^T \langle f \rangle dt$$

or, for an equilibrium ensemble,

$$\langle \bar{f} \rangle = \langle f \rangle \quad , \tag{C8}$$

an example of a very general rule of probability theory; an ensemble average $\langle f \rangle$ is not the same as a measured value $f(t)$ or a measured average $\bar{f}$; but it is equal to the <u>expectations</u> of both of those quantities.

But (C8), like (C3), tells us nothing about whether the prediction is a reliable one; to answer this we must again consider the variance

$$(\Delta \bar{f})^2 \equiv \langle (\bar{f} - \langle \bar{f} \rangle)^2 \rangle$$
$$= \frac{1}{T^2} \int_0^T dt_1 \int_0^T dt_2 [\langle f(t_1) f(t_2) \rangle - \langle f(t_1) \rangle \langle f(t_2) \rangle] \quad . \tag{C9}$$

WHERE DO WE STAND ON MAXIMUM ENTROPY? 277

Only if $|\overline{\Delta f}/\langle \overline{f}\rangle| \ll 1$ is the ensemble making a sharp prediction of the measured average $\overline{f}$. Now, however, the time averaging can help us; for $\overline{\Delta f}$ may become very small compared to Δf, if we average over a long enough time.

Now in an equilibrium ensemble the integrand of (C9) is a function of $(t_2 - t_1)$ only, and defines the <u>covariance function</u>

$$\phi(\tau) \equiv \langle f(t)f(t+\tau)\rangle - \langle f(t)\rangle\langle f(t+\tau)\rangle$$

$$= \langle f(0)f(\tau)\rangle - \langle f\rangle^2 \qquad (C10)$$

from which (C9) reduces to a single integral:

$$(\overline{\Delta f})^2 = \frac{2}{T^2} \int_0^T (T-\tau)\phi(\tau)d\tau . \qquad (C11)$$

A sufficient (stronger than necessary) condition for $\overline{\Delta f}$ to tend to zero is that the integrals

$$\int_0^\infty \phi(\tau)\tau d\tau \quad , \quad \int_0^\infty \phi(\tau)d\tau \qquad (C12)$$

converge; and then the characteristic correlation time

$$\tau_c \equiv \left[\int_0^\infty \phi(\tau)d\tau\right]^{-1}\left[\int_0^\infty \tau\,\phi(\tau)d\tau\right] \qquad (C13)$$

is finite, and we have asymptotically,

$$(\overline{\Delta f})^2 \sim \frac{2}{T}\int_0^\infty \phi(\tau)d\tau . \qquad (C14)$$

$\overline{\Delta f}$ then tends to zero like $1/\sqrt{T}$, and the situation is very much as if successive samples of the function over non-overlapping intervals of length τ_c were independent. However, the slightest positive correlation, if it persists indefinitely, will prevent any sharp prediction of $\overline{f}$. For, if $\phi(\tau) \to \phi(\infty) > 0$, then from (C11) we have

$$(\overline{\Delta f})^2 \to \phi(\infty) \qquad (C15)$$

and the ensemble can never make a sharp prediction of the measured average; i.e., any postulate that the ensemble average equals the time average, violates the mathematical rules of probability theory. These results correspond to (B23).

Now everything we have said about measurable values of f can be repeated <u>mutatis mutandis</u> for the measurable fluctuations $\delta f(t)$; we need only take a step up the hierarchy of successively higher order correlations. For, over the observation time T, the measured mean-square fluctuation in f(t)--i.e., deviation from the measured mean--is

$$(\delta f)^2 \equiv \frac{1}{T} \int_0^T [f(t) - \bar{f}]^2 \, dt \tag{C16}$$

$$= \overline{f^2} - \bar{f}^2 \tag{C17}$$

which corresponds to the choice $G = f^2(t_1) - f(t_1)f(t_2)$ in (C5). The "best" prediction we can make of this from the ensemble, is its expectation, which reduces to

$$\langle(\delta f)^2\rangle = (\Delta f)^2 + (\Delta \bar{f})^2 \tag{C18}$$

as a short calculation using (C4), (C11) will verify. This is in itself a very interesting (and I am sure to many surprising) result. The predicted <u>measurable</u> fluctuation δf is not the same as the <u>ensemble</u> fluctuation Δf unless the ensemble is such, and the averaging time so long, that $\Delta \bar{f}$ is negligible compared to Δf.

But (C18) tells us nothing about whether the prediction $\langle(\delta f)^2\rangle$ is a reliable one; to answer this we must, once more, examine the variance

$$V = \langle(\delta f)^4\rangle - \langle(\delta f)^2\rangle^2 \,. \tag{C19}$$

Unless (C19) is small compared to the square of (C18), the ensemble is not making any definite prediction of $(\delta f)^2$. After some computation we find that (C19) can be written in the form

$$V = \frac{1}{T^4} \int_0^T dt_1 \int_0^T dt_2 \int_0^T dt_3 \int_0^T dt_4 \, \psi(t_1,t_2,t_3,t_4) \tag{C20}$$

where ψ is a four-point correlation function:

$$\psi(t_1,t_2,t_3,t_4) = \langle f(t_1)f(t_2)f(t_3)f(t_4)\rangle - 2\langle f(t_1)f(t_2)f^2(t_3)\rangle$$
$$+ \langle f^2(t_1)f^2(t_2)\rangle - [(\Delta f)^2 + (\Delta \bar{f})^2]^2 \tag{C21}$$

which we have written in reduced form, taking advantage of the symmetry of the domain of integration in (C20).

As we see, the person who supposes that the RMS fluctuation Δf in the ensemble is also the experimentally measurable RMS fluctuation δf, is inadvertently supposing some rather nontrivial mathematical properties of that ensemble, which would seem to require some nontrivial justification! Yet to the best of my knowledge, no existing treatment of fluctuation theory even recognizes the distinction between δf and Δf.

In almost all discussions of random functions in the existing literature concerned with physical applications, it is taken for granted that (C6) holds for all functionals. One can hardly avoid this if one postulates, with Rowlinson, that "the probability of an event is the frequency with which it occurs in a given situation." But if it requires the computation (C20) to justify this for the mean-square fluctuation, what would it take to justify it in general? That is just the "ergodic" problem for this model.

Future progress in a number of areas will, I think, require that the relation between ensembles and physical systems be more carefully defined. The issue is not merely one of "philosphy of interpretation" that practical people may ignore; for not only the quantitative details, but even the qualitative kinds of physical predictions that a theory can make, depend on how these conceptual problems are resolved. For example, as was pointed out in my 1962 Brandeis Lectures, [loc. cit. Eqs. (83)-(93)], one cannot even state, in terms of the underlying ensemble, the criterion for a phase transition, or distinguish between laminar and turbulent flow, until the meaning of that ensemble is recognized.

A striking example of the need for clarifications in fluctuation theory is provided by quantum electrodynamics. Here one may calculate the expectation of an electric field at a point: $<E(x,t)> = 0$, but the expectation of its square diverges: $<E^2(x,t)> = \infty$. Thus $\Delta E = \infty$; in present quantum theory one interprets this as indicating that empty space is filled with "vacuum fluctuations," yielding an infinite "zero-point" energy density. But when we see the distinction between ΔE and δE, a different interpretation suggests itself. If $\Delta E = \infty$ that does not have to mean that any _physical_ quantity is infinite; it means only that the present theory is totally unable to predict the field at a point, i.e., the only thing which is infinite is the uncertainty of the prediction.

It had been thought for 30 years that these vacuum fluctuations had to be real, because they were the physical cause of the Lamb shift; however it has been shown (Jaynes, 1978) that a classical calculation leads to just the same formula for this frequency shift without invoking any field fluctuations. Therefore, it appears that a reinterpretation of the "fluctuation laws" of quantum theory along these lines might clear up at least some of the paradoxes of present quantum theory.

The situation just noted is only one of a wide class of connections that might be called "generalized fluctuation-dissipation theorems," or "fluctuation-response theorems." These include all of the Kubo-type theorems relating transport coefficients to various "thermal fluctuations." I believe that relations of this type will become more general and more useful with a better understanding of fluctuation theory.

Biology. Perhaps the largest and most obvious beckoning new field for application of statistical thermodynamics is biology. At present, we do not have the input information needed for a useful theory, we do not know what simplifying assumptions are appropriate; and indeed we do not know what questions to ask. Nevertheless, molecular biology has advanced to the point where some preliminary useful results do not seem any further beyond us now than the achievement of an integrated circuit computer chip did thirty years ago.

In the case of the simplest organism for which a great deal of biochemical information exists, the bacterium E. coli, Watson (1965) estimated that "one-fifth to one-third of the chemical reactions in E. coli are known," and noted that additions to the list were coming at such a rate that by perhaps 1985 it might be possible to describe "essentially all the metabolic reactions involved in the life of an E. coli cell."

As a pure speculation, then, let us try to anticipate a problem that might just possibly be amenable to the biochemical knowledge and computer technology of the year 2000: Given the structure and chemical composition of E. coli, predict its experimentally reproducible properties, i.e., the range of environmental conditions (temperature, pH, concentrations of food and other chemicals) under which a cell can stay alive; the rate of growth as a function of these factors. Given a specific mutation (change in the DNA code), predict whether it can survive and what the reproducible properties of the new form will be.

Such a program would be a useful first step. It seems, in my very cloudy crystal ball, that (1) its realization might be a matter of decades rather than centuries, (2) success in one instance would bring about a rapid increase in our ability to deal with more complicated problems, because it would reveal what simplifying assumptions are permissible.

At present one could think of several thousand factors that might, as far as we know, be highly relevant for these predictions. If a single cell contains 20,000 ribosomes where protein synthesis is taking place, are they performing 20,000 different functions, each one essential to the life of the cell? This just seems unlikely. I would conjecture that of all the complicated detail that can be seen in a cell, the overwhelmingly greatest part is--like every detail of the hair on our heads, or our fingerprints--accidental to the history of that particular individual; and not at all essential for its biological function.

WHERE DO WE STAND ON MAXIMUM ENTROPY? 281

The problem seems terribly complicated at present, because in all this detail we do not know what is relevant, what is irrelevant. But success in one instance would show us how to judge this. It might turn out that prediction of biological activity requires information about only a dozen separate factors, instead of a million. If so, then one would have both the courage and the insight needed to attack more complicated problems.

This has been stated so as to bring out the close analogy with what has happened in the theory of irreversible processes. In the early 1950's the development of a general formalism for irreversible processes appeared to be a hopelessly complicated program, not to be thought of in the next thousand years, if ever. Thus, van Hove (1956) stated: "... in view of the unlimited diversity of possible nonequilibrium situations, the existence of such a set of equations seems rather doubtful." Yet, as noted in Section A above, the principle which has solved this problem already existed, unrecognized, at that time. And today it seems that our major problem is not the complications of detail, but the conceptual difficulty in understanding how such a complicated problem could have such a (formally) simple solution. The answer is that, while full dynamical information is extremely complicated, the relevant information is not.

Perhaps there is a general principle--which we are conceptually unprepared to recognize today because it is too simple-- that would tell us which features of an organism are its essential, relevant biological features; and which are not.

Of course, applications of statistical mechanics to biology may be imagined at many different levels, so widely separated that they have nothing to do with each other. Thus, while I have been speculating about complexities within a single cell, the contribution of E. H. Kerner to this Symposium goes after the opposite extreme, interaction of many organisms. At that level the relevant information is now so much simpler and more easily obtained that many interesting results are already available.

Basic Statistical Theory. From the standpoint of statistical theory in general, the principle of maximum entropy is only one detail, which arose in connection with the problem of generalizing Laplace's statistical practice from (A6), and we have examined it above only in the finite discrete case. As $n \to \infty$ a new feature is that for some kinds of testable information there is no upper bound to the entropy. For mean-value constraints, the partition function may diverge for all real λ, or the constraint equations (B4) may not have a solution. In this way, the theory signals back to us that we have not put enough information into the problem to determine any definite inferences. In the finite case, the mere enumeration of the possibilities $\{i = 1, 2, \ldots n\}$ specifies enough to ensure that a solution exists. If $n \to \infty$, we have specified far less in the enumeration, and it is hardly surprising that this must be compensated by specifying more information in our constraints.

Rowlinson quotes Leslie Ellis (1842) to the effect that "Mere ignorance is no ground for any inference whatever. Ex nihilo nihil." I am bewildered as to how Rowlinson can construe this as an argument against maximum entropy, since as we see the maximum entropy principle immediately tells us the same thing. Indeed, it is the principle of maximum entropy--and not Leslie Ellis--that tells us precisely how much information must be specified before we have a normalizable distribution so that rational inferences are possible. Once this is recognized, I believe that the case $n \to \infty$ presents no difficulties of mathematics or of principle.

It is very different when we generalize to continuous distributions. We noted that Boltzmann was obliged to divide his phase space into discrete cells in order to get the entropy expression (A14) from the combinatorial factor (A11). Likewise, Shannons uniqueness proof establishing $-\Sigma p_i \log p_i$ as a consistent information measure, goes through only for a discrete distribution. We therefore approach the continuous case as the limit of a discrete one. This leads (Jaynes, 1963b, 1968) to the continuous entropy expression

$$S = - \int p(x) \log \frac{p(x)}{m(x)} dx \qquad (C22)$$

where the "measure function" $m(x)$ is proportional to the limiting density of discrete points (all this theory is readily restated in the notation of measure theory and Stieltjes integrals; but we have never yet found a problem that needs it). So, it is the entropy relative to some "measure" $m(x)$ that is to be maximized. Under a change of variables, the functions $p(x)$, $m(x)$ transform in the same way, so that the entropy so defined is invariant; and in consequence it turns out that the Lagrange multipliers and all our conclusions from entropy maximization are independent of our choice of variables. The maximum-entropy probability density for prescribed averages

$$\int f_k(x) p(x) dx = F_k \quad , \quad 1 \leq k \leq m \qquad (C23)$$

is

$$p(x) = \frac{m(x)}{Z(\lambda_1\ldots\lambda_m)} \exp\left[-\sum_k \lambda_k f_k(x)\right] \qquad (C24)$$

with the partition function

$$Z(\lambda_1\ldots\lambda_m) \equiv \int dx\, m(x)\, \exp\left[-\sum_k \lambda_k f_k(x)\right]. \qquad (C25)$$

An interesting fact, which may have some deep significance not yet seen, is that the class of maximum-entropy functions (C24) is, by the Pitman-Koopman theorem, identical with the class of functions admitting sufficient statistics; that is, if as in (B72)-(B81) we think of (C24) as defining a class of sampling distributions from which, given data $D \equiv$ "N measurements of x yielded the results $\{x_1\ldots x_N\}$," we are to estimate the $\{\lambda_1\ldots\lambda_m\}$ by applying Bayes' theorem, we find that the posterior distribution of the λ's depends on the data only through the observed averages:

$$\bar{f}_k \equiv \frac{1}{N}\sum_{r=1}^N f_k(x_r) \qquad (C26)$$

all other aspects of the data being irrelevant. This seems to strengthen the point of view noted before in (B72)-(B81). For many more details, see Huzurbazar (1976).

But now let us return to our usual viewpoint, that (C24) is not a sampling distribution but a prior distribution from which we are to make inferences about x, which incorporate any testable prior information. If the space S_x in which the continuous variable x is defined, is not the result of any obvious limiting process, there seems to be an ambiguity; for what now determines m(x)?

This problem was discussed in some detail before (Jaynes, 1968). If there are no constraints, maximization of (C22) leads to $p(x) = Am(x)$ where A is a normalization constant; thus m(x) has the intuitive meaning that it is the distribution representing "complete ignorance" and we are back, essentially, to Bernoulli's problem from where it all started. In the continuous case, then, before we can even apply maximum entropy we must deal with the problem of complete ignorance.

Suppose a man is lost in a rowboat in the middle of the ocean. What does he mean by saying that he is "completely ignorant" of his position? He means that, if he were to row a mile in any direction he would still be lost; he would be just as ignorant as before. In other words, ignorance of one's location is a state of knowledge which is not changed by a small change in that location. Mathematically, "complete ignorance" is an invariance property.

The set of all possible changes of location forms a group of translations. More generally, in a space S_x of any structure, one can define precisely what he means by "complete ignorance" by specifying some group of transformations of S_x onto itself under which an element of probability $m(x)dx$ is to be invariant. If the group is transitive on S_x (i.e., from any point x, any other point x' can be reached by some element of the group), this determines $m(x)$, to within an irrelevant multiplicative constant, on S_x.

This criterion follows naturally from the basic desideratum of consistency: In two problems where we have the same state of knowledge, we should assign the same probabilities. Any transformation of the group defines a new problem in which a "completely ignorant" person would have the same state of prior knowledge. If we can recognize a group of transformations that clearly has this property of transforming the problem into one that is equivalent in this sense, then the ambiguity in $m(x)$ has been removed. Quite a few useful "ignorance priors" have been found in this way; and in fact for most real problems that arise the solutions are now well-- if not widely--known.

But while the notion of transformation groups greatly reduces the ambiguity in $m(x)$, it does not entirely remove it in all cases. In some problems no appropriate group may be apparent; or there may be more than one, and we do not see how to choose between them. Therefore, still more basic principles are needed; and active research is now underway and is yielding promising results.

One of these new approaches, and the one on which there is most to report, is the method of marginalization (Jaynes, 1979). The basic facts pointing to it were given already by Jeffreys (1939; §3.8), but it was not realized until 1976 that this provides a new, constructive method for defining what is meant by "ignorance," with the advantage that everything follows from the basic rules (A8), (A9) of probability theory, with no need for any such desiderata as entropy or group invariance. We indicate the basic idea briefly, using a bare skeleton notation to convey only the structure of the argument.

Marginalization. We have a sampling distribution $p(x|\theta)$ for some observable quantity x, depending on a parameter θ, both multidimensional. From an observed value x we can make inferences about θ; with prior information I_1, prior probability distribution $p(\theta|I_1)$ Bayes' theorem (B24) yields the posterior distribution

$$p(\theta|xI_1) = A p(\theta|I_1) p(x|\theta) \quad . \tag{C27}$$

In the following, A always stands for a normalization constant, not necessarily the same in all equations.

But now we learn that the parameter θ can be separated into two components: $\theta = (\zeta,\eta)$ and we are to make inferences only about ζ. Then we discover that the data x may also be separated: $x = (y,z)$ in such a way that the sampling distribution of z depends only on ζ:

$$p(z|\zeta\eta) = \int p(yz|\eta\zeta) dy = p(z|\zeta) \quad . \tag{C28}$$

Then, writing the prior distribution as $p(\theta|I_1) = \pi(\zeta)\pi(\eta)$, (C27) gives for the desired marginal posterior distribution

$$p(\zeta|y,z\, I_1) = A\, \pi(\zeta) \int p(y,z|\zeta\eta)\pi(\eta) d\eta \tag{C29}$$

which must in general depend on our prior information about η. This is the solution given by a "conscientious Bayesian" B_1.

At this point there arrives on the scene an "ignorant Bayesian" B_2, whose knowledge of the experiment consists only of the sampling distribution (C28); i.e., he is unaware of the existence of the components (y,η). When told to make inferences about ζ, he confidently applies Bayes' theorem to (C28), getting the result

$$p(\zeta|z\, I_2) = A\, \pi(\zeta) p(z|\zeta) \quad . \tag{C30}$$

This is what was called a "pseudoposterior distribution" by Geisser and Cornfield (1963).

B_1 and B_2 will in general come to different conclusions because B_1 is taking into account extra information about (y,η). But now suppose that for some particular prior $\pi(\eta)$, B_1 and B_2 happen to agree after all; what does that mean? Clearly, B_2 is not incorporating any information about η; he doesn't even know it exists. If, nevertheless, they come to the same conclusions, then it must be that B_1 was not incorporating any information about η either. In other words, a prior $\pi(\eta)$ that leaves B_1 and B_2 in agreement must be, within the context of this model, a <u>completely uninformative</u> prior; it contains no information relevant to questions about ζ.

Now the condition for equality of (C29), (C30) is just a Fredholm integral equation:

$$\int p(y,z|\zeta,\eta)\pi(\eta) d\eta = \lambda(y,z) p(z|\zeta) \tag{C31}$$

where $\lambda(y,z)$ is a function to be determined from (C31). Therefore, the rules of probability theory already contain the criterion for defining what is meant by "completely uninformative."

Mathematical analysis of (C31) proves to be quite involved; and we do not yet know a necessary and sufficient condition on $p(y,z|\zeta\eta)$ for it to have solutions, or unique solutions, although a number of isolated results are now in (Jaynes, 1979). We indicate one of them.

Suppose y and η are positive real, and η is a scale parameter for y; i.e., we have the functional form

$$p(y,z|\zeta,\eta) = \eta^{-1} f(z,\zeta;y/\eta) \tag{C32}$$

for the sampling density function. Then, (C31) reduces to

$$y^{-1} \int f(z,\zeta;\alpha)\left[\frac{y}{\alpha}\pi\left(\frac{y}{\alpha}\right)\right]d\alpha = \lambda(y,z)\int f(z,\zeta;\alpha)d\alpha \tag{C33}$$

where we have used (C28). It is apparent from this that the Jeffreys prior

$$\pi(\eta) = \eta^{-1} \tag{C34}$$

is always a solution, leading to $\lambda(y,z) = y^{-1}$. Thus (C34) is "completely uninformative" for all models in which η appears as a scale parameter; and it is easily shown (Jaynes, 1979) that one can invent specific models for which it is unique.

We have therefore, the result that the Jeffreys prior is uniquely determined as the only prior for a scale parameter that is "completely uninformative" without qualifications. We can hardly avoid the inference that it represents, uniquely, the condition of "complete ignorance" for a scale parameter.

This example shows how marginalization is able to give results consistent with those found before, but in a way that springs directly out of the principles of probability theory without any additional appeal to intuition (as is involved in choosing a transformation group). At the moment, this approach seems very promising as a means of rigorizing and extending the basic theory. However, there are enough complicated technical details not noted here, so that it will require quite a bit more research before we can assess its full scope and power. In fact, the chase is at present quite exciting, because it is still mathematically an open question whether the integral equations may in some cases become overdetermined, so that no uninformative prior exists. If so, this would call for some deep clarification, and perhaps revision, of present basic statistical theory.

D. An Application: Irreversible Statistical Mechanics.
The calculation of an irreversible process usually involves
three distinct stages; (1) Setting up an "ensemble," i.e.,
choosing a density matrix $\rho(0)$, or an N-particle distribution
function, which is to describe our initial knowledge about the
system of interest; (2) Solving the dynamical problem; i.e.,
applying the microscopic equations of motion to obtain the
time-evolution of the system $\rho(t)$; (3) Extracting the final
physical predictions from the time-developed ensemble $\rho(t)$.

Stage (3) has never presented any procedural difficulty; to
predict the quantity F from the ensemble ρ, one follows the
practice of equilibrium theory, and computes the expectation
value $\langle F \rangle = \text{Tr}(\rho F)$. While the ultimate justification of this
rule has been much discussed (ergodic theory), no alternative
procedure has been widely used.

In this connection, we note the following. Suppose we are
to choose a number f, representing our estimate of the physical
quantity F, based on the ensemble ρ. A reasonable criterion
for the "best" estimate is that the expected square of the
error, $\langle (F-f)^2 \rangle$ shall be made a minimum. The solution of this
simple variational problem is: $f = \langle F \rangle$. Thus, if we regard
statistical mechanics, not in the "classical" sense of a means
for calculating time averages in terms of ensemble averages,
but rather as an example of statistical estimation theory
based on the mean square error criterion, the usual procedure
is uniquely determined as the optimal one, independently of
ergodic theory. A justification not depending on ergodic
theory is in any event necessary as soon as we try to predict
the time variation of some quantity F(t); for the physical
phenomenon of interest then consists just of the fact that
the ensemble average $\langle F(t) \rangle$ is not equal to a time average.

The dynamical problem of stage (2) is the most difficult to
carry out, but it is also the one in which most recent progress
has been booked (Green's function methods). While the present
work is not primarily concerned with these techniques, they are
available, and needed, in carrying out the calculations indicated here for all but the simplest problems.

It is curious that stage (1), which must logically precede
all the others, has received such scant attention since the
pioneering work of Gibbs, in which the problem of ensemble
construction was first recognized. Most recent discussions
of irreversible processes concentrate all attention on stage
(2); many fail to note even the existence of stage (1). One
consequence of this is that the resulting theories apply unambiguously only to the case of "response functions," in
which the nonequilibrium state is one resulting from a
dynamical perturbation (i.e., an explicitly given term in the

Hamiltonian), starting from thermal equilibrium at some time
in the past; the initial density matrix is then given by conventional equilibrium theory, and so the problem of ensemble
construction is evaded.

If, however, the nonequilibrium state is defined (as it
usually is from the experimental standpoint) in terms of
temperature or concentration gradients, rate of heat flow,
shearing stress, sound wave amplitudes, etc., such a procedure
does not apply, and one has resorted to various ad hoc devices.
An extreme example is provided by some problems in astrophysics,
in which it is clear that the system of interest has never, in
the age of the universe, been in a state approximating thermal
equilibrium. Such cases have been well recognized as presenting
special difficulties of principle.

We show here that recognition of the existence of the stage
(1) problem, and that its general solution is available, can
remove such ambiguities and reduce the labor of stage (2). In
the case of the nonequilibrium steady state, stage (2) can be
dispensed with entirely if stage (1) has been properly treated.

Background. To achieve a certain unity within the present
volume, we shall take the review article of Mori, Oppenheim,
and Ross (1962)--hereafter denoted MOR--as indicating the
level to which nonequilibrium theory had been brought before
the introduction of Information Theory notions. This work is
virtually unique in that the Stage 1 problem, and even the
term "ensemble construction" appear explicitly. The earlier
work of Kirkwood, Green, Callen, Kubo and others, directly
related to ours, is noted in MOR, Sec. 6.

To fix ideas, consider the calculation of transport properties in systems close to equilibrium (although our final
results will be far more general). In the treatments discussed
by MOR, dissipative-irreversible effects did not appear in the
ensemble initially set up. For example, a system of N particles
of mass m, distributed with macroscopic density $\rho(x)$, local
temperature $T(x)$, is often described in classical theory by
an N-particle distribution function, or Liouville function,
of the form:

$$W_N(x_1 p_1 \cdots x_N p_N) = \prod_{i=1}^{N} \frac{\rho(x_i)}{Nm} [2\pi mkT(x_i)]^{3/2} \exp\left\{-\frac{p_i^2}{2mkT(x_i)}\right\} \quad (D1)$$

where x_i, p_i denote the (vector) position and momentum of the
ith particle. But, although this distribution represents non-
vanishing density and temperature gradients $\nabla \rho$, ∇T, the diffusion current or heat flow computed from (D1) is zero.

Likewise, in quantum theory MOR described such a physical
situation by the "local equilibrium," or "frozen-state"

density matrix:

$$\rho_t = \frac{1}{Z} \exp\left\{-\int d^3x\ \beta(x)[H(x) - \mu(x)n(x)]\right\} \quad (D2)$$

where $H(x)$, $n(x)$ are the Hamiltonian density and number density operators. Again, although (D2) describes gradients of temperature, concentration, and chemical potential, the fluxes computed from (D2) are zero.

Mathematically, it was found that dissipative effects appear in the equations only after one has carried out the following operations: (a) approximate forward integration of the equations of motion for a short "induction time," and (b) time smoothing or other coarse-graining of distribution functions or Heisenberg operators.

Physically, it has always been somewhat of a mystery why either of these operations is needed; for one can argue that, in most experimentally realizable cases, irreversible flows (A) are already "in progress" at the time the experiment is started, and (B) take place slowly, so that the low-order distribution functions and expectation values of measurable quantities must be already slowly-varying functions of time and position; and thus not affected by coarse-graining. In cases where this is not true, coarse-graining would result in loss of the physical effects of interest.

The real nature of the forward integration and coarse-graining operations is therefore obscure; in a correctly formulated theory neither should be required. We are led to suspect the choice of initial ensemble; i.e., that ensembles such as (D1) and (D2) do not fully describe the conditions under which irreversible phenomena are observed, and therefore do not represent the correct solution of the stage (1) problem. [We note that (D1) and (D2) were not "derived" from anything more fundamental; they were written down intuitively, by analogy with the grand canonical ensemble of equilibrium theory.] The forward integration and coarse-graining operations would, on this view, be regarded as corrective measures which in some way compensate for the error in the initial ensemble.

This conclusion is in agreement with that of MOR. These authors never claimed that ρ_t in (D2) was the correct density matrix, but supposed that it differed by only a small amount from another matrix $\rho(t)$, which they designate as the "actual distribution." They further supposed that after a short induction time, ρ_t relaxes into $\rho(t)$, which would explain the need for forward integration.

Such relaxation undoubtedly takes place in the low-order distribution functions derived from ρ, as was first suggested by Bogoliubov for the analogous classical problem. However, this is not possible for the full "global" density matrix; if ρ_t and $\rho(t)$ differ at $t = 0$ and undergo the same unitary transformation in their time development, they cannot be equal at any other time. Furthermore, $\rho(t)$ was never uniquely defined; given two different candidates $\rho_1(t)$, $\rho_2(t)$ for this role, MOR give no criterion by which one could decide which is indeed the "actual" distribution.

For reasons already explained in earlier Sections and in Jaynes (1967), we believe that such criteria do not exist; i.e., that the notion of an "actual distribution" is illusory, since different density matrices connote only different states of knowledge. In the following Section we approach the problem in a different way, which yields a definite procedure for constructing a density matrix which is to replace ρ_t, and will play approximately the same role in our theory as the $\rho(t)$ of MOR.

The Gibbs Algorithm. If the above reasoning is correct, a re-examination of the procedures by which ensembles are set up in statistical mechanics is indicated. If we can find an algorithm for constructing density matrices which fully describe non-equilibrium conditions, we should find that transport and other dissipative effects are obtainable by direct quadratures over the initial ensemble.

This algorithm, we suggest, was given already by Gibbs (1902). The great power and scope of the methods he introduced have not been generally appreciated to this day; until recently it was scarcely possible to understand the rationale of his method for constructing ensembles. This was (loc. cit., p. 143) to assign that probability distribution which, while agreeing with what is known, "gives the least value of the average index of probability of phase," or as we would describe it today, maximizes the entropy. This process led Gibbs to his canonical ensemble for describing closed systems in thermal equilibrium, the grand canonical ensemble for open systems, and (loc. cit., p. 38) an ensemble to represent a system rotating at angular velocity $\vec{\omega}$ in which the probability density is proportional to

$$\exp[-\beta(H - \vec{\omega}\cdot\vec{M})] \tag{D3}$$

where H, M are the phase functions representing Hamiltonian and total angular momentum.

Ten years later, the Ehrenfests (1912) dismissed these ensembles as mere "analytical tricks," devoid of any real significance, and asserted the physical superiority of Boltzmann's methods, thereby initiating a school of thought which dominated

statistical mechanics for decades. It is one of the major
tragedies of science that Gibbs did not live long enough to
answer these objections, as he could have so easily.

The mathematical superiority of the canonical and grand
canonical ensembles for calculating equilibrium properties has
since become firmly established. Furthermore, although Gibbs
gave no applications of the rotational ensemble (D3), it was
shown by Heims and Jaynes (1962) that this ensemble provides
a straightforward method of calculating the gyromagnetic
effects of Barnett and Einstein-de Haas. At the present time,
therefore, the Gibbs methods--like the Laplace methods and
the Jeffreys methods--stand in a position of proven success
in applications, independently of all the conceptual problems
regarding their justification, which are still being debated.

The development of Information Theory made it possible to
see the method of Gibbs as a general procedure for inductive
reasoning, independent of ergodic theory or any other physical
hypotheses, and whose range of validity is therefore not restricted to equilibrium problems; or indeed to physics. In
the following we show that the Principle of Maximum Entropy
is sufficient to construct ensembles representing a wide
variety of nonequilibrium conditions, and that these new
ensembles yield transport coefficients by direct quadratures.
Indeed, we shall claim--for reasons already explained in
Jaynes (1957b), that this is the only principle needed to
construct ensembles which predict any experimentally reproducible effect, reversible or irreversible.

The general rule for constructing ensembles is as follows.
The available information about the state of a system consists
of results of various macroscopic measurements. Let the
quantities measured be represented by the operators $F_1, F_2, \ldots F_m$.
The results of the measurements are, of course, simply a set
of numbers: $\{f_1, \ldots, f_m\}$. These numbers make no reference to
any probability distribution. The ensemble is then a mental
construct which we invent in order to describe the range of
possible microscopic states compatible with those numbers, in
the following sense.

If we say that a density matrix ρ "contains" or "agrees
with" certain information, we mean by this that, if we communicate the density matrix to another person he must be able,
by applying the usual procedure of stage (3) above, to recover
this information from it. In this sense, evidently, the
density matrix agrees with the given information if and only
if it is adjusted to yield expectation values equal to the
measured numbers:

$$f_k = \mathrm{Tr}(\rho F_k) = \langle F_k \rangle \quad , \quad k = 1, \ldots, m \qquad (D4)$$

and in order to ensure that the density matrix describes the <u>full</u> range of possible microscopic states compatible with (D4), and not just some arbitrary subset of them (in other words, that it describes <u>only</u> the information given, and contains no hidden arbitrary assumptions about the microscopic state), we demand that, while satisfying the constraints (D4), it shall maximize the quantity

$$S_I = -\text{Tr}(\rho \log \rho) \ . \tag{D5}$$

A great deal of confusion has resulted from the fact that, for decades, the single word "entropy" has been used interchangeably to stand for either the quantity (D5) or the quantity measured experimentally (in the case of closed systems) by the integral of dQ/T over a reversible path. We shall try to maintain a clear distinction here by following the usage introduced in my 1962 Brandeis lectures (Jaynes, 1963b); referring to S_I as the "information entropy" and denoting the experimentally measured entropy by S_E. These quantities are different in general; in the equilibrium case (the only one for which S_E is defined in conventional thermodynamics) the relation between them was shown (<u>loc. cit</u>.) to be: for all density matrices ρ which agree with the macroscopic information that defines the thermodynamic state; i.e., which satisfy (D4),

$$kS_I \leq S_E \tag{D6}$$

where k is Boltzmann's constant, with equality in (D6) if and only if S_I is computed from the canonical density matrix

$$\rho = \frac{1}{Z(\lambda_1 \ldots \lambda_m)} \exp[\lambda_1 F_1 + \ldots + \lambda_m F_m] \tag{D7}$$

where the λ_k are unspecified real constants. In the nonequilibrium theory we find it easier to change our sign convention, so that all λ's here are the negative of the usual ones; otherwise, from this point on it would be invariably $(-\lambda)$ rather than λ that we need. For normalization (Tr $\rho = 1$) we have

$$Z(\lambda_1 \ldots \lambda_m) = \text{Tr} \exp[\lambda_1 F_1 + \ldots + \lambda_m F_m] \tag{D8}$$

which quantity will be called the <u>partition function</u>. It remains only to choose the λ_k [which appear as Lagrange multipliers in the derivation of (D7) from a variational principle] so that (D4) <u>is</u> satisfied. This is the case of

$$f_k = \langle F_k \rangle = \frac{\partial}{\partial \lambda_k} \log Z \ , \qquad k = 1, 2, \ldots, m \ . \tag{D9}$$

WHERE DO WE STAND ON MAXIMUM ENTROPY?   293

If enough constraints are specified to determine a normalizable
density matrix, it will be found that these relations are just
sufficient to determine the unknowns λ_k in terms of the given
data $\{f_1 \ldots f_m\}$; indeed, we can then solve (D9) explicitly for
the λ_k as follows. The maximum attainable value of S_I is, from
(D7), (D8),

$$(S_I)_{max} = \log Z - \sum_{k=1}^{m} \lambda_k <F_k> \ . \tag{D10}$$

If this quantity is expressed as a function of the given data,
$S(f_1 \ldots f_m)$, it is easily shown from the above relations that

$$\lambda_k = - \frac{\partial S}{\partial f_k} \ . \tag{D11}$$

It has been shown (Jaynes, 1963b, 1965) that the second law
of thermodynamics, and a generalization thereof that tells
which nonequilibrium states are accessible reproducibly from
others, follow as simple consequences of the inequality (D6)
and the dynamical invariance of S_I.

We note an important property of the maximum entropy ensemble,
which is helpful in gaining an intuitive understanding of this
theory. Given any density matrix ρ and any ε in $0 < \varepsilon < 1$, one
can define a "high-probability linear manifold" (HPM) of finite
dimensionality $W(\varepsilon)$, spanned by all eigenvectors of ρ which
have probability greater than a certain amount $\delta(\varepsilon)$, and such
that the eigenvectors of ρ spanning the complementary manifold
have total probability less than ε. Viewed in another way, the
HPM consists of all state vectors ψ to which ρ assigns an "array
probability" as defined in Jaynes (1957b), Sec. 7, greater than
$\delta(\varepsilon)$. Specifying the density matrix ρ thus amounts to asserting
that, with probability $(1-\varepsilon)$, the state vector of the system
lies somewhere in this HPM. As ε varies, any density matrix
ρ thus defines a nested sequence of HPM's.

For a macroscopic system, the information entropy S_I may be
related to the dimensionality $W(\varepsilon)$ of the HPM in the following
sense: if N is the number of particles in the system, then as
$N \to \infty$ with the intensive parameters held constant, $N^{-1} S_I$ and
$N^{-1} \log W(\varepsilon)$ approach the same limit independently of ε. This is
a form of the asymptotic equipartition theorem of Information
Theory, and generalizes Boltzmann's $S = k \log W$. The process of
entropy maximization therefore amounts, for all practical purposes,
to the same thing as finding the density matrix which, while
agreeing with the available information, defines the <u>largest</u>
possible HPM; this is the basis of the remark following (D4).
An analogous result holds in classical theory (Jaynes, 1965),
in which $W(\varepsilon)$ becomes the phase volume of the "high-probability

region" of phase space, as defined by N-particle distribution function.

The above procedure is sufficient to construct the density matrix representing equilibrium conditions, provided the quantities F_k are chosen to be constants of the motion. The extension to nonequilibrium cases, and to equilibrium problems in which we wish to incorporate information about quantities which are not intrinsic constants of the motion (such as stress or magnetization) requires mathematical generalization which we give in two steps.

It is a common experience that the course of a physical process does not in general depend only on the present values of the observed macroscopic quantities; it depends also on the past history of the system. The phenomena of magnetic hysteresis and spin echoes are particularly striking examples of this. Correspondingly, we must expect that, if the F_k are not constants of the motion, an ensemble constructed as above using only the present values of the $\langle F_k \rangle$ will not in general suffice to predict either equilibrium or nonequilibrium behavior. As we will see presently, it is just this fact which causes the error in the "local equilibrium" density matrix (D2).

In order to describe time variations, we extend the F_k to the Heisenberg operators

$$F_k(t) = U^{-1}(t) F_k(0) U(t) \tag{D12}$$

in which the time-development matrix $U(t)$ is the solution of the Schrödinger equation

$$i\hbar \dot{U}(t) = H(t) U(t) \tag{D13}$$

with $U(0) = 1$, and $H(t)$ is the Hamiltonian. If we are given data fixing the $\langle F_k(t_i) \rangle$ at various times t_i, then each of these must be considered a separate piece of information, to be given its Lagrange multiplier λ_{ki} and included in the sum of (D7). In the limit where we imagine information given over a continuous time interval, $-\tau < t < 0$, the summation over the time index i becomes an integration and the canonical density matrix (D7) becomes

$$\rho = \frac{1}{Z} \exp\left\{ \sum_{k=1}^{m} \int_{-\tau}^{0} \lambda_k(t) F_k(t) \, dt \right\} \tag{D14}$$

where the partition function has been generalized to a partition functional

$$Z[\lambda_1(t) \ldots \lambda_m(t)] \equiv \text{Tr} \, \exp\left\{ \sum_{k=1}^{m} \int_{-\tau}^{0} \lambda_k(t) F_k(t) \, dt \right\} \tag{D15}$$

and the unknown Lagrange multiplier functions $\lambda_k(t)$ are determined from the condition that the density matrix agree with the given data $\langle F_k(t)\rangle$ over the "information-gathering" time interval:

$$\langle F_k(t)\rangle = \text{Tr}[\rho F_k(t)] = f_k(t) \quad , \quad -\tau \leq t \leq 0 \quad . \tag{D16}$$

By the perturbation methods developed below, we find that (D16) reduces to the natural generalization of (D9):

$$\langle F_k(t)\rangle = \frac{\delta}{\delta \lambda_k(t)} \log Z \quad , \quad -\tau \leq t \leq 0 \tag{D17}$$

where δ denotes the functional derivative.

Finally, if the operators F_k depend on position as well as time, as in (D2), Eq. (D12) is changed to

$$F_k(x,t) = U^{-1}(t) F_k(x,0) U(t) \tag{D18}$$

and the values of these quantities at each point of space and time now constitute the independent pieces of information, which are coupled into the density matrix via the Lagrange multiplier function $\lambda_k(x,t)$. If we are given macroscopic information about $F_k(x,t)$ throughout a space-time region R_k (which can be a different region for different quantities F_k), the ensemble which incorporates all this information, while locating the largest possible HPM of microscopic states, is

$$\rho = \frac{1}{Z} \exp\left\{\sum_k \int_{R_k} dt\, d^3x\, \lambda_k(x,t) F_k(x,t)\right\} \tag{D19}$$

with the partition functional

$$Z = \text{Tr}\, \exp\left\{\sum_k \int_{R_k} dt\, d^3x\, \lambda_k(x,t) F_k(x,t)\right\} \tag{D20}$$

and the $\lambda_k(x,t)$ determined from

$$\langle F_k(x,t)\rangle = \frac{\delta}{\delta \lambda_k(x,t)} \log Z \quad , \quad (x,t) \text{ in } R_k \quad . \tag{D21}$$

Prediction of any quantity $J(x,t)$ is then accomplished by calculating

$$\langle J(z,t)\rangle = \text{Tr}[\rho J(x,t)] \tag{D22}$$

The form of equations (D19)-(D22) makes it appear that stages (1) and (2), discussed in the Introduction, are now fused into a single stage. However, this is only a consequence of our using the Heisenberg representation. According to the usual conventions, the Schrödinger and Heisenberg representations coincide at time t = 0; thus we may regard the steps (D19)-(D21) equally well as determining the density matrix $\rho(0)$ in the Schrödinger representation; i.e., as solving the stage (1) problem. If, having found this initial ensemble, we switch to the Schrodinger representation, Eq. (D22) is then replaced by

$$<J(x)>_t = Tr[J(x)\rho(t)] \qquad (D23)$$

in which the problem of stage (2) now appears explicitly as that of finding the time-evolution of $\rho(t)$. The form (D23) will be more convenient if several different quantities $J_1, J_2, \ldots$ are to be predicted.

Discussion. In equations (D19)-(D23) we have the generalized Gibbs algorithm for calculating irreversible processes. They represent the three stages: (1) finding the ensemble which has maximum entropy subject to the given information about certain quantities $\{F_k(x,t)\}$; (2) utilizing the full dynamics by working out the time evolution from the microscopic equations of motion; (3) making those predictions of certain other quantities of interest $\{J_i(x,t)\}$ which take all the above into account, and minimize the expected square of the error. We do not claim that the resulting predictions must be correct; only that they are the best (by the mean-square error criterion) that could have been made from the information given; to do any better we would require more initial information.

Of course this algorithm will break down, as it should, and refuse to give us any solution if we ask a foolish, unanswerable question; for example, if we fail to specify enough information to determine any normalizable density matrix, if we specify logically contradictory constraints, or if we specify space-time variations incompatible with the Hamiltonian of the system.

The reader may find it instructive to work this out in detail for a very simple system involving only a (2 x 2) matrix; a single particle of spin 1/2, gyromagnetic ratio γ, placed in a constant magnetic field $\vec{B}$ in the z-direction, Hamiltonian $H = -(1/2)\hbar \gamma (\vec{\sigma} \cdot \vec{B})$. Then the only dynamically possible behavior is uniform precession about $\vec{B}$ at the Larmor frequency $\omega_0 = \gamma B$. If we specify any time variation for $<\sigma_x>$ other than sinusoidal at this frequency, the above equations will break down; while if we specify $<\sigma_x(t)> = a \cos \omega_0 t + b \sin \omega_0 t$, we find a unique solution whenever $(a^2 + b^2) \leq 1$.

Mathematically, whether the ensemble ρ is or is not making a sharp prediction of some quantity J is determined by whether the variance $<J^2> - <J>^2$ is sufficiently small. In general, information about a quantity F would not suffice to predict some other quantity J with deductive certainty (unless J is a function of F). But in inductive reasoning, Information Theory tells us the precise extent to which information about F is relevant to predictions of J. In practice, due to the enormously high dimensionality of the spaces involved, the variance $<J^2> - <J>^2$ usually turns out to be very small compared to any reasonable mean-square experimental error; and therefore the predictions are, for all practical purposes, deterministic.

Experimentally, we impose various constraints (volume, pressure, magnetic field, gravitational or centrifugal forces, sound level, light intensity, chemical environment, etc.) on a system and observe how it behaves. But only when we reach the degree of control where reproducible response is observed, do we record our data and send it off for publication. Because of this sociological convention, it is not the business of statistical mechanics to predict everything that can be observed in nature; only what can be observed reproducibly. But the experimentally imposed macroscopic constraints surely do not determine any unique microscopic state; they ensure only that the state vector is somewhere in the HPM. If effect A is, nevertheless, reproducible, then it must be that A is characteristic of each of the overwhelming majority of possible states in the HPM; and so averaging over those states will not change the prediction.

To put it another way, the macroscopic experimental conditions still leave billions of microscopic details undetermined. If, nevertheless, some result is reproducible, then those details must have been irrelevant to the phenomenon; and so with proper understanding we ought to be able to eliminate them mathematically. This is just what Information Theory does for us; it removes irrelevant details by averaging over them, while retaining what is relevant to the particular question being asked [i.e., the particular quantity $J(x,t)$ that we want to predict].

It is clear, then, why the maximum entropy prescription works in such generality. If the constraints used in the calculation are the same as those actually operative in the experiment, then the maximum-entropy density matrix will locate the same HPM as did the experimental conditions; and will therefore make sharp predictions of any reproducible effect, provided that our assumed microscopic physics (enumeration of possible states, equations of motion) is correct.

For these reasons--as stressed in Jaynes (1957b),--if the class of phenomena predictable from the maximum entropy principle is found to differ in any way from the class of reproducible phenomena, that would constitute evidence for new microscopic laws of physics, not presently known. Indeed (Jaynes, 1968) this is just what did happen early in this Century; the failure of Gibbs' classical statistical mechanics to predict the correct heat capacities and vapor pressures provided the first clues pointing to the quantum theory. Any successes make this theory useful in an "engineering" sense; but for a research physicist its failures would be far more valuable than its successes.

We emphasize that the basic physical and conceptual formulation of the theory is complete at this point; what follows represents only the working out of various mathematical consequences of this algorithm.

Perturbation Theory. For systems close to thermal equilibrium, the following general theorems are useful. We denote an "unperturbed" density matrix ρ_o, by

$$\rho_o = \frac{e^A}{Z_o} \quad , \qquad Z_o \equiv \text{Tr}\left(e^A\right) \quad , \tag{D24}$$

a "perturbed one by

$$\rho = \frac{e^{A+\varepsilon B}}{Z} \quad , \qquad Z \equiv \text{Tr}\left(e^{A+\varepsilon B}\right) \tag{D25}$$

where A, B are Hermitian. The expectation values of any operator C over these ensembles are respectively

$$\langle C \rangle_o = \text{Tr}(\rho_o C) \quad , \qquad \langle C \rangle = \text{Tr}(\rho C) \quad . \tag{D26}$$

The cumulant expansion of $\langle C \rangle$ to all orders in ε is derived in Heims and Jaynes (1962), Appendix B. The n'th order term may be written as a covariance in the unperturbed ensemble:

$$\langle C \rangle - \langle C \rangle_o = \sum_{n=1}^{\infty} \varepsilon^n \left[\langle Q_n C \rangle_o - \langle Q_n \rangle_o \langle C \rangle_o \right] \quad . \tag{D27}$$

Here Q_n is defined by $Q_1 \equiv S_1$, and

$$Q_n \equiv S_n - \sum_{k=1}^{n-1} \langle Q_k \rangle_o S_{n-k} \quad , \qquad n > 1 \tag{D28}$$

in which S_n are the operators appearing in the well-known expansion

$$e^{A+\varepsilon B} = e^A \left[1 + \sum_{n=1}^{\infty} \varepsilon^n S_n \right] \quad . \tag{D29}$$

More explicitly,

$$S_n = \int_0^1 dx_1 \int_0^{x_1} dx_2 \cdots \int_0^{x_{n-1}} dx_n \, B(x_1) \cdots B(x_n) \quad \text{(D30)}$$

where

$$B(x) \equiv e^{-xA} B e^{xA} \quad . \quad \text{(D31)}$$

The first-order term is thus

$$<C> - <C>_o = \varepsilon \int_0^1 dx \left[<e^{-xA} B e^{xA} C>_o - <B>_o <C>_o \right] \quad \text{(D32)}$$

and it will appear below that all relations of linear transport theory are special cases of (D32).

For a more condensed notation, define the average of any operator B over the sequence of similarity transformations as

$$\overline{B} \equiv \int_0^1 dx \, e^{-xA} B e^{xA} \quad \text{(D33)}$$

which we will call the <u>Kubo transform</u> of B. Then (D32) becomes

$$<C> - <C>_o = \varepsilon \, K_{CB} \quad \text{(D34)}$$

in which, for various choices of C, B, the quantities

$$K_{CB} \equiv <\overline{B}C>_o - <\overline{B}>_o <C>_o \quad \text{(D35)}$$

are the basic covariance functions of the linear theory.

We list a few useful properties of these quantities; in all cases, the result is proved easily by writing out the expressions in the representation where A is diagonal. Let F, G be any two operators; then

$$<\overline{F}>_o = <F>_o \quad \text{(D36)}$$

$$K_{FG} = K_{GF} \quad . \quad \text{(D37)}$$

If F, G are Hermitian, then

$$K_{FG} \text{ is real} \, , \quad K_{FF} \geq 0 \quad . \quad \text{(D38)}$$

If ρ_o is a projection operator representing a pure state, then $K_{FG} \equiv 0$. If ρ_o is not a pure state density matrix, then with Hermitian F, G,

$$K_{FF}K_{GG} - K_{FG}^2 \geq 0 \tag{D39}$$

with equality if and only if $F = qG$, where q is a real number. If G is of the form

$$G(u) = e^{-uA} G(0) e^{uA} \tag{D40}$$

then

$$\frac{d}{du} K_{FG} = \langle [F,G] \rangle_o \quad . \tag{D41}$$

This identity, with u interpreted as a time, provides a general connection between statistical and dynamical problems.

<u>Near-Equilibrium Ensembles</u>. A closed system in thermal equilibrium is described, as usual, by the density matrix

$$\rho_o = \frac{e^{-\beta H}}{Z_o(\beta)} \tag{D42}$$

which maximizes S_I for prescribed $\langle H \rangle$, and is a very special case of (D19). The thermal equilibrium prediction for any quantity F is, as usual,

$$\langle F \rangle_o = \text{Tr}(\rho_o F) \quad . \tag{D43}$$

But suppose we are now given the value of $\langle F(t) \rangle$ throughout the "information-gathering" interval $-\tau \leq t < 0$. The ensemble which includes this new information is of the form (D19), which maximizes S_I for prescribed $\langle H \rangle$ and $\langle F(t) \rangle$. It corresponds to the partition functional

$$Z[\beta, \lambda(t)] = \text{Tr} \exp\left[-\beta H + \int_{-\tau}^{0} \lambda(t) F(t) dt\right] \quad . \tag{D44}$$

If, during the information-gathering interval, this new information was simply $\langle F(t) \rangle = \langle F \rangle_o$, it is easily shown from (D17) that we have identically

$$\int_{-\tau}^{0} \lambda(t) F(t) dt = 0 \quad . \tag{D45}$$

In words: if the new information is redundant (in the sense that it is only what we would have predicted from the old information), then it will drop out of the equations and the ensemble is unchanged. This is a general property of the formalism here presented. In applications it means that there is never any need, when setting up an ensemble, to ascertain whether the different pieces of information used are independent; any redundant parts will drop out automatically.

If, therefore, we treat the integral in (D44) as a small perturbation, we are expanding in powers of the departure from equilibrium. For validity of the perturbation scheme it is not necessary that $\lambda(t)F(t)$ be everywhere small; it is sufficient if the integral is small. First-order effects in the departure from equilibrium, such as linear diffusion or heat flow, are then predicted using the general formula (D32), with the choices $A = -\beta H$, and

$$\varepsilon B = \int_{-\tau}^{0} \lambda(t) F(t) dt \quad . \tag{D46}$$

With constant H, the Heisenberg operator $F(t)$ reduces to

$$F(t) = \exp(iHt/\hbar) F(0) \exp(-iHt/\hbar) \tag{D47}$$

and its Kubo transform (D33) becomes

$$\overline{F}(t) = \frac{1}{\beta} \int_{0}^{\beta} du\, F(t - i\hbar u) \quad , \tag{D48}$$

the characteristic quantity of the Kubo (1957, 1958) theory.

In the notation of (D34), the first-order expectation value of any quantity $C(t)$ will then be given by

$$<C(t)> - <C>_0 = \int_{-\tau}^{0} K_{CF}(t,t')\, \lambda(t')\, dt' \tag{D49}$$

where K_{CF} is now indicated as a function of the parameters t, t' contained in the operators:

$$K_{CF}(t,t') \equiv <\overline{F}(t')C(t)>_0 - <F>_0 <C>_0 \tag{D50}$$

Remembering that the parameters t, t' are part of the operators C, F, the general reciprocity law (D37) now becomes

$$K_{CF}(t,t') = K_{FC}(t',t) \quad . \tag{D51}$$

When H is constant, it follows also from (D47) that

$$K_{CF}(t,t') = K_{CF}(t - t') \tag{D52}$$

and (D41) becomes

$$i\hbar \frac{\partial}{\partial t} K_{CF}(t,t') = <[C(t), F(t')]>_0\, \beta^{-1} \tag{D53}$$

Integral Equations for the Lagrange Multipliers. We wish to find the $\lambda_k(x,t)$ to first order in the given departures from equilibrium, $<F_k(x,t)> - <F_k(x,t)>_0$. This could be done by direct application of the formalism; by finding the perturbation expansion of log Z to second order in the λ's and taking the functional derivative explicitly according to (D21). It will be sufficient to do this for the simpler case described by Equations (D44)-(D53); but on carrying through this calculation we discover that the result is already contained in our perturbation-theory formula (D49). This is valid for any operator $C(t)$; and therefore in particular for the choice $C(t) = F(t)$. Then (D49) becomes

$$\int_{-\tau}^{0} K_{FF}(t,t') \lambda(t') dt' = <F(t)> - <F>_0 \qquad (D54)$$

If t is in the "information-gathering interval" $(-\tau \leq t \leq 0)$ this is identical with what we get on writing out (D17) explicitly with log Z expanded to second order. In lowest order, then, taking the functional derivative of log Z has the effect of constructing a linear Fredholm integral equation for $\lambda(t)$, in which the "driving force" is the given departure from equilibrium.

However, from that direct manner of derivation it would appear that (D54) applies <u>only</u> when $(-\tau \leq t \leq 0)$; while the derivation of (D49) makes it clear that (D54) has a definite meaning for all t. When t is in $-\tau \leq t \leq 0$, it represents the integral equation from which $\lambda(t)$ is to be determined; when $t > 0$, it represents the <u>predicted</u> <u>future</u> of $F(t)$; and when $t < -\tau$, it represents the <u>retrodicted</u> <u>past</u> of $F(t)$.

If the information about $<F(t)>$ is given in the entire past, $\tau = \infty$, (D54) becomes a Wiener-Hopf equation. Techniques for the solution, involving matching functions on strips in the complex fourier transform space are well known; we remark only that the solution $\lambda(t)$ will in general contain a δ-function singularity at $t = 0$; it is essential to retain this in order to get correct physical predictions. In other words, the upper limit of integration in (D54) must be taken as (0+). The case of finite τ, where we generally have δ-functions at both end-points, is discussed by Middleton (1960).

For example, with a Lorentzian correlation function

$$K_{FF}(t-t') = \frac{a}{2} \exp[-a|t-t'|] \qquad (D55)$$

and $\tau = \infty$, the solution is found to be

$$\lambda(t) = \left(1 - \frac{1}{a^2} \frac{d^2}{dt^2}\right) f(t) + \frac{1}{a^2} f(0)[a\delta(t) - \delta'(t)] \qquad (D56)$$

WHERE DO WE STAND ON MAXIMUM ENTROPY?

where
$$f(t) \equiv \begin{cases} <F(t)> - <F>_o & t \leq 0 \\ 0 & t > 0 \end{cases} \tag{D57}$$

Then we find
$$\int_{-\infty}^{0+} K_{FF}(t-t')\lambda(t')dt' = \begin{cases} f(t), & t < 0 \\ f(0)e^{-at}, & t > 0 \end{cases} \tag{D58}$$

in which it is necessary to note the δ-functions in $f''(t)$ at the upper limit. The nature and need for these δ-functions becomes clear if we approach the solution as the limit of the solutions for a sequence $\{f_n\}$ of "good" driving functions each of which satisfies the same boundary conditions as K_{FF} at the upper limit:

$$\left[f_n'(t')K_{FF}(t-t') - f_n(t')\frac{\partial}{\partial t'}K_{FF}(t-t')\right]_{t'=0} = 0, \quad t < 0. \tag{D59}$$

The result (D58) thus predicts the usual exponential approach back to equilibrium, with a relaxation time $\tau = a^{-1}$. The particular correlation function (D55) is "Markoffian" in that the predicted future decay depends only on the specified departure from equilibrium at $t = 0$; and not on information about its past history. With other forms of correlation function we get a more complicated prediction, with in general more than one relaxation time.

<u>Relation to the Wiener Prediction Theory</u>. This problem is so similar conceptually to Wiener's (1949) problem of optimal prediction of the future of a random function whose past is given that one would guess them to be mathematically related. However, this is not obvious from the above, because the Wiener theory was stated in entirely different terms. In particular, it contained no quantity such as $\lambda(t)$ which enables us to express both the given past and predicted future of $F(t)$ in a single equation (D54). To establish the connection between these two theories, and to exhibit an alternative form of our theory, we may eliminate $\lambda(t)$ by the following purely formal manipulations.

If the resolvent $K_{FF}^{-1}(t,t')$ of the integral equation (D54) can be found so that

$$\int_{-\tau}^{0} K_{FF}(t,t'')K_{FF}^{-1}(t'',t')dt'' = \delta(t-t'), \quad -\tau \leq t, t' \leq 0 \tag{D60}$$

$$\int_{-\tau}^{0} K_{FF}^{-1}(t,t'')K_{FF}(t'',t')dt'' = \delta(t-t'), \quad -\tau \leq t, t' \leq 0 \tag{D61}$$

then
$$\lambda(t) = \int_{-\tau}^{0} K_{FF}^{-1}(t,t')[<F(t')> - <F>_o]dt' , \quad -\tau \leq t \leq 0 \qquad (D62)$$

and the predicted value (D49) of any quantity $C(t)$ can be expressed directly as a linear combination of the given departures of F from equilibrium:

$$<C(t)> - <C>_o = \int_{-\tau}^{0} R_{CF}(t,t')[<F(t')> - <F>_o]dt' \qquad (D63)$$

in which

$$R_{CF}(t,t') \equiv \int_{-\tau}^{0} K_{CF}(t,t'')K_{FF}^{-1}(t'',t')dt'' \qquad (D64)$$

will be called the <u>relevance function</u>.

In consequence of (D61), the relevance function is itself the solution of an integral equation:

$$K_{CF}(t) = \int_{-\tau}^{0} R_{CF}(t,t')K_{FF}(t')dt' , \quad -\infty < t < \infty \qquad (D65)$$

so that, in some cases, the final prediction formula (D63) can be obtained directly from (D65) without the intermediate step of calculating $\lambda(t)$.

In the Wiener theory we have a random function $f(t)$ whose past is known. For any "lead time" $h > 0$, we are to try to predict the value of $f(t+h)$ by a linear operation on the past of $f(t)$, i.e., the prediction is

$$\hat{f}(t+h) = \int_{0}^{\infty} f(t-t')W(t')dt' \qquad (D66)$$

and the problem is to find that $W(t)$ which minimizes the mean square error of the prediction:

$$I[W] = \lim_{T \to \infty} \frac{1}{2T} \int_{-T}^{T} |f(t+h) - \hat{f}(t+h)|^2 dt . \qquad (D67)$$

We readily find that the optimal W satisfies the Wiener-Hopf integral equation

$$\phi(t+h) = \int_{0}^{\infty} \phi(t-t')W(t')dt' , \quad t \geq 0 \qquad (D68)$$

where

$$\phi(t) \equiv \lim_{T \to \infty} \frac{1}{2T} \int_{-T}^{T} f(t+t')f(t')dt' \qquad (D69)$$

is the autocorrelation function of $f(t)$, assumed known.

Evidently, the response function $W(t)$ corresponds to our relevance function $R_{FF}(t,t')$; and to establish the formal identity of the two theories, we need only show that R also satisfies the integral equation (D68). But, with the choice $C(t) = F(t)$, this is included in (D65) making the appropriate changes in notation; our "quantum covariance function" $K_{FF}(t)$ corresponding to Wiener's autocorrelation function $\phi(t)$. In the early stages of this work, the discovery of the formal identity between Wiener's prediction theory and this special case of the maximum-entropy prediction theory was an important reassurance.

The relevance function $R_{CF}(t,t')$ summarizes the precise extent to which information about F at time t' is relevant to prediction of C at time t. It is entirely different from the physical impulse-response function $\phi_{CF}(t-t')$ discussed, for example, by Kubo (1958), Eq. (2.18). The latter represents the dynamical response $<C(t)> - <C>_0$ at a time $t > t'$, to an impulsive force term in the Hamiltonian applied at $t = t'$: $H(t) = H_0 + F \delta(t-t')$, while in (D63) the "input" $<F(t')> - <F>_0$ consists only of <u>information</u> concerning what the system, with a fixed Hamiltonian but in a nonequilibrium state, was doing in the interval $-\tau \leq t' \leq 0$. This distinction is perhaps brought out most clearly by emphasizing again that (D63) is valid for an arbitrary time t, which may be before, within, or after this information-gathering interval. Thus, while our conception of causality is based on the postulate that a force applied at time t' can exert physical influences only at <u>later</u> times, there is no such limitation in (D63). It therefore represents an explicit statement of the fact that, while <u>physical in-</u><u>fluences</u> propagate only forward in time, <u>logical inferences</u> propagate equally well in either direction; i.e., new information about the present affects our knowledge of the past as well as the future. Although relations such as (D63) have been rather rare in physics, the situation is, of course, commonplace in other fields; sciences such as geology depend on logical connections of this type.

<u>Space-Time Variations</u>. Suppose the particle density $n(x,t)$ departs slightly from its equilibrium value in a space-time region R. Defining $\delta n(x,t) \equiv n(x,t) - <n(x)>_0$, the ensemble containing this information corresponds to the partition functional

$$Z[\beta,\lambda(x,t)] = \text{Tr } e\left[-\beta H + \int_R \lambda(x,t)\delta n(x,t)d^3xdt\right] \qquad (D70)$$

and (D34) becomes

$$<\delta n(x,t)> = \int_R K_{nn}(x-x';t-t')\lambda(x',t')d^3x'dt' \quad . \qquad (D71)$$

When (x,t) are in R this represents the integral equation determining $\lambda(x',t')$; when (x,t) are outside R it gives the predicted nonequilibrium behavior of $<n(x,t)>$ based on this information, and the predicted departure from equilibrium of any other quantity $J(x,t)$ is

$$<J(x,t)> - <J>_0 = \int_R K_{Jn}(x-x';t-t')\lambda(x',t')d^3x'dt' \quad . \tag{D72}$$

To emphasize the generality of (D72), note that it contains no limitation on time scale or space scale. Thus it encompasses both diffusion and ultrasonic propagation.

In (D71) we see the deviation $<\delta n>$ expressed as a linear superposition of basic relaxation functions $K_{nn}(x,t) = <\delta \bar{n}(0,0)\delta n(x,t)>_0$ with $\lambda(x',t')$ as the "source" function. The class of different nonequilibrium ensembles based on information about $<\delta n>$ is in 1:1 correspondence with different functions $\lambda(x,t)$. In view of the linearity, we may superpose elementary solutions in any way, and while to solve a problem with specific given information would require that we solve an integral equation, we can extract the general laws of nonequilibrium behavior from (D71), (D72) without this, by considering $\lambda(x,t)$ as the independent variable.

For example, let J be the α-component of particle current, and for brevity write the covariance function in (D72) as

$$<\delta\bar{n}(x',t')J_\alpha(x,t)>_0 = K_\alpha(x-x';t-t') \quad . \tag{D73}$$

Now choose R as all space and all time $t < 0$, and take

$$\lambda(x,t) = \mu(x)\dot{q}(t) \quad , \quad t < 0 \tag{D74}$$

where $\mu(x)$, $q(t)$ are "arbitrary" functions (but of course, sufficiently well-behaved so that what we do with them makes sense mathematically). In this ensemble, the current is

$$<J_\alpha(x,t)> = \int d^3x' \, \mu(x')\int_{-\infty}^0 dt' \, \dot{q}(t')K_\alpha(x-x',t-t') \quad . \tag{D75}$$

Integrate by parts on t and use the identity $\dot{n}+\nabla\cdot J = 0$: the RHS of (D75) becomes

$$\int d^3x' \, \mu(x')\left[q(0)K_\alpha(x-x',0) + \frac{\partial}{\partial x'_\beta}\int_{-\infty}^0 dt' \, q(t)K_{\alpha\beta}(x-x',t-t')\right] \tag{D76}$$

where $K_{\alpha\beta}$ is the current-current covariance:

$$K_{\alpha\beta}(x-x',t-t') \equiv <\bar{J}_\beta(x',t')J_\alpha(x,t)>_0 \quad . \tag{D77}$$

But from symmetry $K_\alpha(x-x',0) \equiv 0$. Another integration by parts then yields

$$\langle J_\alpha(x,t)\rangle = -\int_{-\infty}^{0} dt' q(t') \int d^3x' \, K_{\alpha\beta}(x-x',t-t') \frac{\partial \mu}{\partial x'_\beta} \tag{D78}$$

and thus far no approximations have been made.

Now let us pass to the "long wavelength" limit by supposing $\mu(x)$ so slowly varying that $\partial\mu/\partial x'_\beta$ is essentially a constant over distances in which $K_{\alpha\beta}$ is appreciable:

$$\langle J_\alpha(x,t)\rangle \cong -\frac{\partial \mu}{\partial x_\beta} \int_{-\infty}^{0} dt' \, q(t') \int d^3x' \, K_{\alpha\beta}(x-x',t-t') \tag{D79}$$

and in the same approximation (D71) becomes

$$\langle \delta n(x,t)\rangle \simeq q(0)\mu(x) \int d^3x' \, K_{nn}(x',0) . \tag{D80}$$

Therefore, the theory predicts the relation

$$\langle J_\alpha \rangle = -D_{\alpha\beta} \frac{\partial}{\partial x_\beta} \langle \delta n \rangle \tag{D81}$$

with

$$D_{\alpha\beta} \equiv \frac{\int_{-\infty}^{0} dt' \, q(t') \int d^3x' \, K_{\alpha\beta}(x-x',t-t')}{q(0) \int d^3x' \, K_{nn}(x-x',0)} . \tag{D82}$$

If the ensemble is also quasi-stationary, $q(t)$ very slowly varying, only the value of $q(t)$ near the upper limit matters, and the choice $q(t) = \exp(\varepsilon t)$ is as good as any. This leads to just the Kubo expression for the diffusion coefficient.

If instead of taking the long-wavelength limit we choose a plane wave: $\mu(x) = \exp(ik \cdot x)$, (D75), (D71) become

$$\langle J_\alpha(x,t)\rangle = e^{ik\cdot x} \int_{-\infty}^{0} dt' \, \dot{q}(t') K_\alpha(k;t-t') \tag{D83}$$

$$\langle \delta n(x,t)\rangle = e^{ik\cdot x} \int_{-\infty}^{0} dt' \, \dot{q}(t') K_{nn}(k,t-t') \tag{D84}$$

where $K_\alpha(k,t)$, $K_{nn}(k,t)$ are the space fourier transforms. These represent the decay of sound waves as linear superpositions of many characteristic decays $\sim K_\alpha(k,t)$, $K_{nn}(k,t)$ with various "starting times" t'. If we take time fourier transforms, (D84) becomes

$$\langle \delta n(x,t)\rangle = e^{ik\cdot x} \int \frac{d\omega}{2\pi} K_{nn}(k,\omega) Q(\omega) e^{-i\omega t} \tag{D85}$$

which shows how the exact details of the decay depend on the method of preparation (past history) as summarized by $Q(\omega)$. Now, however, we find that $K_{nn}(k,\omega)$ usually has a sharp peak at some frequency $\omega = \omega_0(k)$, which arises mathematically from a pole near the real axis in the complex ω-plane. Thus if $\omega_1 = \omega_0 - i\alpha$ and $K_{nn}(k,\omega)$ has the form

$$K_{nn}(k,\omega) = \frac{-iK_1}{\omega-\omega_1} + \hat{K}(k,\omega) \tag{D86}$$

where $\hat{K}(k,\omega)$ is analytic in a neighborhood of ω_1, this pole will give a contribution to the integral (D86) of

$$K_1 Q(\omega_1) e^{i(k \cdot x - \omega_0 t)} e^{-\alpha t} , \quad t > 0 . \tag{D87}$$

Terms which arise from parts of $K_{nn}(k,\omega)Q(\omega)$ that are not sharply peaked as a function of ω, decay rapidly and represent short transient effects that depend on the exact method of preparation. If α is small, the contribution (D87) will quickly dominate them, leading to a long-term attenuation and propagation velocity essentially independent of the method of preparation.

Thus, the laws of ultrasonic dispersion and attenuation are contained in the location and width of the sharp peaks in $K_{nn}(k,\omega)$.

Other Forms of the Theory. Thus far we have considered the application of maximum entropy in its most general form: given some arbitrary initial information, to answer an arbitrary question about reproducible effects. Of course, we may ask any question we please; but maximum entropy can make sharp predictions only of reproducible things (that is in itself a useful property; for maximum entropy can tell us which things are and are not reproducible, by the sharpness of its predictions). Maximum entropy separates out what is relevant for predicting reproducible phenomena, and discards what is irrelevant (we saw this even in the example of Wolf's die where, surely, the only reproducible events in his sample space of $6^{20,000}$ points were the six face frequencies or functions of them; just the things that maximum entropy predicted).

Likewise, in the stage 2 techniques of prediction from the maximum entropy distribution, if we are not interested in every question about reproducible effects, but only some "relevant subset" of them, we may seek a further elimination of details that are irrelevant to those particular questions. But this kind of problem has come up before in mathematical physics; and Dicke (1946) introduced an elegant projection operator technique for calculating desired external elements of a scattering matrix while discarding irrelevant internal details. Our present problem, although entirely different in physical and mathematical

WHERE DO WE STAND ON MAXIMUM ENTROPY? 309

details, is practically identical formally; and so this same technique must be applicable.

Zwanzig (1962) introduced projection operators for dealing with two interacting systems, only one of which is of interest, the other serving only as its "heat bath." Robertson (1964) recognized that this will work equally well for any kind of separation, not necessarily spatial; i.e., if we want to predict only the behavior of a few physical quantities $\{F_1...F_m\}$ we can introduce a projection operator P which throws away everything that is irrelevant for predicting those particular things; allowing, in effect, everything else to serve as a kind of "heat bath" for them.

In the statistical theory this dichotomy may be viewed in another way: instead of "relevant" and "irrelevant" read "systematic" and "random." Then, referring to Robertson's presentation in this volume, it is seen that Eq. (9.3), which could be taken as the definition of his projection operator P(t), is formally just the same as the solution of the problem of "reduction of equations of condition" given by Laplace for the optimal estimate of systematic effects. A modern version can be found in statistics textbooks, under the heading: "multiple regression." Likewise, his formulas (9.8), (9.9) for the "subtracted correlation functions" have a close formal correspondence to Dicke's final formulas.

Of course, this introduction of projection operators is not absolutely required by basic principles; it is in the realm of art, and any work of art may be executed in more than one way. All kinds of changes in detail may still be thought of; but needless to say, most of them have been thought of already, investigated, and quietly dropped. Seeing how far Robertson has now carried this approach, and how many nice results he has uncovered, it is pretty clear that anyone who wants to do it differently has his work cut out for him.

Finally, I should prepare the reader for his contribution. When Baldwin was a student of mine in the early 1960's, I learned that he has the same trait that Lagrange and Fermi showed in their early works: he takes delight in inventing tricky variational arguments, which seem at first glance totally wrong. After long, deep thought it always developed that what he did was correct after all. A beautiful example is his derivation of (4.3), where most readers would leave the track without this hint from someone with experience in reading his works: you are not allowed to take the trace and thus prove that $a = 1$, invalidating (4.4), because this is a formal argument in which the symbol <F> stands for $Tr(F\sigma)$ even when σ is not normalized to $Tr(\sigma) = 1$. For a similar reason, you are not allowed to protest that if $F_0 \equiv 1$, then $\delta <F_0> \equiv 0$. One of my mathematics professors once threw our class into an uproar by the same trick; evaluating an integral, correctly, by differentiating with respect to π. For those with a taste for subtle trickery, variational mathematics is the most fun of all.

References
T. Bayes (1763), Phil. Trans. Roy. Soc. 330-418. Reprint, with biographical note by G. A. Barnard, in Biometrika 45, 293 (1958) and in Studies in the History of Statistics and Probability, E. S. Pearson and M. G. Kendall, editors, C. R. Griffin and Co. Ltd., London (1970). Also reprinted in Two Papers by Bayes with Commentaries, W. E. Deming, editor, Hafner Publishing Co., New York (1963).

E. T. Bell (1937), Men of Mathematics, Simon and Schuster, N.Y.

L. Boltzmann (1871), Wien. Ber. 63, 397, 679, 712.

G. Boole (1854), The Laws of Thought, Reprinted by Dover Publications, Inc., New York.

R. T. Cox (1946), Am. J. Phys. 17, 1. See also The Algebra of Probable Inference, Johns Hopkins University Press (1961); reviewed by E. T. Jaynes in Am. J. Phys. 31, 66 (1963).

E. Czuber (1908), Wahrscheinlichkeitsrechnung und Ihre Anwendung auf Fehlerausgleichung, Teubner, Berlin. Two Volumes.

P. and T. Ehrenfest (1912), Encykl. Math. Wiss. English translation by M. J. Moravcsik, The Conceptual Foundations of the Statistical Approach in Mechanics. Cornell University Press, Ithaca, N. Y. (1959).

P. L. Ellis (1842), "On the Foundations of the Theory of Probability", Camb. Phil. Soc. vol. viii. The quotation given by Rowlinson actually appears in a later work: Phil. Mag. vol. xxxvii (1850). Both are reprinted in his Mathematical and other Writings (1863).

K. Friedman and A. Shimony (1971), J. Stat. Phys. 3, 381. See also M. Tribus and H. Motroni, ibid. 4, 227; A. Hobson, ibid. 6, 189; D. Gage and D. Hestenes, ibid. 7, 89; A. Shimony, ibid. 9, 187; K. Friedman, ibid. 9, 265.

J. W. Gibbs (1902), Elementary Principles in Statistical Mechanics, Reprinted in Collected works and commentary, Yale University Press (1936), and by Dover Publications, Inc. (1960).

I. J. Good (1950), Probability and the Weighing of Evidence, C. Griffin and Co., London.

D. Heath and Wm. Sudderth (1976), "de Finetti's Theorem on Exchangeable Variables," The American Statistician, 30, 188.

d. ter Haar (1954), Elements of Statistical Mechanics, Rinehart and Co., New York.

S. P. Heims and E. T. Jaynes (1962), Rev. Mod. Phys. 34, 143.

V. S. Huzurbazar (1976), Sufficient Statistics (Volume 19 in Statistics Textbooks and Monographs Series) Marcel Dekker, Inc., New York.

E. T. Jaynes (1957a,b), Phys. Rev. 106, 620; 108, 171.

E. T. Jaynes (1963a), "New Engineering Applications of Information Theory," in Proceedings of the First Symposium on Engineering Applications of Random Function Theory and Probability, J. L. Bogdanoff and F. Kozin, editors; J. Wiley and Sons, New York, pp. 163-203.

E. T. Jaynes (1963b), "Information Theory and Statistical Mechanics," in Statistical Physics (1962 Brandeis Lectures) K. W. Ford, ed. W. A. Benjamin, Inc., New York, pp. 181-218.

E. T. Jaynes (1965), "Gibbs vs. Boltzmann Entropies," Am. J. Phys. 33, 391.

E. T. Jaynes (1967), "Foundations of Probability Theory and Statistical Mechanics," in Delaware Seminar in the Foundations of Physics, M. Bunge, ed., Springer-Verlag, Berlin; pp. 77-101.

E. T. Jaynes, (1968), "Prior Probabilities," IEEE Trans. on Systems Science and Cybernetics, SSC-4, pp. 227-241. Reprinted in Concepts and Applications of Modern Decision Models, V. M. Rao Tummala and R. C. Henshaw, eds., (Michigan State University Business Studies Series, 1976).

E. T. Jaynes (1971), "Violation of Boltzmann's H-theorem in Real Gases," Phys. Rev. A 4, 747.

E. T. Jaynes (1973), "The Well-Posed Problem," Found. Phys. 3, 477.

E. T. Jaynes (1976), "Confidence Intervals vs. Bayesian Interv Intervals," in Foundations of Probability Theory, Statistical Inference, and Statistical Theories of Science, W. L. Harper and C. A. Hooker, eds., D. Reidel Publishing Co., Dordrecht-Holland, pp. 175-257.

E. T. Jaynes (1978), "Electrodynamics Today," in Proceedings of the Fourth Rochester Conference on Coherence and Quantum Optics, L. Mandel and E. Wolf, eds., Plenum Press, New York.

E. T. Jaynes (1979), "Marginalization and Prior Probabilities," in Studies in Bayesian Econometrics and Statistics, A. Zellner, ed., North-Holland Publishing Co., Amsterdam (in press).

H. Jeffries (1939), Theory of Probability, Oxford University Press.

J. M. Keynes (1921), A Treatise on Probability, MacMillan and Co., London.

R. Kubo (1957), J. Phys. Soc. (Japan) 12, 570.

R. Kubo (1958), "Some Aspects of the Statistical-Mechanical Theory of Irreversible Processes," in Lectures in Theoretical Physics, Vol. 1, W. E. Brittin and L. G. Dunham, eds, Interscience Publishers, Inc., New York; pp. 120-203.

S. Kullback (1959), Information Theory and Statistics, J. Wiley and Sons, Inc.

J. Lindhard (1974), "On the Theory of Measurement and Its Consequences in Statistical Dynamics," Kgl. Danske. Vid. Selskab Mat.-Fys. Medd. 39, 1.

J. C. Maxwell (1859), "Illustrations of the Dynamical Theory of Gases," Collected Works, W. D. Niven, ed., London (1890), Vol. I, pp. 377-409.

J. C. Maxwell (1876), "On Boltzmann's Theorem on the Average Distribution of Energy in a System of Material Points," ibid. Vol. II, pp. 713-741.

D. Middleton (1960), Introduction to Statistical Communication Theory, McGraw-Hill Book Company, Inc., New York, pp. 1082-1102.

WHERE DO WE STAND ON MAXIMUM ENTROPY? 313

H. Mori, I. Oppenheim, and J. Ross (1962), "Some Topics in Quantum Statistics," in Studies in Statistical Mechanics, Volume 1, J. de Boer and G. E. Uhlenbeck, eds., North-Holland Publishing Company, Amsterdam; pp. 213-298.

G. Polya (1954), Mathematics and Plausible Reasoning. Princeton University Press, two volumes.

B. Robertson (1964), Stanford University, Ph.D. Thesis.

J. S. Rowlinson (1970), "Probability, Information, and Entropy," Nature, 225, 1196.

L. J. Savage (1954), The Foundations of Statistics, John Wiley and Sons, Inc., Revised Edition by Dover Publications, Inc., 1972.

E. Schrodinger (1948), Statistical Thermodynamics, Cambridge University Press.

C. Shannon (1948), Bell System Tech. J. 27, 379, 623. Reprinted in C. E. Shannon and W. Weaver, The Mathematical Theory of Communication, University of Illinois Press, Urbana (1949).

C. Truesdell (1960), "Ergodic Theory in Classical Statistical Mechanics," in Ergodic Theories, Proceedings of the International School of Physics "Enrico Fermi," P. Caldirola, ed., Academic Press, New York, pp. 21-56.

J. Venn (1866), The Logic of Chance

R. von Mises (1928), Probability, Statistics and Truth, J. Springer. Revised English Edition, H. Geiringer, ed., George Allen and Unwin Ltd., London (1957).

A. Wald (1950), Statistical Decision Functions, John Wiley and Sons, Inc., New York.

J. D. Watson (1965), Molecular Biology of the Gene, W. A. Benjamin, Inc., New York. See particularly pp. 73-100.

N. Wiener (1949), Extrapolation, Interpolation, and Smoothing of Stationary Time Series, Technology Press, M.I.T.

S. S. Wilks (1961), Forward to Geralomo Cardano, The Book on Games of Chance, translated by S. H. Gould. Holt, Rinehart, and Winston, New York.

R. Zwanzig (1960), J. Chem. Phys. 33, 1338; Phys. Rev. 124.
983 (1961).

Additional Reference

Dicke, R. H., J. Applied Phys. 18 (1947), p. 873.

11. CONCENTRATION OF DISTRIBUTIONS AT ENTROPY MAXIMA (1979)

After sending off the 'Where Do We Stand?' manuscript for the MIT Proceedings Volume, my thoughts returned to Rudolph Wolf's dice data. The hastily concocted argument for five degree of freedom instead of three bothered me a bit, and I decided to calculate how much Chi-squared could have been reduced if the parameters λ_1, λ_2 had been chosen to minimize it, rather than from the *Maxent* rule. But first, as a check on the old computer program, I recalculated the entropies of those various distributions. Quite by accident, I noticed that the numerical values of $N\Delta H$ were in every case just half the values of Chi-squared in the MIT article.

The light suddenly dawned: my calculations of Chi-squared were foolish. There was no need to work out the squares of the residuals and compute their weighted sum; it seemed that the entropy of a *Maxent* distribution, determined immediately from the partition function, already contained that information.

Then a little analytical work, to understand why this connection existed, brought forth the Concentration Theorem. Result: all the tedious analysis of Wolf's dice data in the MIT volume could now be reproduced, more accurately and more meaningfully, in a few lines. To test any number of theories concerning which constraints were present, against Wolf's data, we need only compare their predicted entropies to the entropy of the data.

But this better method also revealed that I had been wrong in arguing for five degrees of freedom. Although it was correct to say that the parameters were not chosen by the criterion of minimizing Chi-squared, nevertheless the analysis showed unequivocally that the distribution with three degrees of freedom was the correct one determining the distribution of entropy over class C — chalk up another for Fisher. Therefore, instead of reporting that Wolf's data were not quite significant at the 5% level, I must now report that they were slightly significant, and do contain evidence for a third, very tiny, imperfection.

In September 1981 I visited Zürich, saw the observatory where Rudolph Wolf had worked, and thanks to the kindness of Dr. Lucien Preuss and Dr. Hans Primas, came away with more detailed evidence concerning this now famous die. It will be reported in due course.

Nevertheless, it is fortunate that I did not know of this more efficient method in the MIT volume. The plodding, familiar type of analysis given there will be far more convincing to almost every reader. The efficient procedure described below is just too slick and subtle to be grasped immediately; it gives that impression of 'getting something for nothing' to anyone who does not really understand what is happening. Knowing how many years it took for the bare *Maxent* principle to be comprehended by persons who accused me of trying to get something for nothing out of it, I expect that some time will pass before the principle of hypothesis testing by comparing entropies will be usable in public.

CONCENTRATION OF DISTRIBUTIONS AT ENTROPY MAXIMA[†]

E. T. Jaynes

Department of Physics, Washington University
St. Louis, Missouri 63130, U.S.A.

1. INTRODUCTION

It has long been recognized, or conjectured, that the notion of entropy defines a kind of measure on the space of probability distributions such that those of high entropy are in some sense favored over others. The basis for this was stated first in a variety of intuitive forms: that distributions of high entropy represent greater "disorder," that they are "smoother," that they are "more probable," that they "assume less" according to Shannon's interpretation of entropy as an information measure, etc. While each of these doubtless expresses an element of truth, none seems explicit enough to lend itself to a quantitative demonstration. This alone, however, has not prevented the useful exploitation of this property of entropy.

In many statistical problems we have information which places some kind of restriction on a probability distribution without completely determining it. If, given two distributions that agree equally well with the information at hand, we prefer the one with greater entropy, then the distribution with the maximum entropy compatible with our information will be the most favored of all. Thus conversion of prior information into a definite prior probability assignment becomes a variational problem in which the prior information plays the role of constraint.

[†]To be presented at the 19'th NBER-NSF Seminar on Bayesian Statistics, Montreal, October 1979.

But while this Principle of Maximum Entropy has an established usefulness in a variety of applications, it has left an unanswered question in the minds of many. Granted that the distribution of maximum entropy has a favored status, in exactly what sense, and how strongly, are alternative distributions of lower entropy ruled out?

Probably most information theorists have considered it obvious that, in some sense, the possible distributions are concentrated strongly near the one of maximum entropy; i.e., that distributions with appreciably lower entropy than the maximum are atypical of those allowed by the constraints.

Likewise, Schrödinger (1948) noted that this is the reason why, in statistical mechanics, the Darwin-Fowler method and the Boltzmann "method of the most probable distribution" lead to the same result in the limit $N \to \infty$, where N is a suitable "size" parameter (i.e., in statistical mechanics the number of particles in a system; in communication theory the number of symbols in a message; in statistical inference the number of trials of a random experiment). A general proof of this limiting form (i.e., a generalized Darwin-Fowler theorem) is given by van Campenhout and Cover (1979).

But these results, pertaining only to the limiting distribution, leave us in the same unsatisfactory state as did the original limit theorem of Jacob Bernoulli (1713): {as $N \to \infty$, the observable frequency f = r/N of successes converges in probability to p}. This said nothing about how large N must be for a given accuracy. For applications one needed the more explicit de Moivre-Laplace theorem: {Asymptotically, $f \sim N(p,\sigma)$ where $\sigma^2 = N^{-1} p(1-p)$}.

CONCENTRATION OF DISTRIBUTIONS 319

Similarly, in our present problem it would be desirable to have a quantitative demonstration of this entropy concentration phenomenon for finite N, so that one can see just how the limit is approached. This is so particularly because there are still some who, apparently unaware or unconvinced of the reality of the phenomenon, reject the Principle of Maximum Entropy as a method of inference.

This problem was discussed at the M.I.T. Maximum Entropy Formalism Conference of May 1978, in connection with some alternative solutions that had been proposed for maximum entropy problems. The result was a lengthy but awkward and unsatisfactory analysis (Jaynes, 1978) in which real insight into the problem had not yet been achieved. We give here a simpler, more accurate, and more general treatment of entropy concentration.

The general Principle of Maximum Entropy is applicable to any problem of inference with a well-defined hypothesis space but incomplete information, whether or not it involves a repetitive situation such as a random experiment. However, we consider below only the special applications where we use entropy as a criterion for (1) estimating frequencies in a random experiment about which incomplete information is available; or (2) testing hypotheses about systematic effects in experiments where frequency data are available.

The second application is illustrated by analyzing the famous dice data of R. Wolf. We show how entropy analysis enables one to draw conclusions about the specific physical imperfections that must have been present (not knowing whether those dice are still in existence, so that our conclusions might be checked directly).

2. ENTROPY CONCENTRATION THEOREM

A random experiment has n possible results at each trial; thus in N trials there are n^N conceivable outcomes (we use the word "result" for a single trial, while "outcome" refers to the experiment as a whole; thus one outcome consists of an enumeration of N results, including their order). Each outcome yields a set of sample numbers $\{N_i\}$ and frequencies $\{f_i = N_i/N, 1 \leq i \leq n\}$, with an entropy

$$H(f_1 \ldots f_n) = - \sum_{i=1}^{n} f_i \log f_i . \qquad (1)$$

Consider the subclass C of all possible outcomes that could be observed in N trials, compatible with m linearly independent constraints (m < n) of the form

$$\sum_{i=1}^{n} A_{ji} f_i = D_j , \quad (1 \leq j \leq m) . \qquad (2)$$

The conceptual interpretation is that m different "physical quantities" have been measured, the matrix A_{ji} defines their "nature," and D_j are the particular "data" for the case under study. These data tell us that the actual outcome must have been in class C, but are insufficient to determine the frequencies $\{f_i\}$. We examine the combinatorial basis for using--and the consequences of failing to use--the entropy (1) as a criterion for estimating the $\{f_i\}$.

Although it is not needed for this purpose, we note that in a real application one will wish, if possible, to choose the constraint matrix A_{ji} so that the resulting quantities D_j represent systematic physical influences, real or conjectured, (for example, eccentric position of the center of gravity of a die), which constrain the frequencies to be different from the uniform distribution of absolute maximum entropy $H_0 = \log n$. In using entropy analysis for hypothesis testing, the mathematical relations are used in the other direction, considering the $\{f_i\}$ as known experimentally. A successful hypothesis about the systematic influences is then one for which the experimentally observed entropy (1) is sufficiently close to the maximum H_{max} permitted by the assumed constraints (2), "sufficiently close" being defined by the following concentration theorem.

A certain fraction F of the outcomes in class C will yield an entropy in the range

$$H_{max} - \Delta H \leq H(f_1 \ldots f_n) \leq H_{max} \tag{3}$$

where H_{max} may be determined by the following algorithm: define the partition function

$$Z(\lambda_1 \ldots \lambda_m) \equiv \sum_{i=1}^{n} \exp\left(-\sum_{j=1}^{m} \lambda_j A_{ji}\right) . \tag{4}$$

Then

$$H_{max} = \log Z + \sum_{j=1}^{m} \lambda_j D_j \tag{5}$$

in which the Lagrange multipliers $\{\lambda_j\}$ are found from

$$\frac{\partial}{\partial \lambda_j} \log Z + D_j = 0 , \quad (1 \leq j \leq m) \tag{6}$$

a set of m simultaneous equations for m unknowns. The frequency distribution which has this maximum entropy is then

$$f_i = Z^{-1} \exp\left(-\sum_j \lambda_j A_{ji}\right) , \quad (1 \leq i \leq n) . \tag{7}$$

Other distributions $\{f'_i\}$ allowed by the constraints (2) will have various entropies less than H_{max}. Their concentration near this upper bound (i.e., the functional relation connecting F and ΔH) is given by the <u>Concentration</u> <u>Theorem:</u> Asymptotically, $2N \Delta H$ is distributed over class C as Chi-squared with $k = n - m - 1$ degrees of freedom, independently of the nature of the constraints. That is, denoting the critical Chi-squared for k degrees of freedom at the 100 P % significance level by $\chi_k^2(P)$, ΔH is given in terms of the upper tail area (1-F) by

$$2N \Delta H = \chi_k^2(1-F) \qquad (8)$$

The proof is relegated to the Appendix, since it consists of little more than repeating <u>mutatis</u> <u>mutandis</u> Karl Pearson's original derivation of the Chi-squared distribution, taking note of the reduction of dimensionality due to constraints. Note that the theorem is combinatorial, expressing only a counting of the <u>possibilities;</u> it does not become a statement of <u>probabilities</u> unless one assigns equal probability to each outcome in class C.

3. EXAMPLES: FREQUENCY ESTIMATION

We illustrate the meaning and use of this result by a much-discussed example. Suppose a die is tossed N = 1000 times and we are told only that the average number of spots up was not 3.5 as we might expect from a "true" die, but 4.5, i.e.,

$$\sum_{i=1}^{6} i\, f_i = 4.5 \qquad (9)$$

which is a special case of (2). Given this information and nothing else (i.e., not making use of any additional information that you or I might get from inspection of the die or from past experience with dice in general), what estimates should we make of the frequencies $\{f_i\}$ with which the different faces appeared? This is a kind of caricature of a class of real problems that arises constantly in physical applications.

The distribution which has maximum entropy subject to the constraint (9) is given by (4)-(7) with $n = 6$, $m = 1$, $A_{ji} = i$, $Z(\lambda) = (e^{-\lambda} + \ldots + e^{-6\lambda})$, $\lambda = -0.37105$. The result, derived in more detail before (Jaynes, 1978), is

$$\{f_1 \ldots f_6\} = \{0.0543, 0.0788, 0.1142, 0.1654, 0.2398, 0.3475\} \qquad (10)$$

and it has entropy

$$H_{max} = 1.61358 \qquad (11)$$

as compared to the value $\log_e 6 = 1.79176$, corresponding to no constraint and a uniform distribution.

Applying the concentration theorem, we have $6-1-1=4$ degrees of freedom; entering the Chi-squared tables at the conventional 5% significance level, we find that 95% of all <u>possible</u> outcomes allowed by the constraint (9) have entropy in the range (3) of width $\Delta H = (2N)^{-1} \chi_4^2(0.05) = 0.00474$; or, to sufficient accuracy,

$$1.609 \leq H \leq 1.614. \tag{12}$$

Thus on the "null hypothesis" which supposes that no further systematic influence is operative in the experiment other than the one taken into account (i.e., which assigns equal probability to all outcomes in class C), there is less than a 5% chance of seeing a frequency distribution with entropy outside the interval (12).

A remarkable feature is that the "95% concentration range"

$$H_{max} - \frac{4.74}{N} \leq H \leq H_{max} \tag{13}$$

is valid asymptotically for any random experiment with four degrees of freedom, although the value of H_{max} may vary widely with other details.

More interesting numerical results are found at more extreme significance levels. Thus, in any experiment with 1000 trials and four degrees of freedom, 99.99% of all outcomes allowed by the constraints have entropy in a range of width $\Delta H = (2N)^{-1} \chi_4^2(0.0001) = 0.012$. In the above example this is

$$1.602 \leq H \leq 1.614 \tag{14}$$

and only one in 10^8 of the possible outcomes has entropy below the range

$$1.592 \leq H \leq 1.614 \tag{15}$$

Thus, given certain incomplete information, the distribution of maximum entropy is not only the one that can be realized in the greatest number of ways; in fact, for large N the overwhelming majority of all possible distributions compatible with our information have entropy very close to the maximum.

Note that the width of this region of concentration goes down like N^{-1}; and not like $N^{-1/2}$ as one might have guessed. Thus, in 20,000 tosses agreeing with (9), 95 percent of the possible outcomes have entropy in the interval $(1.61334 < H < 1.61358)$ and only one in 10^8 has $H < 1.61253$. As $N \to \infty$, any frequency distribution other than the one of maximum entropy thus becomes highly atypical of those allowed by the constraints.

Even more interesting numbers are readily found. Rowlinson (1970) rejected the principle of maximum entropy for this problem, and proposed as an alternative solution in place of (10) the binomial distribution

$$f'_i = \binom{5}{i-1} p^{i-1} (1-p)^{6-i} \quad , \quad 1 \le i \le 6 \qquad (16)$$

which also satisfies the constraint (9) if $p = 0.7$. But the distribution (16) has entropy $H' = 1.4136 = H_{max} - 0.200$, far below the limit (15). We now have $2N \, \Delta H = 400 = \chi_4^2(1-F)$; or from (A8),

$$1 - F \simeq 2.94 \times 10^{-84} \quad . \qquad (17)$$

This indicates that in 1000 tosses, less than one in 10^{83} of the outcomes compatible with the constraint (9) have entropy as low as H'.

But the concentration theorem is valid only asymptotically, because of the approximation (A4) made in its derivation; and even for N = 1000 we might distrust its numerical accuracy that far out in the tail of the distribution. However, we can check the magnitude of (17) by direct counting.

The number of ways W in which a specific set of sample numbers $\{N_1 \ldots N_6\}$ can be realized is given by the multinomial coefficient (A1). The asymptotic formula (A3) for the ratio W/W' (which is free from any errors that might result from the aforementioned approximation) says that, for every way in which the binomial distribution (16) can be realized, there are about $\exp(N \Delta H) \simeq \exp(200)$, or more than 10^{86} ways, in which the maximum-entropy distribution (10) can be realized (about 10^{62} ways for every microsecond in the age of the universe). While this result does not take into account the volume element factors (r^{k-1} dr) of the full concentration theorem, it does indicate that (17) did not mislead us.

Even if we come down to N = 50, we find the following. The sample numbers which agree most closely with (10), (16) while summing to $\Sigma N_k = 50$ are $\{N_k\} = \{3,4,6,8,12,17\}$ and $\{N_k'\} = \{0,1,7,16,18,8\}$ respectively. With such small numbers, we no longer need asymptotic formulas; for every way in which the distribution $\{N_k'\}$ can be realized, there are exactly W/W' = (7!16!18!)/(3!4!6!12!17!) = 38,220 ways in which the maximum-entropy distribution $\{N_k\}$ can be realized.

Such numbers illustrate rather clearly just what we are accomplishing when we maximize entropy. If our data do not fully determine a distribution $\{f_i\}$, it is prudent to adopt, for purposes of inference, that distribution which has maximum entropy subject to the data we do have.

CONCENTRATION OF DISTRIBUTIONS

4. HYPOTHESIS TESTING: WOLF'S DICE DATA

The Swiss astronomer Rudolph Wolf (1816-1893; best known today as the discoverer of the correlation between terrestrial magnetic disturbances and sunspot activity) performed a number of random experiments, conducted with great care, presumably to check the validity of statistical theory. An account with references is given by Czuber (1908).

In one of these experiments, a red and white die were tossed together 20,000 times in a way that precluded any systematic favoring of any face over any other. The resulting 36 joint sample numbers are given in Table 1 (taken from Czuber).

Table 1. Wolf's Dice Data

		White Die						Row Total
		1	2	3	4	5	5	
Red Die	1	547	587	500	462	621	690	3407
	2	609	655	497	535	651	684	3631
	3	514	540	468	438	587	629	3176
	4	462	507	414	413	509	611	2916
	5	551	562	499	506	658	672	3448
	6	563	598	519	487	609	646	3422
	Column Total	3246	3449	2897	2841	3635	3932	20000

These are the sample numbers $\{N_i, 1 \leq i \leq n\}$ of a random experiment with n = 36 possible results at each trial. On the null hypothesis which assigns uniform probabilities $p = n^{-1} = 1/36$, the expectation and standard deviation of any sample number are $Np = 555.55$, $\sigma = [Np(1-p)]^{\frac{1}{2}} = 23.24$ respectively.

Czuber, writing in the days when commonly understood statistical inference consisted of little more than fitting by least squares, compared σ with the observed mean-square deviation

$$\left[n^{-1} \Sigma(N_i - Np)^2\right]^{\frac{1}{2}} = 76.87 \tag{18}$$

and concluded only that the null hypothesis must have been wrong; "die Würfelseiten nicht als gleichmögliche Fälle sich darstellen."

Keynes (1921) also cited Wolf's dice data, but did even less with them. Noting only that agreement with predictions of the null hypothesis was atrocious, he concludes: "This, then, is the sole conclusion of these immensely laborious experiments, -- that Wolf's dice were very ill made. Indeed, the experiments could have had no bearing except upon the accuracy of his dice." This appears to be an outstanding example of blindness to an important result staring one in the face. Why did he not see in such "bad" data a golden opportunity for further analysis, that would have been lost had Wolf worked with perfect dice and produced the kind of data expected of him?

Today, another sixty years have passed, and to the best of the writer's knowledge no statistician has ever attempted to draw any specific inferences about the imperfections in Wolf's dice from these data. Yet to a physicist those data tell us something very clear and simple about his dice; informatic that can be extracted by a straightforward entropy analysis that does not require us to go into complicated mechanical details.

Ludwig Boltzmann, writing thirty years before Czuber and about six years before Wolf's experiment, had given the principle by which this analysis may be carried out; and J. Willard Gibbs, writing six years before Czuber, had developed the resulting mathematical apparatus to a high degree of perfection. Yet today, 100 years after Boltzmann's work, it still seems generally believed that the principles of statistical mechanics apply only to molecules; and not to dice.

We do not expect, and Wolf's data do not give evidence for, any correlations between the results of the two dice. Therefore, the import of the data for our purposes is contained in the marginal totals. The observed frequencies $\{f_i\}$ and their deviations $\{\Delta_i = f_i - 1/6\}$ from the null hypothesis prediction are given in Table 2.

Table 2. Wolf's Marginal Frequencies

i	Red Die f_i	Δ_i	White Die f_i	Δ_i
1	0.17035	+0.00368	0.16230	-0.00437
2	0.18155	+0.01488	0.17245	+0.00578
3	0.15880	-0.00787	0.14485	-0.02182
4	0.14580	-0.02087	0.14205	-0.02464
5	0.17240	+0.00573	0.18175	+0.01508
6	0.17110	+0.00443	0.19960	+0.02993

On the null hypothesis that the dice were true, the standard deviations of $\{f_i\}$ from $p = 1/6$ should be $\sigma = [p(1-p)/N]^{1/2} = 0.0026$. The observed deviations Δ_i are many times this amount.

Now let us judge the deviation by the entropy criterion, considering only the white die. The entropy of the observed distribution lies below the maximum, log 6, by

$$\log 6 = 1.791\ 759$$
$$H_{Wolf} = 1.784\ 990$$
$$\Delta H = 0.006\ 769$$

which looks rather small; but this is for N = 20,000 trials. As a "quick and dirty" estimate based on (A3) we find $\exp(N\Delta H) \simeq 6 \times 10^{58}$, indicating an unmistakably strong constraint (i.e., systematic influence) keeping the frequencies away from the uniform distribution that could happen in the greatest number of ways if the die were equally free to settle in all positions.

The more precise concentration theorem gives

$$2N \ \Delta H = 270.1 = \chi_5^2(1 - F) \qquad (19)$$

and therefore, from (A8),

$$1 - F = 1.07 \times 10^{-56} \ . \qquad (20)$$

Only one in 10^{56} of the 6^N conceivable outcomes has an entropy as low as Wolf's data give.

In Jaynes (1978) we considered what specific imperfections one might expect to find in a die, that might tend to make the frequencies nonuniform. The two most obvious are (1) a shift of the center of gravity due to the mass of ivory excavated from the spots, which being proportional to the number of spots on any side, should make the quantity $\{f_1(i) \equiv i - 3.5, 1 \leq i \leq 6\}$ have a nonzero expectation; and (2) errors in trying to machine a perfect cube, which will tend to make one dimension (the last side cut) slightly different from the other two. It is clear from the data that Wolf's white die gave a lower frequency for the faces (3,4); and therefore that the (3-4) dimension was undoubtedly greater than the (1-6) or (2-5) ones. The effect of this is that the function

$$f_2(i) = \begin{cases} +1, & i = 1,2,5,6 \\ -2, & i = 3,4 \end{cases} \qquad (21)$$

has a non-zero expectation. The strength of these two systematic influences is indicated by Wolf's measured averages for them:

$$\overline{f}_1 = 0.0983 , \qquad \overline{f}_2 = 0.1393 \qquad (22)$$

Now if these are the only two imperfections present, we expect that the die will be equally free to yield any outcome compatible with the constraints (22). Therefore the observed frequencies should be the ones that can be realized in the greatest number of ways while agreeing with (22); i.e., which has maximum entropy subject to these two constraints. On the other hand, if the entropy of the observed distribution is appreciably below the maximum allowed by (22), that would be evidence that there is still another imperfection present; i.e., a third systematic influence not yet taken into account.

The maximum entropy H_{max} allowed by (22) was calculated in Jaynes (1978) by the algorithm (4)-(7), with the result indicated below:

$$H_{max} = 1.785\ 225$$
$$\underline{H_{Wolf} = 1.784\ 990}$$
$$\Delta H = 0.000\ 235$$

The discrepancy is reduced by nearly a factor of thirty. The concentration theorem now gives

$$2N\ \Delta H = 9.38 = \chi_3^2(1 - F) \qquad (23)$$

or

$$1 - F = 0.025 \qquad (24)$$

The result appears just barely significant. That is, 97.5 percent of all outcomes compatible with (22) have an entropy greater than observed by Wolf. To assume a further very tiny imperfection [the (2-3-6) corner chipped off] we could make even this discrepancy disappear; but in view of the great number of trials one will probably not consider the result (24) as sufficiently strong evidence for this.

5. CONCLUSION

In Jaynes (1978) we gave a much more lengthy analysis, using the conventional Chi-squared test but arriving at less detailed and less accurate conclusions. At that time, in ignorance of the concentration theorem, it was not realized that there is no need to carry out the laborious computation of Chi-squared from the observed deviations Δ_i; the discrepancy between the observed entropy and that allowed by the hypothesis is already a more precise measure of significance.

We now see that the single maximum entropy formalism defined by (1) - (7) provides not only the procedure for predicting frequencies when incomplete data are available, that is optimal by a certain well-defined criterion; but also the criterion for testing hypotheses about systematic influences when frequency data are at hand.

APPENDIX

In N trials of the aforementioned random experiment, the i'th result occurs $N_i = N f_i$ times, $1 \leq i \leq n$. Out of the n^N conceivable outcomes, the number which yield a particular set of frequencies $\{f_i\}$ is

$$W(f_1 \ldots f_n) \equiv \frac{N!}{(Nf_1!) \ldots (Nf_n)!} \tag{A1}$$

and as $N \to \infty$ we have by the Stirling approximation

$$N^{-1} \log W \longrightarrow H(f_1 \ldots f_n) \quad , \tag{A2}$$

the entropy function (1). Given two sets of frequencies $\{f_i\}$ and $\{f_i'\}$, the ratio (number of ways f_i can be realized)/(number of ways f_i' can be realized) is asymptotically

$$\frac{W}{W'} \sim A\, e^{N(H-H')} \left[1 + \frac{B}{12N} + O(N^{-2})\right] \tag{A3}$$

where

$$A \equiv \prod_i (f_i'/f_i)^{\frac{1}{2}}$$

$$B \equiv \sum_i (f_i - f_i')/f_i f_i' \tag{A4}$$

represent corrections from the higher terms in the Stirling approximation. Their variation with $\{f_i\}$ is, of course, completely overwhelmed by that of the factor exp $N(H-H')$.

The conceivable frequencies $\{f_1 \ldots f_n\}$ may be regarded as cartesian coordinates of a point P in an n-dimensional space, restricted to $\{S: 0 \leq f_i, \Sigma f_i = 1\}$, an (n-1)-dimensional convex set whose vertices are the n points

$\{f_i = 1, 1 \leq i \leq n\}$. On S, the entropy (1) varies continuously, taking on all values in $(0 \leq H(P) \leq \log n)$ as P moves from a vertex to the center.

But now we obtain information that imposes the m linearly independent constraints (2), which define an (n-m)-dimensional hyperplane M. P is now confined to the intersection $S' = M \cap S$, a closed set comprising a bounded portion of hyperplane of dimensionality $k = n - m - 1$.

On S' the entropy attains a maximum $H_{max} \leq \log n$. That this is attained at a unique point of S' may be proved analytically, but is perhaps made obvious as follows. Since any "mixing" increases the entropy, the set $\{S_x: P \in S, H(P) \geq x\}$ is strictly convex. Entropy maximization with constraints linear in $\{f_i\}$ thus amounts to finding the value of $x = H_{max}$ for which S' is a supporting tangent plane to S_x.

After these preliminaries, our argument follows slavishly the original derivation by Karl Pearson, as recalled by Lancaster (1969). In S' we may define new coordinates $\{x_1 \ldots x_k\}$ as appropriate linear functions of $\{f_1 \ldots f_n\}$ such that the new origin is at the maximum-entropy point, and there is a distance $r = (\Sigma x_i^2)^{1/2}$ such that near the origin a power series expansion yields

$$H(P) = H_{max} - a r^2 + \ldots , \qquad a > 0 . \tag{A4}$$

We then have a volume element in S' proportional to r^{k-1} dr. The domain of all possible frequency distributions $\{f_1 \ldots f_n\}$ which satisfy the constraints and whose entropy is in the range (3) is a k-sphere of radius R, given by $aR^2 = \Delta H$.

In N trials this sphere contains a fraction F of all possible outcomes in class C. From (A2), (A4) this is given asymptotically by

$$F \sim I(R)/I(\infty) \tag{A5}$$

where

$$I(R) \equiv \int_0^R e^{-Nar^2} r^{k-1} dr . \tag{A6}$$

But, setting $NaR^2 = N \Delta H = (1/2)\chi^2$, this is just the cumulative Chi-squared distribution with k degrees of freedom; in conventional notation the relation between ΔH and F is given by Eq. (4).

In our applications we are generally concerned with numerical values for large $N \Delta H$, beyond the range of tables. The Chi-squared distribution $F(N \Delta H)$ may be expressed analytically as

$$F(x) = \frac{1}{s!} \int_0^x t^s e^{-t} dt \tag{A7}$$

where $s = (k/2) - 1$. For large $x = N \Delta H$, this yields the asymptotic expansion

$$1 - F(x) \sim (s!)^{-1} x^s e^{-x} [1 + sx^{-1} + s(s-1)x^{-2} + ...] \tag{A8}$$

When s is an integer (k even) this terminates and gives the exact result. Most of the numerical results cited in the text have been obtained from (A8).

REFERENCES

E. Czuber (1908), Wahrscheinlichkeitsrechnung, Teubner, Berlin; Vol. I, pp. 149-151.

E. T. Jaynes (1978), in The Maximum Entropy Formalism, R. D. Levine and M. Tribus, Editors, M.I.T. Press, Cambridge, Mass.

H. O. Lancaster (1969), The Chi-squared Distribution, J. Wiley & Sons, Inc., N. Y. pp. 7-8.

J. S. Rowlinson (1970), "Probability, Information, and Entropy," Nature, 225, 1196.

E. Schrödinger (1948), Statistical Thermodynamics, Cambridge University Press.

J. van Campenhout & T. M. Cover, "Maximum Entropy and Conditional Probability," I.E.E.E. Information Theory Trans. (in press).

12. MARGINALIZATION AND PRIOR PROBABILITIES (1980)

This article calls for an unusually long commentary, because a great deal of misinformation about improper priors is now propagating through the statistical literature, inspiring wasted effort on non-problems. Some further explanations are needed.

At the London meeting in 1973 I first met Dennis Lindley and learned from him about the startling new 'Marginalization Paradox' of Dawid, Stone, and Zidek, which had been presented at a meeting of the Royal Statistical Society about a month earlier. The consensus was that the culprit was the use of improper (i.e. non-normalizable) prior probability distributions; and that henceforth Bayesians must avoid them.

The neat — and clearly intuitively right — results that follow from the Jeffreys improper priors made me doubt this; but it was the eagerness with which Oscar Kempthore seized upon marginalization as sounding the death-knell of all Bayesianity, that clinched my resolve to study the matter for myself. But various misfortunes intervened, and it was not until September 1976 that I finally located the DSZ article and gave it some intensive study.

It required two weeks of increasing frustration at not being able to find the suspected error in the DSZ calculations, before the comprehension dawned; the difficulty was not in the calculations at all, but in the faulty logic that one is tricked into by an inadequate notation. Once seen, the resolution was trivial. As I had expected from the start, there is no fault in the Jeffreys priors (defined, of course, as limits of proper priors). Indeed, it was just the failure to follow Jeffreys in notation that made one see a paradox where none existed.

The Bayesian posterior probability of some hypothesis H is conditional on both the data D and the prior information I. In principle, if you fail to specify one of them, the problem is just as ill-posed and indeterminate as if you had failed to specify the other. Jeffreys had of course recognized this from the start, and his probability symbols $p(H \mid D, I)$ always indicated which prior information was intended. But orthodox statistics, which recognized the existence only of sampling probabilities of the form $p(D \mid H)$, did not use even the term 'prior information', much less a symbol for it.

Now statisticians trained from the orthodox literature still clung to the habit of orthodox notation after they became Bayesians. That is, they denoted posterior probabilities by $p(H \mid D)$, leaving out explicit mention of the prior information. One can get along for a while in this way, without disaster, understanding the prior information from the surrounding context. But eventually a problem will arise in which probabilities conditional on two different pieces of prior information I_1 and I_2, appear; and then we are in danger of misreading our equations.

In the DSZ work two Bayesians, B_1 and B_2, are given the same data but come to different conclusions. This is held to be a paradoxical inconsistency. Now B_1 had used an improper prior, while B_2 did not; and DSZ draw the conclusion that B_1 is at fault for using an improper prior.

But if they used different prior probabilities, they were necessarily taking into account different prior information. That they came to different conclusions is no more paradoxical than if they had come to different conclusions from different data. Indeed, the prior information used by B_2 did not include even the existence of the parameter to which B_1 had assigned an improper prior! If DSZ had used the full Jeffreys notation $p(H \mid D, I_1)$, $p(H \mid D, I_2)$ for the two calculations, there could never have been any appearance of paradox.

However, if this were the whole story, the matter could not have erupted into controversy. The trouble was that a subtle mathematical point lay, so to speak, right on top of this not so subtle logical point, and they were not easy to distinguish. Since some workers in this field have not yet recognized the double nature of the confusion, let us emphasize once more what is explained more fully in the following article.

The reason why DSZ placed the blame on improper priors was their purported proof that the 'paradox' (i.e., difference in conclusions) cannot arise if B_1 uses a proper prior. However, that proof rested on two assumptions that are in general mutually contradictory. But — and this is the extremely subtle point — in the problems they considered, those assumptions were contradictory for proper priors, but they became compatible just for the class of improper priors under discussion!

So they got their 'paradox' with improper priors, without noticing that they had proved nothing for the case of proper priors. The result of this double confusion was: improper priors were blamed for causing a 'paradox' which they did not cause and which was not a paradox.

In the following article I demonstrated that improper priors are not the cause of the difference by showing that one can reproduce the same 'paradox' to arbitrarily great accuracy by use of proper priors.

MARGINALIZATION AND PRIOR PROBABILITIES 339

Indeed, that something is amiss is clear from results in the DSZ article itself. As noted in my article, their Example 5 already provides a case in which B_1 may use an arbitrary proper prior, yet the 'paradox' is still present, a counter-example to their proof.

It is unfortunate that we Bayesians must dissipate so much of our efforts in putting out fires, that more important matters suffer. This commentary has not touched at all on the constructive, useful aspects of marginalization theory in pointing to another avenue by which uninformative priors may be defined and constructed. However, perhaps a few readers may push their way through to the final Sections of the following article, in which the prior probability problem takes on a new and different mathematical form, thanks to the work of DSZ.

A great deal more work is needed on the purely mathematical problems of finding solutions and existence proofs for these simultaneous integral equations. In my opinion, new results important for statistical theory will reward the person who masters them.

In the volume where this article was originally published, it is followed by some comments by DSZ (actually, I believe, by Merwyn Stone alone) and a rejoinder by me. While this discussion is a major contribution to statistical polemics, it does not contribute to the technical development of statistics, and is not reprinted here.

MARGINALIZATION AND PRIOR PROBABILITIES*

EDWIN T. JAYNES
Washington University

1. Introduction

A recent article of Dawid, Stone and Zidek (1973) notes two Bayesian calculation methods- the first using an improper prior, the second avoiding it – that one feels intuitively ought to lead to the same result, but in general do not. This "marginalization paradox" has been interpreted widely as revealing a fundamental inconsistency in the common Bayesian practice of using improper priors to express prior ignorance.

We argue that, on the contrary, resolution of the paradox is very simple, the discrepancy arising not from any defect of improper priors, but from a rather subtle failure of the second method to take into account all the relevant information. This situation, far from revealing an inconsistency in Bayesian methods, shows that to violate them in seemingly harmless ways can generate paradoxes, i.e. it is only by strict adherence to the Bayesian principles expounded by Jeffreys in 1939, that one can avoid inconsistencies in statistical reasoning.

The marginalization process is then turned to advantage by showing that it leads to a new means for defining what is meant by "uninformative" and for constructing noninformative priors, as the solution of an integral equation. This method draws only upon the universally accepted principles of probability theory, making no appeal to such additional desiderata as entropy, group invariance, or Fisher information. However, its range of applicability is still largely unexplored.

2. The paradox

A conscientious Bayesian B_1 studies a problem with parameters θ which he partitions into two sets, $\theta = (\eta, \zeta)$, being interested only in inferences about ζ.

*A preliminary account of this work was given at the 14th NBER–NSF Seminar on Bayesian Inference, Holmdel, N.J., June 1977

MARGINALIZATION AND PRIOR PROBABILITIES

Dawid, Stone and Zidek (1973: hereafter denoted by DSZ) note that, in several examples where B_1 uses an improper prior for η, the data x may also be partitioned into two sets, $x = (y, z)$ in such a way that B_1's marginal posterior distribution for ζ "is a function of z only", while the sampling distribution of z depends only on ζ.

A lazy Bayesian B_2 then tries to derive the posterior distribution $p(\zeta|x) = p(\zeta|z)$ more easily by applying Bayes's theorem directly to the sampling distribution $p(z|\zeta)$; and finds that he cannot reproduce B_1's result whatever prior $\pi(\zeta)$ he assigns.

DSZ then point the accusing finger at B_1 thus: "B_2's intervention has revealed the paradoxical unBayesianity of B_1's posterior distribution for ζ." In the ensuing discussion there was near unanimity of all opinions expressed, holding that B_1 is the party at fault, his transgression lying in his use of an improper prior.

A group-theoretical analysis by DSZ showed that if the sampling distribution $p(dydz|\eta\zeta)$ has a certain group structure (invariant under the combined action of coupled homomorphic groups G, $\bar{G}$ which are exact and transitive on the spaces $S(y)$, $S(\eta)$), the paradox can be avoided by choosing the prior as the right-invariant measure on $S(\eta)$. This procedure has indeed been advocated by a long list of writers starting with Poincaré (1912). However, as soon as we pass beyond the case of location and scale parameters it is rather exceptional to find a problem with all that group structure; and the paradox persists even in problems that have no group structure at all. In general, therefore, no way emerged of avoiding the paradox.

Predictably, some have seized upon this as a new tool for the abrogation of Bayesian statistics in general. However unimportant the practical consequences may be, it is imperative for Bayesian theory that this puzzle be cleared up.

In what follows we argue these points:

(1) Resolution of the paradox is far too simple a problem to be in need of group-theoretical analysis. That it can be made to appear and disappear by different choices of the η-prior, shows immediately where the difficulty lies.

(2) The real cause of the paradox is not B_1's use of improper priors, or indeed any transgression on B_1's part. On the contrary, it appears only when B_2 violates elementary Bayesian principles. B_2's transgression was concealed from view by concise notation.

(3) Nevertheless, the prior $\pi(\eta)$ that "avoids the paradox" has a useful interpretation as being, in a certain sense, "completely uninformative".

(4) Recognizing this, marginalization leads to a new means for constructing noninformative priors, via a set of simultaneous integral equations. This method is consistent with, but appears more general than, the group analysis.

3. The resolution

We must be careful to note exactly what the first quoted statement of DSZ means. From the mathematics it is clear that to say the posterior distribution of ζ

"is a function of z only" means that it depends *on the data* x only through the value of z. But of course, any posterior distribution depends not only on the data, but also on the prior information. As Jeffreys (1939) stressed, to avoid ambiguities the prior information (or hypotheses) on which our probabilities are conditional, ought to be stated explicitly to the right of the stroke in our probability symbols $p(A|B)$.

B_1's prior information includes the whole structure of the model, the qualitative fact of the *existence* of the components η and y, and the prior distribution of η. How, then, can one be sure that B_2 is justified in considering only the reduced problem in which (η, y) never appear at all? According to Bayesian principles, one may not disregard any part of either the data or the prior information, unless that part is shown to be irrelevant in the sense that it cancels out mathematically.

B_2's reduction appears, at first glance, to be reasonable; but so did a multitude of *ad hoc* procedures of non-Bayesian statistics, which were found eventually to contain defects. Surely, there is no room for personal opinions about this; the mathematical rules of probability theory are quite competent to tell us whether B_2's reduction is or is not justified.

As a constant reminder of the presence of prior information, we extend the notation of DSZ by introducing the symbols I_1 and I_2 to stand for the totality of prior information used by B_1 and B_2, respectively. The quoted first statement is then, more precisely,

$$p(\zeta|xI_1) = p(\zeta|zI_1). \tag{1}$$

Now the rules of probability theory tell us that

$$p(\zeta|xI_1) = \int d\eta \; p(\zeta|\eta xI_1)p(\eta|xI_1). \tag{2}$$

If, given I_1 and all the data x, additional knowledge of η would be irrelevant for inference about ζ, i.e. if

$$p(\zeta|\eta xI_1) = p(\zeta|xI_1), \tag{3}$$

then η integrates out of (2) trivially. But if (3) does not hold, then η *is* relevant, and the posterior distribution $p(\eta|xI_1)$ intrudes itself inevitably into the problem, bringing with it a dependence on the prior $\pi(\eta|I_1)$.

In this case we have to expect that the separation property (1) cannot hold for all η-priors. If (1) holds for some class C of priors, then while $p(\zeta|xI_1)$ is, in a sense, "a function of z only", it is a different function of ζ for different η-priors in C. But since B_2's posterior distribution $p(\zeta|zI_2)$ is independent of $\pi(\eta|I_1)$, it appears that we have at hand all the material needed to manufacture paradoxes. In other words, we suggest that this paradox has, fundamentally, nothing to do with improper priors; B_1 and B_2 obtain different results when, and only when, B_2 ignores relevant prior information (about η and/or the model), that B_1 is taking into account.

It remains to be shown that the mechanism just suggested is the one actually

MARGINALIZATION AND PRIOR PROBABILITIES

operative in the examples of DSZ. Since (3) is equivalent to

$$p(\eta, \zeta | xI) = p(\eta | xI) p(\zeta | xI), \tag{4}$$

we examine some of the DSZ examples for this factorization property.

Example 1. The model is described in DSZ. For present purposes we need note only that the raw data $x = \{x_1 \ldots x_n\}$ are partitioned into $y \equiv x_1$, and $z \equiv \{z_i = x_i/x_1, 1 \leq i \leq n\}$. The joint posterior distribution is

$$p(\eta \zeta | xI_1) \propto \pi(\zeta | I_1) c^{-\zeta} \eta^n \exp(-\eta y Q) \pi(\eta | I_1), \tag{5}$$

where

$$Q(\zeta, z) \equiv \sum_1^\zeta z_i + c \sum_{\zeta+1}^n z_i \tag{6}$$

is a function that is known from the data. Here B_1 has helped B_2's prospects as much as possible by assigning independent priors to η, ζ. Nevertheless, the likelihood function mixes them up and we find the conjectured lack of factorization (4). B_1's marginal posterior distribution is

$$p(\zeta | xI_1) \propto \pi(\zeta | I_1) c^{-\zeta} \int_0^\infty \eta^n e^{-\eta y Q} \pi(\eta | I_1) d\eta, \tag{7}$$

from which we note several things.

(a) As predicted, the dependence on the prior $\pi(\eta | I_1)$ is manifest. Prior information about η is clearly relevant to inference about ζ; and B_2's reduction violates Bayesian principles by throwing it away.

(b) The dependence on y drops out on normalization, leading to (1), if and only if the η-prior is in the class $\{C: \pi(\eta | I_1) \propto \eta^k, -n-1 < k < \infty\}$, which includes all those considered by DSZ.

(c) For any prior in class C there are no convergence problems. It is therefore difficult to see how use of an improper prior can in itself be grounds for reproach; all of B_1's conclusions can be approximated to any accuracy we please (e.g. one part in 10^{1000}) by use of a proper prior (as shown explicitly below, eq. (24)).

(d) On the other hand, use of a proper prior, $\int \pi(\eta | I_1) d\eta = 1$, will take us out of the class C. But then the statistic y cannot be disentangled, and remains relevant; the separation property (1) is lost, and B_2 becomes superfluous.

(e) The proof of DSZ that use of proper priors avoids the paradox, rested on two assumptions: that B_1 uses a proper prior, and (DSZ, eq. (1.20)) that the separation property still holds. But for this model, those assumptions are contradictory. DSZ supposed that, with proper priors, the paradox would disappear because B_1 and B_2 then agree. We now see that, at least in this example, the paradox disappears rather because the comparison disappears; B_2 can no longer play his game at all.

For efficient verbalization at this point, we need to coin a new term. A prior

$\pi(\eta)$ that leads to the separation property (1) nullifies the effect of the data y for inference about ζ. Let us call such a prior *nullifying* (more precisely: y-nullifying within the context of a particular model). What DSZ proved is then: if a proper prior is also nullifying, then it necessarily leaves B_1 and B_2 in agreement. However, except in the trivial case of complete independence: $p(dy dz|\eta\zeta) = p(dy|\eta)p(dz|\zeta)$ one cannot assume *ohne weiteres* the existence of such a prior, as this example illustrates.

Example 2. We have parameters $\theta = (\mu_1, \mu_2, \sigma)$ and data $x = (u_1, u_2, s)$ with sampling density function

$$p(u_1 u_2 s | \mu_1 \mu_2 \sigma) = A(s^{\nu-1}/\sigma^{\nu+2})\exp(-Q), \tag{8}$$

where A is a normalizing constant and

$$Q \equiv \frac{1}{2\sigma^2}[(u_1 - \mu_1)^2 + (u_2 - \mu_2)^2 + vs^2]. \tag{9}$$

However, we are interested in inference only about

$$\zeta \equiv \frac{\mu_1 - \mu_2}{\sigma\sqrt{2}} \tag{10}$$

and the sampling distribution of

$$z = \frac{u_1 - u_2}{s\sqrt{2}} \tag{11}$$

is found to depend only on ζ:

$$p(z|\mu_1\mu_2\sigma) = p(z|\zeta) = \sqrt{(2\pi)} A \int_0^\infty \omega^\nu e^{-R} d\omega, \tag{12}$$

where

$$R(z, \zeta, \omega) \equiv \tfrac{1}{2}[v\omega^2 + (\omega z - \zeta)^2]. \tag{13}$$

Making the additional change of variables

$$\mu = \tfrac{1}{2}(\mu_1 + \mu_2); \quad u = \tfrac{1}{2}(u_1 + u_2), \tag{14}$$

the unwanted components are $\eta = (\mu, \sigma)$; $y = (u, s)$, and B_1's posterior distribution of ζ is

$$p(\zeta|xI_1) \propto \pi(\zeta|I_1) \int_0^\infty \omega^\nu d\omega \, e^{-R} f(u, \sigma), \tag{15}$$

where

$$f(u, \sigma) \equiv \frac{1}{\sqrt{\pi}} \int_{-\infty}^\infty d\mu \, \pi(\mu, \sigma) \exp\left[-\left(\frac{u-\mu}{\sigma}\right)^2\right], \tag{16}$$

and in the integration over $\omega = s/\sigma$, s is held constant while σ varies. Again, a certain class C of priors $\pi(\eta) = \pi(\mu, \sigma)$ is found to be nullifying, leading to the separation property (1). This includes the class

$$\{C': d\mu_1 d\mu_2 \sigma^{-k} d\sigma = \sqrt{(2)}\, d\zeta d\mu\, \sigma^{1-k} d\sigma\} \tag{17}$$

considered by DSZ, or indeed any prior $\pi(\mu, \sigma)$ independent of μ, for which (15) is independent of y:

$$p(\zeta|xI_1) \propto \pi(\zeta|I_1) \int_0^\infty \omega^{\nu-1}\, d\omega\, e^{-R} \pi(\sigma). \tag{18}$$

From this we confirm eq. (1.3) of DSZ (with $k=1$) *and comparing with* (12) *it is seen that* B_1 *and* B_2 *will agree if* $k=2$. Once again it is clear from (15) and (18) that in general their conclusions will differ because B_1 is taking into account relevant prior information about $\eta = (\mu, \sigma)$ that B_2 is ignoring.

Rather than continuing with a rather tedious, but still superficial, inspection of more examples, which would only reconfirm the mechanism already established, we can get a better understanding by returning to a second look at example 1.

4. A reinterpretation

We may take a more charitable view of B_2 if DSZ will grant a similar courtesy to B_1. In these examples, independently of all questions of priors, it is true that the marginal sampling distribution of z depends only on ζ. Suppose that we now regard B_2, not as a lazy fellow who "always arrives late on the scene of inference" and tries to simplify B_1's analysis; but merely, through no fault of his own, an uninformed fellow whose knowledge about the experiment consists only of the sampling distribution

$$p(z|\zeta I_1) = p(z|\zeta I_2) \tag{19}$$

and is unaware of the existence of the components (η, y). Then B_2 is following strict Bayesian principles, and

$$p(\zeta|zI_2) \propto \pi(\zeta|I_2) p(z|\zeta I_2) \tag{20}$$

will always represent the best inferences that can be made on the information he has – whether or not B_1's posterior distribution has the separation property ($p(\zeta|yzI_1)$ independent of y) that initiated all this.

It then makes sense to compare B_2's results with B_1's in all cases, whether B_1's prior for η is proper or improper; and in all cases the comparison will reveal just how much difference B_1's extra information has made. For the most meaningful comparison, we suppose they have the same prior information about ζ:

$$\pi(\zeta|I_1) = \pi(\zeta|I_2) = \pi(\zeta). \tag{21}$$

Returning now to example 1, from eq. (1.2) of DSZ, B_2's conclusions are given by

$$p(\zeta|zI_2) \propto \pi(\zeta) c^{-\zeta} [Q(\zeta, z)]^{-n}, \tag{22}$$

while B_1's are given by our eq. (7). Let B_1 assign a proper η-prior of the conjugate form

$$\pi(\eta|I_1) \propto \eta^{k-1} e^{-t\eta}; \quad t > 0, \tag{23}$$

which as $t \to 0$ goes into the family of improper priors used by DSZ. Then B_1's result is

$$p(\zeta|yzI_1) \propto \pi(\zeta) c^{-n} \left[\frac{y}{t + yQ(\zeta, z)} \right]^{n+k} \tag{24}$$

which, we note, goes smoothly and continuously into B_2's result (22) as $t \to 0$ and $k \to 0$.

But no "paradoxical unBayesianity" or "impropriety" is apparent. Strictly speaking, the dependence on y drops out, leading to the separation property (1), only when $t = 0$, but for $t \ll yQ$ there is virtually no y-dependence, even though the prior is still proper. There is no discontinuous change; as t becomes smaller and the prior (23) becomes more nearly nullifying, the y-statistic just becomes less and less informative.

If then B_1, having noted that $t \ll yQ$, decides to simplify (24) by setting $t = 0$, this now appears, not as a paradox-creating transgression into the realm of improper priors, but rather as a perfectly harmless and reasonable approximation – indeed, an approximation far better justified than many that are accepted without question in non-Bayesian statistics.

If B_1 has very little prior information about η (i.e. if (t, k) are small), then there is virtually no difference between his conclusions and B_2's, whether his prior is proper or improper. If, on the other hand, (t, k) are large, then B_1 is in possession of additional, highly cogent, information relevant to inference about ζ; and it is only right and proper that his conclusions deviate from B_2's. Any statistical method that failed to make use of this information although it was available to the user, would then be deserving of the epithet, "impropriety".

5. Improper priors – discussion

In view of the great emphasis on the issue of improper priors in DSZ and in the ensuing discussion – almost to the exclusion of all else – and subsequent attempts to use this as an argument against all Bayesian methods, some further exegesis defending the use of improper priors is needed.

A sequence $\{\pi_i\}$ of proper priors defines a corresponding sequence $\{P_i\}$ of posterior distributions. Often, even though the limit of $\{\pi_i\}$ is improper, the limit of $\{P_i\}$ is a proper, well-behaved, and analytically simple distribution. The Bayesian will often take that limit for mathematical convenience, *after* it is clear –

MARGINALIZATION AND PRIOR PROBABILITIES 347

whether by specific calculation in the manner of (24) or through past experience with similar problems – that this will make no practical difference in the results. Often, the experimental data are so much more informative than the prior information that to carry along all the details of any particular proper prior, although in principle the correct thing to do, would in practice only increase the amount of computation without yielding anything of value for the purposes at hand. Usually, it is so clear when we have this situation that there is no need to construct specific sequences of the type (23), (24); one proceeds immediately to this simpler limit.

In a similar way, a person using the χ^2 test knows in advance, from common sense and the past experience of statisticians in general, about how much data he needs, and how many categories, to give a test that is good enough for his purposes. Beyond that, further refinements, although correct in principle, would only increase the amount of computation without useful return. In orthodox statistics, use of a little practical common sense in applying a method is not regarded as an inconsistency. Perhaps, when his methods are more widely understood, the Bayesian may hope to be granted an equal dispensation.

Now, as noted by several participants in the discussion following the DSZ paper and discussed at greater length in Jaynes (1976), it is just the Bayesian results based on noninformative improper priors that correspond closely – often exactly – with those obtained by orthodox methods. In these cases it is difficult to see how one can reject the Bayesian use of an improper prior, without thereby rejecting with equal force the orthodox method which yields the same result.

On the other hand, in some cases the attempted passage to an improper prior may fail, by yielding a non-normalizable posterior distribution in the limit. This is symptomatic that the experiment is so uninformative that our prior information is, necessarily, still highly relevant to any inference that can be made; and in such a case we had *better* take that prior information explicitly into account by using the appropriate proper prior. In this case an orthodox method, by its nature incapable of taking prior information into account, is virtually guaranteed to produce absurd or dangerously misleading results (for a specific example, see Jaynes (1976); reply to Kempthorne's comments).

The actual equations both in DSZ and in the present work are not in any way changed by our reinterpretation of B_2's role; but the analysis is seen in a different and perhaps more constructive light. We are not merely exhibiting the folly of a defective Bayesian procedure – ignoring information or using improper priors. We are comparing two entirely correct Bayesian procedures, making inferences about the same quantity ζ, at two different stages of knowledge. The parameter η is not merely an "unwanted" complication; it represents new information relevant to the desired inference about ζ.

The comparison is, in effect, a microcosm of an often-occurring real life phenomenon, the effect of advancing knowledge on a scientific inference.

For example, from the known rate ($z = 2 \times 10^{20}$ MW) of radiation of energy from the sun, estimate its future lifespan (ζ); how much longer can it continue pouring out energy at that rate? The datum (z) was known 120 years ago about as well as it is today, and on the basis of the laws of physics as then known, B_2 (better known as Lord Kelvin (1862)) estimated a future life of $\zeta = $ a few million years;[1] an entirely valid conclusion from the information he had. But today we know of a new parameter ($\eta = $ energy release from nuclear reactions) that has an important bearing on the question, and we have new data ($y = $ abundance of various elements in the sun, and energy release of a number of nuclear reactions). As a result, B_1 (Gamow (1945)) re-estimated the future life of the sun to be vastly greater, about 10^{10} years.[2]

Doubtless, an econometrician could give much more immediate examples; e.g. the effect of new knowledge (the role of oil prices) on prediction of economic activity from models in which, prior to 1973, oil price did not appear as a factor.

6. The integral equations

Can we extract something of positive value from all this, leaving Bayesian theory with a net gain? As is now clear, there is no reason to be surprised when B_1 and B_2 disagree; that was only to be expected. What is perhaps surprising, and calls for explanation, is rather that in some cases B_1's extra information was unavailing, and they *did* agree, after all. How is that possible?

Clearly, in all cases B_2 was incorporating no prior information about η. If, nevertheless, they agree in one case, it seems natural to conclude that, in that case, B_1 must not have been incorporating any prior information either; at least, none that was relevant to ζ.

The prior $\pi(\eta)$ that leaves them in agreement should, then, have some close relation to the one describing "complete ignorance" of η, if such exists. Is it possible that marginalization is giving us a new, objective, and above all, *workable* criterion for defining precisely what is meant by "complete ignorance," and for telling us whether and when such priors do or do not exist?

But we must proceed cautiously. It is not clear how marginalization could tell us that a prior is "completely uninformative" without qualifications. But marginalization can and does provide an answer to the question whether, within the context of a given model, any proposed prior $\pi(\eta)$ is or is not "completely uninformative about ζ."

Two further cautions are necessary. A prior $\pi(\eta)$ that is held to be unin-

---

[1] He concludes: "It seems, therefore, on the whole most probable that the sun has not illuminated the earth for 100,000,000 years, and almost certain that it has not done so for 500,000,000 years. As for the future, we may say, with equal certainty, that inhabitants of the earth cannot continue to enjoy the light and heat essential to their life, for many million years longer, unless sources now unknown to us are prepared in the great storehouse of creation." Those unknown sources were revealed 43 years later, when Albert Einstein wrote $E = mc^2$.

[2] The concluding sentence is: "And the year 12,000,000,000 after the Creation of the Universe, or AD 10,000,000,000, will find infinite space sparsely filled with still receding stellar islands populated by dead or dying stars."

formative about η ought, one would suppose, to have the property that it is uninformative *a fortiori* about any other quantity ζ. Clearly, however, the converse need not hold. In any given model, a prior $\pi(\eta)$ might express very great knowledge about η; and still be ζ-uninformative because of the functional form of $p(yz|\eta\zeta)$. On the other hand, if one could prove that a given prior $\pi(\eta)$ is ζ-uninformative for all models in which η appears as a scale parameter, and unique for one, that would seem to be valid grounds for a stronger claim.

Finally, we note that in another respect the property of a prior $\pi(\eta)$ to leave B_1 in agreement with B_2 involves rather more than what one usually means by the term "uninformative". B_1's advantage over B_2 does not lie only in his prior knowledge of η; he has also the additional data y. If a prior $\pi(\eta)$ is to leave him in agreement with B_2, therefore, it is not enough for $\pi(\eta)$ to have the passive property of being ζ-uninformative (i.e. of not in itself providing any information relevant to ζ). It must perform also the active function of rendering the new data y irrelevant to ζ; that is what we have termed "nullifying".

The necessary and sufficient condition for a prior $\pi(\eta)$ to be nullifying independently of $\pi(\zeta)$ is that B_1's quasilikelihood contains y and ζ in separate factors

$$\int p(y, z|\eta, \zeta)\pi(\eta)\mathrm{d}\eta = f(y, z)g(z, \zeta) \tag{25}$$

for some functions f, g. The surprising discovery of DSZ was then that, while a proper prior that is nullifying is also necessarily uninformative [*Proof*: integrating y out of (25), it then reduces to B_2's likelihood $p(z|\zeta)$], an improper prior may be nullifying without being uninformative. In example 1, the prior (23) is nullifying if $t = 0$; but it is not uninformative unless $k = 0$ also.

Let us seek the necessary and sufficient condition for agreement of B_1 and B_2, subject to three assumptions. First is the property (19) without which we should have little reason to compare B_1 and B_2 at all. Secondly, it would make little sense to ask whether an η-prior is uninformative about ζ if it contained ζ; so we assume that B_1 assigns independent priors:

$$\pi(\eta\zeta|I_1) = \pi(\eta|I_1)\pi(\zeta|I_1). \tag{26}$$

Thirdly, DSZ considered whether B_2 could, by any choice of his prior $\pi(\zeta|I_2)$, achieve agreement with B_1's posterior distribution. But for present purposes we cannot allow B_2 that much freedom; for if B_1 and B_2 had different priors for ζ, that would in itself lead to a difference in their conclusions, which really has nothing to do with B_1's prior knowledge about η, although agreement of the posterior distributions might be, fortuitously, restored by a particular η-prior. But in that case it would be very wrong to label such an η-prior as "uninformative about ζ". To avoid this, we must suppose rather that B_1 and B_2 start from the same state of prior knowledge about ζ as in (21):

$$\pi(\zeta|I_1) = \pi(\zeta|I_2) = \pi(\zeta),$$

and end up still in agreement as to the posterior distribution. And by "agreement", we do not mean that they agree for one particular sample or one particular prior. In order to justify saying that B_1's prior for η was completely ζ-uninformative and y-nullifying, they must remain in agreement for all data sets $x = (y, z)$, all sample sizes, and all priors $\pi(\zeta)$.

With these assumptions, B_1's posterior distribution is

$$p(\zeta|y, z, I_1) \propto \pi(\zeta) p(z|\zeta) \int d\eta \ \pi(\eta|I_1) p(y|z\eta\zeta I_1), \tag{27}$$

while B_2's is

$$p(\zeta|zI_2) \propto \pi(\zeta) p(z|\zeta). \tag{28}$$

Evidently, the necessary and sufficient condition for agreement is that

$$\int d\eta \ \pi(\eta|I_1) p(y|z\eta\zeta I_1) = p(y|z\zeta I_1) \tag{29}$$

shall be independent of ζ for all y,z. Denoting the parameter space and our partitioning of it into subspaces by $S_\theta = S_\zeta \times S_\eta$, we may write (29) as

$$\int_{S_\eta} p(y, z|\zeta, \eta) \pi(\eta) d\eta = \lambda(y, z) p(z|\zeta), \qquad \zeta \text{ in } S_\zeta. \tag{30}$$

This is a Fredholm integral equation in which the kernel is B_1's likelihood, $K(\zeta, \eta) = p(y, z|\zeta, \eta)$, the "driving force" is B_2's likelihood $p(z|\zeta)$, and $\lambda(y, z) \equiv p(y|zI_1)$ is an unknown function to be determined from (30).

For each possible data set $x = (y, z)$ we have an equation of the form (30); so if a single prior $\pi(\eta)$ is to suffice for all data sets it must satisfy not just one integral equation, but a large – in general infinite – class of simultaneous integral equations.

Now in other applications we are accustomed to find that a single Fredholm equation has already a unique solution. At first glance, therefore, it seems almost beyond belief that the system of equations (30) could fail to be grossly overdetermined; from which one would be forced to conclude, with the anti-Bayesian skeptics, that uninformative priors do not exist.

Clearly, the consistency of previous Bayesian thought, which presupposed the existence of uninformative priors, is being put here to a severe test. But it is also an eminently fair and "objective" test. The question whether, in a given model, the notion of an uninformative prior is contradictory, ambiguous, or well defined, is removed from the realm of philosophical debate, and reduced to the question whether a set of simultaneous integral equations is overdetermined, underdetermined, or well-posed.

7. An example

We know at least that the system of equations (30) is not *always* overdetermined; for in several examples DSZ were able to recognize particular priors $\pi(\eta)$ which leave B_1 and B_2 in harmony for all samples. Each of the DSZ examples can tell us something about the mathematical structure of (30) and its correspondence with previous group invariance arguments.

Example 1. The sampling distribution is

$$p(yz|\eta\zeta) = \eta^n c^{n-\zeta} y^{n-1} \exp[-\eta y Q(z, \zeta)] \tag{31}$$

with $Q(z, \zeta)$ defined by (6). This gives the marginal sampling distribution

$$p(z|\zeta) = (n-1)! \, c^{n-\zeta} Q^{-n}. \tag{32}$$

S_η is the positive real line, and the family of integral equations (30) becomes: for each possible sample (y, z),

$$\int_0^\infty \pi(\eta)\eta^n e^{-\eta y Q} \, d\eta = (n-1)! \, y\lambda(y, z) [yQ(z, \zeta)]^{-n},$$

$$\zeta = 1, 2, \ldots, n-1. \tag{33}$$

Now choose any two values $\zeta \neq \zeta'$, and write $Q \equiv Q(z, \zeta)$, $Q' \equiv Q(z, \zeta')$. Eq. (33) then requires that for all (y, z)

$$\int_0^\infty \pi(\eta)(\eta y Q)^n e^{-\eta y Q} \, d\eta = \int_0^\infty \pi(\eta)(\eta y Q')^n e^{-\eta y Q'} \, d\eta. \tag{34}$$

or

$$\int_0^\infty \left[\pi(\eta) - \frac{Q}{Q'} \pi\left(\eta \frac{Q}{Q'}\right)\right] \eta^n e^{-\eta y Q} \, d\eta = 0, \qquad 0 < y < \infty. \tag{35}$$

Since the Laplace transform is uniquely invertible, this requires that for all choices of $\{z, \zeta, \zeta'\}$ we must have, setting $a \equiv Q/Q'$,

$$\pi(\eta) = a\pi(a\eta), \qquad 0 < \eta < \infty. \tag{36}$$

But this is the same functional equation that was deduced earlier from the transformation group that expresses "complete ignorance" of a scale parameter. To complete the proof, note that from (6), if $\zeta' < \zeta$, eq. (36) must hold in the continuous range $(1 \leq a \leq c)$, and so the only possibility is, to within a constant factor,

$$\pi(\eta) = \eta^{-1}. \tag{37}$$

This argument shows that no $\pi(\eta)$ other than (37) can satisfy (33). Conversely, on substitution we see that (33) is indeed satisfied for all $\{y, z, \zeta\}$, with $y\lambda(y, z) = 1$. Again, we note several things.

(a) This argument made no use of the separation property (1). The solution (37) implies this as a necessary, if obvious, condition for agreement of B_1 and B_2.

(b) The problem turned out to be well-posed; there is one unique prior $\pi(\eta)$ that is "completely uninformative about ζ", and it is just the one that Jeffreys anticipated, on partly intuitive grounds, some forty years ago (as the prior representing "complete ignorance" of a parameter known to be positive). It follows also from the fact that η is a scale parameter, by some transformation group methods (for example, Jaynes (1968); one of several quite different approaches all called "the transformation group method" or "the group invariance principle", although they utilize different groups which operate in different spaces, are chosen by different criteria, and yield different results. For further comments, see Appendix A.)

(c) But the prior (37) is now derived in a way that is completely independent of anybody's intuition or any additional desiderata such as entropy, group invariance, or Fisher Information. Given the sampling distribution (31), the result (37) follows by straightforward mathematical steps. [Indeed, on sufficiently fine analysis, it will be seen that the only elements of probability theory used in the transition from (31) to (37) are the product rule $p(AB|C) = p(A|BC)p(B|C)$, and the sum rule $p(A|B) + p(\sim A|B) = 1$].

(d) This, however, recalls the oft-quoted remark of Lindley (1971): "Why should one's knowledge, or ignorance, of a quantity depend on the experiment being used to determine it?" The answer, in our view, is that the prior distribution should, of course, be based on *all* the prior information available. But the role a parameter plays in a sampling distribution is always a part of that information. Indeed, that is the irreducible minimum information without which a problem of inference cannot be formulated. Often, in pedagogical examples, it is the *only* prior information at hand, because (as in all the DSZ examples) the person formulating the problem simply neglects to provide any more. In this case – and only in this case – the prior distribution is, necessarily, determined (not necessarily uniquely) by the sampling distribution. But this is just the case we are solving by (37).

(e) In a real problem a parameter will be, in general, "a physically meaningful quantity about which we know something". But for the mechanics of incorporating that something into our informative prior there are, to the best of the writer's knowledge, only two known principles: Bayes's theorem and maximum entropy; and both of these still require an ignorance preprior like (37) as their starting-point (Jaynes (1968)). Therefore, for any problem of inference we see no way to avoid the notion of "complete ignorance", any more than we could avoid the concept of zero in arithmetic. Nor should we wish to avoid it; for clearly, to ask "What is our state of knowledge after receiving information I?" cannot have any definite answer until we specify: What was our state of knowledge *before* receiving I? And this holds with equal force whether we choose to classify I as part of the data, or part of the prior information (see however, the Appendix for some further comments).

(f) In this example, η was a scale parameter, the sampling distribution (31) having the functional form $p(y, z | \zeta, \eta) = y^{-1} g(z, \zeta; y\eta)$. For any sampling distribution of this form [or equally well, $y^{-1} g(z, \zeta; y/\eta)$] one readily verifies that the Jeffreys prior $\pi(\eta) \sim \eta^{-1}$ satisfies (30), and $y\lambda(y, z)$ is then a constant. Whether this solution is unique depends, of course, on how z, ζ enter into the function g.

8. The one-dimensional case

With the insight gained from the DSZ example 1, we are able to give a more general discussion of the case where y, η are one dimensional. We started cautiously, asking only for a prior $\pi(\eta)$ that is uninformative about ζ within the context of a given model. We now see that for a scale parameter η, the Jeffreys prior is ζ-uninformative for all models, and unique for one. But this is already enough to establish it as the only prior for a scale parameter that is "completely uninformative" without qualifications.

Since the location and scale parameter cases are equivalent by the transformation $\mu = \log \sigma$, it follows that the uniform prior $d\mu$ is similarly general and unique for a location parameter (but in this case the result is so intuitive that it had never been doubted anyway).

The analysis may be generalized in the following way (suggested to the writer by a remark of W. D. Fisher). Consider any sampling distribution with the functional form

$$p(yz|\eta\zeta) = g[z, \zeta; h(y, \eta)] \frac{\partial h}{\partial y}, \tag{38}$$

for which the property $p(z|\eta\zeta) = p(z|\zeta)$ underlying marginalization theory follows at once. The integral equation (30) for an uninformative prior becomes

$$\int g(z\zeta; h) \frac{\partial h}{\partial y} \pi(\eta) d\eta = \lambda(y, z) \int g(z\zeta; h) dh. \tag{39}$$

If this is to hold without further assumptions about the functional form of $g(z\zeta; h)$, it is necessary that $\lambda(y, z) = \lambda(y)$ be independent of z, and that

$$\frac{\partial h}{\partial y} \pi(\eta) = \lambda(y) \frac{\partial h}{\partial \eta}. \tag{40}$$

But then, making the change of variables $(y, \eta) \to (\bar{y}, \bar{\eta})$ where

$$\bar{y} \equiv \exp \int \lambda(y) dy; \qquad \bar{\eta} \equiv \exp \int \pi(\eta) d\eta, \tag{41}$$

eq. (40) reduces h to a function of $(\overline{y\eta})$:

$$h(y, \eta) = \bar{h}(\overline{y\eta}) \tag{42}$$

and (38) assumes the functional form $p(\bar{y}, z|\bar{\eta}, \zeta) = \bar{y}^{-1} \bar{g}(z, \zeta; \overline{y\eta})$ of remark (f) above, where $\bar{g}(z\zeta; \alpha) \equiv g(z, \zeta; \bar{h}(\alpha))$. Thus, the class of functions $h(y, \eta)$ for which $\pi(\eta)$ and $\lambda(y)$ can be constructed as in (40) takes us back, to within the change of variables (41), to the scale parameter case.

For example, if

$$h(y, \eta) = \tanh \sqrt{(y^n + \eta^m - a)}, \qquad 0 < \eta < \infty,$$

we have at once from (40) that the uninformative prior is

$$\pi(\eta) = \eta^{m-1}.$$

Likewise, if

$$h(y, \eta) = f(y)[\tan \alpha\eta]^{3/2}, \qquad 0 < \alpha\eta < \frac{\pi}{2}$$

the uninformative prior is

$$\pi(\eta) = \csc(2\alpha\eta).$$

and if

$$h(y, \eta) = \log\left[\frac{(\eta y - 1)(\eta + y)}{\eta y}\right]$$

the uninformative prior is

$$\pi(\eta) = 1 + \eta^{-2}.$$

Now, although the result (40) is rather special in the class of all problems with one-dimensional (y, η), it is easily seen to exhaust the possibilities of the DSZ group analysis for that class. They took the sampling distribution as (DSZ, eq. (2.6)) $p(dydz|\eta\zeta) = f(yz|\eta\zeta)\mu_G(dy)dz$, where μ_G is "a fixed general measure element" and defined the group structure by (DSZ, eq. (2.7)):

$$f(y, z|\eta, \zeta) = f(gy, z|\bar{g}\eta, \zeta), \tag{43}$$

where $g, \bar{g}$ are corresponding elements of the groups $G, \bar{G}$ mentioned in section 2 above. Evidently, if (y, η) are one dimensional, eq. (43) says only that we have the functional form (compare (38)):

$$f(yz|\eta\zeta) = g[z, \zeta; h(y, \eta)], \tag{44}$$

the "combined action of the groups" signifiying a kind of hydrodynamic flow in the (y, η) plane, whose streamlines are the contours $h(y, \eta) = $ const. But just as our eq. (40) cannot be satisfied for all functional forms of $h(y, \eta)$, so the group structure (43) restricts the form of $h(y, \eta)$ in (44).

The form of that restriction can be anticipated at once by the following argument. A continuous exact group of mappings of the real line onto itself is necessarily a one-parameter group (for in $y' = gy$ with fixed y, each group element g is represented by one and only one value of y'; thus y' parameterizes the group). But a one-parameter continuous group is isomorphic with the group of simple translations $(x' = x + a)$. We infer that the group structure (43) must restrict us to problems that are equivalent, to within a change of variables, to the location/

MARGINALIZATION AND PRIOR PROBABILITIES

scale parameter case. Indeed, on following through the analysis (Hamermesh (1962)) we find that the condition imposed on (44) by the group structure is just our eq. (40), i.e. the functions denoted $h(y, \eta)$ in (38) and (44) are identical.

The condition found here is the same as that given by Lindley (1958) for agreement of a Bayesian posterior with a fiducial distribution; such relations were noted also by Fraser (1961) and Villegas (1971).

Now we arrive at the really interesting question: What happens in the one-dimensional case if we try to go beyond the class of problems just discussed? Do we continue to find uninformative priors from (30) beyond those obtainable by group analysis; or do we come up against that threatening overdetermination? This opens up a wide class of new mathematical problems, interesting in their own right and of obvious importance for the future of Bayesian statistical theory. At the time of writing (January 1978) progress on them is far from complete, consisting mostly of isolated results.

The following example, due to C. L. Mallows, shows that further solutions do exist beyond those resulting from the group structure (43), and that the apparent overdetermination is not always real.

Let y, z be non-negative integers, and

$$p(yz|\eta\zeta) \propto \frac{\zeta^z \eta^y (1-\eta)^{z-y}}{y!(z-y)!}, \quad \begin{array}{l} 0 \leq \zeta, \eta < \infty \\ 0 \leq y \leq z \end{array}. \tag{45}$$

Then the marginal sampling distributions are Poisson:

$$p(z|\zeta\eta) = p(z|\zeta) = e^{-\zeta} \frac{\zeta^z}{z!} \tag{46}$$

independent of η, as required by marginalization theory and

$$p(y|\zeta, \eta) = e^{-\zeta\eta} \frac{(\eta\zeta)^y}{y!}, \tag{47}$$

depending on both parameters; thus seemingly leading to a nontrivial marginalization problem. This example lacks the group structure (43), since y is discrete, η continuous. But we now find the peculiarity that

$$p(y|z, \zeta, \eta) \tag{48}$$

is independent of ζ, and as a consequence the integral equations (30) are satisfied trivially; all priors $\pi(\eta)$ are nullifying and uninformative about ζ.

That the opposite behavior can also occur, is shown by example 5 of DSZ. The concluding message of their section 1 was that all is well as long as B_1 uses proper priors. Later, they consider the model

$$p(yz|\eta\zeta) \propto \int_0^\infty t^{2n-1} \exp\{-\tfrac{1}{2}[t^2 + n(zt-\zeta)^2 + n(yt-\eta))^2]\}dt \tag{49}$$

and note that, if $y=0$, the posterior distributions of B_1 and B_2 are

$$p(\zeta|zI_1) \propto \pi(\zeta) \int_0^\infty t^{2n-1} \exp\{-\tfrac{1}{2}[t^2+n(zt-\zeta)^2]\}dt. \tag{50}$$

$$p(\zeta|zI_2) \propto \pi(\zeta) \int_0^\infty t^{2n-2} \exp\{-\tfrac{1}{2}[t^2+n(zt-\zeta)^2]\}dt. \tag{51}$$

But if $z>0$, the ratio of the integrals is (by a Schwarz inequality) a monotonic increasing function of ζ; and so B_1 and B_2 cannot agree unless they assign a singular prior $\pi(\zeta)=\delta(\zeta-\zeta_0)$, in which case their posterior distribution is independent of the data.

DSZ (Appendix 2) term this situation, "*The Inevitable Paradox of Example 5*". It is, perhaps, even more inevitable and more paradoxical than they intended; for it is clear from (49) that this situation arises for all priors $\pi(\eta)$, proper or improper! What, then, are we to make of their proof in Section 1, that this discrepancy "could not have arisen if B_1 had employed proper prior distributions"?

Passing over this query, the DSZ example 5 is particularly instructive, just because at first glance the trouble appears so acute. The only nullifying prior is the uniform one $\pi(\eta)=$ const.; and it leads us back to (50) for all y. Surely, we have now run up against that overdetermination; it is simply a mathematical fact that there is *no* prior $\pi(\eta)$ that can leave B_1 and B_2 in agreement for all data sets (y, z) and all priors $\pi(\zeta)$.

Yet we would argue that there is still no real paradox here. This situation should not be disconcerting to anyone who has noted, in other Bayesian problems, that the effective sample number n often drops by one unit when we integrate out an unwanted parameter; or who, in using the χ^2 test, has reduced the number of degrees of freedom by one unit to take account of a parameter estimated from the data.

In fact, the explanation was noted in our section 3 above; the mere qualitative fact of the *existence* of the components (η, y), i.e. the knowledge that other parameters are present in our model beyond those of interest, *already* constitutes prior information relevant to B_1's inferences, that B_2 is ignoring. For further discussion, see section 11 below.

These examples demonstrate that two opposite extremes of behavior are possible; presumably, many or all of the conceivable intermediate cases are also possible. It is evident that a great deal more insight into the content of the integral equations (30) will be needed before any overall understanding of marginalization and its implications for Bayesian theory can be reached. In the writer's judgment, the remaining space available here is best used, not in communicating a mass of further isolated results like the above (which the reader can easily invent for himself), but by giving a preliminary survey of a more general attack on the structure of those integral equations, not restricted to the one-dimensional case. But before turning to that, we note some further pertinent clues from the DSZ examples with higher dimensionality.

9. Higher dimensionality

It appears from the foregoing that the case of a single location or scale parameter – or one that can be reduced to this by a change of variables – is disposed of once and for all; the only remaining function of the integral equations (30) is to determine whether, in a given model, the result is unique. Mathematically, this is the question whether the kernel of the integral equation is complete.

If the parameter η is two-valued, comprising both a location and scale parameter, i.e. if $\eta = (\mu, \sigma)$ and the corresponding data y can be separated into two components $y = (u, s)$ such that the sampling distribution has the form

$$p(zus|\zeta\mu\sigma) = s^{-2} g\left(z, \zeta; \frac{u-\mu}{\sigma}, \frac{s}{\sigma}\right), \tag{52}$$

then we can verify that the element of prior probability

$$\pi(\mu, \sigma)d\mu d\sigma = \frac{d\mu d\sigma}{\sigma} \tag{53}$$

will satisfy (30) with $s\lambda(s, u, z)$ a constant. Clearly, then, whether or not (53) is uniquely determined by (30), no disagreement between B_1 and B_2 can arise from its use. Yet DSZ produce apparent counterexamples, in which a prior of the form (53) *does* lead to disagreement! The DSZ paradoxes must, then, have been in part illusory. In the following examples we will see just how this has come about.

Example 2. Here we appear to be in the aforementioned difficulty, for DSZ note that the "paradox" (i.e. disagreement between B_1 and B_2) does not disappear for the "widely recommended prior" $d\mu_1 d\mu_2 d\sigma/\sigma$; but it does for $d\mu_1 d\mu_2 d\sigma/\sigma^2$ for which "no recommendations appear to exist". Of course, in a problem with two parameters the prior $d\mu d\sigma/\sigma$ is indeed widely – and as we have just seen, justifiably – recommended; but that is a very different problem. In the Appendix we discuss the present problem from the standpoint of the transformation group method recommended by the writer (Jaynes (1968)) and show that either result may be obtained depending on further details of the "real-life" situation which are not conveyed by the mere words "location parameter" or "scale parameter".

Our sampling distribution is, in the notation of eqs. (9)–(18),

$$p(usz|\mu\sigma\zeta) = A\frac{s^\nu}{\sigma^{\nu+2}} \exp\left[-\left(\frac{u-\mu}{\sigma}\right)^2 - R\left(z, \zeta, \frac{s}{\sigma}\right)\right]. \tag{54}$$

358 E. T. JAYNES

Note particularly that from (14), $\mu = \frac{1}{2}(\mu_1 + \mu_2)$. Since (54) has the form (52), the prior $d\mu d\sigma/\sigma$ must be a solution of (30). To see this directly, and to see whether the result is unique, we can write (30), using (12), in the form

$$\int_0^\infty \omega^\nu\, e^{-R}[f(u, \sigma) - s\lambda(u, s, z)]d\omega = 0, \qquad \begin{array}{l} 0 < s < \infty, \\ -\infty < u, z, \zeta < \infty, \end{array} \tag{55}$$

where $\lambda(usz) = p(us|zI_1)$, $f(u, \sigma)$ is given by (16), and in the integration over $\omega = s/\sigma$, s is held constant.

But (55) is an integral equation with complete kernel, since e^{-R} is the generating function for a complete set of (Hermite) functions;

$$e^{-R} = \exp[-\tfrac{1}{2}\omega^2(\nu + z^2)] \sum_{n=0}^\infty H_n\left(\frac{\omega z}{\sqrt{2}}\right)\frac{(\zeta/\sqrt{2})^n}{n!}. \tag{56}$$

Substituting this into (55), it is apparent that each term of the summand must vanish separately. A function orthogonal to a complete set must vanish almost everywhere, and so the necessary and sufficient condition (NASC) for an uninformative prior reduces to

$$f(u, \sigma) = s\lambda(u, s, z), \qquad \begin{array}{l} 0 < s, \sigma < \infty, \\ -\infty < u, z < \infty, \end{array} \tag{57}$$

from which we infer that $f(u, \sigma)$ is independent of σ, and the undetermined function

$$g(u) \equiv s\lambda(u, s, z) \tag{58}$$

is independent of s, z. Using (16), the NASC is then

$$\frac{1}{\sqrt{\pi}}\int_{-\infty}^\infty d\mu \exp\left[-\left(\frac{\mu - u}{\sigma}\right)^2\right]\pi(\mu, \sigma) = g(u), \qquad \begin{array}{l} 0 < \sigma < \infty, \\ -\infty < u < \infty. \end{array} \tag{59}$$

Evidently, for any $\pi(\mu, \sigma)$ that could be taken seriously as a prior, the integral (59) converges so well that $g(u)$ must be an entire function. But then appealing again to completeness and generating function relations of the form (56), the most general function satisfying (59) is

$$\pi(\mu, \sigma) = \sum_{n=0}^\infty a_n \sigma^{n-1} H_n\left(\frac{\mu}{\sigma}\right), \tag{60}$$

where a_n are arbitrary constants, and H_n are the Hermite polynomials. We then find $g(u) = \Sigma a_n(2u)^n$. Conversely, on substituting (60) back into (15), we find that B_1's posterior distribution reduces to B_2's, all the arbitrary constants a_n cancelling out upon normalization.

Of course, not all functions of the form (60) satisfy the further requirement $\pi(\mu, \sigma) \geq 0$; but (60) includes many non-negative priors. For example, if $\omega(q)$ is any non-negative function with moments of all orders, the choice

$$a_n = \frac{1}{n!} \int_{-\infty}^{\infty} \omega(q) q^n \, dq \tag{61}$$

leads to a non-negative prior

$$\pi(\mu, \sigma) = \sigma^{-1} \int dq\, \omega(q) \exp(2q\mu - q^2\sigma^2) \tag{62}$$

for which B_1 and B_2 will be in agreement. As a special case, if $\omega(q)$ goes into a delta function $\delta(q-t)$ we get $a_n = t^n/n!$ which in turn yields the anticipated Jeffreys prior $d\mu d\sigma/\sigma$ in the special case $t = 0$.

Also in this example, then, our early fears for the poverty of overdetermination disappear in an embarrassment of riches; from a mathematical standpoint the problem is grossly underdetermined. Nevertheless, out of the many different solutions of (59) the Jeffreys prior $\pi(\mu, \sigma) \sim \sigma^{-1}$ still appears to hold a favored position. Out of the class of solutions (62) it is the only one that does not become exponentially large as $|\mu| \to \infty$. We conjecture that some further restriction on the allowable behavior at infinity (for example that $\pi(\mu, \sigma)$ shall be at most $0(|\mu|^N)$ for some $N < \infty$) may lead, after all, to the Jeffreys prior as the unique solution.

Example 3. We have n independent observations of a bivariate (x_1, x_2) with model structure

$$x_1 = \sigma_1 e_1; \quad x_2 = \gamma x_1 + \sigma_2 e_2, \tag{63}$$

where e_1, e_2 are independent and $N(0, 1)$. We require inference about the correlation coefficient $\zeta = \gamma\sigma_1/(\gamma^2\sigma_1^2 + \sigma_2^2)^{1/2}$. The prior that avoids the paradox should then express complete ignorance about those components of the parameter space that can be varied with ζ held constant. DSZ note that the "recommended" prior

$$d\gamma \frac{d\sigma_1}{\sigma_1} \frac{d\sigma_2}{\sigma_2} \tag{64}$$

yields a posterior distribution identical with Fraser's structural distribution; but that the marginalization paradox is still present.

However, from the standpoint of the writer's transformation group method, the difficulty with (64) is obvious. For the model equation (63) is written in such a way that the parameter γ is not decoupled (i.e. it gets entangled in the change of scale transformations which express the fact that σ_1, σ_2 are scale parameters); and

360   E. T. JAYNES

so of course it cannot be assigned an independent prior. If we rewrite (63) as

$$x_1 = \sigma_1 e_1, \qquad x_2 = \sigma_2(e_2 + \tau e_1), \tag{65}$$

then $\tau \equiv \gamma \sigma_1/\sigma_2$ is decoupled, an arbitrary change of scale $\sigma_1' = a_1\sigma_1$, $\sigma_2' = a_2\sigma_2$ inducing no change in τ. But since the parameter of interest ζ is a function of τ (if $\tau = \tan\alpha$, then $\zeta = \sin\alpha$), the prior assigned to τ should not have anything to do with the paradox. Re-examining the equations of DSZ we find, as expected, that use of the prior

$$f(\tau)\,d\tau\,\frac{d\sigma_1}{\sigma_1}\,\frac{d\sigma_2}{\sigma_2} \tag{66}$$

avoids the paradox, where $f(\tau)$ is an arbitrary function.

The same result can be reasoned out without introducing τ. For in the model equation (63), γ appears not as a location parameter, but as a scale parameter (note that the product or ratio of two scale parameters is still a scale parameter). Complete ignorance of all three parameters should then be represented by a product of three Jeffreys priors. But again the prior assigned to the quantity of interest ζ should not matter; so we should be able to insert an arbitrary function $g(\zeta)$ without disrupting the agreement of B_1 and B_2. Indeed, one can verify that the prior

$$g(\zeta)\frac{d\gamma}{\gamma}\frac{d\sigma_1}{\sigma_1}\frac{d\sigma_2}{\sigma_2} \tag{67}$$

leads to agreement, and is equivalent to (66). The prior which DSZ noted as "avoiding the paradox" is a special case of (67), corresponding to $g(\zeta) = \zeta/(1-\zeta^2)^{1/2}$.

The joint likelihood function, which forms the kernel of the integral equations, can be read off from eq. (1.8) of DSZ. However, to avoid cumbersome expressions we introduce the notation

$$\beta \equiv \log\frac{\sigma_1}{\sigma_2}\sqrt{(1-\zeta^2)}, \tag{68}$$

$$b \equiv \log\sqrt{\left(\frac{S_{11}}{S_{22}}\right)}, \tag{69}$$

$$\omega \equiv \frac{\sqrt{(S_{11}S_{22})}}{\sigma_1\sigma_2}, \tag{70}$$

where S_{11}, S_{22} are the sample moments as defined by DSZ. The joint likelihood is then

$$L(\zeta, \sigma_1, \sigma_2) = \omega^n\, e^{-\omega T}, \tag{71}$$

where

$$T(z, b; \zeta, \beta) \equiv \frac{\cosh(\beta - b) - z\zeta}{\sqrt{(1 - \zeta^2)}} \qquad (72)$$

and $z \equiv S_{12}/\sqrt{(S_{11}S_{22})}$ is the sample correlation coefficient, whose sampling distribution depends only on ζ (DSZ, eq. (1.10)). The "unwanted" components of the parameter and sample spaces may then be taken as $\eta = (\sigma_1, \sigma_2)$; $y = (S_{11}, S_{22})$; and the NASC that a prior $\pi(\sigma_1, \sigma_2)$ shall be completely uninformative about ζ is

$$\int_0^\infty d\omega \int_{-\infty}^\infty d\beta [\pi(\sigma_1, \sigma_2) - \omega\lambda]\omega^{n-2}e^{-\omega T} = 0, \qquad \begin{array}{c} -1 < z, \zeta < 1 \\ 0 < S_{11}, S_{22} < \infty \\ n = 2, 3, \ldots \end{array}, \quad (73)$$

where $\lambda(z, S_{11}, S_{22})$ is an undetermined function, and $\{S_{11}, S_{22}, \zeta\}$ are held constant in the integrations over ω, β.

Evidently, if the kernels $\omega^{n-2}\exp(-\omega T)$ are complete on the domain of integration we shall be led to the Jeffreys prior $(d\sigma_1/\sigma_1)(d\sigma_2/\sigma_2)$ as the unique solution. We conjecture that this is the case; however, we have not succeeded in finding a fully rigorous proof of this, or a counterexample. Therefore, in view of the writer's astonishment at discovering the nonuniqueness of (59) after long believing it unique but being unable to prove it, we leave this an open question which others may perhaps answer.

Example 4a. At this point in the DSZ narrative, the sense of paradox increases sharply; for they produce two versions of a problem that appear not only paradoxical, but unavoidably inconsistent with each other. We obtain the sample $\{x_{11}, x_{12}, \ldots, x_{1n}\}$ from $N(\mu_1, \sigma)$ and $\{x_{21}, x_{22}, \ldots, x_{2n}\}$ from $N(\mu_2, \sigma)$. In version (1) B_1 is interested in inference about $\zeta \equiv \{\zeta/\zeta_1 = \mu_1/\sigma, \zeta_2 = \mu_2/\sigma\}$. The "unwanted" component is $\eta = \sigma$, and the corresponding data separation is in part $\{z_i = (ns)^{-1} \Sigma_j x_{ij}, i = 1, 2]$; y was not specified. Using the class of priors

$$\sigma^p d\mu_1 d\mu_2 d\sigma, \qquad (74)$$

DSZ show that B_1 and B_2 cannot agree unless $p = -3$.

But then in version (2) B_2 "asserts his interest in ζ_1 alone". Now ζ_2 becomes part of the "unwanted" parameters: $\eta = \{\sigma, \mu_2/\sigma\}$, and the paradox is resurrected; B_1 and B_2 cannot agree unless $p = -2$. Not only do the two priors contradict the Jeffreys rule; they contradict each other!

In view of our earlier demonstrations, something must be very amiss; yet we can find no error in the DSZ calculations. So the resolution must be – and is – very much simpler. We have here a case of paradox by optical illusion.

B_1 and B_2 are free to partition the parameter space $S_\theta = S_\zeta \times S_\eta$ in any way they please; but having chosen any such partition, the mathematical problem is: what prior $\pi(\eta)$ in the space S_η, i.e. *with ζ held constant*, is uninformative about ζ? The trouble was simply that, after choosing a partition, DSZ continued to write their prior (74) in terms of the old variables (μ_1, μ_2, σ), thereby failing to make the condition (ζ = const.) visible. Had DSZ transformed their priors to the new variables $\pi(\zeta)\pi(\eta)$ they would have found, in all these examples, that the "paradox" disappears just for the priors $\pi(\eta)$ recommended by Jeffreys. Far from suggesting any inconsistency in Bayesian principles, marginalization thus demonstrates again the power and basic soundness of the notions introduced into this field by Jeffreys some forty years ago.

10. Singular solutions: knowledge is ignorance

In the DSZ example 3, the correlation coefficient was considered the quantity of interest, $\rho = \zeta$, and we found that the prior $\pi(\eta)d\eta = d\sigma_1/\sigma_1)(d\sigma_2/\sigma_1)$ was completely uninformative about ζ. Can we reverse our viewpoint and find an uninformative prior $\pi(\rho)d\rho$ for the correlation coefficient? Most people, facing the problem of expressing ignorance of ρ, have chosen the form $\pi(\rho) \sim (1-\rho^2)^{-k}$ more or less instinctively; but complete agreement on the value of k still eludes us.

We would like to make the choice: $\{\zeta = (\sigma_1, \sigma_2), \eta = \rho\}$, and our method then requires that we find a separation of data $x = (y, z)$ for which $p(z|\zeta\eta) = p(z|\zeta)$. Perhaps this is possible, but our first guess: $\{z = (S_{11}, S_{22}); y = r = S_{12}/(S_{11}S_{22})^{1/2}\}$ does not work. The joint sampling distribution of (S_{11}, S_{22}) still depends on ρ, containing as a factor the modified Bessel function

$$I_{n/2-1}\left[\frac{\rho}{\sigma_1\sigma_2}\sqrt{\left(\frac{S_{11}S_{22}}{1-\rho^2}\right)}\right].$$

However, the sampling distribution of S_{11} depends only on σ_1; so let us marginalize using the choice: $\{\zeta = \sigma_1, \eta = (\sigma_2, \rho), z = S_{11}, y = (S_{22}, r)\}$. In view of what we found above, we shall perhaps be willing to take from the start $\pi(\eta) = \pi(\sigma_2, \rho) = \sigma_2^{-1}\pi(\rho)$. From eq. (1.8) of DSZ, we are led to the integral equation

$$\int_{-1}^{1}\pi(\rho)d\rho\int_{0}^{\infty}\frac{d\sigma_2}{\sigma_2^{n+1}}\exp\left\{-\frac{1}{2}\left[\frac{S_{11}}{(1-\rho^2)\sigma_1^2} - \frac{2\rho S_{12}}{(1-\rho^2)^{1/2}\sigma_1\sigma_2} + \frac{S_{22}}{\sigma_2^2}\right]\right\}$$

$$= \lambda \exp\left(-\frac{S_{11}}{2\sigma_1^2}\right), \tag{75}$$

where λ must be independent of σ_1. For the class of samples: $\{S_{22} = 2, S_{12} = 0\}$, (75) collapses to

$$\left(\frac{n-2}{2}\right)!\int_{-1}^{1}\exp\left\{-\frac{\rho^2 S_{11}}{2(1-\rho^2)\sigma_1^2}\right\}\pi(\rho)d\rho = \lambda(S_{11}), \tag{76}$$

MARGINALIZATION AND PRIOR PROBABILITIES   363

But this cannot be independent of σ_1 unless $\pi(\rho)$ is singular:

$$\pi(\rho) = \delta(\rho), \tag{77}$$

i.e. the prior information must be that $\rho = 0$ with certainty! Conversely, the prior (77) satisfies (75) for all samples.

On further reflection, we see that this result does, after all, make sense. B_2 is (in our reinterpretation) given data $\{x_1 \ldots x_n\}$ whose sampling distribution depends only on σ_1, and uses it to make the standard Bayesian inference about σ_1. In addition, B_1 has the other data components $\{y_1 \ldots y_n\}$. But if B_1 also knows that $\rho = 0$, then these additional data cannot help him to estimate σ_1; uncorrelated normal distributions are independent. B_1 and B_2 will then agree, not because B_1 is totally ignorant of ρ, but for the opposite reason, i.e. that his perfect knowledge of ρ makes his extra data irrelevant.

To recognize this puts a new dimension into the marginalization game. A prior $\pi(\eta)$ that is uninformative about ζ does not necessarily express ignorance about η; it depends on the structure of the model. If B_1 did not know ρ, his extra data $\{y_1 \ldots y_n\}$ would always be relevant and he would always revise B_2's conclusions about σ_1; there is no ignorance prior $\pi(\rho) \sim (1 - \rho^2)^{-k}$ that can avoid this. But if B_1 had far greater knowledge, he might throw away the new data and accept B_2's conclusions after all!

This is not paradoxical, but is a natural and necessary part of consistent plausible reasoning. We can see this phenomenon in generality already in eq. (2). If for some particular value $\eta = \eta_0$ we should have

$$\frac{\partial}{\partial y} p(\zeta | \eta_0, y, z, I_1) = 0, \tag{78}$$

then the singular prior $\pi(\eta | I_1) = \delta(\eta - \eta_0)$ will bring about agreement between B_1 and B_2 by making the data y irrelevant. Whenever the property (78) exists, the integral equation will have singular solutions representing ignorance of ζ due to perfect knowledge of η.

Now, at long last, we have enough clues in hand to commence a general attack on the integral equations.

11. Structure of the integral equations

For any fixed data set $x = (y, z)$, (30) is an integral equation which we write, for suggestiveness, in the form

$$\int_{S_\eta} K(\zeta, \eta) \pi(\eta) d\eta = \lambda f(\zeta), \qquad \zeta \in S_\zeta. \tag{79}$$

Already at this stage it is possible to have "ζ-*overdetermination*". The set of all functions on S_ζ forms a Hilbert space $\mathcal{H}_\zeta$. As η ranges over S_η the functions $K(\zeta,$

η) span a certain subspace $\mathcal{H}'_\zeta$ of $\mathcal{H}_\zeta$. If $f(\zeta)$ does not lie in $\mathcal{H}'_\zeta$, there can be no solution of (79). In these cases the mere qualitative fact of the existence of the components (η, y) – irrespective of their numerical values – already constitutes prior information relevant to B_1's inferences (because introducing them restricts the space of B_1's possible posterior distributions to $\pi(\zeta)H'_\zeta$). We saw an example in (49). In this case, the shrinkage of $\mathcal{H}_\zeta$ cannot be restored by any prior on S_η and the integral equations (79) ask an ill-advised question. In what follows we consider only the case where the problem is free from ζ-overdetermination.

If we think of $\pi(\eta)$ as a vector in a Hilbert space $\mathcal{H}$ of functions on S_η, then for each ζ, eq. (79) specifies the inner product of $\pi(\eta)$ with the function $K(\zeta, \eta)$. If as ζ varies over S_ζ these functions span the full space of $\mathcal{H}$ then the kernel $K(\zeta, \eta)$ may be said to be "complete", and the function $\pi(\eta)$ is defined "uniquely"; i.e. almost everywhere.

On the other hand, if the functions $\{K(\zeta, \eta):\zeta \in S_\zeta\}$ are not complete on S_η, they span some subspace $\mathcal{H}_0 \subset \mathcal{H}$, and (79) determines only the projection $\pi_0(\eta)$ of $\pi(\eta)$ onto $\mathcal{H}_0$, i.e. $\pi(\eta) = \pi_0(\eta) + \pi_1(\eta)$, where $\pi_1(\eta)$ is orthogonal to $\pi_0(\eta)$ but is otherwise undetermined. Since the coefficient λ is at this stage arbitrary, $\pi_0(\eta)$ is determined to within a multiplicative constant.

But all this has referred to only one particular data set x. For every different data set we can have a different kernel

$$K_x(\zeta, \eta) = p(yz|\eta,\zeta), \tag{80}$$

a different "driving force"

$$f_x(\zeta) = p(z|\zeta) = \int dy\, p(yz|\eta\zeta), \tag{81}$$

and a different coefficient λ_x. The equations

$$\int_{S_\eta} K_x(\zeta, \eta)\pi(\eta)d\eta = \lambda_x f_x(\zeta), \quad \zeta \in S_\zeta, \tag{82}$$

will, for two different data sets x, x', determine the projections $\pi_x(\eta)$, $\pi_{x'}(\eta)$ of $\pi(\eta)$ onto two different subspaces $\mathcal{H}_x$, $\mathcal{H}_{x'}$ of $\mathcal{H}$. If $\mathcal{H}_x$, $\mathcal{H}_{x'}$ are disjoint, the two integral equations (82) determine no relation between these solutions, i.e. the arbitrary constants C_x in $[\pi_x(\eta), \lambda_x]$ and $C_{x'}$ in $[\pi_{x'}(\eta), \lambda_{x'}]$ may be specified independently. But if $\mathcal{H}_x$ and $\mathcal{H}_{x'}$ are not disjoint (i.e. they have a common linear manifold $\mathcal{M}$), then there are several possibilities.

Case I. If $\mathcal{M}$ has dimensionality greater than one, the two integral equations for x and x' may determine different (i.e. linearly independent) projections of $\pi(\eta)$ onto $\mathcal{M}$. If these are both nonzero, then formally we can still escape overdetermination by setting $\lambda_x = \lambda_{x'} = 0$; and then hoping that some other data point x'' will allow $\lambda_{x''} \neq 0$. If one of the projections (say of $\pi_{x'}$) vanishes, then we need set only $\lambda_x = 0$. But in either case there will be an embarrassing situation; since λ_x has the meaning: $\lambda(y, z) = p(y|zI_1)$, we are escaping overdetermination only by assigning a prior $\pi(\eta)$ which says that the trouble-making data set $x = (y, z)$ is impossible!

One would be very reluctant to accept such a prior as "uninformative"; indeed, it would seem to be a rather obvious minimum requirement of any prior deserving of that name that it should not exclude in advance any data set permitted by the sampling distribution $p(y, z|\eta\zeta)$. A fully acceptable solution ought to lead to $\lambda > 0$ over the entire sample space. Case I thus represents a kind of moral – even if not formal mathematical – overdetermination. If it should occur for many pairs of data points, we could have also mathematical overdetermination, the only solution of (82) being $\lambda_x \equiv 0$, $\pi(\eta) \equiv 0$.

Case II. The two integral equations for x, x' agree that $\pi(\eta)$ is orthogonal to $\mathcal{M}$. Then the situation is basically as if the subspaces $\mathcal{H}_x$, $\mathcal{H}_{x'}$ were disjoint, i.e. no connection is established, and λ_x, $\lambda_{x'}$ may still be specified independently. As far as x, x' are concerned, the "unused" manifold $\mathcal{M}$ could be removed from the Hilbert space with no essential change in the problem (of course, if some other data point x'' should determine a nonzero projection onto $\mathcal{M}$, we are back to case I).

Case III. The two integral equations agree in assigning nonzero projections of $\pi(\eta)$ onto $\mathcal{M}$ that are the same within a multiplicative constant. Then the existence of a single function $\pi(\eta)$ demands that these multiplicative constants be equal. A connection is thus established so that, given λ_x, $\lambda_{x'}$ is determined. This can happen whatever the dimensionality of $\mathcal{M}$. If we can find a third data set x'' for which $\lambda_{x'}$ and $\lambda_{x''}$ have a common manifold, then $\lambda_{x''}$ is in turn determined by λ_x.

In this way, by a sequence of points $\{x \to x' \to x'' \to \ldots\}$ with overlapping manifolds the constants λ_x, originally arbitrary and independent at each point of the sample space, become tied together by the requirement of a single solution $\pi(\eta)$, into a function $\lambda(x)$ defined at many points. The existence or nonexistence of unique and "morally acceptable" noninformative priors then depends on whether by this process a single function $\lambda(x) > 0$ can be set up over the entire sample space.

Let us call any sequence $\{x_1, x_2, x_3, \ldots\}$ such that $\mathcal{H}_{x_i}$ and $\mathcal{H}_{x_{i+1}}$ overlap, a continuation path P. Then for any two points x, x' that can be connected by a continuation path, the ratio $(\lambda(x')/\lambda(x))$ is determined by the integral equations (82). The process is somewhat analogous to analytic continuation (but very different topologically, i.e. a sequence $\mathcal{H}_x$, $\mathcal{H}_{x'}$, ... of overlapping manifolds does not in general correspond to a continuous path in the sample space).
Case III is thus in turn comprised of three possibilities.

Case IIIa, "Nonintegrability". Two points x, x' can be connected by two different continuation paths P_1, P_2; but they yield different ratios $[\lambda(x')/\lambda(x)]_1 \neq [\lambda(x')/\lambda(x)]_2$. Then there is no single-valued nonvanishing function $\lambda(x)$ and, as in case I, the problem is morally – and, depending on how much of the sample space is infected with this disease, perhaps also mathematically – overdetermined. The avoidance of this case is analogous to a condition of integrability (but again, very different topologically!).

Case IIIb, "Intransitivity". The sample space S_x can be decomposed into two subspaces $S_x^{(1)}$, $S_x^{(2)}$ in such a way that there is no continuation path from any point in $S_x^{(1)}$ to any point in $S_x^{(2)}$. Then no connection is established between $\lambda^{(1)}(x)$ and $\lambda^{(2)}(x)$, i.e. they can be assigned independent arbitrary multiplicative constants. The problem is then underdetermined, and more than one "noninformative prior" exists. If there are K disconnected subspaces $\{S_x^{(1)} \ldots S_x^{(K)}\}$, then the prior $\pi(\eta)$ determined by (30) will contain K arbitrary constants.

Case IIIc, "Integrable transitivity". Any two points x, x' in the sample space can be connected by a continuation path, and if more than one such path exists, all paths assign the same ratio $[\lambda(x')/\lambda(x)]$. Then a single-valued function $\lambda(x) > 0$ exists over the entire sample space, and eqs. (82) define one unique noninformative prior $\pi(\eta)$, to within a normalization constant.

Previous Bayesian thought (including the writer's) which simply took for granted the existence of unique noninformative priors, has thus in effect assumed that we always have case IIIc. But looked at in this new way, it seems astonishing that such a thing could be true. If for any two points x, x' in our sample space we should have case I or case IIIa, then it is all over with our search for a "morally acceptable" noninformative prior. Yet we have the counterexamples of DSZ where such solutions *do* exist. What, in the structure of the problem, prevents these cases from occurring?

For enlightenment let us turn back, still another time, to our faithful DSZ Example 1, which has never yet failed to give us an interesting and useful answer to any question we have put to it.

12. Example 1 – a fourth look

We have seen already, in eq. (37), that the integral equations determine the Jeffreys prior $\pi(\eta) = \eta^{-1}$ uniquely; now we want to examine in minute detail the mechanism by which this is accomplished. Introducing the Laplace transform of $\eta^n \pi(\eta)$:

$$F(a) \equiv \int_0^\infty \eta^n \pi(\eta) e^{-a\eta} d\eta, \tag{83}$$

the set of integral equations (33) becomes

$$F[yQ(z, \zeta)] = \frac{(n-1)! \, y\lambda(y, z)}{[yQ(z, \zeta)]^n}, \quad \zeta = 1, 2, \ldots, (n-1). \tag{84}$$

MARGINALIZATION AND PRIOR PROBABILITIES 367

For a fixed data set $x=(y, z)$ this determines the value of $F(a)$ to within a multiplicative factor, at $(n-1)$ discrete points $a_\zeta = yQ(z, \zeta)$. The set of points $\{a_1(y, z) \ldots a_{n-1}(y, z)]$ will be called the spectrum of x. We can suppose without loss of generality that $c > 1$. From (6), the a_i are nonincreasing: $a_1 \geq a_2 \geq \ldots \geq a_{n-1}$, and all lie within a factor c of each other, i.e.

$$1 \leq a_1/a_{n-1} \leq c. \tag{85}$$

The subspace $\mathcal{H}_x$ is the one spanned by the set of functions

$$\phi_i(\eta) = \eta^n e^{-a_i \eta}, \quad i = 1, 2, \ldots, (n-1) \tag{86}$$

linearly independent if the a_i are all distinct.

Clearly, for $n > 3$ we can in general find another data set $x' = (y', z')$ such that

$$a_1(y, z) = a_2(y', z'), \tag{87a}$$

$$a_2(y, z) = a_3(y', z'), \tag{87b}$$

and the subspaces $\mathcal{H}_x, \mathcal{H}_{x'}$ then have a two-dimensional linear manifold $\mathfrak{M}$ in common, consisting of all functions of the form

$$\phi(\eta) = C_1 \phi_1(\eta) + C_2 \phi_2(\eta), \tag{88}$$

with arbitrary coefficients C_1, C_2.

Therefore unless both data sets x, x' determine the same projection of $\pi(\eta)$ onto this manifold:

$$\frac{F(a_1)}{F(a_2)} = \frac{F(a_2')}{F(a_3')}, \tag{89}$$

we shall have case I, and the problem will be morally overdetermined. Now from the data set x we have

$$\frac{F(a_1)}{F(a_2)} = \left[\frac{Q(z, \zeta_1)}{Q(z, \zeta_2)}\right]^n, \tag{90}$$

and from $x' = (y'z')$

$$\frac{F(a_2')}{F(a_3')} = \left[\frac{Q(z', \zeta_2')}{Q(z', \zeta_3')}\right]^n. \tag{91}$$

But (87) is, more explicitly,

$$yQ(z, \zeta_1) = y'Q(z', \zeta_2'), \tag{92a}$$

$$yQ(z, \zeta_2) = y'Q(z', \zeta_3'), \tag{92b}$$

from which we see that (90) and (91) are indeed equal. We have escaped overdetermination only because of the connection (87) required to produce a common linear manifold.

368   E. T. JAYNES

Likewise, we could have a three-dimensional common manifold by adding to (87) the condition

$$a_3(y, z) = a_4(y', z') \tag{93}$$

which is generally possible if $n > 4$. But again the three conditions (87a), (87b) and (93) are just sufficient to bring about equality of the three ratios $F(a_i)/F(a_j)$; and so on. We continue to have case IIIc.

We see now how different this problem is from the usual theory of integral equations with complete kernel. It is just the very great incompleteness of our kernel $K(\zeta, \eta)$ that, so to speak, creates room for agreement so that all the integral equations (82) can be satisfied simultaneously. Because of this incompleteness the subspaces $\mathcal{H}_x$ are so small, and scattered about so widely in the full Hilbert space $\mathcal{H}$ like stars in the sky, that it requires a very special relation between x, x' to bring about any overlapping manifold at all.

But it still seems magical that the relation required to produce overlapping should also be just the one that brings about agreement in the projections. So we still have not found the real key to understanding how case I is avoided.

Since there is a unique solution (37), a single-valued function $\lambda_x > 0$ must have been determined over the sample space S_x. Evidently, then, our integral equations must be transitive on S_x, i.e. there must exist a continuation path connecting any two points x, x'. What are these paths? Are there more than one for given x, x'? If so, how was nonintegrability (case IIIa) avoided?

We suppose the spectra $\{a_1, a_2, \ldots, a_{n-1}\}$, $\{a'_1, a'_2, \ldots, a'_{n-1}\}$ of x, x' to have no point in common (otherwise $\mathcal{H}_x$, $\mathcal{H}_{x'}$ have already a common manifold $\mathcal{M}$ and there is no need for a continuation path). If any point a_i is within a factor c of some point a'_j we define a new data point $x'' = (y'', z'')$ by

$$y'' = \frac{ca'_j - a_i}{c - 1}; \quad z''_2 = \frac{a_i - a'_j}{ca'_j - a_i}, \tag{94}$$

and $z''_3 = z''_4 = \ldots = z''_n = 0$. Then the first two points of the spectrum of x'' are, from (6),

$$a''_1 = a_i; \quad a''_2 = a'_j, \tag{95}$$

and $x \to x'' \to x'$ is a continuation path. If the spectra of x, x' are more widely separated (i.e. if $c^{k-1} < a_{n-1}/a''_1 < c^k$), then (because of the restriction (85) on the spectrum of any one point x'') it will require a continuation path with at least k intermediate points to connect them; but this can always be done by repetition of the above process. The reason for the transitivity is thus clear.

Now, how does this determine the function $\lambda(y, z)$? Writing the family of integral equations (33) as

$$G(a) \equiv \int_0^\infty d\eta \, \pi(\eta)(\eta y Q)^n \, e^{-\eta y Q} = (n - 1)! \, y\lambda(y, z) \tag{96}$$

for any given data point x, the necessary and sufficient condition that $\pi(\eta)$ be uninformative about ζ was that the integral in (96) take on equal values at $(n-1)$ discrete values of ζ, or $G(a_1) = G(a_2) = \ldots = G(a_{n-1})$. Introducing new data points x', x'', ... connected by continuation paths, this equality is extended to further values a', a'', ... Now $G(a)$ is a continuous function. As we continue to all points of the sample space S_x, if the set of spectral points a' where $G(a') = G(a_1)$ becomes everywhere dense on $0 < a' < \infty$, then $G(a) = $ const. is the only possibility. Eq. (96) then reads

$$\int_0^\infty d\eta\, \pi(\eta)\eta^n e^{-\eta a} = (\text{const.}) \times a^{-n}, \quad 0 < a < \infty \tag{97}$$

and, on inverting the Laplace transform, we have again the unique solution (41). On setting $\pi(\eta) = \eta^{-1}$, (97) reduces to

$$\lambda(y, z) = y^{-1} > 0. \tag{98}$$

Our questions have now been answered. Uniqueness of the solution requires that the set of spectral points a' be everywhere dense on the positive real line, and nonintegrability was avoided because the extension of λ_x along any continuation path connecting two points took the eminently satisfactory form that a function of (x, ζ) was a constant. So, by study of example 1 we see how all the conditions can be met, leading to the case IIIc most pleasing to a Bayesian.

The structure thus revealed will, of course, generalize readily to other problems. But our story has already grown too long, and the next Chapter must be told elsewhere.

13. Conclusion

While the full implications of marginalization for Bayesian statistical theory are still far from explored, the analysis given here represents at least the necessary beginnings. However, in research of this type, more than half the game usually lies in the slow process of recognizing the *existence* of an important solvable problem, and learning how to reduce vague conceptual questions to definite, clearly formulated mathematical ones. After that, further progress to the limit of our mathematical capabilities generally comes rapidly.

Viewed in this way, one is encouraged to think that the slow initial stages are now over, and we may hope in the near future to see major advances in the determination of prior probabilities by logical analysis.

The integral equations introduced here may or may not prove to be more widely useful, in practice, than previous desiderata for uninformative priors. At present, they seem to have at least the advantage of being general and noncontroversial, i.e. they express only the universally accepted principles of probability theory, making no use of intuitive ideas (symmetry, entropy, indifference, group invariance, "letting the data speak for themselves", etc.) which appeal differently to different people. Of course, with full understanding, those integral equations may in time be seen as stepping-stones to a still more appropriate and useful method, as yet unimagined.

The most encouraging sign of all is simply that, at last, the first prerequisite for progress in Bayesian theory is now an accomplished fact. The blind alleys have been tracked to their ends, and after decades of neglect and worse – even from some who professed to be Bayesians – the program started by Jeffreys is recognized as the true road to progress. Mathematical problems that might have been solved by Wald or Fisher in 1940 are, at last, being taken seriously and actually worked on.

At present, the crucial problem before us is: *What is the necessary and sufficient condition on the functional form of $p(y, z|\eta, \zeta)$ for the integral equations (30) to possess nontrivial and "morally acceptable" solutions?* Our analysis in section 11 above does not yet answer this; only the future will tell how close it has come to that goal.

Appendix A: comments on group analysis

The explicit mathematical use of group invariance as a criterion for assigning probability distributions goes back to Poincaré (1912), although of course the intuitive recognition of symmetry in gambling devices was present from the very beginnings (Cardano and Pascal). It appears to the writer that, in the final analysis, all applications of probability theory are based necessarily on such considerations, however much those motivations have been disavowed.

Since the term "group analysis" has several different meanings, we try here to indicate how they are related to each other and to the general problem of inference.

In the group structure (43) considered by DSZ the sampling distribution is invariant under two groups G, $\overline{G}$ operating simultaneously on the sample space and the parameter space. The status of that approach can be seen as follows. (a) Whatever group structure of this kind a problem may possess, is determined by the functional form of the sampling distribution. (b) Hence, whatever results may be deduced from that group structure, must also be deducible directly from the functional form. (c) Since group analysis cannot be more general than a "functional form" analysis – and is easily seen to be less general – the question of method reduces to whether, in a problem where it is applicable, group analysis leads to a more efficient calculation, or a better intuitive understanding, of the result. It seems clear that group analysis does accomplish both – often brilliantly. Therefore, by all means, let us take advantage of the DSZ group analysis whenever we can. (d) Nevertheless, whether or not any group structure exists, the necessary and sufficient condition for agreement between B_1 and B_2 is always the set of integral equations (30). For a general understanding of marginalization, then, it appears that we should appeal to the integral equations rather than the group structure.

Now an entirely different kind of group analysis (Jaynes (1968; 1971; 1976)) has also been proposed and illustrated in several applications. Since I believe it to be closer to the spirit of what one means intuitively by "ignorance", and also more widely applicable mathematically, let us look at it in the context of the DSZ example 2 (eqs. (54)–(62) above).

What prior probability element $\pi(\mu_1\mu_2\sigma)d\mu_1 d\mu_2 d\sigma$ expresses "complete ignorance" of two location parameters associated with a common scale parameter? We have a sampling distribution of the form

$$p(\mathrm{d}x\mathrm{d}y|\mu_1\mu_2\sigma) = h\left(\frac{x-\mu_1}{\sigma}; \frac{y-\mu_2}{\sigma}\right)\frac{\mathrm{d}x}{\sigma}\frac{\mathrm{d}y}{\sigma} \tag{A.1}$$

and we consider

Problem 1. Given the data $D \equiv \{(x_1, y_1); (x_2, y_2), \ldots, (x_n, y_n)\}$, estimate (μ_1, μ_2, σ).

Complete initial ignorance means, intuitively, that having no other basis for inference, our estimates are obliged to follow the data, i.e. a noninformative prior is the means by which one achieves Fisher's goal of letting the data speak for themselves. As noted in the text, it is also the necessary starting point for the construction of an informative prior.

Of course, a mere verbal statement such as "complete initial ignorance" is too vague to determine any mathematically well-posed problem. However, there is a rather obvious and basic desideratum of consistency: *in two problems where we have the same prior information we should assign the same prior probabilities.* Surely, any method for assigning priors which was found to violate this requirement would be rejected as self-contradictory.

Yet, as noted before (Jaynes (1968)), in many cases this desideratum is already sufficient to determine a unique solution. For, given the above problem 1, we can carry out a transformation of all variables: $\{x_i, y_i, \mu_1\mu_2\sigma\} \to \{x'_i y'_i \mu'_1 \mu'_2 \sigma'\}$ which involves a mapping $\theta \to \theta'$ of the parameter space onto itself, and consider:

Problem 2. Given the data D', estimate $(\mu'_1, \mu'_2, \sigma')$.

Any proposed prior $f(\mu_1 \mu_2 \sigma)d\mu_1 d\mu_2 d\sigma$ will be transformed into $g(\mu'_1 \mu'_2 \sigma')d\mu'_1 d\mu'_2 d\sigma'$ according to the Jacobian of the transformation

$$g(\mu'_1\mu'_2\sigma') = J^{-1}f(\mu_1\mu_2\sigma), \tag{A.2}$$

where $J(\mu_1\mu_2\sigma) = \partial(\mu'_1\mu'_2\sigma')/\partial(\mu_1\mu_2\sigma)$; and of course the transformation rule (A.2) will hold whatever the function $f(\mu_1\mu_2\sigma)$. But now the transformation may be such that we recognize problems 1 and 2 as entirely equivalent *problems*, i.e. they have the same sampling distribution and if initially we were "completely ignorant" of $(\mu_1\mu_2\sigma)$ in problem 1 – whatever that means – we are at least in the *same* state of knowledge about $(\mu'_1\mu'_2\sigma')$ in problem 2. But our desideratum of consistency then demands that f and g must be the same function, i.e. the prior representing complete ignorance must satisfy the functional equation

$$f(\mu_1'\mu_2'\sigma') = J^{-1}f(\mu_1\mu_2\sigma), \tag{A.3}$$

which determines the ratio of prior density at any two points θ, θ' of the parameter space that are connected by the mapping.

If then the mapping $\theta \to \theta'$ is one of a group of transformations that is transitive on the parameter space (i.e. from any point θ any other point θ' can be reached by some transformation of the group), then (A.3) determines the prior, to within a multiplicative constant, everywhere.

Note that, in this method the prior is determined by the Jacobian of the transformation *on the parameter space*; and this remains true whether the group is Abelian or non-Abelian, compact or noncompact. Therefore, considerations of right Haar measure or left Haar measure do not arise. Haar measure is defined on the group manifold; and not on our parameter space.

Furthermore, this method is more general; for if by any means we can recognize the group on the parameter space that transforms our prior state of knowledge into an equivalent one, the same result (A.3) will follow whether there is or is not an image group on the sample space. Thus, the Mallows example (45) has no group structure of the DSZ type; yet there is a natural group induced on S_n by Bayes' theorem (Jaynes (1968)) which leads to the uninformative prior $\pi(\eta) \propto [\eta(1-\eta)]^{-1}$; and let me acknowledge (correcting an erroneous statement in Jaynes (1968)) that, unknown to me at the time, this result, too, had been anticipated by Jeffreys and Haldane.

This will perhaps make clearer the distinction between our method and other group invariance arguments which do not appear to be motivated by the desideratum of consistency; or at least, to the best of the writer's knowledge, do not explicitly invoke it.

In this method a noninformative prior is not in general determined merely by the form of the sampling distribution; it is determined by specifying the invariance group on the parameter space. Furthermore, even if we do choose a group by the form of the sampling distribution, a given sampling distribution may be invariant under more than one group.

For the sampling distribution (A.1) perhaps the simplest transformation group is given by

$$\begin{aligned}
\mu' &= a\sigma, & 0 &< a < \infty, \\
\mu_1' &= \mu_1 + b_1, & -\infty &< b_1 < \infty, \\
\mu_2' &= \mu_2 + b_2, & -\infty &< b_2 < \infty,
\end{aligned} \tag{A.4}$$

with

$$\begin{aligned}
(x_i' - \mu_1') &= a(x_i - \mu_1), \\
(y_i' - \mu_2') &= a(y_i - \mu_2).
\end{aligned} \tag{A.5}$$

MARGINALIZATION AND PRIOR PROBABILITIES 373

The new sampling distribution $p(\mathrm{d}x'\mathrm{d}y'|\mu_1'\mu_2'\sigma')$ is then identical to (A.1). If our state of prior knowledge is such that this transformation results in a problem 2 that is entirely equivalent to problem 1, then from the Jacobian $J=a^{-1}$ of (A.4) the uninformative prior must satisfy the functional equation

$$af(\mu_1+b_1, \mu_2+b_2, a\sigma)=f(\mu_1\mu_2\sigma), \qquad (A.6)$$

from which we obtain the "widely recommended" prior element

$$f(\mu_1\mu_2\sigma)\mathrm{d}\mu_1\mathrm{d}\mu_2\mathrm{d}\sigma=\mathrm{d}\mu_1\mathrm{d}\mu_2\frac{\mathrm{d}\sigma}{\sigma}. \qquad (A.7)$$

However, when two "location" parameters are present, we may in some cases feel that this does not represent our prior knowledge. In (A.4) a change of scale $\sigma'=a\sigma$ affects only the accuracy of the x_i, y_i measurements. It may be that for other reasons not discernible in the sampling distribution (A.1), we know that the parameter σ not only sets the scale for the "measurement errors" $(x_i-\mu_1)$, $(y_i-\mu_2)$, but it also sets the scale on which the difference of means $(\mu_2-\mu_1)$ is to be measured.

As a concrete – if oversimplified – example, a spectroscopist may wish to determine the difference in magnetic moment of two atomic states by observation of the Zeeman effect, but the available magnet has uncontrollable field fluctuations. Here σ corresponds to the magnetic field strength, and μ_1, μ_2 to the resonant frequencies one is trying to measure. Doubling σ doubles the probable error in the measurements; but it also doubles the measurable difference $(\mu_2-\mu_1)$. On the other hand, the crystalline environment of the atoms affects both their frequencies in the same unknown way independent of σ. All this prior information is in the mind of an experimenter E_1, but it does not appear at all in the sampling distribution (A.1). Because of it, E_1 replaces (A.4) by

$$\begin{aligned} \sigma'&=a\sigma, & 0<a<\infty, \\ (\mu_2'-\mu_1')&=a(\mu_2-\mu_1)+c, & -\infty<c<\infty, \\ \tfrac{1}{2}(\mu_1'+\mu_2')&=\tfrac{1}{2}(\mu_1+\mu_2)+b, & -\infty<b<\infty. \end{aligned} \qquad (A.8)$$

This leads to the functional equation

$$a^2f(q\mu_1+p\mu_2+b-c, p\mu_1+q\mu_2+b+c, a\sigma)=f(\mu_1\ \mu_2\ \sigma), \qquad (A.9)$$

where $2q\equiv 1+a$, $2p\equiv 1-a$. But the left-hand side can be independent of both b and c only if $f(\mu_1\mu_2\sigma)=f(\sigma)$. The functional equation then collapses to $a^2f(a\sigma)=f(\sigma)$, or $f(\sigma)\propto \sigma^{-2}$, the prior that DSZ found to avoid the paradox in example 2.

We are far from having exhausted the number of transitive groups under which the sampling distribution (A.1) is invariant. For example,

$$\begin{aligned} \sigma'&=a\sigma, & 0<a<\infty, \\ \mu_1'&=a\mu_1+b, & -\infty<b<\infty, \\ \mu_2'&=\mu_2+c, & -\infty<c<\infty, \end{aligned}$$

leads again to $f(\mu_1\mu_2\sigma)\propto\sigma^{-2}$; while

$\sigma' = a\sigma,$
$\mu_1' = a\mu_1 + b,$
$\mu_2' = a\mu_2 + c,$

leads to $f(\mu_1\mu_2\sigma)\propto\sigma^{-3}$, which DSZ noted as avoiding the paradox in example 4a, version 1.

All of these correspond to different possible kinds of prior knowledge about the physical meaning of the parameters. These differences cannot be seen in the sampling distribution, which describes only the measurement errors. Thus, when we pass beyond pedagogical examples to real life problems, a further aspect of the quoted remark of Lindley (1971) becomes apparent.

As we see from this, group analysis does not answer questions of uniqueness. A given group leads to a definite prior, but there may be more than one group; and in any event group analysis – at least in any form yet visualized – does not tell us whether other solutions of the integral equations (30) may exist beyond those resulting from the group structure. However, it may be that new theorems bearing on this are waiting to be discovered.

Appendix B: historical note

Since statistical theory is returning to the original viewpoint of Laplace on the relation of inference and probability, we follow Laplace's example also in concluding with two remarks on the background of the marginalization problem, in addition to those noted by DSZ.

The mathematical facts underlying marginalization were fully recognized – and in the writer's view correctly interpreted – by Geisser and Cornfield (1963). Their eqs. (3.10) and (3.26) are just what we now call B_1's result and B_2's result; but instead of seeing a paradox in the difference, they very wisely termed the latter a "pseudoposterior distribution".

And inevitably, when we search for the origin of a Bayesian result, we turn to Jeffreys (1939). His section 3.8 considers the bivariate normal case, and although the sample correlation coefficient r is a sufficient statistic for ρ, the posterior distributions (10) and (24) again reveal the slight difference in B_1's conclusions caused by different prior information about the location parameters (a, b). He then gives B_2's result in (28), the agreement with (24) showing that a uniform prior for (a, b) is uninformative about ρ. Thus Jeffreys had unearthed essentially all the pertinent facts on which marginalization theory is based.

References

Dawid, A. P., M. Stone and J V. Zidek (1973) "Marginalization Paradoxes in Bayesian and Structural Inference", *Journal of the Royal Statistical Society* B35, 189–233.

Fraser, D. A. S. (1961) "On Fiducial Inference", *Annals of Mathematical Statistics* 32, 661–676.
Gamow, G. (1945) *The Birth and Death of the Sun* (Penguin Books, New York).
Geisser, S. and J Cornfield (1963). "Posterior Distributions for Multivariate Normal Parameters", *Journal of the Royal Statistical Society* B25, 368–376.
Hamermesh, M. (1962) *Group Theory* (Addison-Wesley, Reading, Mass.) pp. 239–295
Jaynes, E. T. (1968) "Prior Probabilities", *IEEE Transactions, Systems Science and Cybernetics* SSC-4, 227–241. Reprinted in V. M. Rao Tummala and R C. Henshaw, eds., *Concepts and Applications of Modern Decision Models* (Michigan State University Business Studies Series).
Jaynes, E. T. (1971) "The Well-Posed Problem", in: V. P. Godambe and D. A. Sprott, eds., *Foundation of Statistical Inference* (Holt, Rinehart & Winston, Toronto).
Jaynes, E T. (1976) "Confidence Intervals vs. Bayesian Intervals", in: W. L. Harper and C. A. Hooker, eds., *Foundations of Probability Theory, Statistical Inference, and Statistical Theories of Science*, vol. II (Reidel, Dordrecht) pp. 175–257.
Jeffreys, H. (1939) *Theory of Probability* (Clarendon Press, Oxford).
Kelvin (Lord) (1862) "On the Age of the Sun's Heat", *Macmillan's Magazine*, March.
Lindley, D. V. (1958) "Fiducial Distributions and Bayes' Theorem", *Journal of the Royal Statistical Society* B20, 102.
Lindley, D. V. (1971) *Bayesian Statistics: A Review* (Society of Industrial and Applied Mathematics, Philadelphia).
Poincaré, H. (1912) *Calcul des Probabilites*, pp. 118–130.
Villegas, C. (1971) "On Haar Priors", in: V. P. Godambe and D. A. Sprott, eds., *Foundations of Statistical Inference* (Holt, Rinehart & Winston, Toronto).

13. WHAT IS THE QUESTION? (1981)

At the International Meeting on Bayesian Statistics at Valencia, Spain in June 1979, two papers presented by Professors Arnold Zellner of the University of Chicago and Jose Bernardo of the University of Valencia, considered similar problems of hypothesis testing. In both, the point at issue was whether some conjectured new parameter λ was needed to represent the data. The problems were interesting because Bayesian theory allows more than one way of formulating them. One may define a 'null hypothesis' H_0 which asserts that $\lambda = 0$, and ask for its posterior probability; or one may ask simply for an estimate of λ, and from that decide the issue. The procedures ask different questions, yet either should be adequate for the practical purpose.

In Bernardo's problem λ was the only parameter involved, while Zellner's was more complicated in that there was also an unknown scale parameter σ. However, both chose the null hypothesis method and became involved in the question which prior probability $p(d\lambda \mid I)$ will lead to realistic conclusions. Bernardo used an information theory argument, asking essentially, 'For which prior would we learn the most from the data?' and was led to a prior that had a (to most of us) disconcerting dependence on the amount n of data. Zellner, following Jeffreys, used a Cauchy prior $p(d\lambda \mid \sigma I) \propto \sigma \, d\lambda/(\sigma^2 + \lambda^2)$ expressing a cautious kind of prior ignorance, that had a (to me at the time) disconcerting dependence on σ.

But from these very different appearing beginnings they emerged with quite similar results. One could interpret both as saying: the criterion that the data give significant evidence for a non-zero value of λ is that its estimated value $\hat{\lambda}$ be large compared to the probable error $\delta \lambda$ of that estimate. A relation to the parameter estimation method thus seemed to force its way into both solutions.

Now Laplace had considered very similar problems in the 18th Centrury, to decide whether astronomical data indicated the existence of some new systematic effect not accounted for in his calculations. He used the parameter estimation method, and got essentially the same conclusions that Zellner and Bernardo did, except that he did not need any tricky arguments about prior probabilities; the uniform priors that would have caused trouble for Zellner and Bernardo were just the ones that led Laplace to neat results without any trouble. So the relation between hypothesis testing and parameter estimation seemed closer than one would imagine from the conventional statistical literature.

In my role as discussant for these papers, it struck me that the presentation by Professor R. T. Cox at the M.I.T. Maximum Entropy Formalism Conference a year earlier, in which he recognized that answers to different questions do not necessarily convey different information, had a fundamental bearing on just this situation. Indeed, on some analysis it developed that if there are sufficient statistics, then all questions from whose answers the sufficient statistics may be recovered, elicit just the same information from the data, however different the questions may seem.

The following article was written partly out of evangelistic zeal to promote Cox's ideas, and partly out of a feeling that modern Bayesians, while certainly wise to pay attention to Jeffreys, ought also to pay some attention to Laplace, who started it all. He made many important contributions to scientific inference, not all of which are remembered in a literature that has done him so much injustice for over a Century.

The final conclusion comes out rather heavily in favor of parameter estimation over null hypothesis formulation; and I still believe that is where most of the truth lies, most of the time. But in subsequent discussions Zellner has persuaded me that there is after all, inherent in the nature of some real problems, some natural merit in the null hypothesis way of looking at things, and that the dependence of his prior on σ is, afterall, reasonable.

Pragmatically, one of the main advantages of Bayesian over orthodox methods is that Bayesians recognize the relevance of, and take into account, prior information even though it does not consist of frequencies. But Bayesians in turn need to be vigilant that we are not falling into the same error on a higher level and ignoring relevant information that does not fit into a preconceived pattern. In the last analysis, each problem must be considered separately, on its own merits; and if there are additional considerations that I have ignored here, one will quite properly come to a different conclusion than I did.

378   E. T. JAYNES

International Meeting on Bayesian Statistics

Valencia, Spain, May-June, 1979

WHAT IS THE QUESTION?

(Discussion by E. T. Jaynes on the Zellner-Siow and Bernardo presentations)

It is always interesting to recall the arguments that Jeffreys used to find priors. The case recounted by Zellner is a typical example where it appears at first glance that we have nothing to go on; yet by thinking more deeply, Jeffreys finds something. He shows an uncanny ability to see intuitively the right thing to do, although the rationalization he offers is sometimes, as Laplace said of Bayes' argument, "fine et très ingénieuse, quoiqu'un peu embarrassée." It was from studying these flashes of intuition in Jeffreys that I became convinced that there must exist a general formal theory of determination of priors by logical analysis of the prior information--and that to develop it is today the top priority research problem of Bayesian theory.

Pragmatically, the actual results of the Jeffreys-Zellner-Siow and Bernardo tests seem quite reasonable; without considerable analysis one could hardly say how or whether we should want them any different. Likewise, there is little to say about the mathematics, since once the premises are accepted, all else seems to follow in a rather straightforward and inevitable way. So let us concentrate on the premises; more specifically, on the technical problems encountered in both these works, caused by putting that lump of prior probability on a single point $\lambda = 0$.

WHAT IS THE QUESTION? 379

1. THE PROBLEM

In most Bayesian calculations the same prior appears in numerator and denominator, and any normalization constant cancels out. Usually, passage to the limit of an "uninformative" improper prior is then uneventful; i.e., our conclusions are very robust with respect to the exact prior range. But in Jeffreys' significance test this robustness is lost, since $K = p(D|H_0)/p(D|H_1)$ contains in the denominator an uncancelled factor which is essentially the prior density $\pi(\lambda)$ at $\lambda = \bar{x}$. Then in the limit of an improper prior we have $K \to \infty$ independently of the data D, a result given by Jeffreys (1939, p. 194, Eq. 10), and since rediscovered many times. Note that the difficulty is not due solely to the different dimensionality of the parameter spaces; it would appear in any problem where we think of H_0 as specifying a definite, fixed prior range, but fail to do the same for H_1.

Jeffreys (1961) dealt with this and other problems by using a Cauchy prior $\pi(\lambda|\sigma)$ scaled on σ in the significance test, although he would have used a uniform prior $\pi(\lambda) = 1$ in the same model H_1 had he been estimating λ. But then a question of principle rears up. To paraphrase Lindley's rhetorical question: Why should our prior knowledge, or ignorance, of λ depend on the question we are asking about it? Even more puzzling: why should it depend on another parameter σ, which is itself unknown? One feels the need for a clearer rationalization.

Furthermore, the difficulty was not really removed, but only concealed from view, by Jeffreys' procedure. All his stated conditions on the prior would have been met equally well had he chosen a Cauchy distribution with interquartile span 4σ instead of σ; but then all his K-values would have been quadrupled, leading to indifference at a very different value of the t-statistic [see Eq. (5-13) below]. We do not argue that Jeffreys made a bad choice; quite the contrary. Our point is rather that in his choice there were elements of arbitrariness, arising from a still unresolved question of principle. Pending that resolution, one is not in a position to say much about the "uniqueness" or "objectivity" of the test beyond the admitted virtue of yielding results that seem reasonable.

Bernardo comes up against just the same problem, but deals with it more forthrightly. Finding again that the posterior probability P_0 of the null hypothesis H_0 increases with the prior variance σ_1 in a disconcerting way, he takes what I should describe as a meat-axe approach to the difficulty and simply chops away at its prior probability p until $P_0 = pK/(pK+1-p)$ is reduced to what he considers reasonable (from the Jeffreys-Zellner-Siow standpoint he chops a bit too much, since his P_0 tends only to 1/2 on prolonged sampling when H_0 is true). This approach has one great virtue: whereas the Jeffreys results tended to be analytically messy, calling for tedious approximations, Bernardo emerges triumphantly (in the limit of large σ_1) with a beautifully neat expression (Eq. (11)) which has also, intuitively, a clear ring of truth to it.

But for this nice result, Bernardo pays a terrible price in unBayesianity. He gets it only by making p vary with the sample size n, calling for another obvious paraphrase of Lindley. This elastic quality of his prior is rationalized by an information-theoretic argument; it is, in a sense, the prior for which one would expect (before seeing the data) to learn the most from the experiment. But is this the property one wants?

WHAT IS THE QUESTION? 381

If a prior is to incorporate the <u>prior information</u> we had about λ before the sample was observed, it cannot depend on the sample. The difficulty is particularly acute if the test is conducted sequentially; must we go back to the beginning and revise our prior as each new data point comes in? Yet after all criticisms I like the general tone of Bernardo's result, and deplore only his method of deriving it.

The common plot of these two scenarios is: we (1) start to apply Bayes' theorem in what seems a straightforward way; (2) discover that the result has an unexpected dependence on the prior; (3) patch things up by tampering with the prior until the expected kind of result emerges. The Jeffreys and Bernardo tamperings are similar in effect, although they offer very different rationalizations for what they do. But in both cases the tampering has a mathematical awkwardness and the rationalization a certain contrived quality, that leads one to ask whether some important point has been missed.

Now, why should that first result have been unexpected? If, according to H_1, we know initially only that λ is in some very wide range $2\sigma_1$, and we then receive data showing that it is actually within $\pm\sigma/\sqrt{n}$ of the value predicted by H_0--as a physicist would put it, "the data agree with H_0 to within experimental error"--that is indeed very strong evidence in favor of H_0. Such data ought to yield a likelihood ratio $K = \sqrt{n}\sigma_1/\sigma$ increasing with σ_1, just as Bernardo finds. This first result is clearly the correct answer to the question Q_1 that was being asked.

If we find that answer disconcerting, it can be only because we had in the back of our minds a different, unenunciated question Q_2. On this view, the tampering is seen as a mutilation of equations originally designed to answer Q_1, so as to force them to answer instead Q_2.

The higher-level question: "Which question should we ask?" does not seem to have been studied explicitly in statistics, but from the way it arises here, one may suspect that the answer is part of the necessary "software" required for proper use of Bayesian theory. That is, just as a computer stands ready to perform any calculation we ask of it, our present theory of Bayesian inference stands ready to answer any question we put to it. In both cases, the machine needs to be programmed to tell it which task to perform. So let us digress with some general remarks on question-choosing.

2. LOGIC OF QUESTIONS

For many years I have called attention to the work on foundations of probability theory by R. T. Cox (1946, 1961) which in my view provides the most fundamental and elegant basis for Bayesian theory. We are familiar with the Aristotelian deductive logic of propositions; two propositions are equivalent if they say the same thing, from a given set of them one can construct new propositions by conjunction, disjunction, etc. The probability theory of Bernoulli and Laplace included Aristotelian logic as a limiting form, but was a mathematical extension to the intermediate region $(0 < p < 1)$ between proof and disproof where, of necessity, virtually all our actual reasoning takes place. While orthodox doctrine was rejecting this as arbitrary, Cox proved that it is the only consistent extension of logic in which degrees of plausibility are represented by real numbers.

Now we have a new work by Cox (1978) which may prove to be of even more fundamental importance for statistical theory. Felix Klein (1939) suggested that questions, like propositions, might be used as logical elements. Cox shows that in fact there is an exactly parallel logic of questions: two questions are equivalent if they ask the same thing, from a given set of them one can construct new questions by conjunction (ask both), disjunction (ask either), etc. All the "Boolean algebra" of propositions may be taken over into a new symbolic algebra of questions. Every theorem of logic about the "truth value" of propositions has a dual theorem about the "asking value" of questions.

Presumably, then, besides our present Bayesian statistics--a formal theory of optimal inference telling us which propositions are most plausible-- there should exist a parallel formal theory of optimal inquiry, telling us which questions are most informative. Cox makes a start in this direction, showing that a given question may be defined in many ways by the set of its possible answers, but the question possesses an entropy independent of its defining set, and the entropies of different questions obey algebraic rules of combination much like those obeyed by the probabilities of propositions.

The importance of such a theory, further developed, for the design of experiments and the choosing of procedures for inference, is clear. For over a century we have argued over which ad hoc statistical procedures ought to be used, not on grounds of any demonstrable properties, but from nothing more than ideological committments to various preconceived positions. There is still a great deal of this in my exchanges with Margaret Maxfield and Oscar Kempthorne in Jaynes (1976), and even a little in the exchange with Dawid, Stone, and Zidek over marginalization in Jaynes (1979). A formal theory of optimal inquiry might resolve differences of opinion in a way that Wald-type decision theory and Shannon-type information theory have not accomplished.

Our present problem involves a special case of this. If, seeing the answer to question Q_1 we are unhappy with it, what alternative question Q_2 did we have, unconsciously, in the back of our minds? Is there a question Q_3 that is the optimal one to ask for the purpose at hand? Since the conjectured formal theory of inquiry is still largely undeveloped, we try to guess some of its eventual features by studying this example.

Note that the issue is not which question is "correct." We are free to ask of the Bayesian formalism any question we please, and it will always give us the best answer it can, based on the information we have put into it. But still, we are in somewhat the position of a lawyer at a courtroom trial. Even when he has on the stand a witness who knows all the facts of the case and is sworn to tell the truth, the information he can actually elicit from this witness still depends on his adroitness in asking the right questions.

If his witness is unfriendly, he will not extract any information at all unless he knows the right questions to force it out, phrasing them as sharp leading questions and demanding unequivocal "yes" or "no" answers. But if a witness is friendly and intelligent, one can get all the information desired more quickly by asking simply, "Please tell us in your own words what you know about the case?" Indeed, this may bring out unexpected new facts for which one could not have formulated any specific question.

Significance tests which specify a sharply defined hypothesis and preassigned significance level, and demand to know whether the hypothesis does or does not pass at that level, therefore in effect treat probability theory as an unfriendly witness and automatically preclude any possibility of getting more information than that one bit demanded.

WHAT IS THE QUESTION? 385

Suppose we try instead the opposite tactic, and regard Bayesian formalism as a friendly witness, ready and willing to give us all the pertinent information in our problem--even information that we had not realized was pertinent--if we only allow it the freedom to do so. Instead of demanding the posterior probability of some sharply formulated null hypothesis H_0, suppose we ask of it only, "Please tell us in your own words what you know about λ?" Perhaps by asking a less sharp and restrictive question, we shall elicit more information.

3. INFORMATION FROM QUESTIONS

Evidently, to deal with such problems one ought to be an information theorist--and not only in the narrow sense of One-Who-Uses-Entropy. In the present problem we are concerned not only with the range of possible answers, as measured by the entropy of a question, but also with the specific kind of information that the question can elicit. In the following we use the word "information" in this specific sense rather than the entropy sense.

All statistical procedures are in the last analysis prescriptions for information processing: what information have we put into our mathematical machine, and what information are we trying to get out of it? In these terms, what is the difference--if any-- between significance testing and estimation? Having put certain information (model, prior, and data) into our hopper, we may carry out either, by asking different questions. But the answers to different questions do not necessarily convey different information.

The tests considered by Zellner and Bernardo sought information that can help us decide whether to adopt a new hypothesis H_1 with a value of λ different from its currently supposed value $\lambda = 0$. Presumably, any procedure which yields the same information would be equally acceptable for this purpose, even though current pedagogy might not call it a "significance test."

Now this information criterion establishes an ordering of different procedures, or "tests," rather like the notion of admissibility. If test B (which answers question Q_B) always gives us the same information as test A, and sometimes more, then B may be said to dominate A in the sense of information yield; or question Q_B dominates Q_A in "asking power." And if B requires no more computation, on what grounds could one ever prefer A?

In my work of 1976 (p. 185 and p. 219), I showed that the original Bayesian significance test of Laplace, which asks for the posterior probability P_1 of a one-sided alternative hypothesis, dominates the traditional orthodox t-test and F-test in just this sense. That is, given P_1 we know what the verdict would be, at any significance level, for all three of the corresponding orthodox tests (one equal-tails and two one-sided but the verdict of any one orthodox test is far from determining P_1. Thanks to Cox, we have now a much broader view of this phenomenon.

Let us call a question simple if its answer is a single real number; or in Cox's terminology, if its irreducible defining set is a set of real numbers. For example: "What is the probability that λ, or some function of λ lies in a certain region R?"

In any problem involving a single parameter λ for which there is a single sufficient statistic u, then given any simple question Q_A about λ, the answer will be, necessarily, some function a(u). Given any two such questions Q_A, Q_B and any fixed prior information, the answers $a(u)$, $b(u)$, being functions of a single variable u, must obey some functional relation $a = f(b)$. If $f(b)$ is single-valued, then the answer to Q_B tells us everything that the answer to Q_A does. As Cox puts it, "An assertion answering a question answers every implicate of that question." If the inverse function $b = f^{-1}(a)$ is not single-valued, then Q_B dominates Q_A.

In the case of a single sufficient statistic, then, any simple question whose answer is a strict monotonic function of u, yields all the information that we can elicit about λ, whatever question we ask; and it dominates any simple question whose answer is not a strict monotonic function of u. But this is just the case discussed by Bernardo; he considers σ known, and consequently $\bar{x}$ is a sufficient statistic for λ. Since his odds ratio $K(\bar{x})$ is not a strict monotonic function of $\bar{x}$, we know at once that Bernardo's test is dominated by another.

The Jeffreys-Zellner-Siow tests are more subtle in this respect, since σ is unknown, and consequently there are two jointly sufficient statistics $(\bar{x}, s)$. Given two simple questions Q_A, Q_B with answers $a(\bar{x}, s)$, $b(\bar{x}, s)$, the condition that they ask essentially the same thing, leading to a functional relation $a = f(b)$, is that the Jacobian $J = \partial(a,b)/\partial(\bar{x},s)$ should vanish. If $J \neq 0$, then neither question can dominate the other and no simple question can dominate both. But any two simple questions for which $(\bar{x}, s)$ are uniquely recoverable as single-valued functions $\bar{x}(a,b)$, $s(a,b)$ will jointly elicit all the information that any question can yield, and thus their conjunction dominates any simple question.

We may, therefore, conclude the following. Since Jeffreys' test asks a simple question, whose answer is the odds ratio $K(\bar{x},s)$, it can be dominated by a compound question, the conjunction of two simple questions. Indeed, since K depends only on the magnitude of the statistic t, it is clear that Jeffreys' question is dominated by any one simple question whose answer is a strict monotonic function of t.

These properties generalize effortlessly to higher dimensions and arbitrary sets. Whenever sufficient statistics exist, the most searching questions for any hypothesis test are those (simple or compound) from whose answers the sufficient statistics may be recovered; and all such questions elicit just the same information from the data.

As soon as I realized this, it struck me that this is exactly the kind of result that Fisher would have considered intuitively obvious from the start; however, a search of his collected works failed to locate any passage where such an idea is stated. Perhaps others may recall instances where he made similar remarks in private conversation; it is difficult to believe that he was unaware of it.

With these things in mind, let us re-examine the rationale of the Jeffreys-Zellner-Siow and Bernardo tests.

4. WHAT IS OUR RATIONALE?

In pondering this--trying to see where we have confused two different questions and what the question Q_2 is--I was struck by the contrast between the reasoning used in the proposed tests and the reasoning that physicists use, in everyday practice, to decide such matters. We cite one case history recent memory would yield a dozen equally good, which make the same point.

WHAT IS THE QUESTION? 389

In 1958, Cocconi and Salpeter proposed a new theory H_1 of gravitation, which predicted that the inertial mass of a body is a tensor. That is, instead of Newton's F = Ma, one had $F_i = \Sigma M_{ij} a_j$. For terrestrial mechanics the principal axes of this tensor would be determined by the distribution of mass in our galaxy, such that with the x-axis directed toward the galactic center, $M_{xx}/M_{yy} = M_{xx}/M_{zz} = (1+\lambda)$. From the approximately known galactic mass and size, one could estimate (Weisskopf, 1961) a value $\lambda \simeq 10^{-8}$.

Such a small effect would not have been noticed before, but when the new hypothesis H_1 was brought forth it became a kind of challenge to experimental physicists: devise an experiment to detect this effect, if it exists, with the greatest possible sensitivity. Fortunately, the newly discovered Mössbauer effect provided a test with sensitivity far beyond one's wildest dreams. The experimental verdict (Sherwin, et al, 1960) was that λ, if it exists, cannot be greater than $|\lambda| < 10^{-15}$. So we forgot about H_1 and retained our null hypothesis: H_0 = Einstein's theory of gravitation, in which $\lambda = 0$.

From this and other case histories in which other conclusions were drawn, we can summarize the procedure of the physicist's significance test as follows: (A) Assume the alternative H_1, which contains a new parameter λ, true as a working hypothesis. (B) On this basis, devise an experiment which can measure λ with the greatest possible precision. (C) Do the experiment. (D) Analyze the data as a pure estimation problem--Bayesian, orthodox, or still more informal, but in any event leading to a final "best" estimate and a statement of the accuracy claimed: $(\lambda)_{est} = \lambda' \pm \delta\lambda$. It is considered good form to claim an accuracy $\delta\lambda$ corresponding to at least two, preferably three, standard deviations. (E) Let λ_0 be the correct value according to the null hypothesis H_0 (we supposed $\lambda_0 = 0$ above, but it is now best to bring it explicitly into view), and define the "statistic" $t \equiv (\lambda' - \lambda_0)/\delta\lambda$. Then there are three possible outcomes:

If $|t| < 1$, retain H_0 STATUS QUO

If $|t| \gg 1$, accept H_1 AWARD NOBEL PRIZES

If $1 < |t| < 3$, withhold judgment   SEEK BETTER EXPERIMENTS

That is, to within the usual poetic license, the reasoning format in which the progress of physics takes place.

You see why I like the actual results reported here by Zellner and Bernardo, although I find their rationalizations puzzling. They did indeed find, as the criterion for accepting H_1, that the estimated deviation $|\lambda' - \lambda_0$ should be large compared to the accuracy of the measurement, considered known ($\sigma/\sqrt{n}$) in Bernardo's problem, and estimated from the data in the usual way ($s/\sqrt{n}$) in Zellner's.

It is in the criterion for accepting H_0 that we seem to differ; contra the physicist's rationale with that usually advanced by statisticians, Bayesian or otherwise. When we retain the null hypothesis, our reason is not that it has emerged from the test with a high posterior probability, or even that it has accounted well for the data. H_0 is retained for the totally different reason that if the most sensitive available test fails to detect its existence, the new effect $(\lambda - \lambda_0)$ can have no observable consequences. That is, we are still free to adopt the alternative H_1 if we wish to; but then we shall be obliged to use a value of λ so close to the previous λ_0 that all our resulting predictive distributions will be indistinguishable from those based on H_0.

WHAT IS THE QUESTION?   391

In short, our rationale is not probabilistic at all, but simply pragmatic; having nothing to gain in predictive power by switching to the more complicated hypothesis H_1, we emulate Ockham. Note that the force of this argument would be in no way diminished even if H_0 had emerged from some significance test with an extremely low posterior probability; we would still have nothing to gain by switching. Our acceptance of H_1 when $|t| \gg 1$ does, however, have a probabilistic basis, as we shall see presently.

Today, most physicists have never heard the term "significance test." Nevertheless, the procedure just described derives historically from the original tests devised by Laplace in the 18'th Century, to decide whether observational data indicate the existence of new systematic effects. Indeed, the need for such tests in astronomy was the reason why the young Pierre Simon developed an interest in probability theory, forty-five years before he became the Marquis de Laplace. This problem is therefore the original one, out of which "Bayesian statistics" grew.

As noted also by E. C. Molina (1963) in introducing the photographic reproduction of Bayes' paper, even the result that we call today "Bayes' theorem" was actually given not by Bayes, but by Laplace (the only valid reason I have found for calling it "Bayes' theorem" was provided at this meeting: "There's no theorem like Laplace's theorem" does not set well to Irving Berlin's music). Molina also offers some penetrating remarks about Boole's work, showing that those who have quoted Boole in support of their criticisms of Bayes and Laplace may have mistaken Boole's intention.

Now, although Laplace's tests were thoroughly "Bayesian" in the sense just elucidated, they encountered no such difficulty as those found by Jeffreys and Bernardo; he always got clear-cut decisions from uniform priors without tampering. To see how this was managed, let us examine the simplest of all Laplacian significance tests.

As soon as fairly extensive birth records were kept, it was noticed that there were almost always slightly more boys than girls, the ratio for large samples lying usually in the range $1.04 < (n_b/n_g) < 1.06$. Today we should, presumably, reduce this to some hypothesis about a difference in properties of X and Y chromosomes (for example, the smaller Y chromosome, leading to a boy, would be expected to migrate more rapidly). But for Laplace, knowing nothing of such things, the problem was much simpler. Making no reference to any causal mechanism, he took the model of Bernoulli trials with parameter λ = probability of a boy.

His problem was then: given specific data $D = \{n_b, n_g\}$, do these data indicate the existence of some systematic cause favoring boys? Always direc and straightforward in his thinking, for him the proper question to ask of the theory was simply: Q_L = "Conditional on the data, what is the probability that $\lambda > (1 - \lambda)$?" With uniform prior, the answer was

$$P_L = \frac{(n + 1)!}{n_b! \, n_g!} \int_{\lambda_0}^1 \lambda^{n_b} (1 - \lambda)^{n_g} \, d\lambda$$

with $n = n_b + n_g$, $\lambda_0 = 1/2$. In his <u>Essai Philosophique</u> Laplace reports many results from this, and in the <u>Theorie Analytique</u> (Vol. 2, Chap. 6) he gives the details of his rather tedious methods for numerical evaluation.

Needless to say, Laplace was familiar with the normal approximation to $p(d\lambda|D)$, the inverse of the de Moivre-Laplace limit theorem. But Laplace also realized that the normal approximation is valid only within a few standard deviations of the peak, and when the numbers n_b, n_g become very large, it can lead easily to errors of a factor of 10^{100} in $P_L/(1 - P_L)$; hence his tedious methods.

WHAT IS THE QUESTION? 393

Bernardo's example of Mrs. Stewart's telepathic powers, where the null hypothesis value $\lambda_0 = 0.2$ is about 24 standard deviations out, is another instance where the normal approximation leads to enormous numerical errors in K (many millions, by my estimate).

But pragmatically, once it is estimated that an odds ratio is about 10^{130}, it hardly matters if the exact value is really only 10^{120}. Once it is clear that the evidence is overwhelmingly in favor of H_1, nobody cares precisely how overwhelming it is. After Laplace's time, physicists lost interest in his accurate but tedious evaluations of P_L; for the criterion that we have overwhelming evidence in favor of a positive effect $(\lambda > \lambda_0)$, is just that the overwhelmingly greater part of the mass of the posterior distribution $p(d\lambda|D)$ shall lie to the right of λ_0. In the above example, the peak and standard deviation of $p(d\lambda|D)$ are $\lambda' = n_b/n$, $\delta\lambda = [\lambda'(1-\lambda')/n]^{\frac{1}{2}}$ and this criterion reduces to the aforementioned $t = (\lambda' - \lambda_0)/\delta\lambda \gg 1$, of the modern physicist's significance test--just the same criterion that Jeffreys and Bernardo arrive at in their different ways.

We have noted above that the orthodox t-test and F-test are dominated by Laplace's, and argued that the Jeffreys and Bernardo tests must also be dominated by some other. Let us now compare their specific tests with the ones Laplace would have used in their problems.

5. COMPARISONS WITH LAPLACE

In Bernardo's problem we have a normal sampling distribution $p(dx|\lambda,\sigma) \sim N(\lambda,\sigma)$ with σ known. Hypothesis H_0 specifies $\lambda = \lambda_0$, H_1 a normal prior $\pi(d\lambda|H_1) \sim N(\mu_1,\sigma_1)$, leading to a normal posterior distribution $p(d\lambda|DH_1) \sim N(\lambda',\delta\lambda)$ where

$$(\delta\lambda)^{-2} = n\,\sigma^{-2} + \sigma_1^{-2} \tag{5-1}$$

$$\lambda' = n(\delta\lambda/\sigma)^2\,\bar{x} + (\delta\lambda/\sigma_1)^2\,\mu_1 \tag{5-2}$$

Laplace, asking for the probability of a positive effect, would calculate

$$P_L = p(\lambda > \lambda_0 | DH_1) = \Phi(t) \tag{5-3}$$

where $\Phi(t)$ is the cumulative normal distribution, and as always, $t \equiv (\lambda'-\lambda_0)/\delta\lambda$

Bernardo (Eq. 9) finds for the posterior odds ratio

$$K_B = p(H_0|D)/p(H_1|D) = \exp(-R/2) \tag{5-4}$$

where

$$R = \frac{(\bar{x}-\lambda_0)^2}{\sigma^2/n} - \frac{(\bar{x}-\mu_1)^2}{\sigma_1^2 + \sigma^2/n} \tag{5-5}$$

But by algebraic rearrangement, we find this is equal to

$$R = t^2 - w^2 \tag{5-6}$$

where $w \equiv (\mu_1 - \lambda_0)/\sigma_1$ is independent of the data and drops out if $\mu_1 = \lambda_0$ or if $\sigma_1 \to \infty$. Bernardo would then find for the posterior probability of the null hypothesis

$$P_B = p(H_0|D) = [\exp(t^2/2) + 1]^{-1} \tag{5-7}$$

and comparing with (5-3) we have, as anticipated, a functional relation $P_B = f(P_L)$. To see the form of it, I plotted P_B against P_L and was surprised to find a quite accurate semicircle, almost as good as one could make with a compass. To all the accuracy one could use in a real problem,

WHAT IS THE QUESTION?

the functional relation is simply

$$P_B = [P_L(1-P_L)]^{\frac{1}{2}} \quad , \quad 0 \leq P_L < 1 \tag{5-8}$$

The error in (5-8) vanishes at five points in $(0 \leq P_L \leq 1)$.

Since $P_B = f(P_L)$ is single-valued while the inverse function is not, we have the result that Laplace's original significance test does, indeed, dominate Bernardo's. As stressed in Jaynes (1976), one-sided tests always dominate two-sided ones; given P_L we know everything that Bernardo's K or P_B can tell us; and if $|t| \gg 1$ we know in addition whether $\lambda > \lambda_0$ or $\lambda < \lambda_0$, which P_B does not give.

Of course, in this case one can determine that extra bit of information from a glance at the data; so the mere fact of domination is hardly a strong selling point. What is important is that Laplace's method achieves this without any elements of arbitrariness or unBayesianity.

In Jeffreys' problem we have the same sampling distribution, with the standard likelihood function $L(\lambda,\sigma) = \sigma^{-n} \exp[-ns^2 Q^2(\lambda)/2\sigma^2]$, where

$$Q(\lambda) \equiv [1 + (\lambda-\bar{x})^2/s^2]^{\frac{1}{2}} \quad . \tag{5-9}$$

H_0 and H_1 assign common priors $d\sigma/\sigma$, but H_0 specifies $\lambda = \lambda_0$, while H_1 assigns the Cauchy prior $p(d\lambda|\sigma H_1) = \pi(\lambda|\sigma)d\lambda$ with the density

$$\pi(\lambda|\sigma) = \frac{a\sigma}{\pi(a^2\sigma^2 + \lambda^2)} \tag{5-10}$$

scaled on σ (Jeffreys takes $a = 1$, $\lambda_0 = 0$, but we define the problem thus to bring out some points noted in Sec. 1). To analyze the import of the data, Jeffreys then calculates the likelihood ratio

$$K_J(\bar{x},s) = \frac{p(D|H_0)}{p(D|H_1)} = M^{-1} \int_0^\infty L(\lambda_0,\sigma)d\sigma/\sigma \tag{5-11}$$

while Laplace (if he used the same prior) would calculate instead the probability of a positive effect, given H_1:

$$P_L(\bar{x},s) = p(\lambda>\lambda_0|DH_1) = M^{-1} \int_{\lambda_0}^{\infty} d\lambda \int_0^{\infty} d\sigma \, \sigma^{-1} \, \pi(\lambda|\sigma)L(\lambda,\sigma) \qquad (5\text{-}12)$$

These expressions have a common denominator M, equal to the integral in (5-12) with $\lambda_0 = -\infty$.

It is straightforward but lengthy to verify that Jeffreys and Laplace do not ask exactly the same question; i.e., $J \equiv \partial(K_J, P_L)/\partial(\bar{x},s) \neq 0$. However, they are not very different, as we see on making the same approximation (large n) that Jeffreys makes. Doing the σ-integration in (5-12) approximately, the other integrals may be done exactly, leading to the approximate form

$$K_J \simeq [\pi(n-1)/2]^{\frac{1}{2}} a(1+q^2)/Q^n(\lambda_0) \qquad (5\text{-}13)$$

where $q \equiv (\bar{x}/as)$. This reduces to Jeffreys' result [Zellner's Eq. (2.7) in this volume] when $a = 1$, $\lambda_0 = 0$. In the same approximation, Laplace's result is the tail area of a t-distribution with (n-2) degrees of freedom:

$$P_L \simeq A_n \int_{\lambda_0}^{\infty} d\lambda/Q^{n-1}(\lambda) \qquad (5\text{-}14)$$

where A_n is a normalization constant. Of course, if Laplace used a uniform prior for λ, he would find instead the usual "Student" result with (n-1) degrees of freedom.

In the limit of an improper prior ($a \to \infty$), K_J diverges as noted in Sec. 1, the original motivation for both the Jeffreys and Bernardo tampering but the arbitrary parameter "a" cancels out entirely from Laplace's leading term, appearing only in higher terms of relative order n^{-1}.

WHAT IS THE QUESTION?

Had we been estimating λ instead, we should find the result $(\lambda)_{est} = \lambda' \pm \delta\lambda$, where $\lambda' = \bar{x}$, $\delta\lambda = s/\sqrt{n}$. But Laplace's result (5-14) is a function only of the statistic $t = (\lambda' - \lambda_0)/\delta\lambda$, and Jeffreys' (5-13) is too for all practical purposes (exactly so if $\lambda_0 = 0$, as Jeffreys assumes). Therefore, while considering σ unknown has considerably complicated the mathematics, it does not lead to any real difference in the conclusions. Again, Laplace's test yields the same information as that of Jeffreys, and in addition tells us the sign of $(\lambda - \lambda_0)$. In all cases--Jeffreys, Bernardo, Laplace, and the modern physicist's test--the condition that the data indicate the existence of a real effect is that $|t| \gg 1$.

6. WHERE DOES THIS LEAVE Q_1?

In summary it should not, in my view, be considered "wrong" to ask the original question Q_1 = "What is the relative status of H_0 and H_1 in the light of the data?" But the correct answer to that question depends crucially on the prior range of λ according to H_1; and so the question appears in the retrospect awkward.

Now the original motivation for asking Q_1, stated very explicitly by Jeffreys, was to provide a probabilistic justification for the process of induction in science, whereby sharply defined laws are accepted as universally valid. But as both Jeffreys and Bernardo note, H_0 can never attain a positive posterior probability unless it is given some to start with; hence that "pump-priming" lump of prior probability on a single point $\lambda = 0$. It seems usually assumed that this step is the cause of the difficulty.

However, the question Q_1 is awkward in another, and I think more basic, respect. The experiment cannot distinguish differences in λ smaller than its "resolving power" $\delta\lambda = s/\sqrt{n}$. Yet Q_1 asks for a decision between H_0 and H_1 even when $|\lambda-\lambda_0| < \delta\lambda$. On the other hand, the experiment is easily capable of telling us whether λ is probably greater or less than λ_0 (Laplace's question), but Q_1 does not ask this. In short, Q_1 asks for something which the experiment is fundamentally incapable of giving; and fails to ask for something that the experiment can give.

[Incidentally, a "reference prior" based on the Fisher information $i(\lambda)$ is basically a description of this resolving power $\delta\lambda$ of the experiment. That is, the reference prior could be defined equally well as the one which assigns equal probabilities to the "equally distinguishable" subregions of the parameter space, of size $\delta\lambda$. This property is quite distinct from that of being "uninformative," although they happen to coincide in the case of single location and scale parameters.]

But what we noted in Sec. 4 above suggests a different view of this. Why does induction need a probabilistic justification if it has already a more compelling pragmatic one? It is for the departures from the previous line of induction (i.e., switching to H_1) that we need--and Laplace gave--a probabilistic justification. Bernardo seems to have sensed this also, in being content with the fact that his $p(H_0|D)$ tends only to 1/2 when H_0 is true. Once we see that maintenance of the status quo requires no probabilist justification, the original reason for asking Q_1 disappears.

7. CONCLUSION

What both the Jeffreys and Bernardo tamperings achieved is that they managed to extricate themselves from an awkward start and, in the end, succeeded in extracting the same information from the data (but for the sign of $\lambda - \lambda_0$) that Laplace's question Q_L = "What is the probability that there is a real, positive effect?" elicited much more easily. What, then, was that elusive question Q_2? It was not identical with Q_L, and perhaps does not need to be stated explicitly at all; but in Cox's terminology we may take Q_2 as <u>any implicate of Laplace's question whose answer is a strict monotonic function of $|t|$</u>.

We have seen how the answers to seemingly very different questions may in fact convey the same information. Laplace's original test elicits all the information that can be read off from Jeffreys' $K_J(\bar{x},s)$ or Bernardo's $K_B(\bar{x})$. And for all purposes that are useful in real problems, Laplace's P_L may in turn be replaced by the λ' and $\delta\lambda$ of a pure estimation problem. Because of this, I suggest that the distinction between significance testing and estimation is artificial and of doubtful value in statistics--indeed, negative value if it leads to needless duplication of effort in the belief that one is solving two different problems.

REFERENCES

G. Cocconi and E. E. Salpeter (1958); Nuovo Cimento 10, p. 646.

R. T. Cox (1978); "Of Inference and Inquiry," in The Maximum-Entropy Formalism, R. D. Levine and M. Tribus, Editors, M.I.T. Press, Cambridge, Mass.; pp. 119-167.

E. T. Jaynes (1976); "Confidence Intervals vs. Bayesian Intervals," with discussion, in Foundations of Probability Theory, Statistical Inference, and Statistical Theories of Science, W. L. Harper and C. A. Hooker, Editors, D. Reidel Publishing Co., Dordrecht-Holland; pp. 175-257.

E. T. Jaynes (1979); "Marginalization and Prior Probabilities," with discussion, in Bayesian Analysis in Econometrics and Statistics; Essays in Honor of Sir Harold Jeffreys, A. Zellner, Editor; North Holland Publishing Company, Amsterdam; pp. 43-87.

F. Klein (1939); The Monist, 39, pp. 350-364.

E. C. Molina (1963); "Some Comments on Bayes' Essay," in Two Papers by Bayes W. E. Deming, Editor, Hafner Publishing Co., New York; pp. vii-xii.

C. W. Sherwin, et al (1960); Phys. Rev. Lett. 4, pp. 399-400.

V. F. Weisskopf (1961); "Selected Topics in Theoretical Physics," in Lectures in Theoretical Physics, Vol. III, W. Brittin et al, Editors; Interscience Publishers, Inc., New York; pp. 54-105.

14. THE MINIMUM ENTROPY PRODUCTION PRINCIPLE (1980)

Written in 1979, this work is still too new to have acquired much hindsight-inspired criticism, at least from me. Its rambling, piecemeal character is the result of last-minute radical surgery to meet the length limitation of 10 000 words (the article originally written being over 40 000). Therefore these comments now break precedent, to describe what this article does not contain, but should have.

Some analysis of the logic of Onsager's theory, and of the de Groot–Mazur approach, was sacrificed. Rather than giving a lengthy criticism of that thinking, it seemed more important to describe Tykodi's correction of it which, although applying only to the steady state, was logically unassailable and had been systematically ignored by the Establishment. Clearly, if I did not point it out, it would remain unknown.

It would be hard to overestimate the importance of the work of Truesdell and his co-workers on the reformulation of the phenomenological theory to include arbitrary time variations and fading memory effects. This will surely be the format of thermodynamics (in the true sense of that word) in the next Century. It seems to me almost miraculous that they could reason out, without any appeal to molecular details, almost the identical mathematical formalism that my students and I arrived at laboriously, by many years of statistical analysis. This work too has been systematically ignored by the Establishment, and I regret not having said more about it here.

Some comments on Gibbs' *Heterogeneous Equilibrium* (pointing out that the 'Gibbs Paradox' about entropy of mixing was explained and resolved already in this early work; but this was missed by later writers who had read only his unfinished *Statistical Mechanics*) were also deleted, and will appear elsewhere.

But the great lacuna is the one only crudely plugged up by the paragraph on page 421, referring to Mitchell's macroscopic source theory. Many pages, with some of the most important results of Irreversible Statistical Mechanics, lay on the cutting-room floor at this point. When gathered up again they will make another major article, in a sense the culmination of the whole *Maxent* approach in the application originally visualized.

THE MINIMUM ENTROPY PRODUCTION PRINCIPLE

E. T. Jaynes
Arthur Holly Compton Laboratory of Physics, Washington University, St. Louis, Missouri 63130

INTRODUCTION

It seems intuitively reasonable that Gibbs' variational principle determining the conditions of heterogeneous equilibrium can be generalized to nonequilibrium conditions. That is, a nonequilibrium steady state should be the one that makes some kind of generalized entropy production stationary; and even in the presence of irreversible fluxes, the condition for migrational equilibrium should still be the equality of some generalized chemical potentials.

We summarize progress to date toward this goal, reviewing (*a*) the early history, (*b*) work of Onsager and first attempts at generalization, (*c*) the new direction the field took after 1967 with the work of Tykodi and Mitchell, and (*d*) the present situation and prospects. Our conclusion will be, briefly, that the outlook is good in that the basic principles are believed known; but we do not yet know whether they can be reduced to simple rules immediately useful in practice, in the way that the Gibbs phase rule is useful. For this, we need more experience in the technique of applying them to particular cases, and more data to test some conjectures.

EARLY HISTORY

In 1848, Kirchhoff (1) generalized Ohm's law to three dimensions, and noted an interesting fact. If the electric field is $E = -\nabla\phi$, the conductivity $\sigma(x)$, then when a steady state is reached the potential $\phi(x)$ must cause no accumulation of electric charge at any point:

$$\nabla \cdot (\sigma \nabla \phi) = 0. \qquad 1.$$

But this is just the Euler-Lagrange equation stating that the rate of

ENTROPY PRODUCTION 403

production of Joule heat in a volume V

$$\int_V \sigma(\nabla\phi)^2 \, dV \qquad \qquad 2.$$

is stationary with respect to variations $\delta\phi(x)$ that vanish on the boundary of V. Thus the current distributes itself so as to dissipate the least possible heat for given voltages applied on its boundary. This is probably the first example of a steady nonequilibrium state determined by a variational principle.

In this respect, quantitative nonequilibrium thermodynamics may claim an earlier origin even than our conventional equilibrium theory, for Kirchhoff's discovery antedated by 27 years Gibbs' announcement (2) of the general variational principle for heterogeneous equilibrium, and even preceded Clausius' introduction (3) of the word "entropy" by 17 years. Yet 125 years after Kirchhoff's result, Girardeau & Mazo (4) state: "Variational methods for nonequilibrium statistical mechanics are virtually nonexistent." Why, after such a promising head start, has nonequilibrium theory lagged so far behind thermostatics?

It was evident that Kirchhoff's result could be generalized, and quickly other laws of "least dissipation of energy" and the almost equivalent reciprocal relations were found. In particular, an 1859 work of Helmholtz (5), which contained some of his greatest mathematical achievements, gave the acoustical reciprocity theorem, later extended by Rayleigh (6) and Lorentz (7) to mechanics and electrodynamics.

These first applications (where the thermal aspect, although in the picture, was not in the foreground) all involved variational principles for energy dissipation. Gibbs surely had first-hand knowledge of them, for he had spent a post-doctoral year (1868–1869) with Kirchhoff and Helmholtz in Heidelberg. But in Gibbs' own work, which began to appear four years later, the thermal aspect was the primary thing, and he gave instead a variational principle for entropy.

Gibbs lived another 25 years after completing his monumental work on heterogeneous equilibrium. Why then, with his seemingly perfect background for it, did not Gibbs himself generalize the Kirchhoff-Helmholtz results, and announce the principle of minimum entropy production 100 years ago? Perhaps Gibbs saw at once the difficulty.

Anyone familiar with Kirchhoff's work might simplify the arrangement to this: two resistors R_1, R_2 are in thermal contact with two heat reservoirs at temperatures T_1, T_2. Connecting the resistors in parallel, we send a total current $I = I_1 + I_2$ through them. How does it divide?

When a steady state is reached, the rates of production of heat and entropy are $\dot{Q} = R_1 I_1^2 + R_2 I_2^2$, $\dot{S} = (R_1/T_1)I_1^2 + (R_2/T_2)I_2^2$. The entropy production is a minimum when the current distribution satisfies $R_1 I_1/T_1 = R_2 I_2/T_2$. We know, of course, that the actual distribution will satisfy $R_1 I_1 = R_2 I_2$, which is the condition for minimum heat production.

The example is admittedly oversimplified; but one can invent arbitrarily complicated networks with the resistors at different temperatures and again, given the existence of a potential field $\phi(x)$ and the phenomenological laws $\Delta\phi_i = R_i I_i$ connecting current and potential difference for the individual elements, the steady-state current distribution for any applied voltages or currents is completely determined by Kirchhoff's condition of charge conservation at the nodes; there is logically no room for any further principle.

Now there is nothing special about electric current; what is true for fluxes of electrons is surely true for fluxes of any kind of stable particles, or of anything else that is conserved (energy, momentum, etc). Given the phenomenological relations connecting fluxes and forces, the steady state is determined by the conservation laws, leaving no room for any other principle; but then, what are we to make of the recent discussions of it?

Prigogine (8) postulates the existence of fluxes J_i and forces X_i, connected by the phenomenological relations $J_i = L_{ij} X_j$ (summation over repeated indices understood), so defined that the rate of entropy production is $\dot{S} = J_i X_i = L_{ij} X_i X_j$. Considering some of the forces to be fixed and others to be free, the condition that $\dot{S}$ be a minimum with respect to a free variable X_m is $\partial \dot{S}/\partial X_m = (L_{mj} + L_{jm}) X_j = 0$, if the L_{ij} are constants. But if the reciprocal relations $L_{ij} = L_{ji}$ hold, this is the same as $J_m = 0$, which is considered synonymous with "stationary state." This is the entire content of his theorem.

de Groot & Mazur (9) generalize Prigogine's treatment by taking spatial variations (but not convection currents) into account. They undertake to show that in heat conduction, "the stationary state is characterized by a minimum of the entropy production, compatible with the imposed temperature distribution at the walls of the system." Their proof is a paraphrase of Kirchhoff's, and it requires the assumption that the phenomenological coefficient L_{qq} defined by the heat current expression $J_q = L_{qq} \nabla(T^{-1})$ is independent of temperature; i.e. that the thermal conductivity λ defined by $J_q = -\lambda \nabla T$ varies with temperature as T^{-2}.

Since there is no known substance obeying this relation, there is no real situation involving heat conduction where the stationary state would be predicted quantitatively by minimizing entropy production. If $\lambda \propto T^b$, the steady state is the condition for minimum rate of production of the quantity $F = \int T^{b+2} dS$. But for all b, the steady state is predicted correctly by energy conservation, $\nabla \cdot J_q = \nabla \cdot (T^b \nabla T) = 0$. The same difficulty would have invalidated Kirchhoff's theorem if the electric conductivity σ varied with the potential ϕ.

de Groot & Mazur then give a more general example involving simultaneous heat conduction, diffusion, and chemical reactions. Their argument must now assume all the phenomenological coefficients L_{ij} involved to be independent of both temperature and the concentrations of the participating substances.

In all the examples given in (8, 9), after these restrictive assumptions are made the final Euler-Lagrange equations expressing minimum entropy production reduce simply to the conservation laws, which were valid exactly without any restrictive assumptions. So if we have enough information to apply the principle with any confidence, then we have more than enough information to solve the steady-state problem without it. This same criticism was made by Klein (10).

Gibbs surely would not have given any principle unless it met his standards of logical precision and was of some constructive use; so we are no longer surprised at his failure to give this one.

Yet after all criticisms, there remains a feeling that the principle does at least hint at an important truth, however imperfectly expressed. If the principle had nothing in it but misdirection, there would be no reason to write a review article about it.

REORIENTATION

There is a major part missing from our theoretical structure: On the one hand, the Kirchhoff-Helmholtz principles call out for generalization to thermodynamics; on the other, Gibbs' variational principle calls out for generalization to nonequilibrium cases. Surely, this gap can be filled; i.e. there must exist an exact variational principle for steady irreversible processes. It should include Gibbs' principle as a special case and be also (a) precise and general, requiring no restrictive assumptions like the above, and (b) constructive, yielding useful information that we would not have without it. But to find such a principle we must reorient our thinking in two respects.

First, we note the backward direction of the logic in the aforementioned examples. One assumed phenomenological forms which were only approximate; then stated a principle which could be only an approximate substitute for the conservation laws. We should rather take the conservation laws as exact and given, and seek a principle which gives the correct phenomenological relations without our having to assume them. It is reasoning in this direction that might lead to a precise, constructive principle.

But reversing the direction of the logic ought to reverse the principle. If the conservation laws represent the approximate condition of minimum entropy production for prescribed approximate phenomenological laws, then perhaps the exact phenomenology is the one that has maximum entropy production for prescribed exact conservation laws. Indeed, such a reversed principle would be much closer to the spirit of Gibbs' work.

Second, we need a verbal reorientation. The main difficulties that have retarded progress for a century are not mathematical, but conceptual; and these in turn are mainly artifacts of semantics. The words "irreversible," "entropy," "probability" are used indiscriminately with many different meanings, and the fact that the same word is used prevents many from seeing that the meanings are different.

Thus such a common phrase as "the paradox of how to reconcile the irreversibility of the second law with the reversibility of the equations of motion" records not a paradox but an abuse of language, the term "reversible" being used with two entirely different meanings. It is impossible to think and communicate rationally about these problems unless we use different words and symbols to convey different ideas.

By far the most abused word in science is "entropy." Confusion over the different meanings of this word, already serious 35 years ago, reached disaster proportions with the 1948 advent of Shannon's information theory, which not only appropriated the same word for a new set of meanings; but even worse, proved to be highly relevant to statistical mechanics. So it is necessary to insert at this point a short lexicon.

ENTROPY

As befits a word with many mutually contradictory meanings, "entropy" has also a rich and varied folklore concerning its etymology. According to Prigogine (8) it comes "from the Greek $\epsilon \nu \tau \rho o \pi \eta$ meaning 'evolution'." According to Clausius (3) it comes from $\tau \rho o \pi \eta$, meaning "a turning" or "a turning point" (the same root that appears in isotropic, phototropic, troposphere, etc). Clausius states that he added the *en-* only to make the word look and sound like "energy," although he might have noted that *en-* is in Greek, as in German and English, a standard modifying prefix, and $\epsilon \nu \tau \rho o \pi \eta$ which (according to three Greek dictionaries and two Greek friends) means "to turn one's head aside," rather neatly expresses the one-sided character of S that he had discovered. Because every

German noun is required to have a gender he also determined, by means unexplained, that "Die Entropie" is feminine.

Prigogine & Mayné (11) consider a quantity S_{PM} which they call "entropy," so defined that only near equilibrium can one express it in terms of macroscopic quantities. Their "second law" $\dot{S}_{PM} > 0$ is then to be a theorem in dynamics, and not in phenomenological physics.

The "entropies" with which we shall be concerned here are of a totally different nature. First is the experimental entropy S_E of Clausius, Gibbs, and G. N. Lewis, which is by construction a function $S_E(T, P, M, ...)$ of the observed macroscopic quantities. For us, as for them, the term "second law" refers to a property of S_E observed in laboratory experiments. It is therefore, by definition, a proposition of macroscopic phenomenology. Whether it might be also a theorem in dynamics was answered in the negative already by Gibbs (2) with a very vivid example of gas diffusion.

Second, we use the information entropy $S_I = -\Sigma p_i \log p_i$, a property of any probability distribution. In quantum theory, the $\{p_i\}$ are eigenvalues of a density matrix ρ, and $S_I(\rho) = -Tr(\rho \log \rho)$.

If S_I is maximized subject to certain constraints $\{A_1 ... A_n\}$, the maximum attained defines a third entropy $S(A_1 ... A_n) = (S_I)_{max}$, which is a function of those constraints. Since we may choose the constraints in many different ways, there are many different quantities $S(A)$, with different meanings. Just as Clausius' S_E is undefined until we specify which macroscopic variables are to be used, one must also indicate in each case which constraints are used—and therefore become the independent variables—in defining $S(A)$. In our applications, the $\{A_i\}$ may be any macroscopic quantities about which we have some information.

To keep the distinctions clear, our S_E is, as in conventional thermodynamics, a numerical multiple of Boltzmann's constant k, while S_I and $S(A)$ are dimensionless, following Shannon. Being defined as the maximum in a constrained variational problem, $S(A)$ will have, like S_E, a tendency to increase whenever a constraint is removed, thus paralleling in our mathematics what is observed in the laboratory (12).

Many other entropies appear in the literature, among which we note the Boltzmann and Gibbs S_B, S_G defined from the single-particle and N-particle distribution functions, and the quantity $S_{BEP} = k \log W$ of Boltzmann, Einstein, and Planck. The relations between these have been discussed in detail elsewhere (13a,b).

As should be evident, there is no possibility of finding the correct relations for irreversible processes unless one understands clearly the distinctions in meaning and the different properties of S_E, S_I, $S(A)$, S_B, S_G, and S_{BEP}. We can hardly expect that the variational principle we seek can hold for all of them. While the properties of S_I

and $S(A)$ are mathematical theorems, those of S_E are summaries of experimental facts.

For a closed system, Clausius defined S_E by the integral of dQ/T over a reversible path and stated that, in an adiabatic process from an initial equilibrium state (T_1, V_1) to a final one (T_2, V_2),

$$S_E(2) \geq S_E(1) \qquad 3.$$

with equality if and only if the process is reversible. Of all the statements of the second law made by Clausius and Planck, only Eq. 3 meets our requirements of logical precision; given certain provisos that we have stressed before (13), its truth or falsity can be determined in the laboratory to an accuracy limited only by the accuracy of our measurements, and not by the accuracy of definition of the terms in the equation. But in applications it tells us only in what general direction a change of state will go—not how far, how fast, or along what path.

Gibbs (2) generalized this to open systems and showed that a stronger statement is more useful in practice, telling us precisely "how far" and thus leading to quantitative predictions of the final equilibrium state reached. Let us call Eq. 3 the Clausius weak form of the second law, and append to it the Gibbs strong form: S_E not only "tends" to increase; it *will* increase, to the maximum value permitted by the constraints imposed. The exact constraints for which this is asserted (essentially the conservation laws) involve some standard technical discussion.

In the strong form we see entropy rising above its obscure beginnings and, so to speak, "presiding over" all of thermostatics; i.e. it determines, by its variational properties $\delta S = 0$, the set of all possible equilibrium states. In a similar way, the Lagrangian L presides over all of mechanics and electrodynamics, determining by its variational properties $\delta \int L dt = 0$ all the equations of motion, in any coordinate system.

We seek a generalization of entropy with properties more like a Lagrangian, which can by its variational properties generate our "equations of motion," telling us how fast, and along what path, an irreversible process will take place. The first general attack on this problem was made by Onsager (14a, b), whose work we now survey.

ONSAGER'S THEORY

Irreversible thermodynamics had its historical origins in Thomson's analysis of the thermocouple in 1854. For the effect of transporting a charge q around the circuit, he assumed that one might apply Carnot's principle in the form $\Sigma Q_i / T_i = 0$ to the reversible Peltier and Thomson heat effects even though irreversible heat conduction was also present.

ENTROPY PRODUCTION   409

Indeed, unless something like this were true, there would be few real applications in which one could ever apply Carnot engine arguments with any confidence. His assumption went beyond the principles of thermostatics and yielded, for the interaction of heat flow and electric current, the first example of an "Onsager reciprocal relation."

By 1931 many such relations had been noted, and Onsager (14) sought a general theoretical justification for them. His argument is still worth recalling because the formal relations survive, generalized and reinterpreted, in our present theory. We summarize it briefly, noting the four "serious" assumptions by Roman numerals and limiting comments to the square brackets.

A closed system is characterized by certain parameters $\{a_1 \ldots a_n\}$, so defined that they vanish in the equilibrium state of maximum entropy. Then in a neighborhood of equilibrium we may expand:

$$S = S_0 - (1/2)\Sigma G_{ij} a_i a_j + \ldots \qquad 4.$$

where G is a positive definite, symmetric matrix, $G = G^T$. The system is displaced from equilibrium by means unspecified, then released to find its way back to equilibrium. The derivatives

$$X_i \equiv \partial S / \partial a_i = -\Sigma G_{ij} a_j \qquad 5.$$

are thought of as the "forces" which drive the system back according to

I. $\dot{a}_i = \Sigma_j L_{ij} X_j$ \qquad 6.

where the L_{ij} are the "Onsager phenomenological coefficients." Thus the a's relax to zero along a trajectory given in matrix notation by $\dot{a} = -LGa$, or

$$a(t+\tau) = \exp(-LG\tau)a(t), \quad \tau > 0. \qquad 7.$$

Now we turn to situations very close to equilibrium and examine the small thermal fluctuations in the a's (which were neglected above). We postulate that the same entropy function $S(a_i)$ that supplied the forces X_i is also to supply the probability distribution of these fluctuations, i.e. the equilibrium distribution of the a's at equal times is given by a density function

II. $f(a_1 \ldots a_n) \propto \exp\left[k^{-1} S(a_i) \right]$ \qquad 8.

where k is Boltzmann's constant [at this point it appears that Onsager's entropy is most closely related to the S_{BEP} noted above]. Denoting averages over this distribution by angular brackets, we have $\langle a_i \rangle = 0$, while the matrix of second moments, $K_{ij} \equiv \langle a_i a_j \rangle$ is essentially the inverse of G: $KG = GK = kI$, where I is the unit matrix; K_{ij} is a covariance indicating how far, but not how rapidly, the a_i may be expected to fluctuate about zero.

We now make an assumption about this: that the average regression of these spontaneous fluctuations follows the same law, Eq. 7, as that assumed for forced deviations from equilibrium. That is, given the event $a(t)$, the conditional average of $a(t+\tau)$ at a later time, over many repetitions of the event, shall be

III. $\langle a(t+\tau)\rangle = \exp(-LG\tau)a(t), \quad \tau > 0.$ 9.

[This step is characteristic of the logic of stochastic theories; instead of asking what the microscopic equations of motion have to say about the matter, one simply ignores them and introduces intuitive "stochastic assumptions" at the macroscopic level.]

With this assumption we can define a time-dependent covariance matrix:

$$K_{ij}(\tau) \equiv \langle\!\langle a_i(t+\tau)a_j(t)\rangle\!\rangle \quad 10.$$

in which the double average is over the different motions averaged in Eq. 9, and then over the distribution, Eq. 8. Inserting Eq. 9 into Eq. 10, this means that the covariance matrix must also decay according to the macroscopic law, Eq. 7:

$$K(\tau) = \exp(-LG\tau)K(0) = K(0)\exp(-GL\tau), \quad \tau > 0 \quad 11.$$

where $K(0) = kG^{-1}$ is the same matrix that we denoted by K above, and we used an identity of any matrix function: $f(LG)G^{-1} = G^{-1}f(GL)$. $K(\tau)$ as defined by Eq. 10 is independent of t; $K(-\tau) = K^T(\tau)$; or from Eq. 11,

$$K(-\tau) = K(0)\exp(-GL^T\tau), \quad \tau > 0 \quad 12.$$

since the transposed matrix function is $f^T(LG) = f(G^TL^T) = f(GL^T)$.

Finally, we invoke the famous assumption that Onsager called "microscopic reversibility":

IV. $K(-\tau) = K(\tau).$ 13.

Comparing Eq. 11 and 12 we have the grand result

$$L = L^T. \quad 14.$$

Onsager's argument showed a remarkable instinct for sensing the right formal relations, which have stood the test of fifty years. But he chose a thorny path to them, ignoring the smooth path made by his predecessor at Yale. The relation he needed was Eq. 11; given that, the rest of the derivation is a two-line triviality. But to reach it he (*a*) assumed a phenomenological form that was (I*a*) linear, (I*b*) without memory; (b) assumed that the average regression of fluctuations follows that same phenomenological law; (c) from these deduced Eq. 11: the covariance function $K(t)$ also follows that phenomenological law.

But had he taken the path of a Gibbsian statistical theory instead of a stochastic one, this result—including space dependences and all memory effects—would have been present from the start with no need to assume any phenomenological form or to mention regression of fluctuations at all. For in such a theory, the predicted space-time dependence of any macroscopic process is given by a covariance function $K(x, t; x', t')$.

For example, in acoustics the sound pressure $\delta P(x, t)$ due to a source distribution $s(x', t')\sec^{-1}$ (i.e. cm^3 sec^{-1} per cm^3) is given by a linear superposition

$$\delta P(x,t) = \int d^3x' \int_{-\infty}^{t} dt' G(x,t; x', t') s(x', t'). \qquad 15.$$

At thermal equilibrium, Gibbsian statistical theory gives for the Green's function

$$G(x, t; x', t') = (1/kT)\langle \delta P(x, t) \delta P(x', t') \rangle, \qquad 16.$$

i.e. just $(kT)^{-1}$ times the covariance of the thermal pressure fluctuations. This linear response kernel contains all memory effects, including propagation time delays, reflection from walls, "ringing" due to multiple scatterers and resonators, ultrasonic dispersion and attenuation due to relaxation in the medium, etc. Its obvious symmetry is just the Helmholtz-Rayleigh reciprocity theorem.

Onsager's viewpoint fits in nicely with our conjectured reorientation. If, as stated by Eq. 5, the force driving the system back to equilibrium is the entropy gradient, then instead of minimizing entropy production, the system is maximizing it, trying to get to equilibrium as rapidly as it can, subject to whatever restraints are preventing this. But looking at the relations in this way suggests an additional conjecture.

It appears to us that Onsager might have obtained more useful results by making a different assumption, which seems no stronger than Eq. 13. Since G is real and symmetric, it can be diagonalized by an orthogonal matrix O. In the coordinate system of the new variables $a'_k = \Sigma_i O_{ki} a_i$, the matrix $G' = OGO^{-1}$ is diagonal, and so the force X'_k is merely a numerical multiple of a'_k. The a'_k are uncorrelated at equal times and the entropy function $S = \Sigma S(a'_k)$ is the same as if we had n separate, noninteracting systems. So it seems plausible that in the absence of magnetic or Coriolis forces the a'_k should relax independently; in other words, that the new phenomenological matrix $L' = OLO^{-1}$ should also be diagonal. If so, then L and G must commute in the original coordinate system $[a_i]$.

But $LG = GL$ is a stronger condition than the Onsager symmetries $L = L^T$. For example, with $n = 3$ fluxes, the Onsager relations reduce the number of independent phenomenological coefficients from $n^2 = 9$ to $n(n+1)/2 = 6$. The condition $LG = GL$ yields these and three additional relations, leaving only $n = 3$ independent coefficients. If the matrix G were known from equilibrium measurements, one would then need only three nonequilibrium measurements: for example the self-conductances L_{11}, L_{22}, L_{33}; whereupon all six coupling coefficients L_{ij}, $i \neq j$, would be determined. In the case $n = 2$, the coupling coefficients would reduce to $L_{12} = L_{21} = G_{12}(L_{11} - L_{22})/(G_{11} - G_{22})$.

We point this out in the hope that some readers may be in possession of enough experimental data to check the relation $LG = GL$. If this conjecture should be confirmed, irreversible thermodynamics would become more useful, since one could predict considerably more about irreversible processes from equilibrium data.

INTERLUDE

In the 1940s and 1950s some attempts were made to generalize Onsager's treatment to a macroscopic continuum theory based on the notions of local equilibrium and local rate of entropy production. In 1962 this approach was summarized in the book of de Groot & Mazur (9), where references to the vast literature it generated can be found.

This approach postulates the existence of a local entropy density $s(x, t)$ which plays the role of a field variable. It is to have also a flow rate J_s and source strength $\sigma(x, t) \geq 0$, so as to obey the field equation $\dot{s} + \nabla \cdot J_s = \sigma(x, t)$. Entropy is thus conceived of as a kind of fluid which, once created, is conserved forever after.

Mathematically, the notion of entropy can be generalized to nonequilibrium conditions in many different ways. Basically, the issue is not which is "correct," but which ones have demonstrable and useful physical properties. We agree that a useful theory should be set up as a continuum field theory; but if we allow entropy to degrade into no more than one of many field variables, we shall lose just those properties that made entropy uniquely useful in the work of Gibbs and Onsager.

Therefore we shall seek, rather, to elevate entropy to a functional $S[A_1(x, t) \ldots A_n(x, t)]$ over the thermokinetic history of the field variables so that it can retain those properties, while acquiring a new generating power like a Lagrangian; only thus do we see the possibility of reaching our goal.

In any event, de Groot & Mazur use, without defining, a local entropy density in an inhomogeneous nonequilibrium state. In addition they suppose that the equilibrium expressions for temperature and chemical potentials can be used as local field variables, obeying the Gibbs equilibrium relation $TdS = dU + PdV - \Sigma \mu_i dn_i$ even when gradients and irreversible fluxes are present.

Now one expects that procedures of this kind should, like Thomson's, meet with some success very close to equilibrium; and of course de Groot and Mazur did not claim any more than this. But a "local equilibrium" approach has no criterion for judging its range of validity and provides no basis for further development, since it contains scarcely any quantity that has a precise meaning in a nonequilibrium state.

This approach, therefore, reached a dead end. The logic of using equilibrium relations in nonequilibrium situations was hardly an advance over that used by Thomson in 1854; indeed, we are unable to see wherein they differ at all. To make further progress beyond this point, it was necessary to go back to first principles and reason things out all over again, much more carefully, The coup de grace and final benedictions were administered by Wei (15) and Truesdell (16).

RESURRECTION

In 1967, Tykodi (17) showed how entropy production theories might be not only salvaged, but made in a sense exact, using logic so simple and direct that one could not question any part of it without at the same time questioning a considerable part of established equilibrium theory. He simply abandoned altogether the notions of local equilibrium and local entropy production, and reasoned as follows.

There is one case where logically impeccable inferences about an irreversible process were drawn from the relations of equilibrium theory: the Joule-Thomson porous plug experiment of 1852. The inflowing gas is at thermal equilibrium with temperature and pressure (T_1, P_1), and we measure the outflowing gas far enough downstream from the plug so that it has come back to thermal equilibrium, with new values (T_2, P_2). By a simple argument given in all the textbooks we are persuaded at once that, however violent the irreversible process taking place in the plug (it might, for example, involve locally supersonic velocities, shock waves, chemical reactions catalyzed by the plug, etc), if the plug cannot communicate directly with the outside world, so it does no work and all the heat generated must be carried off by the effluent gas, then when a steady state is reached, the (enthalpy + kinetic energy of mass flow) of the incoming and outgoing gases must be the same.

In other words, established equilibrium theory does enable us to draw rigorous inferences about steady irreversible processes that begin and end in states of complete thermal equilibrium. This is just the conclusion we noted already in Eq. 3, and which had been stressed in the writer's pedagogical article (13). But as soon as we recognize this the road is straight and we can see for miles, for the Joule-Thomson example can be generalized endlessly.

In the first place, the barrier need not be a simple "plug." It may contain apparatus of any complexity, and even if conditions in it never come to a steady state, but go into limit-cycle oscillations, if the apparatus contains suitable "mufflers" so that there is eventually uniform inflow and outflow, the conclusion still holds.

Furthermore, nothing restricts us to a system with only two channels, one for inflow and one for outflow. We can have any number of inflow channels, containing different chemical substances, or mixtures of them, at different pressures and temperatures, flowing at different rates; and any number of similar outflow channels. And nothing restricts us to gases; a channel could transport liquid, solids, plasma, electrons, radiation, etc. There need not be a single reaction region; the plumbing might be arranged to carry any number of substances to any number of reaction vessels in any sequence. In short, we may imagine an arbitrary continuous-flow processing plant.

For any such arrangement we can define an energy flux $H_i' =$ (enthalpy + kinetic energy of mass flow) transported from the reaction region per unit time or per oscillation cycle in the ith channel, and a similar entropy flux S_i. The reaction region may communicate directly with the outside world, doing work W per unit time or per oscillation cycle. Under these conditions the energy balance requirement gives rigorously $\Sigma H_i' + W = 0$, while at the same time the total rate of entropy production ΣS_i is now unambiguously defined by equilibrium theory.

Only at this point is one in a position to discuss entropy production principles in a meaningful way. All ambiguities about the definition of temperature and entropy in a nonequilibrium state have been eliminated, since however such notions may or may not be defined eventually, at least in a steady state they are not changing. And we are not limited to near-equilibrium regimes with linear phenomenological laws; nor have we neglected fading memory effects.

If J_i is the flux in the ith channel in moles (grams) per second, then the rate of entropy production is

$$\dot{S}_E = k \Sigma_i \lambda_i J_i \qquad 17.$$

where $k\lambda_i$ is the entropy per mole (gram) of the ith substance. If it is a pure chemical substance, then $\lambda_i = -\mu_i/kT$ is essentially the chemical potential. The quantities λ_i, which we call simply the "potentials," are, however, the fundamental quantities of our theory.

Although Eq. 17 looks at first glance like the Onsager expression $\dot{S} = \Sigma X_i J_i$, it has a different meaning. In the first place, Eq. 17 is not a quadratic approximation holding near equilibrium; it is the exact rate of entropy production for any departure from equilibrium. Secondly, there are more terms in Eq. 17 and they are not independent. If particles of type k enter via channel 3 and emerge unchanged but for pressure and temperature in channel 7, in Eq. 17 this contributes two terms $k(\lambda_3 J_3 + \lambda_7 J_7)$, constrained by the conservation law $J_3 + J_7 = 0$, but only one $X_k J_k$ in the Onsager form. Where Onsager took his forces as derivatives, $X_k = \partial S/\partial a_k$, we see that the exact "force" should be $X'_k = k(\lambda_7 - \lambda_3)$, a finite difference of potentials.

If we eliminate fluxes determined by the conservation laws and rewrite Eq. 17 in terms of independently variable fluxes we obtain the Onsager form $\dot{S}_E = \Sigma X'_k J_k$. In these terms, Tykodi states a minimum entropy production principle that, close to equilibrium, is equivalent to the Onsager relations. He conjectures that this principle (varying X_m while holding the other forces constant, minimum $\dot{S}$ occurs at $J_m = 0$) should hold also far from equilibrium. It would be interesting to have experimental data which could check this.

Of course, other conjectures may be made. If we restate the phenomenology in differential form, $dJ_k = \Sigma_j L'_{kj} dX_j$, then the symmetries $L'_{kj} = L'_{jk}$ will hold in the nonlinear regime if and only if there exists a function $f(X_1 \ldots X_m)$ such that $J_k = \partial f/\partial X_k$. Because it appears that this form may be obtained from a Gibbsian statistical theory, experiments to check the symmetry of L'_{kj} far from equilibrium would be of great interest.

In summary, progress to this point consists of some conjectured principles that, thanks to Tykodi, can at least be stated in precise and experimentally meaningful terms so that their correctness or incorrectness can be determined in the laboratory. But we set for ourselves a more ambitious goal than this.

Since the methods of analysis reviewed above were not powerful enough to guide us to the missing theoretical principle, we are driven finally to recognize what should have been obvious from the start. Only the Gibbs standards of logical reasoning were powerful enough to give us the first variational principle, on which physical chemistry has been feeding for a century; and only a Gibbsian statistical analysis is powerful enough to extend that principle to irreversible processes. But in recent years the field that is now called "statistical mechanics," with its reversion to kinetic theory, stochastic equations, and ergodicity, has deviated so widely from the program for which Gibbs introduced that term, that we need to coin a new name for Gibbs' program if we are not to propagate still more semantic confusion. We now explain briefly an extension of Gibbs' work currently underway, set apart by a new descriptive word.

PREDICTIVE STATISTICAL MECHANICS

Predictive statistical mechanics is not a physical theory, but a form of statistical inference. As such, it is equally applicable in other fields than physics (e.g. engineering, econometrics, etc). In fact, it is having its greatest current success in the new techniques for image reconstruction in optics and radio astronomy (18a,b). We emphasize the sharp distinction in purpose and content between these two methods of reasoning.

A physical theory asks bluntly, "How does the system behave?" and seeks to answer it by deductive reasoning from the known laws of physics. But, for example, the Onsager reciprocal relations cannot be proved by deductive logic from the equations of motion (they are not true for every possible initial state). Therefore, to obtain them in the manner of a physical theory requires that one make extra physical assumptions of an "ergodic" or "stochastic" nature, beyond what is contained in the equations of motion.

Predictive statistical mechanics, instead of seeking the unattainable, asks a more modest question: "Given the partial information that we do, in fact, have, what are the best predictions we can make of observable phenomena?" It does not claim deductive certainty for its predictions, but to ensure the "objectivity" of the predictions we do make, it explicitly forbids the use of extraneous assumptions beyond the data at hand. The formal device which accomplishes this is that we shall draw inferences only from that probability distribution whose sample space represents what is known about the structure of microstates, and that maximizes S_I subject to the macroscopic data.

By this device, the probability is distributed as uniformly as possible over the class C of microstates compatible with our information. Therefore, we shall make sharp predictions only of those phenomena which are characteristic of each of the vast majority of the states in C. But those are just the reproducible phenomena which a physical theory had sought to predict.

Our aim is not to "explain irreversibility," but to describe and predict the observable facts. If one succeeds in doing this correctly from first principles, he will find that philosophical questions about the "nature of irreversibility" will either have been answered automatically, or else will be seen as ill-considered and irrelevant.

The background and technical details of this approach have been explained in another recent review article (19). We recall here only what is needed for the immediate purpose.

On the space Γ of all possible microstates there is defined a measure $d\Gamma$ which may be classical phase volume: $d\Gamma = dq_1 \ldots dp_N$, or some appropriate generalization of this for quantum theory or any other microscopic theory that we might consider. Choosing some set of macroscopic variables $\{A_1 \ldots A_n, n \ll N\}$, the set of their possible values defines a macrospace Ω. The mapping of Γ onto Ω defines a measure on Ω by projection:

$$d\Omega = W(A_1 \ldots A_n) dA_1 \ldots dA_n = \int_R d\Gamma \qquad 18.$$

where the region R of integration is all microstates for which A_i is in dA_i, $1 \leq i \leq n$.

Microscopic properties are relevant to macroscopic predictions only to the extent that certain aspects of the microstates "leak through" and appear at the macroscopic level. Most evident are the conservation laws for mass, energy, and momentum, which made it possible to discover the principles of mechanics at the macroscopic level long before they were recognized as equally valid microscopically, leading to the first law. Next in importance is the above measure W; through this the fantastically great variations in number of microscopic possibilities of realization manifest themselves at the macroscopic level, as the second law. At sufficiently low energies, $\log W$ becomes essentially independent of other parameters, leading to the third law.

These are the only microscopic properties involved in conventional equilibrium thermodynamics; the content of Gibbs' variational principle is that, given the measure W as a function of certain macroscopic quantities (energy, volume, mole numbers, etc) the equilibrium properties of a system are determined. As a procedure for inference, his principle amounts to this: We shall predict that behavior that can happen in the greatest number of ways, consistent with our data.

Predictive statistical mechanics seeks to do no more than this, but only to do it more generally. All its mathematical formalism is nothing but a kind of bookkeeping system by which we may "count the number of ways" in which various conceivable events can happen, consistent with whatever macroscopic data we may have. If our data are of the kind considered by Gibbs (constant in time, piecewise homogeneous in space), then our principle will reduce to his. It is more general in that we must be prepared to deal, both in the information used and in the predictions made, with arbitrary space-time dependences. Mathematically, this means that the functions of Gibbs are promoted to functionals.

Any probability distribution $w(q_1 \ldots p_N)$ over microstates defines a macroscopic distribution $P(A_1 \ldots A_n)$ on Ω by $w \, d\Gamma = P dA$. Its information entropy is then

$$S_I = - \int w \log w \, d\Gamma = - \int dA \, P(A) \log[P(A)/W(A)] \qquad 19.$$

and so, given the measure $W(A)$, we may carry out the maximization in either space.

Direct evaluation of W would be very difficult; much more manageable and equally informative is its n-fold Laplace transform, called the partition function:

$$Z(\lambda_1 \ldots \lambda_n) = \int_\Omega W e^{-\lambda \cdot A} dA = \int_\Gamma e^{-\lambda \cdot A} d\Gamma \qquad 20.$$

where we used the abbreviations $dA = dA_1 \ldots dA_n$, $\lambda \cdot A = \Sigma_i \lambda_i A_i$. When the integral converges, it is because the rapidly increasing factor W is overpowered by an even more rapidly decreasing factor $\exp(-\lambda \cdot A)$, so that the integrand $W \exp(-\lambda \cdot A)$ has an enormously sharp peak at some point $\{\hat{A}_i\}$. Most of the contribution to the integral then comes from the immediate neighborhood of this peak.

Now the probability density $P(A)$ which maximizes S_I subject to prescribed mean values $\langle A_i \rangle$ is just the canonical distribution $P(A) = Z^{-1} W(A) \exp(-\lambda \cdot A)$, of which Gibbs gave several examples. The peak of this density in the macrospace Ω is so sharp that for all practical purposes the mode $\hat{A}_i$ and mean $\langle A_i \rangle$ are the same. Therefore we need only choose the $\{\lambda_i\}$ so as to place that peak at the experimentally observed values $\{A'_1 \ldots A'_n\}$. The simplest way of doing this is to note that the first moments of $P(A)$ are given by

$$\langle A_i \rangle = -\partial \log Z / \partial \lambda_i, \qquad 1 \leq i \leq n \qquad 21.$$

so setting these equal to the experimental values $\langle A_i \rangle = A'_i$, gives n simultaneous equations for the n unknowns λ_i.

In fact, all moments of $P(A)$ are determined by derivatives of $\log Z$; differentiating Eq. 21 with respect to λ_j, we find a combined reciprocity-covariance law:

$$\langle A_i A_j \rangle - \langle A_i \rangle \langle A_j \rangle = -\partial \langle A_i \rangle / \partial \lambda_j = -\partial \langle A_j \rangle / \partial \lambda_i \qquad 22.$$

and we suspect already that reciprocal relations are going to appear rather trivial in this theory.

Note that these relations are perfectly general, whatever microscopic theory we imagine as underlying the macroscopic one. This is a point that was stressed by Einstein many years ago, and it is the reason that he was able to move so confidently in the transition from classical to quantum theory. He knew that Eq. 22 was trustworthy whatever our microscopic theory; as long as conservation of mass and energy were not being called into question, the only thing that could change was the underlying measure: $W_{class} \to W_{quantum}$. So he applied Eq. 22 to determine the energy fluctuations $(\Delta E)^2$ of black-body radiation from the empirical Planck law $\partial \langle E \rangle / \partial T$, noted a term identical with the fluctuations of an ideal gas, and inferred the existence of photons.

Having noted this generality, we may equally well use the notation of quantum theory; the A_i are then operators, the canonical density matrix $\rho = Z^{-1} \exp(-\lambda \cdot A)$ maximizes $S_I = -Tr(\rho \log \rho)$ subject to given values of $A'_i = \langle A_i \rangle = Tr(\rho A_i)$, where the partition function is $Z(\lambda) = Tr \exp(-\lambda \cdot A)$.

For a system of macroscopic size the measure $\log W(A'_i)$ is (13, 19) essentially the maximum of S_I thus attained: $(S_I)_{max} = S(A'_1 \ldots A'_n) = \log Z + \lambda \cdot A'$. For all purposes that could be relevant experimentally, $S(A')$ may be taken as the logarithm of the number of microstates compatible with the macroscopic data A'_i. If this function is known, then the λ's (which arose as Lagrange multipliers in the maximization of S_I) are given simply by $\lambda_i = \partial S/\partial A'_i$. They are, therefore, just the "potentials" appearing in Tykodi's entropy production rate, Eq. 17.

The potential λ_i thus measures the rate at which the number of microscopic possibilities would change if A'_i were slightly different. According to Onsager's interpretation, Eq. 5, the "statistical force" that drives a system back to equilibrium is essentially a change in λ, given near equilibrium by the matrix G of second derivatives of $S(A')$. Tykodi's Eq. 17 suggests that this may be, in fact, exact.

All the formal properties noted above—although perhaps not the interpretation we have just made—have been well known for many years; if the A_i are energy and mole numbers, $P(A)$ reduces to the grand canonical ensemble of Gibbs. Predictive statistical mechanics applies this same formalism, with more general choices of the A_i than Gibbs made. Two different stages of generalization, and therefore two different generalized entropies $S(A')$, are useful in present applications.

The quantity A_i might be observed at different positions $A_i(x_j)$; for each such datum there would be a Lagrange multiplier λ_{ij}. In the limit as the points x_j become dense, the scalar product $\lambda \cdot A$ then goes into $\lambda \cdot A \to \sum_i \int \lambda_i(x) A_i(x) d^3x$. If all this pertains to one time t, we indicate this by a subscript t: the partition function and entropy then become functionals $Z_t = Z_t[\lambda_i(x)]$, $S_t = S_t[A'_i(x)]$.

The density matrix $\rho_t = Z_t^{-1} \exp(-\lambda \cdot A)$ is then, for certain choices of the As, formally identical with what has been called a "local equilibrium" density matrix, but its meaning is here entirely different; in particular, it has nothing to do with equilibrium. ρ_t represents information about the space distribution of the fields at one instant of time, and no other information whatsoever, because it has maximum S_I subject to that constraint. The functional S_I measures the number of microstates compatible with that information, and it generates the potential fields $\lambda_i(x)$ by what has now become functional differentiation: $\lambda_i(x) = \delta S_I/\delta A_i'(x)$.

Note that S_I measures the total number of all microstates compatible with the macrostate $A_i'(x)$ at time t, regardless of the thermokinetic history by which the system came into that state. Thus it contains, with various relative weightings, a kind of mixture of every conceivable history. It is obvious, then, that in general ρ_t cannot contain enough information to predict other quantities B, or the future evolution of the system; for the characteristic feature of irreversible processes is the appearance of fading memory effects, and in ρ_t all memory of the past has been thrown away. This is the logical defect that makes any "local equilibrium" approach inadequate.

In 1964, Robertson (20) showed how, in spite of this, one can make predictions of later irreversible behavior from ρ_t by adding corrective memory terms that accumulate as one integrates the equations of motion forward in time from t. This work developed and applied the continued fraction expansion, later given by Mori. If the important relaxation times are short compared to the time over which one can trust second-order perturbation theory, then one reaches a "plateau" at which transport coefficients may be calculated, as was indeed shown by Green and Kubo in the 1950s. Robertson's recent review (21) gives an extensive list of the many works to 1978 based on this approach.

But there is a more elegant and general way of incorporating memory effects into this theory. Let the $A_i(x)$ now become time-dependent operators in the Heisenberg representation, and suppose we add information about their values at various times t_j. Each of these will now acquire its Lagrange multiplier $\lambda_{ij}(x)$, and again in the limit of dense t_j we have an integral over time. The dot product now goes into

$$\lambda \cdot A \to \Sigma_i \int_{R_i} d^3x dt \lambda_i(x,t) A_i(x,t) \qquad 23.$$

in which R_i is the space-time region in which we have information about $A_i'(x,t)$. The new entropy functional $S[A_i'(x,t)]$ is over all the known thermokinetic history of the system, and it measures the number of microstates consistent with that specific history.

ENTROPY PRODUCTION   421

Analogous to the world-lines of relativity theory, the evolution of a microstate may be visualized as a world-line in "phase space-time," and S is the cross-section of a tube formed of all world-lines by which the given history could have been realized. Let us then call S for any particular history the *caliber* of that history.

We have indicated recently (19) some of the technical details and results of this space-time theory, and applications to hydrodynamics are given by Grandy (22). If the specified history $\{A'_i(x, t), x, t \text{ in } R_i\}$ includes all that is relevant in the laboratory for determining reproducible behavior, then the new ensemble based on Eq. 23 automatically includes all memory effects; the plateau phenomenon is eliminated and one now obtains transport coefficients by direct quadratures over the initial ensemble; they are the full "renormalized" ones.

The theory is freed from previous limitations to the quasi-stationary, long-wavelength case; when all memory effects are included, there is no longer any limitation on time scale or space scale. Thus, as shown in (19), a single equation for the predicted space-time dependence of particle density encompasses both static diffusion and ultrasonic dispersion and attenuation.

An important addition to the technique of applying this theory was added in 1967 by Mitchell (23) in his theory of macroscopic sources, which was identical in philosophy with Schwinger's source theory for quantum fields. From Mitchell's viewpoint, the acoustic Green's function formula Eq. 16 appears as an obvious triviality. He went on to some elegant theorems showing how variational properties of the caliber S of a process determine the conditions for migrational equilibrium in nonequilibrium states, and reciprocity-response theorems about the effect of imposing a new constraint, by which any "renormalization" effects may be analyzed. In the course of this, he formulated what is now called "mode-mode coupling theory." We hope to present elsewhere a detailed account of the kind of results that may be obtained by Mitchell's methods.

The caliber S of a space-time history determines by its variational properties most of the relevant physical information one would like to have. Its first variations determine the conditions of migrational equilibrium, while its second variations generate the "equations of motion." To see why this is so, suppose we have information I_A from one space-time region R_A, which determines a caliber S_A; and we wish to predict—or retrodict—events in some other region R_B. Now we could imagine that someone had given to us a conjectured answer I_B to this, so that we had the total information $I = I_A + I_B$. What would be the caliber S of the combined process? Since S is the result of maximization subject to a further constraint I_B, we shall have $S \le S_A$ with equality if and only if the new information is redundant; i.e. if it is what the theory would have predicted from the old information I_A. Thus the theory always predicts those events in R_B for which the total thermokinetic process will have maximum caliber—an obvious generalization of Gibbs' variational principle.

Although the principle itself evidently holds far from equilibrium, the explicit form of the equations of motion is easily found close to equilibrium where we expect them to be linear. Let S_0 be the caliber corresponding to thermal equilibrium (it is just $k^{-1}S_E$); and let $\delta A = \{\delta A'_1(x, t) \ldots \delta A'_n(x, t)\}$ be some small departures from equilibrium conditions in R_A, while we wish to predict the similar departures δB of some quantities B (which may be the same as the As) in R_B. Then the caliber determined by I_A will be given by an expansion $S_A = S_0 - (1/2)\delta A \cdot G_{AA} \cdot \delta A$, generalizing Onsager's Eq. 4. However, this is compact notation; we remind ourselves that $\delta A \cdot G_{AA} \cdot \delta A$ actually stands for

$$\sum_{jk} \int_{R_j} d^3x\, dt \int_{R_k} d^3x'\, dt'\, \delta A'_j(x, t) G_{jk}(x, t; x', t') \delta A'_k(x', t'). \qquad 24.$$

Now if we add a small variation δB, the caliber acquires more terms: $S = S_A - (1/2)\delta B \cdot G_{BB} \cdot \delta B - \delta B \cdot G_{BA} \cdot \delta A$, where we have used $G_{AB} = G_{BA}$. For fixed δA, the caliber is maximum when

$$G_{BB} \cdot \delta B + G_{BA} \cdot \delta A = 0 \qquad 25.$$

which is a set of simultaneous linear integral equations determining the δB. Had we been given δB and predicted δA, the result would have been $G_{AA} \cdot \delta A + G_{AB} \cdot \delta B = 0$, and $G_{AB} = G_{BA}$ implies a mass of reciprocal relations. Thus the Gs generated by second variations of S are the kernels of the equations of motion.

S usually possesses a convexity property expressed by the inequality of any two neighboring ensembles: $\delta \lambda \cdot \delta A' \le 0$. This is a generalization of the condition given by Gibbs (Reference (2), Eq. 171) from which he deduced all his stability conditions, and leads in the present theory to the positive definite character of G. Then G can be inverted, and the inverse kernels $K = G^{-1}$ are the set of space-time covariance functions generalizing Onsager's Eq. 10, of which the acoustic Green's function Eq. 16 is an example. When the convexity fails, the theory predicts bifurcations or other instabilities, a generalization of Gibbs' condition for phase transitions.

CONCLUSION

As the reader will have sensed, our title is a play on words; logical economy minimizes the principles, not the entropy production. We started by seeking an exact variational property characterizing the nonequilibrium steady state. One such is now apparent, although there may be others more useful. Consider a system evolving according to its equations of motion. Because the caliber S of its history up to time t embodies further constraints beyond those defining the local equilibrium S_t, we have $S \leq S_t$, with equality if and only if that history is the one retrodicted from ρ_t. Now at each instant $t' < t$ there is an $S_{t'|t}$ defined as was S_t by maximizing S_I, but subject to the retrodicted values $A'_i(x, t')$. For reasons explained before (13), a retrodicted history could not be reproduced in the laboratory unless $S_{t'|t} \leq S_t$; but it is a theorem (invariance of S_I under unitary transformations) that $S_{t'|t} \geq S_t$.

These inequalities yield the theorem: Of all reproducible histories terminating at a given state, that one which corresponds to constant S_t throughout the past has the greatest caliber: $S = S_t$. At present it is not known whether this is a pragmatically useful principle in applications; it is, however, of some theoretical importance.

Readers of Truesdell's fresh and fascinating new approach to thermodynamics (16) will resonate at once to this statement. It is a paraphrase of what he calls a "major assertion" in need of proof, from which many other desired results will follow [Reference (16), pp. 22,43]. There is, evidently a close correspondence between these approaches; but to understand it fully and combine them into a single unified theory is a task for the future.

ACKNOWLEDGMENT

It is a pleasure to thank Professor W. Güttinger, Director, for the hospitality of the Institut für Informationsverarbeitung, Universität Tübingen, where a part of this article was written—and where it finally became possible to unravel the mystery about the historical origin of the word "entropy."

Literature Cited

1. Kirchhoff, G. D. 1848. *Ann. Phys.* 75:189
2. Gibbs, J. W. 1876–1878. In *The Scientific Papers of J. Willard Gibbs*. New York: Longmans, Green, 1906; New York: Dover, 1961
3. Clausius, R. 1865. Memoir read at the Philos. Soc. Zürich, April 24. *Pogg. Ann.* 125:353
4. Girardeau, M. D., Mazo, R. M. 1973. *Adv. Chem. Phys.* 24:187–255
5. Helmholtz, H. 1859. *Crelle's J.* 57:1132
6. Lord Rayleigh. 1877. *Theory of Sound.* London: Macmillan
7. Lorentz, H. A. 1895. *Amsterdam Akad. Wetens.* 4:176
8. Prigogine, I. 1961. *Thermodynamics of*

Irreversible Processes. New York: Interscience
9. de Groot, S. R., Mazur, P. 1962. *Non-Equilibrium Thermodynamics.* Amsterdam: North-Holland
10. Klein, M. J. 1960. In *Rend. Sc. Int. Fis. Enrico Fermi.* 10:198–204
11. Prigogine, I., Mayné, F. 1974. In *Transport Phenomena*, Lecture Notes in Physics, Vol. 31, ed. G. Kirczenow, J. Marro, Berlin: Springer
12. Jaynes, E. T. 1963. In *Statistical Physics*, ed. K. W. Ford, 3:181–218. New York: Benjamin
13a. Jaynes, E. T. 1965. *Am. J. Phys.* 33:391–98
13b. Jaynes, E. T. 1971. *Phys. Rev. A* 4:747–50
14a. Onsager, L. 1931. *Phys. Rev.* 37:405–26
14b. Onsager, L. 1931. *Phys. Rev.* 38:2265–79
15. Wei, J. 1966. *Ind. Eng. Chem.* 58:55–60
16. Truesdell, C. 1969. *Rational Thermodynamics.* New York: McGraw-Hill
17. Tykodi, R. J. 1967. *Thermodynamics of Steady States.* New York: Macmillan
18a. Gull, S. F., Daniell, G. J. 1978. *Nature* 272:686–90
18b. Frieden, B. R. 1980. In *Computer Graphics and Image Processing.* In press
19. Jaynes, E. T. 1978. In *The Maximum Entropy Formalism* ed. R. D. Levine, M. Tribus, pp. 15–118. Cambridge: MIT Press
20. Robertson, B. 1964. PhD thesis. Stanford Univ., Calif.
21. Robertson, B. 1978. See Ref. 19, pp. 289–320
22. Grandy, W. T. 1980. *Phys. Rep.* In press
23. Mitchell, W. C. 1967. PhD thesis. Washington Univ., St. Louis, Mo.

SUPPLEMENTARY BIBLIOGRAPHY

A. BACKGROUND

Cox, R. T.: 1946, 'Probability, Frequency, and Reasonable Expectation', *Amer. J. Phys.* 14, 1–13.
Cox, R. T.: 1961, *The Algebra of Probable Inference*, Johns-Hopkins Press, Baltimore, Md..
de Finetti, B.: 1972, *Probability, Induction, and Statistics*, John Wiley and Sons, New York.
Elsasser, W. M.: 1937, 'On Quantum Measurements and the Role of Uncertainty Relations in Statistical Mechanics', *Phys. Rev.* 52, 987–999.
Gibbs, J. W.: 1902, *Elementary Principles in Statistical Mechanics*, Yale University Press, New Haven, Conn.. Reprinted in *The Collected Works of J. Willard Gibbs*, Vol. 2, Yale University Press, 1948, and by Dover Publications, New York, 1960.
Good, I. J.: 1950, *Probability and the Weighing of Evidence*, C. Griffin and Co., London.
Jeffreys, H.: 1939, *Theory of Probability*, Oxford University Press (2nd edition, 1948; 3rd edition, 1961).
Laplace, Pierre Simon de: 1812, *Theorie analytique des probabilités*, Paris (2nd edition, 1814; 3rd edition, 1820). Reprinted in *Oeuvres Complètes*, Vol. VII, Paris, 1847.
Poincaré, H.: 1912, *Calcul des probabilités*, Paris.
Shannon, C. and Weaver, W.: 1949, *The Mathematical Theory of Communication*, The University of Illinois Press, Urbana, Ill..

B. DISCUSSIONS OF ENTROPY MAXIMIZATION AND RELATED TOPICS

Box, G. E. P. and Tiao, G. E.: 1973, *Bayesian Inference and Statistical Analysis*, Addison-Wesley, Reading, Mass..
Csiszar, I.: 1975, 'I-Divergence Geometry of Probability Distributions and Minimization Problems', *Ann. Prob.* 3, 146–158.
Denbigh, K.: 1981, 'How Subjective is Entropy?', *Chemistry in Britain* 17, 168–185.
Dias, P. and Shimony, A.: 1981, 'A Critique of Jaynes' Maximum Entropy Principle', *Advances in Applied Mathematics* 2, 172–211.
Domotor, Z.: 1980, 'Probability Kinematics and Representation of Belief Change', *Phil. Sci*, 47, 384–403.
Domotor, Z.: 1981, 'Probability Kinematics, Conditionals, and Entropy Principles', manuscript, forthcoming in *Synthese*.
Field, H.: 1978, 'A Note on Jeffrey Conditionalization', *Phil. Sci.* 45, 361–367.
Friedman, K. and Shimony, A.: 1971, 'Jaynes Maximum Entropy Prescription and Probability Theory', *J. Statist. Physics* 3, 381–384.

SUPPLEMENTARY BIBLIOGRAPHY

Friedman, K.: 1973, 'Replies to Tribus and Motroni and to Gage and Hestenes', *J. Statist. Physics* 9, 265.
Gage, D. W. and Hestenes, D.: 1973, 'Comments on the paper "Jaynes' Maximum Entropy Prescription and Probability Theory"', *J. Statist. Physics* 7, 89-90.
Hartigan, J.A.: 1964, 'Invariant Prior Distributions', *Ann. Math. Stat.* 35, 836.
Hobson, A.: 1972, 'The Interpretation of Inductive Probabilities', *J. Statist. Phys.* 6, 189-193.
Hobson, A. and Cheng, B.: 1973, 'A Comparison of the Shannon and Kullback Information Measures', *J. Statist. Phys.* 7, 301-310.
Huzurbazar, V. S.: 1976, *Sufficient Statistics*, Marcel Dekker, New York.
Jeffrey, R. C.: 1965, *The Logic of Decision*, McGraw-Hill, New York.
Jeffreys, Sir Harold: 1946, 'An Invariant Form for the Prior Probability in Estimation Problems', *Proc. Royal Society* A186, 453-461.
Johnson, R. W.: 1979a, 'Comments on "Prior Probability and Uncertainty"', *IEEE Trans. on Information Theory* IT-25, 129-132.
Johnson, R. W.: 1979b, 'Axiomatic Characterization of the Directed Divergences and Their Linear Combinations', *IEEE Trans. on Information Theory* IT-25, 709-716.
Kashyap, R. L.: 1971, 'Prior Probabilities and Uncertainty', *IEEE Trans. on Information Theory* IT-17, 641-650.
May, S. and Harper, W. L.: 1976, 'Toward an Optimization Procedure for Applying Minimum Change Principles in Probability Kinematics', in W. L. Harper and C. A. Hooker (eds.), *Foundations of Probability Theory, Statistical Inference, and Statistical Theories of Science*, D. Reidel, Dordrecht, Vol. I, pp. 137-166.
Rosenkrantz, R. D.: 1977, *Inference, Method and Decision*, D. Reidel, Dordrecht, Chapter 3.
Rosenkrantz, R. D.: 1981, *Foundations and Applications of Inductive Probability*, Ridgview Publ. Co., Atascadero, California, Chapter 4.
Rowlinson, J. S.: 1970, 'Probability, Information, and Entropy', *Nature* 225, 1196-1198.
Seidenfeld, T.: 1979, 'Why I am not an Objective Bayesian: Some Reflections Prompted by Rosenkrantz', *Theory and Decision* 11, 413-440.
Shimony, A.: 1973, 'Comment on the Interpretation of Inductive Probabilities', *J. of Statist. Phys.* 9, 187-191.
Shimony, A.: 1981, 'The Status of the Principle of Maximum Entropy', manuscript, forthcoming in *Synthese*.
Shore, J. E. and Johnson, R. W.: 1980, 'Axiomatic Derivation of the Principle of Maximum Entropy and the Principle of Minimum Cross-Entropy', *IEEE Trans. on Information Theory* IT-26, 26-37.
Shore, J. E. and Johnson, R. W.: 1981, 'Properties of Cross Entropy Minimization', *IEEE Trans. on Information Theory* IT-27, 472-482.
Skilling, J.: 1982, 'Prior Probabilities', manuscript, forthcoming in *Synthese*.
Skyrms, B.: 1980, 'Higher Order Degrees of Belief', in D. H. Mellor (ed.), *Prospects for Pragmatism*, Cambridge University Press, Cambridge, pp. 109-137.
Skyrms, B.: 1982, 'Maximum Entropy Inference as a Special Case of Conditionalization', manuscript, forthcoming in *Synthese*.
Tribus, M. and Motroni, H.: 1972, 'Comments on the paper "Jaynes' Maximum Entropy Prescription and Probability Theory"', *J. Statist. Phys.* 4, 227-228.
van Campenhout, J. and Cover, T. M.: 1981, 'Maximum Entropy and Conditional Probability', *IEEE Trans. on Information Theory* IT-27, 483-489.

van Fraassen, B.: 1980, 'Rational Belief and Probability Kinematics', *Phil. Sci.* 47, 165-178.
Villegas, C.: 1971, 'On Haar Priors', in V. P. Godambe and D. A. Sprott (eds.), *Foundations of Statistical Inference*, Holt, Rinehart and Winston of Canada, Toronto, pp. 409-416.
Williams, P. M.: 1980, 'Bayesian Conditionalization and the Principle of Minimum Information', *Brit. J. Phil. Sci.* 31, 131-144.

C. APPLICATIONS

(a) *General Works*

Levine, R. D. and Tribus, M. (eds.): 1978, *The Maximum Entropy Formalism*, M.I.T. Press, Cambridge, Mass..
Grandy, T. W. and Smith, C. R. (eds.): *Proceedings of the Maximum Entropy Workshop*, University of Wyoming Press, Laramie, Wyoming, forthcoming.

(b) *Biology*

Kerner, E. H.: 1978, 'The Gibbs Grand Ensemble and the Eco-genetic Gap', in Levine and Tribus, 1978, pp. 469-476.
Rothstein, J.: 1978, 'Generalized Entropy, Boundary Conditions, and Biology', in Levine and Tribus, 1978, pp. 423-468.

(c) *Chemistry*

Keck, J. C.: 1978, 'Rate-Controlled Constrained Equilibrium Method for Treating Reactions in Complex Systems', in Levine and Tribus, 1978, pp. 219-246.
Levine, R. D.: 1978, 'Maximal Entropy Procedures for Molecular and Nuclear Collisions', in Levine and Tribus, 1978, pp. 247-272.

(d) *Communication Theory*

Gray, R. M., Gray, A. H., Rebolledo, G., and Shore, J. E.: 1980, 'Rate-Distortion Speech Coding with a Minimum Discrimination Information Distortion Measure', manuscript, forthcoming in *IEEE Trans. on Information Theory*.
Jaynes, E. T.: 1959, 'Note on Unique Decipherability', *I.R.E. Trans. on Information Theory* IT-5, 98.

(e) *Decision Making*

Cozzolino, J., 'Maximum Entropy Methods in Management and Decision Making', forthcoming in Grandy and Smith.
Jaynes, E. T.: 1963, 'New Engineering Applications of Information Theory', in J. Bogdonoff and F. Kozin (eds.), *Proceedings of the First Symposium on Engineering Applications of Random Function Theory and Probability*, John Wiley and Sons, New York, pp. 163-203.
Tribus, M.: 1969, *Rational Descriptions, Decisions and Designs*, Pergamon Press, Oxford.

SUPPLEMENTARY BIBLIOGRAPHY

(f) *Econometrics and Forecasting*

Cozzolino, J. and Zahner, M.: 1973, 'The Maximum Entropy Distribution of the Future Market Price of a Stock', *Operations Research* 21, 1200–1211.

Litterman, R. B.: 1979, *Techniques of Forecasting Using Vector Autoregression*, Ph. D. Dissertation, University of Minnesota.

(g) *Engineering and Operations Research*

El-Sayed, Y. M. and Evans, R. B.: 1970, 'Thermoeconomics and the Design of Heat Systems', *J. of Engineering for Power*, pp. 27–35.

Tribus, M.: 1962, 'The Use of the Maximum Entropy Estimate in Reliability Engineering', in R. E. Machol and P. Gray (eds.), *Recent Developments in Decision and Information Processes*, Macmillan, New York, pp. 102–140.

Tribus, M.: 1969, *Rational Descriptions, Decisions and Designs*, Pergamon Press, Oxford.

(h) *Geology and Geophysics*

Currie, R. G.: 1980, 'Detection of the 11-year Sunspot Cycle Signal in Earth Rotation', *Geophys. J. Roy. Astron. Soc.* 46, 513–520.

Robinson, E., 'Fundamentals of Seismic Exploration', forthcoming in Grandy and Smith.

(i) *Maximum Entropy Spectral Analysis (MESA)*

Burg, J.: 1975, 'Maximum Entropy Spectral Analysis', Ph. D. Dissertation, Stanford University (University Microfilms, No. 75–25, p. 499).

Shore, J. E.: 1979, 'Minimum Cross-Entropy Spectral Analysis', *IEEE Trans. Acoust. Speech and Signal Processing* ASSP-29, 230–237.

MESA has been applied in several fields, including geophysics (see Currie, 1980, above), radio astronomy, and optical image reconstruction. Additional references are:

Daniell, G. J. and Gull, S. F.: 1980, 'The Maximum Entropy Algorithm Applied to Image Enhancement', *IEE Proc.* (E) 5, 170–173.

Frieden, B. R.: 1980, 'Statistical Models for the Image Restoration Problem', *Computer Graphics and Image Processing* 12, 40–59.

Skilling, J., Strong, A. W., and Bennett, K.: 1979, 'Maximum Entropy Image Processing in Gamma Ray Astronomy', *Monthly Notices Roy. Astron. Soc.* 187, 145–152.

(j) *Statistics*

Gokhale, D. V. and Kullback, S.: 1978, *The Information in Contingency Tables*, Marcel Dekker, New York.

Good, I. J.: 1967, 'Maximum Entropy for Hypothesis Formulation, especially for Multidimensional Contingency Tables', *Ann. Math. Statist.* 34, 911–934.

Jaynes, E. T.: 1958, *Probability Theory in Science and Engineering*, Vol. 4 in Colloquium Lectures in Pure and Applied Science, Socony-Mobil Oil Company, Inc., Dallas, Texas.

Kullback, S.: 1959, *Information Theory and Statistics*, John Wiley and Sons, New York. Reprinted by Dover Publications, N.Y., 1968.

(k) *Thermodynamics*

Baierlein, R.: 1971, *Atoms and Information Theory: an introduction to statistical mechanics*, W. H. Freeman Co., San Francisco.
Katz, A.: 1967, *Principles of Statistical Mechanics – the Information Theory Approach*, W. H. Freeman Co., San Francisco.
Mead, C. A.: 1978, 'The Special Role of Maximum Entropy in the Application of "Mixing Character" to Irreversible Processes in Macroscopic Systems', in Levine and Tribus, 1978, pp. 273–288.
Robertson, B.: 1978, 'Application of Maximum Entropy to Nonequilibrium Statistical Mechanics', in Levine and Tribus, 1978, pp. 289–320.
Tribus, M.: 1961, *Thermostatics and Thermodynamics*, van Nostrand, New York.

(l) *Traffic Networks*

Benes, V. E.: 1965, *Mathematical Theory of Connecting Networks and Telephone Traffic*, Academic Press, New York.

D. SOME OTHER PAPERS BY JAYNES NOT INCLUDED HERE

Heims, S. P. and Jaynes, E. T.: 1962, 'Theory of Gyromagnetic Effects and Some Related Magnetic Phenomena', *Rev. Mod. Phys.* 34, 143–165.
Cummings, F. W. and Jaynes, E. T.: 1963, 'Comparison of Quantum and Semi classical Radiation Theories with Application to the Beam Maser', *Proc. IEEE* 51, 89–109.
Stroud, C. P. and Jaynes, E. T.: 1970, 'Long-Term Solutions in Semiclassical Radiation Theory', *Phys. Rev.* A1, 106–121.
Jaynes, E. T.: 1972, 'Survey of the Present Status of Neoclassical Radiation Theory', in L. Mandel and E. Wolf (eds.), *Coherence and Quantum Optics*, Plenum Press, New York, pp. 35–81.
Jaynes, E. T.: 1977, 'Electrodynamics Today', presented at the Fourth Rochester Conference on Coherence and Quantum Optics.
Matthys, D. R. and Jaynes, E. T.: 1980, 'Phase-Sensitive Optical Amplifier', *J. Opt. Soc. Am.* 70, 263–267.
Jaynes, E. T.: 1980, 'Quantum Beats', in A. O. Barnt (ed.), *Foundations of Radiation Theory and Quantum Electrodynamics*, Plenum Press, New York.

INDEX

acceptance tests 161ff.
array
  definition of 20
  probabilities 17, 20
  minimum entropy property 22, 38

Bayes' theorem
  inductive reasoning and 203, 248
  Laplace and 249, 391
  maximum entropy and 249, 250, 271–273, 283
Bayesian
  intervals 171ff.
  statistics 116, 149–169 passim, 340–400 passim
Bernoulli
  definition of probability 213
  law of large numbers 312ff., 318
  trials 122, 123, 126–128, 170–171, 213ff.
Bertrand (see paradox)
Bogoliubov scheme 69–70
Boltzmann
  collision equation 9, 96, 103
  distribution 9, 224ff.
  distribution function 64ff., 97
  entropy 70ff., 79–86 passim
  ergodic hypothesis 97
  H-theorem 2, 72–75, 77, 80, 97
Brandeis dice problem (see dice)

chi squared (see significance test)
concentration theorem (see maximum entropy)
conditionalization (see Bayes' theorem, maximum entropy)
confidence intervals 108, 116, 170–182
  vs. Bayesian intervals 170–173, 178ff., 182
  sufficient statistics and 174, 178ff.

consistency (see principle, maximum entropy)

Darwin–Fowler method 250, 318
de Finetti's theorem 247
density matrix 19, 22–25, 58, 59, 110, 111, 292, 293
dice 41–44, 210, 243ff., 323ff.
distribution(s)
  functions 61–70, 77, 81–84
  fiducial 355
  most probable 225, 250, 318
  structural 359

ensemble (see Gibbs)
  canonical 73, 81ff., 105ff., 109, 121, 245, 306
  conceptual problems of 104ff.
  grand canonical 109, 121, 291
  pressure 7, 15
  rotational 5, 109, 276
entropy (see, also, maximum entropy)
  as anthropomorphic 78, 85–86
  axioms for, Shannon's 16
  combinatorial approach to 52, 120, 320
  continuous 59, 60, 114, 124, 282
  cross 39, 60, 114, 124, 262, 282
  experimental vs. informational 39, 45, 55, 110, 292
  Gibbs vs. Boltzmann 70ff., 77, 79–86 passim
  meanings of 406–408
  production 77, 402ff.
  relative (see cross)
  test of fit 315, 324–326, 329–332
ergodic approach 4, 5, 10, 11, 90ff., 99, 104–107, 230ff.
exchangeable sequences 246–248, 262

431

experiment
  Barnett 5
  broom straws 142-143
  Einstein-de Haas 5, 15
  Hahn 85
  male births 392
  molecular beam 22
  nuclear reflection 118-119
  porous plug 413
  uninformative 206, 207
  Wolf's dice data 210, 258ff., 264, 315, 327ff.

Fisher vs. Neyman-Pearson 198-199
fluctuation theory 274ff.
Forney's question 271
frequency interpretation 12, 65ff., 69, 106, 107, 133, 141, 184ff., 214, 215, 227, 230, 234-236

Gibbs (see paradox)
  algorithm 290, 296
  canonical distribution (see ensemble)
  entropy 70ff., 79-86 passim
  entropy maximization 100, 121
  general formalism 101, 102, 110
  H-function 71-74, 79-82
  nature of his work 98-103, 121, 227, 228
  on ergodicity 99, 229
  on irreversibility 100
  phase rule 402, 422
  rules 87, 101-102, 290ff.
  statistical theory 411, 415
  use of 'ensemble' 227ff.
  variational principle 402, 417, 422
gyromagnetic effects 5, 15

Haar measure 125, 341, 372
heat bath 14, 230
high probability manifold 293

ignorance 114, 125-128, 282-284, 352, 371
improper prior (see prior)
indifference between problems 143-144

information (see entropy, questions)
  discrimination 262
  Fisher 398
  game 17, 28-31
  interpretation of 233ff.
  loss and second law 19
  redundant 121, 300
  testable 118, 240-241
irreversible (see thermodynamics)
integral equations 302-303, 348ff.
interval estimation 155-157
invariance argument 114, 125-129, 131-132, 134ff., 195-196, 370ff.
invariant measure 39, 59-60, 114, 124-125, 282

Jeffreys
  invariance theory 221
  marginalization 221, 362, 374
  prior 125, 129, 159, 161, 176, 204, 352, 359-360
  rules for non-informative priors 114-115, 125, 129, 161, 221, 337
  significance tests 159, 377-379, 399

Kubo
  formulas 111
  transform 299

Laplace (see, also, Bayes, principle of indifference)
  contributions to probability 214-224
  criticisms of 199ff., 217-219
  definition of probability 199-200
  on estimation 156
  rule of succession 126-128, 171, 201, 202
  significance tests 156, 193-194, 204, 216, 376, 386, 391-397
  solution of inversion problem 216
  was right afterall 202ff.
likelihood principle 175, 184-185

macroscopic uniformity 11, 13, 20
marginalization 221, 284-286, 337-370 passim
  and group analysis 370-374

INDEX   433

maximum entropy (see, also, entropy)
  applications 3, 273ff., 416
  Boltzmann's use of 51, 224ff.
  concentration theorem 315, 321–322,
    333–335
  conditioning and 249–250
  consistency and 9, 16
  constraint rule 269–271
  correspondence property (see frequencies and)
  criticisms of 250–273 passim
  density matrix formalism and 24ff.
  direct probabilities and 122–123, 257
  existence of 241
  formalism 7–9, 45–52, 108–111,
    118–119, 241–243, 417ff.
  frequencies and 51–52, 119–121,
    131, 245, 254ff.
  Gibbs' use of 7, 100, 121, 229, 418
  irreversible processes and 238–239,
    287–310
  maximum likelihood and 269ff., 283
  new constraints implied by 13, 20,
    121, 257ff., 264, 267, 298, 328ff.
  Shannon's use of 236
  spectral analysis and (MESA) 3, 416
  sufficiency and 283
  uniqueness of 241
  why it works 10–13, 20, 227, 239–
    240, 267, 296–298
Maxwell's velocity distribution 143, 222–
  223, 226
metric transitivity 10–11, 230, 232
minimum entropy production 401–424
  passim
Mitchell's macroscopic source theory 401,
  421
molecular biology 280–281

neoclassical electrodynamics 231
noninformative prior (see, also, prior)
  consistency desideratum (see principle
    of consistency)
  dependence on question 352, 374,
    376, 379–381
  for Bernoulli trials 126–128

  for location parameters 125–126
  for Poisson rate 126
  for regression problem 195–196
  for scale parameters 125–126
  frequencies and 131, 141ff., 145
  invariance of 114, 124–129, 135ff.,
    370ff.
  Jeffreys' rules for (see Jeffreys)
  marginalization and 221, 284–286,
    337–340, 348ff.
  separation property of 342
nuclear polarization effect 15

Onsager's theory 401, 408–412
orthodox statistics 152, 218–219
  'bad' samples for 174ff., 180
  Bayesian vs. 169–208 passim
  single case and 176ff., 204–205
  sufficient statistics and 169, 174–175,
    182, 197, 199

paradox
  Bertrand 128–129, 134–147
  Gibbs 7, 81, 401
  Loschmidt 65, 97, 406
  marginalization 337–346
  Uhlenbeck 237–238
  Zermelo 65, 97, 406
parameters
  nuisance 169, 219
  of location 125–126
  of scale 125–126, 159
partition
  function 101, 123–124, 242, 292
  functional 17, 57, 62, 111, 294
periodicity in scientific progress 1, 4, 90–
  94
personalism vs. impersonalism 117, 186
philosophy 103–104, 112, 154
Poincaré's invariance argument 135, 370
prior (see, also, noninformative prior)
  Bayes-Laplace 125, 128
  improper 205ff., 346–348
  Jeffreys (see Jeffreys)
  nullifying 344, 356
  overdetermined 146, 365–366

reference 398
underdetermined 146, 365-366
uniqueness of 366-369, 374
principle of
consistency 115, 117, 195, 284, 371
indifference 8, 128, 134-135, 144, 212, 200, 229
insufficient reason (*see* indifference)
statistical complementarity 24
probability (*see* also prior)
frequencies and 120, 131, 141ff., 213-218, 228, 245ff., 274
inverse 230
history of 152-153, 199-204, 212-240 passim
problems (*see also* questions)
estimation vs. testing 389
higher-level 146, 272
ill-posed 146
indifference between 128, 144
overdetermined 146, 365-366
string 132
underdetermined 146, 365-366
water and wine 146
projection operators 308-309

quantum electrodynamics 231, 279
quantum mechanics 20ff., 23-24, 87, 93ff.
Copenhagen interpretation 87
infirmities of 93ff.
superposition principle 93
uncertainty principle 94
questions (*see also* problems)
information from 385-388
logic of 382ff.
optimal 383, 388
simple 386

relaxation process 30-31
relevance function 304
reliability tests 157-168 passim

Shannon theorems
asymptotic equipartition 83
entropy 16
significance tests (*see also* Jeffreys, Laplace)
chi squared 210, 262, 315, 322, 332
exact levels 160
F-test 157-158, 160
form of in physics 388-391
statistical mechanics
classical 7-10
generalized 13-15
predictive 2, 19, 37, 416-423
quantum 20-25 passim, 61ff.
statistics (*see also* Bayesian, orthodox)
ancillary 219
sufficient 26, 169, 174-175, 182, 197, 199, 386-388
subjective 17-18, 23
H-theorem 27ff., 38

thermodynamics
equilibrium 10, 53-57, 109
irreversible 7, 17, 25-28, 39, 75, 111-112, 238-239, 287-310
perturbation theory 298ff.
second law of 19, 74, 80, 82-86, 293, 408
time-dependent phenomena 19
thermometer 14
transformation groups (*see* invariance)

uninformative prior (*see* noninformative prior)

Wiener prediction theory 303-305

Made in the USA
Lexington, KY
17 June 2013